# The Biology of Marsupials

## CONTRIBUTORS

RAYMOND D. BARNES

B. J. BRYANT

DON HUNSAKER II

JOHN IRWIN JOHNSON, JR.

JOHN A. W. KIRSCH

STEPHEN POTKAY

LAWRENCE K. SCHNEIDER

DONALD SHUPE

# The Biology of Marsupials

*Edited by*

**DON HUNSAKER II**
Department of Zoology
San Diego State University
San Diego, California

ACADEMIC PRESS New York San Francisco London 1977
*A Subsidiary of Harcourt Brace Jovanovich, Publishers*

ACADEMIC PRESS, INC.
111 Fifth Avenue, New York, New York 10003

*United Kingdom Edition published by*
ACADEMIC PRESS, INC. (LONDON) LTD.
24/28 Oval Road, London NW1

**Library of Congress Cataloging in Publication Data**

Main entry under title:

The biology of marsupials.

Bibliography: p.
Includes index.
1. Marsupialia. I. Hunsaker, Don. [DNLM: 1. Marsupialia. QL737.M3 B615]
QL737.M3B5 1977 599′.2 76-2946
ISBN 0-12-362250-6

PRINTED IN THE UNITED STATES OF AMERICA

# Contents

# List of Contributors

Numbers in parentheses indicate the pages on which the authors' contributions begin.

**Raymond D. Barnes**[1] (387), Department of Anatomy, School of Veterinary Medicine, University of California, Davis, California

**B. J. Bryant**[2] (349), Department of Pathology, School of Medicine, Wayne State University, Detroit, Michigan

**Don Hunsaker II** (95, 279, 507), Department of Zoology, San Diego State University, San Diego, California

**John Irwin Johnson, Jr.** (157), Department of Biophysics, Michigan State University, East Lansing, Michigan

**John A. W. Kirsch** (1, 521), Department of Biology and the Peabody Museum of Natural History, Yale University, New Haven, Connecticut

**Stephen Potkay** (415), Veterinary Resources Branch, Division of Research Services, National Institutes of Health, Bethesda, Maryland

**Lawrence K. Schneider** (51), Division of Biomedical Sciences, School of Medical Sciences, University of Nevada-Reno, Reno, Nevada

**Donald Shupe** (279), Behavioral Science Department, California State Polytechnic University-Pomona, Pomona, California

[1] *Present address:* Department of Veterinary Biology, College of Veterinary Medicine, University of Minnesota, St. Paul, Minnesota

[2] *Present Address:* Radiobiology Laboratory, University of California, Davis, California

# Preface

This work is intended to be a compilation and analysis of the research conducted on New World marsupials. While H. Tyndale-Biscoe's work "Life of Marsupials" (1973, American Elsevier, New York) was directed principally to the Australian forms with some reference to those of the New World, in this book, some chapters relate to all marsupials; others address themselves primarily to the didelphids with as much collateral information as possible on other forms. The chapters reflect the surprisingly detailed knowledge we have in some areas of the biology of marsupials and the complete lack of basic information available in others of scientific interest.

This volume is a result of the numerous discussions researchers have had on the common problems encountered in their studies of marsupials. We agreed that it was time to collate the results of our research and to survey the available literature. The contributors were carefully selected to provide a wide selection of material on the biology of these animals, and they have done an outstanding job. We hope that the aims of the book—to bring to the scientific community the latest information available on certain aspects of the biology of the marsupial and to encourage other investigators to initiate new programs which will eventually fill some of the voids in the knowledge of marsupiology—have been fulfilled.

The influence of many people is reflected in these pages: Jack Thrasher, Ira Fritz, and Ray Barnes were particularly instrumental in the early planning of the volume. Other researchers whose help and advice are greatly appreciated include L. T. Evans, Jorge Hernandez C., Jerry Boggs, Glenn Cadzow, Frederico Medem, and Ernesto Barriga. I am grateful for the hours of aid and encouragement given by my students and assistants, particularly Lory Minning, Jean Hawthorne, Craig Johnson, Del Overton, and Lorna Boggs. Special thanks are also due Mary Duclos and Martha Hayes for their long hours of typing and for aid with the illustrations. The San Diego Zoo officials were also helpful in enabling me to conduct research with their animals.

Don Hunsaker II

# Introduction

The biology of marsupials is a fascinating study for any one of several reasons. First and foremost, the marsupials represent a diverse group of mammals that are intrinsically interesting. Although they have many primitive characteristics, they have evolved some remarkable specializations and represent a very successful group. Study of the marsupials also contributes to our knowledge and understanding of the evolution of the class Mammalia. Although considered by many to be an ancestral form which gave rise to the placentals, the marsupials were quite distinct from them in the Cretaceous period, and were surely derived from an early common ancestor. Since they did not give rise to the placentals, many of the primitive characteristics they share with the reptiles and monotremes provide us with information from which we can infer the probable mechanisms operating in the ancestral forms of both the present-day marsupials and placentals. In recent years, the marsupials, especially the Virginia opossum and a few South American species, have been utilized as excellent models for biomedical research. The accessibility of the immature young in the pouch for experimental medical and biological research has given us added incentive to utilize these animals.

Many researchers have studied the morphological, physiological, and behavioral characteristics of *Didelphis,* while complacently considering that they were studying a primitive animal that retained all of its Cretaceous characteristics. In the light of our understanding of modern evolutionary theory and natural selection, we can only assume that these studies may or may not reflect an ancestral condition. If the characteristics studied are also shared by the placentals, we can be reasonably sure that they are primitive and did occur in the ancestral forms. Many characteristics such as sleep patterns, locomotion, and the scratching of an animal's head with its hind foot are shared by both groups. If we are considering factors which are present only in one or the other of the major groups, we must look for relationships with the even more primitive reptiles or consider the possibility that it is a new adaptation that has evolved in one group and not in the other, or possibly was shared by both in the past and subsequently lost by one. Most modern investigators of the biology of living marsupials appreciate and consider these factors. This research, in conjunction with the paleontological research of Clements, Simpson, Lillegraven, and others, and the modern systematics of Ride and Kirsch provide major steps forward in our understanding of marsupial relationships to one another as well as to other mammals.

The utilization of the New World marsupials as a model for biological research has provided a surprisingly large amount of literature to investigators during the past few decades. Many researchers, beginning with Hartman and McCrady, have described the embryology and development of the marsupials in great detail. Experimental embryologists are presently utilizing marsupials in a variety of laboratories. Other researchers use them for physiological and immunological research. There is also continuing interest in the anatomy and pathological conditions of these animals. Their relatively small size, availability, and low cost of maintenance have encouraged many investigators to utilize didelphids in their research. A great deal of controversy has arisen as to whether *Didelphis virginiana* is really well adapted to laboratory conditions. Many researchers consider this species one of the most miserable laboratory animals in existence, while others have solved the problems of laboratory maintenance and handling, and consider them an ideal species.

The marsupials in general, and certainly the Virginia opossum, have been controversial animals for as long as people have known of their existence. The literature is rampant with stories of this species copulating through the nose via the bifurcate penis of the male, babies that were born already attached to the nipple of the mother, and others relating how the embryos were expelled into the pouch of the mother by an explosive sneeze. Rigorous scientific investigations have dispatched most of the old wives' tales, but the species is still discussed by both the scientific and lay communities at great length. Even such a mundane thing as the correct spelling of the genetic name, *Didelphis,* is still controversial. There is no doubt in anyone's mind that Linneaus intended the genus to be named in reference to the double uteri of the opossum. In order to do this, the correct spelling of the genus should be *Didelphys.* However, due to an oversight by Linneaus, or a mistake by the printer, the first published spelling was *Didelphis* which refers to two porpoises, or sea mammals. The rules of nomenclature are quite clear that the first published spelling is correct, so the official spelling of the genus is *Didelphis.* It is hoped that marsupials as a group will be as well understood in the future as some of the more popular groups.

During the writing and compilation of the information in this book, Dr. Al Gardner published a paper relating to the systematics of the genus *Didel-*

*phis*. In his work, he answered the question "How many species of *Didelphis* really occur in the western hemisphere?" which had been plaguing various investigators working with didelphids. His answer was that there are three: *Didelphis virginiana* in North America extending down to Central America; *Didelphis marsupialis*, a Central and South American species; and *Didelphis albiventris* occurring in the highlands of South America. Most of our manuscripts were in the process of final editing as his information became available. The task of changing all the references of *Didelphis marsupialis*, the more popular name for the Virginia opossum before Gardner's work, to *Didelphis virginiana* was a huge one.

The same problem with the proper nomenclature occurs with a very popular species for biological research, *Marmosa mitis*, the small mouse opossum from South America. The correct name for this species, again following the rules of nomenclature, is *Marmosa robinsoni*, the name which should be used in the literature. Since many authors use *Marmosa mitis* to describe the animal utilized in their research projects, references to *Marmosa robinsoni* and *Marmosa mitis* should be considered synonymous.

In an effort to clarify the marsupial nomenclature problem, Dr. John A. W. Kirsch, in the Appendix at the end of the book, has collated, in tabular form, the genus and species names currently in use with synonymous forms found in the literature.

Don Hunsaker II

# The Biology of Marsupials

# 1

# The Classification of Marsupials

## With Special Reference to Karyotypes and Serum Proteins

JOHN A. W. KIRSCH



## I. INTRODUCTION

The readers of this treatise, like its authors, will presumably already be prejudiced toward marsupials. There are many reasons for the continuing interest in this group of mammals, including the excellent examples they afford of adaptive radiation and of evolutionary parallelism and convergence. In Australia, and in a significantly different way in America, marsupials evolved into a wide range of forms often strikingly similar morphologically or ecologically to placental counterparts elsewhere. Less appreciated are the similarities and differences between the two geographical groups of marsupials. Moreover, the distribution of marsupials has often been cited as evidence for continental drift; today, however, evidence completely independent of biogeography leaves no doubt that drift did occur, and the question now is not whether marsupial distribution provides evidence for drift, but whether drift accounts for the distribution of marsupials.

There are some 250 species of living marsupials (Kirsch and Calaby, 1976) and there are at least 150 known fossil species. [Turnbull (1971) gives combined figures for living and extinct marsupials.] It is not surprising that a group so diverse should present many taxonomic problems, but it *is* surprising that for all their intrinsic interest marsupials have received less than adequate treatment from classical taxonomists. One reason is that workers on one geographical group have mostly ignored the other group, so that there is no really adequate overall treatment of the order; one result is that marsupial classification continues to be discussed in terms of the few morphological features well known in all members of the group.

A purpose of this treatise is to demonstrate that marsupials provide excellent material for a variety of experimental studies. Among these, and particularly suited to taxonomic work, are cytology and comparative serology. Studies of the chromosomes of marsupials were sporadic and few until Sharman's determined survey of Australian marsupials in the 1950's (summarized in Sharman, 1961). Sharman's work has been repeated and extended by Hayman and Martin [Hayman and Martin (1974) is a summary of their work and provides references to earlier papers] and significant additions to our knowledge of American marsupials have been provided by them, and by Biggers *et al.* (1965), Reig and Bianchi (1969), Reig and Sonnenschein (1970), and Reig *et al.* (1972, 1977). Serological studies also comprise a limited literature: some marsupials were included in the phenomenally comprehensive pioneering work of Nuttall (1904) but were not again studied until Weymss' work in 1953. Except for peripheral inclusion of opossums or kangaroos in studies such as Goodman's (1961, 1962) on primates, my own studies, and more recently those of V. M. Sarich (unpublished) complete the literature.

In spite of this very compact literature, it is fair to say that major insights into marsupial relationships have come from karyological and serological studies. A particularly striking feature of these separate approaches is their complementarity, and it is this point that will be stressed at the expense of a detailed description of techniques or results.

Taxonomic problems in Marsupialia are found at every categorical level. Since this volume is mostly about didelphid marsupials, it seems appropriate to concentrate here on those broad relationships which will place the family Didelphidae in perspective and to consider some details of classification within that family.

The major aim, however, is to present those data on marsupial karyotypes and comparative serology which will illustrate the usefulness of these methods in arriving at a classification of marsupials and the phylogenetic inferences from them. To facilitate this aim, and because it seems appropriate in a volume on marsupials, it is necessary to begin with some description of the group and what we presently know of its evolution. Clearly this can be neither exhaustive nor particularly original in its treatment.

## II. THE MARSUPIALS: AN OVERVIEW

At present the standard classification of marsupials, as of mammals in general, is that of Simpson (1945), although he proposed this arrangement in 1930 and it does have definite antecedents in earlier authors (e.g., Gregory, 1910). Simpson's classification distributes the marsupials among six superfamilies, five living and one wholly extinct (Borhyaenoidea). Recently, however, Simpson (1970a) reduced the extinct group to a family of Didelphoidea, while recognizing the Argyrolagidae as representative of another major group; without committing himself definitely, he indicates (Simpson, 1970b) that this would be appropriate for the Groeberiidae as well. These points are discussed more fully later in this section.

Table I lists Simpson's superfamilies with their currently recognized families, indicating the temporal and geographical distribution of these groups and the known range of diploid chromosome numbers found in each extant family. While neither the living nor fossil marsupials provide a completely smooth sequence illustrating the evolution of the group, the major taxa do to a large degree represent distinct points along the lineages. Didelphoidea provides a very satisfactory starting point: living didelphoids are for the most part small- to medium-sized insectivores and carnivores, terrestrial or scansorial (the aquatic *Chironectes* is an exception), with plantigrade pentadactyl limbs. They retain a high dental formula with up to five pairs of upper incisors, a characteristic shared only with some peramelid marsupials.

The great length and relative richness of didelphoid fossil history also contributes to their being considered satisfactory progenitors for the rest of the marsupials. Among the oldest known mammalian fossils are possibly didelphoid teeth dating from the mid-Cretaceous of Texas and occurring in association with other teeth that may be eutherian [Slaughter, 1968a,b,

**TABLE I**

A Classification of Marsupials by Superfamily[a]

| Taxon | Distribution | Temporal range | Karyotypes |
|---|---|---|---|
| Didelphoidea | | | |
| Didelphidae | North America | Late Cretaceous–Recent, but discontinuously | $2n = 14$, 18, or 22 |
| | South America | Latest Cretaceous–Recent | |
| | Europe | Eocene–early Miocene | |
| †Pediomyidae | North America | Late Cretaceous | |
| | South America | Latest Cretaceous | |
| †Stagodontidae | North America | Late Cretaceous | |
| Microbiotheriidae | South America | Paleocene–Recent | $2n = 14$ |
| †Borhyaenidae | South America | Late Paleocene–Pliocene | |
| †Necrolestidae | South America | Early Miocene | |
| Caenolestoidea | | | |
| Caenolestidae | South America | Early Eocene–Recent | $2n = 14$ |
| †Polydolopidae | South America | Late Paleocene–early Eocene | |
| Argyrolagoidea | | | |
| †Argyrolagidae | South America | Pliocene–Pleistocene | |
| Dasyuroidea | | | |
| Dasyuridae | Australasia | Miocene–Recent | $2n = 14$ |
| Myrmecobiidae | Australia | Recent | $2n = 14$ |
| Thylacinidae | Australasia | Late Miocene–Recent | |
| Notoryctidae | Australia | Recent | $2n = 20$ |
| Perameloidea | | | |
| Peramelidae | Australasia | Late Miocene–Recent | $2n = 14$ + supernumeraries |
| Thylacomyidae | Australia | Late Pliocene–Recent | $2n = 18$ female, 19 male |
| Phalangeroidea | | | |
| Phalangeridae | Australasia | Miocene–Recent[b] | $2n = 14$, 20 |
| Macropodidae | Australasia | Miocene–Recent | $2n = 10$–32 |
| Petauridae | Australasia | Miocene–Recent | $2n = 10$–22 + supernumeraries |
| Burramyidae | Australasia | Late Miocene–Recent[b] | $2n = 14$ |
| †Thylacoleonidae | Australia | Miocene–Pleistocene | |
| †Ektopodontidae | Australia | Miocene | |
| †Diprotodontidae | Australasia | Miocene–Pleistocene[b] | |
| Phascolarctidae | Australia | Miocene–Recent | $2n = 16$ |
| Vombatidae | Australia | Miocene–Recent | $2n = 14$ |

**TABLE I (Cont.)**

| Taxon | Distribution | Temporal range | Karyotypes |
|---|---|---|---|
| Tarsipedidae | Australia | Recent | $2n = 24$ |
| †Wynyardiidae | Australia | Miocene | |
| Marsupialia *incertae sedis* | | | |
| †Groeberiidae | South America | Early Oligocene | |

[a] Simpson 1930, 1945, 1970a; cf. Table III. Includes all families currently recognized (Ride, 1970; Archer and Kirsch, 1977; Woodburne *et al.*, 1977). The geographical and temporal range of each family is given (compiled mostly from Patterson and Pascual, 1972; Tedford, 1974; Woodburne *et al.*, 1977), with the range of chromosome numbers known to occur (from Hayman and Martin, 1974; Calaby *et al.*, 1974). Wholly extinct families are indicated by a dagger (†).

[b] Tedford *et al.* (1975) suggest that these three families may occur as early as the Oligocene.

1971; but see Turnbull (1971) and Lillegraven (1974) for discussions of the alternatives]. The late Cretaceous Lance Formation of Wyoming (Clemens, 1966, 1968) and deposits of similar age in Alberta (Lillegraven, 1969) and Montana (Sahni, 1972) are rich in didelphoid remains; Clemens (1968) regards the abundant and persistent fossil *Alphadon* (Fig. 1), a member of the Didelphinae, as possibly ancestral to all later marsupials. These Cretaceous didelphoids are found in association with relatively primitive eutherian, but abundant multituberculate, fossils. (Multituberculates represent a completely separate subclass of mammals, extinct since the Eocene.) Elsewhere in the world marsupials occur this early only in some latest Cretaceous deposits of Peru, where at least didelphid and possibly also pediomyid remains have been found (Sigé, 1968, 1972). Butler and Kielan-Jaworowska (1973) have recently suggested that *Deltatheridium,* a Mongolian late Cretaceous genus generally regarded as placental, might be marsupial; but it seems more likely that the Deltatheridiidae, with their mixture of metatherian and eutherian characters, represent the persistence of the features of the common ancestor of marsupials and placentals. In Europe didelphids were represented from the Eocene until Miocene by the ubiquitous *Peratherium*; in North America didelphids similarly disappeared in the Miocene, and the presently resident didelphids are recent reinvaders from South America. Tedford (1974) suggested that South America might in fact have been the place of marsupial origin.

South American didelphoids radiated widely, evolving into primarily carnivorous and possibly frugivorous forms. The largest of these didelphoids

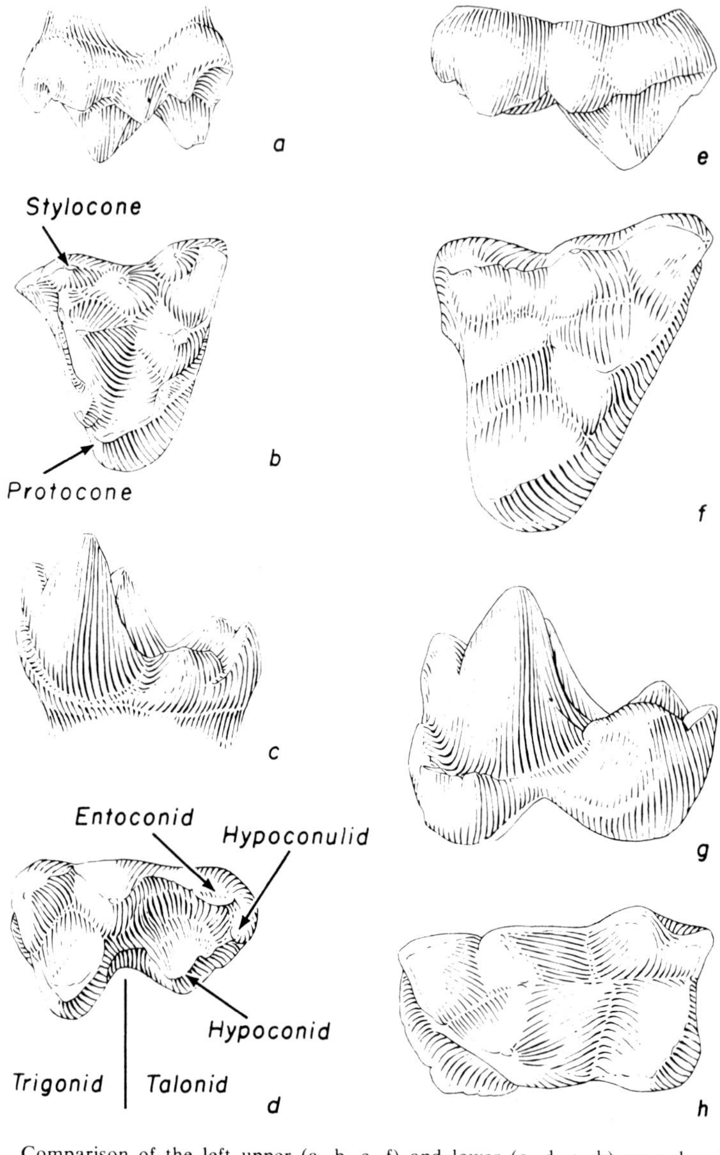


Fig. 1. Comparison of the left upper (a, b, e, f) and lower (c, d, g, h) second molars of *Alphadon marshi* (left, times 10) and *Didelphis marsupialis* (right, times 6). *Alphadon* is thought likely to be the progenitor of all later marsupials. Although there are evident similarities between *Alphadon* and *Didelphis*, the latter genus is distinct and known only from the late Pliocene onward. [From Clemens (1968); reprinted by permission of the editor of *Evolution*.]

Fig. 2. A reconstruction of *Prothylacynus,* one of the largest of the South American carnivorous marsupials. [Painting by Charles Knight, from Scott (1913); reprinted by permission of the American Museum of Natural History.] Lower left: the notoungulate *Interatherium.*

reached panther size, and, along with a number of smaller carnivores, represent a distinct family, the Borhyaenidae (Fig. 2). The borhyaenid dentition was essentially didelphoid, but with reduction in the number of upper incisors to three or four pairs, and as in many Australian carnivores the epipubes, or marsupial bones, were also reduced or lost. One subfamily consisted of sabertooths remarkably like the machairodontine felids (Fig. 3). The single known species of an apparently allied family, Necrolestidae, was a fossorial, molelike form. Many of the borhyaenids became extinct before the invasion of placentals from the north, but there is a persistent notion that the disappearance of *all* the marsupial carnivores was "caused" by competition with eutherians. Patterson and Pascual (1972) suggest that competition from carnivorous birds may have been a factor.*

Australian marsupials include a group, Dasyuroidea, whose members are

* Placental chauvinism, or the anthropocentric derogation of thylacate animals, reaches perhaps its ultimate expression in James Blish's novel, "A Case of Conscience" (1958): the pouched reptiles of Lithia (really more like monotremes with external fertilization than marsupials, despite Father Ruiz-Sachez' statement on page 41 that "we have a few animals with a somewhat similar arrangement on Earth, although they are viviparous") are denied even the possibility of Grace.

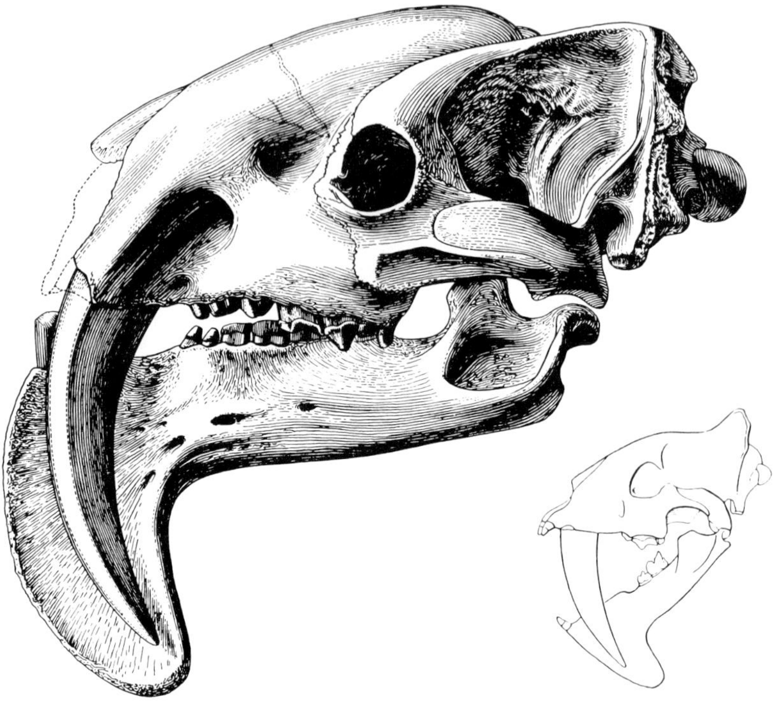

Fig. 3. The skull of *Thylacosmilus atrox,* a panther-sized borhyaenid remarkably similar to the North American saber-toothed felid *Eusmilus* shown in the inset. Not to scale. [Both drawings from Riggs (1934); reprinted by permission of the American Philosophical Association.]

very similar to didelphoids, and in fact one cannot but feel that many fossil (and even some living) marsupials have been assigned to Dasyuroidea or Didelphoidea more because of where they have been found than for any convincing morphological reasons. *Thylacinus,* the Tasmanian tiger (Fig. 4), is so like some of the extinct borhyaenids that its presence in Australia was frequently cited as evidence of a once closer connection between that continent and South America. Sinclair (1906), in fact, referred all the borhyaenids to the family Thylacinidae, and Bensley (1903), one of the most important revisors of Marsupialia, believed *Thylacinus* to be a foreign element in the Australian fauna. Simpson (1941, 1945, 1948) maintained that the resemblances are convergent, and for some time his opinion that the tiger is a dasyurid was generally accepted. However, Ride (1964) stated that *Thylacinus* is not typical of the Australian dasyurids and should at the very least be placed in a separate family, and Archer (1976) has concluded that borhyaenids and thylacinids had a common origin in didelphoids separate from that of dasyuroids, thus approximating Sinclair's view.

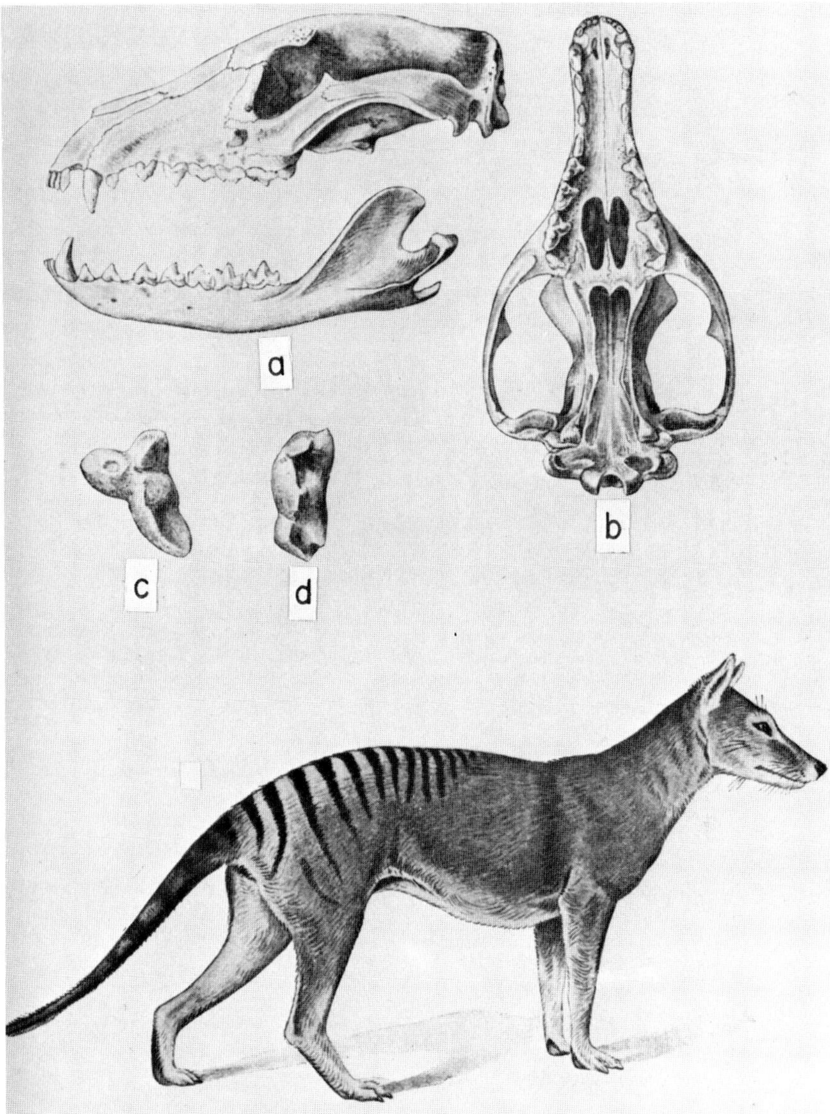


Fig. 4. The Tasmanian tiger, *Thylacinus cynocephalus*, the largest of the dasyuroid marsupials. Note the doglike features of the skull (a, lateral view; b, ventral view), and in particular the specialization of the molars (c,d) for shearing. [From Cabrera's colored drawing, in Cabrera (1919).]

A

B

C

D

E

F

Be that as it may, the case of the tiger does illustrate the similarities between the radiations of Didelphoidea and Dasyuroidea. In some cases, anatomical resemblances are quite close: Fig. 5 (A and B) compares the Chilean didelphid *Marmosa elegans* with the fat-tailed dunnart *Sminthopsis crassicaudata* of Western Australia. Both animals store fat in the tail, although this probably represents adaptations to different climatic conditions in each case. However, there are distinct parallels in the general ecology as well as morphology of members of the two superfamilies; again, in Australia dasyuroids comprise the largest portion of the insectivores and carnivores. Many of them are arboreal, but a number are burrowing desert forms; none, however, is aquatic as is the didelphoid *Chironectes*. Moreover, the arboreal fruit-eating marsupials of Australia belong to a different superfamily, Phalangeroidea, while this niche is partially filled by didelphids (e.g., *Caluromys*) in South America [Fig. 5 (E and F)]. All dasyuroids have a reduced incisor number as do borhyaenids, but the unusual termite-eating *Myrmecobius* has supernumerary molars. There is, as well, an extant family of Australian marsupial moles, Notoryctidae (analogous to Necrolestidae), but it is of very uncertain affinities and may represent a derivative of the Perameloidea.

The Perameloidea, or bandicoots, are a distinctive Australian group. That they probably had a common ancestry with the dasyuroids through a more didelphid-like form is indicated by the retention in many species of five pairs of upper incisors; perameloid molars, in contrast to the triangular molars of didelphids, are square or quadritubercular. In form, most bandicoots are muscular, compact animals weighing about a kilogram or less (but there are some giant New Guinean species), with sharply pointed snouts and insectivorous or omnivorous habits. The front feet are adapted for digging, and some bandicoots, in particular the rabbit-eared bandicoots (*Macrotis*, Fig. 6), are burrowers. *Chaeropus*, a cursorial desert form, has only two functional digits on each forefoot and has accordingly been termed the pig-footed bandicoot.

In fact, it is in the structure of the pes that bandicoots differ as a group most obviously from dasyuroids. In all bandicoots the second and third

Fig. 5. Comparisons of some didelphids (A,C,E) with their ecological and/or morphological counterparts in Australia (B,D,F). (A,B) The southern Chilean *Marmosa* (*Thylamys*) *elegans* and the dasyurid *Sminthopsis crassicaudata*. Note the similarities of these mouse-sized marsupials in head shape and use of the tail for fat storage; (C,D) the Venezuelan *Marmosa* (*Marmosa*) *fuscata* and the arboreal dasyurid *Phascogale tapoatafa*: both animals are about the size of a rat; (E,F) the arboreal, fruit-eating *Caluromys derbianus* and the petaurid *Pseudocheirus peregrinus*: the general ecology and gross morphology of the two animals are much the same, although the Australian marsupial is very different dentally from the polyprotodont dasyurids and didelphids. (All photographs by the author.)

Fig. 6. *Macrotis leucura*, a rabbit-eared bandicoot of Australia. [From a colored plate in Thomas (1888).]

digits are reduced and enclosed in a fold of skin, functioning together as a fur comb. This *syndactyly* is approached in some living didelphids but not in any known dasyurids; on the other hand, syndactyly is entirely characteristic of the Phalangeroidea (see later in this section) which, however, are distinct dentally from bandicoots and other Australian marsupials. Thus the bandicoots have been a troublesome group taxonomically, and the resolution of the conflict is to consider them representative of a group intermediate between Dasyuroidea and Phalangeroidea (Jones, 1923; see Fig. 7). It is this author's belief that the similarity of the pes in perameloids and phalangeroids is convergent [see the discussion in Simpson (1970a)], but Marshall (1972) provides evidence against this view.

The Phalangeroidea (Simpson's sense) is the most diverse of the Australian superfamilies. Phalangeroids are characterized by their syndactyl pes and additionally by reduction of the upper incisors to no more than three pairs, and the functional lowers to a single enlarged and procumbent pair. The adaptation is reminiscent of the dentition of rodents, and the members of at least one family of phalangeroids, Vombatidae (wombats), are much like some rodents in having continually growing teeth. Bensley (1903) believed that this *diprotodonty*, together with the necessary concomitant of an underslung lower jaw, were originally adaptations to grasping insects, and it is true that some of the smallest phalangeroids may use the incisors in this way. At the same time, one can imagine a number of other "uses" for diprotodont teeth—grazing, cropping, browsing, scooping, and, in a strengthened form, piercing—and all of these possibilities have been realized in the Phalangeroidea; for example, the mostly arboreal phalangers and the koala (*Phascolarctos cinereus*) are primarily browsers, while the familiar kangaroos are largely grazing forms. The little honey mouse (*Tarsipes*, Fig. 8) is a nectar and pollen feeder that has practically done away with cheek teeth altogether.

One of the oldest known Australian fossil marsupials, *Wynyardia*, from the early Miocene of Tasmania, is a primitive phalangeroid, but in spite of this provides no real clue to the origins of Phalangeroidea or its relationships with other groups. Similarly, a recently discovered family of extinct marsupials, Ektopodontidae, seems to have been vaguely phalanger-like (Woodburne *et al.*, 1977). The other fossil phalangeroids, like Australian fossil marsupials in general, have been found no earlier than the Oligocene because of the unsuitability of most of Australia for the preservation of Tertiary fossils. As was true of mammals on other continents, many giant forms became extinct at the end of the Pleistocene, among them large browsing kangaroos (Fig. 9) and the Diprotodontidae, quadrupedal forms which included marsupials as large as rhinoceroses (Fig. 10). Perhaps the most striking of the extinct phalangeroids were the Thylacoleonidae, carnivores up to panther size with tusklike incisors and functional cheek teeth

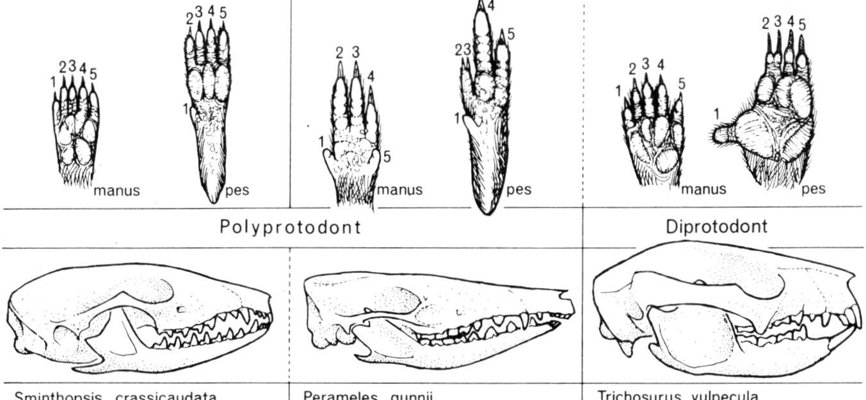


Fig. 7. Diagram illustrating the pes and dental characters upon which the major divisions of Marsupialia are traditionally based, with an example of an Australian marsupial showing each of three of the four possible combinations. Caenolestoids are diprotodont but not syndactylus. [Figure by Frank Knight, from Tyndale-Biscoe (1973); reprinted by permission of Edward Arnold Company.]

Fig. 8. *Tarsipes spencerae,* the honey possum of southwestern Australia. This diprotodont marsupial, which is smaller than a house mouse, is highly adapted for nectar and pollen feeding. (Photograph by the author.)

virtually reduced to a pair of square slicing blades (Fig. 11). It is particularly interesting that a predominately herbivorous group could give rise to such an apparently predaceous animal, and for many years Owen's view (quoted in Broom, 1898) that *Thylacoleo* was "one of the fellest and most destructive of predatory beasts" has been hotly debated.

One group of living marsupials remains to be considered, the South American superfamily Caenolestoidea. They, too, have been something of a taxonomic problem because they are, like the phalangeroids, diprotodont; unlike phalangeroids, they are not syndactyl, and thus, like the bandicoots but with respect to dentition, caenolestoids are taxonomically anomalous.

In spite of this, caenolestoids have frequently been grouped with the phalangeroids, and as with *Thylacinus* have been invoked as evidence for contiguity of the southern continents. Hofer (1952), adducing evidence from Stockard's breeding experiments with dogs (1941), pointed out how easily the two correlated conditions—an underslung jaw and procumbency of the incisors—could arise. Hofer's conclusion that the dental resemblance between phalangeroids and caenolestoids is convergent is strengthened by Ride's (1962) review of the dental embryology of caenolestoids and phalangeroids, which showed that in all probability the diprotodont incisors were not homologous in the two groups; recent work by Kirkpatrick (1969) casts some doubt on Ride's conclusion, however.

It should be said here that there is little doubt as to the monophyly of the Australian diprotodonts; all members of that group share a number of other important characteristics, for example, the superficial thymus gland (Symington, 1898; Johnstone, 1901; Yadav, 1973) and the fasciculus aberrans of the forebrain (see Abbie, 1937, 1941; and Johnson, Chapter 4). Again, caenolestoids possess neither feature, though this would not exclude them as possible ancestors for Phalangeroidea.

In American as in Australian diprotodonts, it is likely that diprotodonty evolved originally as a further adaptation to insectivory—at least this is what one would conclude from a consideration of extant caenolestoids, which are all small, shrewlike marsupials found for the most part in the high, wet, cold paramos of the Andes of Colombia, Ecuador, and Peru (Fig. 12). Analysis of the stomach contents of *Caenolestes obscurus* by Osgood (1921) indicated a diet of insects, but observations the present author made on the same species fed baby murids showed that they can deal effectively with vertebrate prey, using their lower incisors in a quick, rapierlike killing

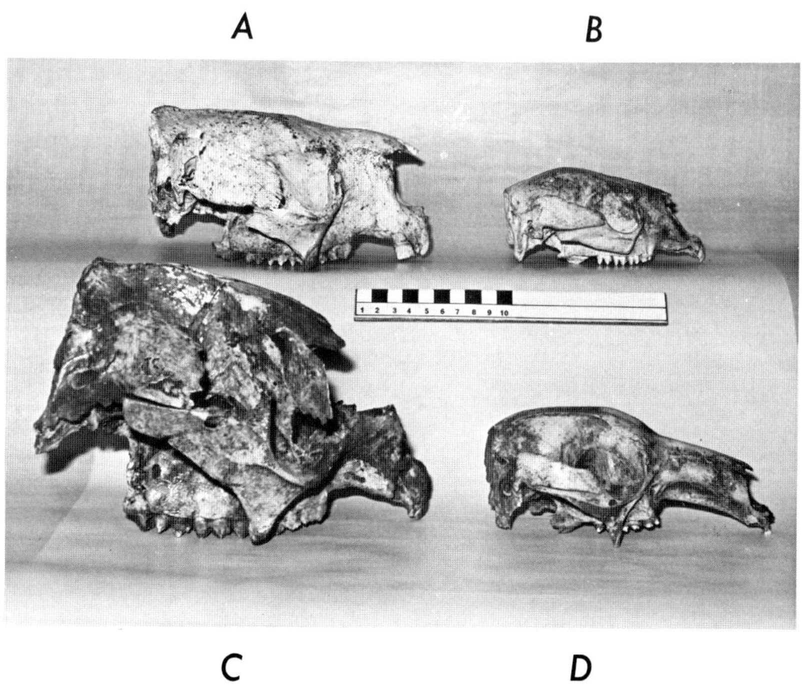


Fig. 9. The skulls of several kangaroos belonging to the Sthenurinae, an extinct subfamily of Macropodidae whose members showed varying degrees of adaptation to browsing. They are compared here with *Macropus fuliginosus*, a grazing macropodine, which is one of the largest of living kangaroos. (A) *Sthenurus* (*Simosthenurus*) sp.; (B) *Sthenurus* (*Sthenurus*) *gilli*; (C) *Procoptodon rapha*; (D) *Macropus fuliginosus*. (Photograph by R. T. Wells.)

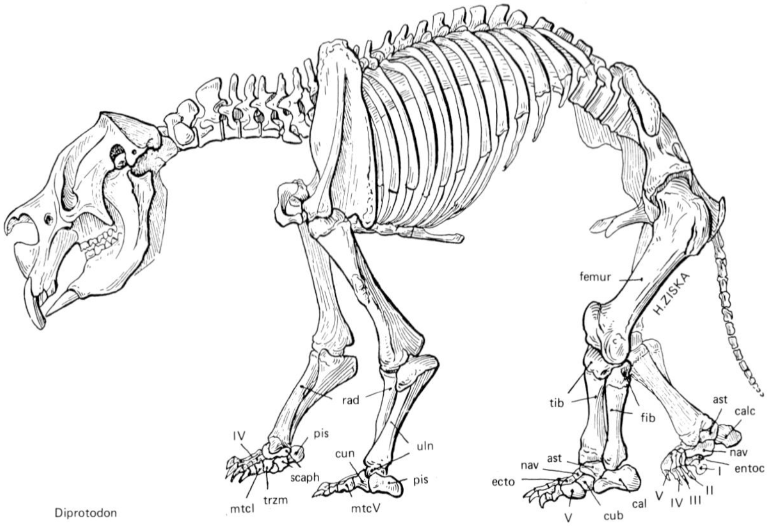


Fig. 10. Reconstruction of the articulated skeleton of the rhinoceros-sized Australian marsupial *Diprotodon australis*, which became extinct in the late Pleistocene, possibly as a result of human predation [Merilees (1968). From Gregory (1951); reprinted by permission of The American Museum of Natural History.]

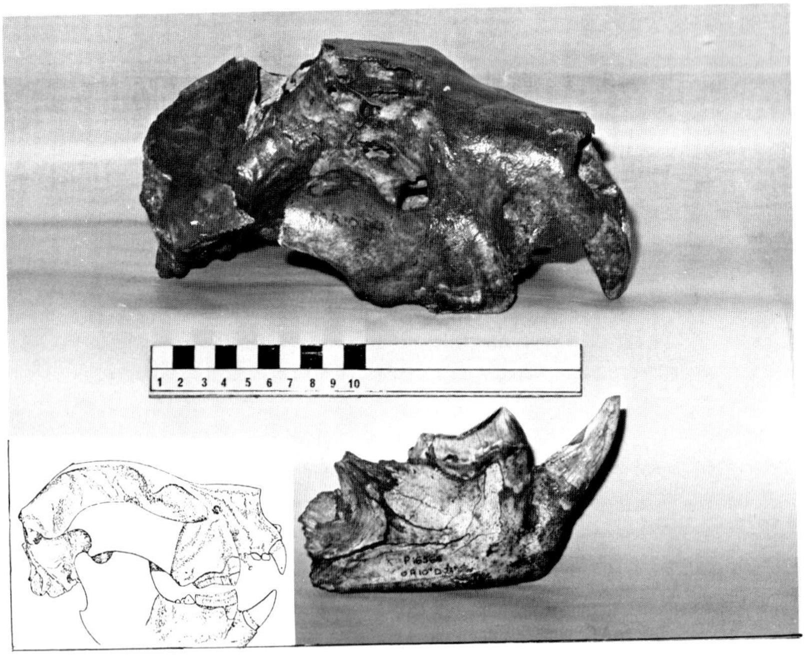


Fig. 11. Skull of *Thylacoleo carnifex*, an extinct phalangeroid of presumably carnivorous habits. [Photograph by R. T. Wells; inset from Flower (1868).]

Fig. 12. The caenolestid marsupial *Lestoros inca*, from near Cuzco, Peru. All caenolestids are about the size of small rats, but are shrewlike in appearance and habits. (Photograph by the author.)

action. Besides showing a wider range of diet in the species, these observations suggest an alternative interpretation of the origin of diprotodonty as a predatory adaptation.

To this point in the discussion there has been no problem in showing relationships of fossil marsupials with a living group, at least at the superfamily level. It is becoming apparent, however, that the South American situation is much more complex than the Australian—not surprisingly, since the relevant part of the Australian fossil record is disappointingly meager. To begin with, the living caenolestoids are hardly a representative sample of their superfamily. The earliest caenolestoids (late Paleocene) are classified as Polydolopidae, a family which is much more specialized dentally than the living Caenolestidae. In particular, polydolopids possess an enlarged shearing cheek tooth characteristic also of some living phalangeroids, including the rat kangaroos and the terrestrial pigmy phalanger *Burramys*. This dental feature is called *plagiaulacoid* after the similarly endowed multituberculate *Plagiaulax*. Pascual and Herrera (1973) have recently mentioned again the possibility that polydolopids may have some special relationship with *Burramys* or other phalangeroids (as opposed, or even in addition, to their relationship with caenolestids), but this does not seem likely, if only because of the tortured phylogeny which must of necessity be involved.

In any event, the polydolopids disappeared completely in the Eocene, although some of the extinct (but later) caenolestids became very polydolopid-like in their convergent development of a shearing tooth. Thus the extant caenolestoids are truly relics of a once greater diversity, and this is all the more impressive in that diprotodonty in caenolestoids probably was directed toward other than the predominately herbivorous ends characteristic of many Australian phalangeroids, owing to the presence in South America of a wide variety of placental herbivores.

To complicate the picture further, Simpson (1970a,b) has shown that two other extinct South American diprotodont marsupial families show no real similarity to caenolestoids beyond this common dental armament. The older of these is Groeberiidae, represented by a few extraordinarily rodent-like mandibles and one skull. Groeberiids appear only in the Oligocene and not afterward, and were possibly a casualty of the invasion of caviomorph rodents (Patterson and Pascual, 1972).

The other group, Argyrolagidae (Fig. 13), appears without obvious progenitors in the Pliocene and disappears without issue or even ecological replacement in the Pleistocene. This short record is surely an artifact of sampling; like many South American Tertiary fossils, argyrolagids are found mainly in Argentina, which may have been the extreme of range for many forms more typical of the poorly known northern part of South America. Argyrolagids were small, saltatory forms astonishingly similar to the North American heteromyid rodent *Dipodomys* or the Australian murid *Notomys*. Thus, in these two marsupial families which, as indicated pre-

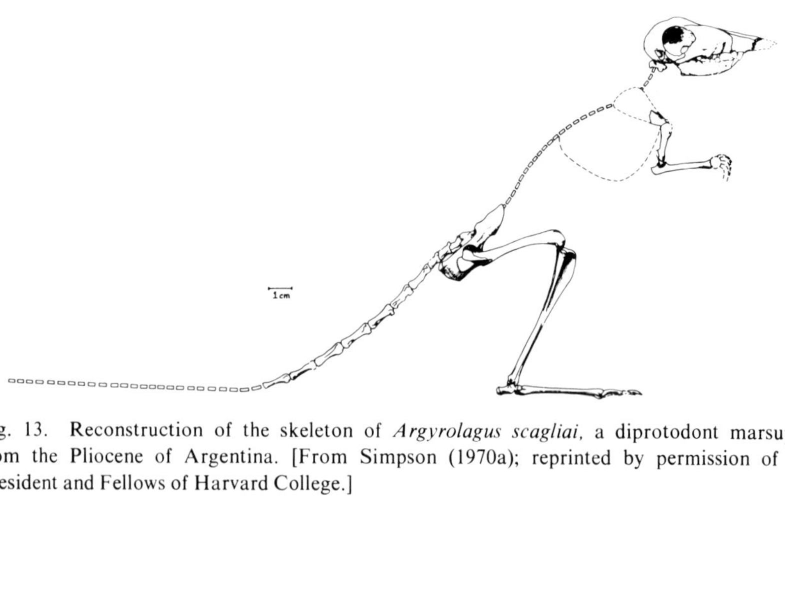


Fig. 13. Reconstruction of the skeleton of *Argyrolagus scagliai*, a diprotodont marsupial from the Pliocene of Argentina. [From Simpson (1970a); reprinted by permission of the President and Fellows of Harvard College.]

viously, Simpson tends to regard as representative of major taxa, we have the closest approach known to the small rodent habitus in marsupials.

To summarize briefly, the fossil history of marsupials is as long as, but probably no longer than, that of placentals; geographically, marsupials were apparently restricted to South America and Australia after the Miocene and until the Panamanian land bridge was reestablished. They proliferated in parallel ways on those two continents, each radiation producing a prodigious variety of carnivorous and diprotodont forms. In spite of the similarities, the two radiations differed significantly in that the Australian group was a complete fauna including herbivores, while in South America marsupials tended to complement the placental herbivores with carnivorous and rodentlike species. The fossil history of South America suggests that the marsupial families there differentiated early, but the Australian record is too poor to indicate if the radiation on that continent paralleled the South American temporally or postdated it. The great distinctness and diversity of the Australian diprotodonts suggest that the major groups were established early in Australia as well, however.

In spite of mooted special relationships between Australian and South American diprotodonts or between *Thylacinus* and the borhyaenids, it is generally assumed that the two radiations proceeded independently of each other. Accordingly, Australian and South American marsupials are implicitly treated by most taxonomists as separate groups. This division is supported by the pairing behavior of South American (but not Australian) marsupial sperm reported by Biggers and Delamater (1966), and is not, as we shall see, contradicted by karyological and serological information.

## III. THE CLASSIFICATION OF AMERICAN MARSUPIALS

Since most of the contributions in this book concern the American marsupials, and especially the Didelphidae, it seems worthwhile to describe the composition of the extant families in a little more detail. Moreover, it will be evident from the review just presented that living American marsupials represent a very attenuated version of the American radiation. The breadth of that radiation suggests that we might expect to find great discontinuities among the present day representatives, and it becomes more important to an interpretation of nontaxonomic studies that we have some idea of the composition of American marsupial genera and of their affinities to each other and the Australian marsupials.

## A. Didelphidae

Following Cabrera (1957), and by present consensus, Didelphidae are allocated to 12 genera. Many of these genera are monotypic, but there is little doubt that the contained species of each are truly generically distinct. On the other hand, over 50 species of *Marmosa* have been described, and there have been suggestions that this assemblage should be considered more than a single genus. Each of the 12 genera is characterized ecologically by Hunsaker (Chapter 3), and the comments here are limited mostly to taxonomy.

### 1. *Didelphis*

There are three species (Gardner, 1973): the North and Central American *D. virginiana* is sympatric in Central America with the predominantly South American *D. marsupialis*; and *D. albiventris*, mostly found at higher altitudes in South America, is readily distinguished from the others by its white ears. All three have well-developed pouches.

### 2. *Philander*

The common species, *P. opossum*, the "gray four-eyed opossum," is found throughout Central and South America, and is somewhat smaller than *Didelphis* species. The fur is short, soft, and gray, lacking the coarse guard hairs of *Didelphis*; a dark form restricted to Peru (*P. mcilhennyi*) is also dentally distinct (Gardner and Patton, 1972). The pouch is present.

### 3. *Metachirus*

*Metachirus nudicaudatus* is the only species now recognized. It is sometimes called the "brown four-eyed opossum" because its markings are similar to those of *Philander opossum*, but the fur is, as indicated, brown or rusty red and much shorter and coarser. Moreover, the female lacks a pouch, and both sexes are posturally and behaviorally distinct from *Philander*.

### 4. *Lutreolina*

This pouchless genus is also monotypic, and the reddish weasellike *L. crassicaudata* is somewhat adapted to an aquatic habitat.

### 5. *Chironectes*

Fully aquatic habits are associated with *C. minimus*, the single species of this genus. The fur is short and individual hairs are flattened, the hind feet are webbed, and the tail is laterally compressed. Not surprisingly, the pouch is well developed. The skull is short and the jaws powerful; the markings are dark brown with lighter patches.

### 6. *Marmosa*

*Marmosa* has more described species than any other marsupial genus, and has been variously subdivided into sections or subgenera. In Tate's 1933 revision—arguably his finest work on marsupials—he recognized five groups with sections in several of them, and a total of 50 species (not 49 as he states). Cabrera (1957) reassessed the status of the South American species and reduced their number considerably, adding one not recognized by Tate, and placing the 37 species in the subgenera *Marmosa* and *Thylamys*. Cabrera's arrangement is the starting point for current lists, and his subgeneric grouping is implicitly accepted by Pine (1972) in describing a third subgenus for the new species *Marmosa* (*Stegomarmosa*) *andersoni*. Pine believes that *Marmosa* probably represents more than a single genus, the recently described *Marmosa* (*Thylamys*) *karimii* being as different (in his opinion) from typical *Marmosa* as is *Didelphis*. Although the present author's acquaintance with living *Marmosa* species is limited to some half-dozen forms, Pine's opinion seems valid. The subgenus *Marmosa* currently includes animals as different as the familiar *M. robinsoni, M. fuscata* (Fig. 5C), and the large *M. cinerea,* which is reminiscent of *Caluromys*. For the moment, pending Pine's further work, we can accept Cabrera's listing of 37 species, with the addition of three North American species (*M. mexicana, M. invicta,* and *M. canescens*; see Hall and Kelson, 1959), all belonging to the subgenus *Marmosa*; and several South American species which have been described since. These include *M. tatei* Handley, 1956, *M. karimii* Petter, 1968, both assignable to the subgenus *Thylamys*, and *M.* (*Stegomarmosa*) *andersoni* Pine, 1972. As Pine (in Collins, 1973) indicates, *M.* (*Thylamys*) *formosa* should also be considered a full species. Thus, there are 44 *Marmosa* species to be reckoned with; a complete list is given in Kirsch and Calaby (1976). *Marmosa* species lack a pouch.

### 7. *Monodelphis*

This genus is also rather diverse, with 11 species currently recognized. Only one *Monodelphis* is found in Central America, a subspecies of *M.*

*adusta*. Reig *et al.* (1977) distinguish *M. orinoci* from *M. brevicaudata*, but Reig (1964) synonomized *M. fosteri* with *M. dimidiata*. Female *Monodelphis* have no pouches.

## 8. *Lestodelphys*

The single species of this genus, *L. halli*, is one of the rarest of the didelphids. It is found only in Patagonia, and was compared by Thomas (1921) with *Marmosa*, though Reig and Simpson (1972) have noted similarities to the extinct sparassocynine marsupials (family Didelphidae). It is not known if there is a pouch.

## 9. *Caluromys*

This genus was long confused nomenclaturally with *Philander*, and contains three species. *Caluromys derbianus* is found throughout Central America and the western part of the Andean countries of South America; the similar *C. lanatus* lives on the eastern slopes of the Andes and in the adjacent Amazon basin, while the much smaller *C. philander* is found in northeastern South America. The pouch is well developed, except in *C. philander* (Pine, 1973), and the common name woolly opossum refers to the luxurious orange and gray fur.

## 10. *Caluromysiops*

As the generic name suggests, the single species *C. irrupta* is similar to *Caluromys*, but the waxy yellow ears and rounder head are distinctive. The species is limited to Peru. The recorded specimens are males, so that it is unknown whether there is a pouch.

## 11. *Glironia*

The two species, *G. venusta* and *G. criniger*, are rare didelphids to judge from the few numbers of recorded specimens (four, although this author inspected two more in Quito). They are apparently also eastern slopes rain-forest species, found in Ecuador, Peru, and Bolivia. The tail is fully furred, and it is unknown whether there is a pouch.

### 12. *Dromiciops*

*Dromiciops australis,* the only species of the genus, is found in southern Chile and adjacent Argentina. The pouch is well formed.

Except for the provocative paper of Reig (1955), there have been few attempts to group the living didelphid genera at the subfamilial level. According to Reig, *Dromiciops* answers closely to the diagnosis of Ameghino's *Microbiotherium*; Reig further maintained that *Caluromys, Caluromysiops,* and *Glironia* partly matched that diagnosis. Hence he considered these four genera to be living members of Microbiotheriinae, if the taxon represented by *Microbiotherium* is ranked as a didelphid subfamily. Reig's hypothesis has provided a base for diverse studies of didelphid taxonomy, as will be shown in Section VI.

## B. Caenolestidae

All caenolestids prefer a densely vegetated, humid microhabitat, typically in the scrub adjacent to the meadows of the high, wet Andean paramos. They appear to be insectivorous, but probably prey on small vertebrates as well, as previously suggested. All lack pouches. The living caenolestids comprise three genera, two of which are monotypic.

### 1. *Caenolestes*

The type genus contains five species, although it seems likely that two or three is a more accurate number. *Caenolestes fuliginosus,* the type species, was originally taken near Quito and occurs as far south as northern Peru; *C. convelatus* is known to occur only near Quito; and *C. obscurus* has been taken near the Venezuelan–Colombian border and throughout the latter country. The other two species, *C. tatei* and *C. caniventer,* known from a single locality each, contrast with the first three in that they are found at decidedly lower altitudes (under 2000 m) and may represent a semitropical adaptation.

### 2. *Lestoros*

The single species *L. inca* (Fig. 12) has only been taken in the vicinity of Cuzco and Machu Picchu, Peru, and is readily distinguishable from *Caenolestes* species by its small canine. This dental characteristic probably indicates a less predaceous habit.

### 3. *Rhyncholestes*

Only three specimens of the one species, *R. raphanurus,* are recorded in the literature. Interestingly, the sexes are dimorphic with respect to the canine, which is large in the male and premolariform in the female. The species is further distinguished by its double-cusped lateral upper incisors (unique among marsupials) and incrassated tail; the other southern Chilean marsupials, *Dromiciops australis* and *Marmosa elegans,* and the Patagonian *Lestodelphys,* also store fat in the tail, and this may well be an adaptation to the cold, wet winters.

The chief taxonomic interest of caenolestids is of course their resemblance to some Australian marsupials, but even if this similarity is accepted as convergent, the question remains of caenolestid relationships to didelphids and the implications for studies of caenolestid biology.

## IV. METHODS

Although the results obtained from other approaches to classification are discussed in this paper, cytological and serological studies are considered in greatest detail. Methodological details will be found in the cited papers; here only an outline will be given of the techniques employed.

### A. Cytology

Most of the earliest chromosome studies of marsupials were done using testes squashes, and this technique was employed by Sharman (1961) as well. This method yields adequate preparations for determination of numbers and general morphology. Tissue culture techniques have become general in recent years, although these are hard to carry out in the field. The procedures used vary from laboratory to laboratory but generally involve culture of leukocytes or organ material in fetal calf serum and culture medium for several days, after which the cells are plated out, fixed, and stained (see Schneider, Chapter 2, for a more detailed description).

Bone marrow and spleen preparations have been used extensively by Reig *et al.* (1977) and Hayman and Martin (1974). In this method, the animal is injected intraperitoneally with dilute colchicine several hours before killing, which has the effect of inhibiting spindle formation and thus accumulating cell divisions in the metaphase stage; accordingly, the chances of finding good chromosome spreads are enhanced. The marrow is aspirated from the dissected long bones and broken up in culture medium; spleen or other organs are similarly treated. After treatment in hypotonic KCl to expand the cells, the preparations are fixed in acetic-alcohol and stained. Conjunctival cells removed from the same animals and stained also provide adequate preparations. While constant temperature baths and centrifugation are needed, the use of such equipment in the field is not beyond possibility. When suitable controls are prepared, accurate DNA estimations are possible as well.

Although for the purposes of this essay chromosome numbers and general morphology are sufficient to make the most important points, it is worth mentioning here the quantitative DNA estimation technique developed by Martin and Hayman (1965), since this promises to increase the usefulness of chromosome studies, particularly in comparing very similar karyotypes and in revealing homologies between distinct ones.

The usual procedure in comparing the karyotypes of different species is to construct an ideogram of the paired metaphase chromosomes, usually in descending order of size, with the sex chromosomes at the end of the series. Next, the lengths of the arms are measured, and their percent lengths relative to the summed lengths of all chromosomes computed. The percent lengths of presumably homologous chromosomes (or arms) from different species are then compared. Equal percents are considered to indicate exact homology; owing to variations in "spread" in different preparations, absolute lengths of arms cannot be compared directly. However, the percent comparisons assume an equal amount of nuclear DNA in each species, which Martin and Hayman (1965) point out may be untrue; marsupial species may differ $1\frac{1}{2}$ times in DNA content, and may differ as much as 30% between species with the same apparent karyotype, but this is not a result of polyploidy. In Martin and Hayman's quantitative method, a microdensitometer is used to measure the total DNA content of a nucleus and, operating on the assumption that the DNA is uniformly distributed throughout the chromosomes, the amount of DNA in each arm is computed. Next, having established the distribution of apparently identical arms (defined in terms of DNA content as well as relative length) over the entire group of species, a hypothesis of relatedness based on the probabilities of co-occurrence of identical chromosomes or arms in any two of them is constructed and tested.

### B. Serology

As defined here, comparative serology is the study of antigens and antibodies and the reactions between them for the purpose of establishing the similarities in antigens from different species. *Antigens* are proteins which, when injected into a foreign vertebrate host, will provoke the production of complementary molecules (*antibodies*) capable of combining with the antigens and inactivating them in one way or another. As antibodies are usually discharged into the host's bloodstream in excess of the need for them, the host's serum (*antiserum*) becomes a reagent for the detection of the antigens in "unknown" samples. The value of this technique for taxonomic comparisons depends on the fact that the antigen–antibody reaction is usually not specific to the original antigens—that is, antibodies are generally capable of combining with antigens that are similar but not identical to the original, stimulating ones. When the antigen sources are a set of related species, the extent of cross-reaction of a particular antiserum with antigens from each of these can be taken as an index of similarity of those proteins, and presumably of the organisms from which they came.

In most comparative serological studies, the antigens used are soluble serum proteins, and the reaction measured is ordinarily the precipitation of these from solution by antibodies. The classic technique is the photoelectric measurement (Boyden, 1942) of the turbidity generated by a mixture of antigens (usually whole serum) and antibodies (antiserum prepared in a rabbit or chicken against the antigens of one of the organisms compared). Since the reaction is inhibited when either reagent is in excess, turbidimetric tests are usually made with a range of antigen dilutions, generating a curve which has its peak at the point of optimum proportions. For any particular antiserum, the curve for each antigen mixture (serum) in the series is compared to the standard (reaction of the stimulating antigens) to give an index of similarity. It is an important logical limitation of this procedure that organisms whose antigens cross-react to about the same extent with a particular antiserum are not necessarily the same; equivalent reactions only mean that they share the same number of serological characters with the standard, and only an antiserum against one or the other of such cross-reacting sera can test whether they are indeed identical. Ideally, one should prepare antisera against the antigens of each of the organisms being compared, but this is often not practical.

Since classic quantitative precipitin tests require large amounts of reagent, semimicroqualitative techniques have, since their introduction, largely superseded turbidimetry. *Double diffusion* visualizes the antigen–antibody reaction in agar gel as a series of lines when a mixture such as native serum is tested. A pattern of wells is punched in a thin layer of agar and charged with antigens and antiserum in such a way that the lines formed by two or more cross-reacting sera can be compared side by side. A related technique is *immunoelectrophoresis,* in which the components of the antigen mixture are first separated in an electric field. Patterns can be compared simply by counting the number of lines formed; additionally, inferences of homology can be made on the basis of similarities in electrophoretic mobilities of the components.

Finally, two newer quantitative techniques should be noted. One, *microcomplement fixation,* has been popularized by Sarich and Wilson (e.g., 1967; Wilson and Sarich, 1969) and applied by V. M. Sarich (unpublished) to marsupial problems. Antigen–antibody reactions depend not only on the presence of these two reagents, but on a series of "auxiliary" proteins collectively known as *complement,* which are normally present in excess, and which may be inactivated by preheating the antiserum so that precipitation cannot occur. The microcomplement fixation test indirectly measures the amount of complement used up in an antigen–antibody reaction in the following way: antiserum with inactivated complement is reacted with antigen in the presence of a known amount of complement, usually in the form of guinea pig serum. Sheep red blood cells (RBC) sensitized with inactivated antiserum against the cells are added to this mixture. The leftover complement will complete the RBC/anti-RBC reaction and cause hemolysis, the degree of which is dependent on the amount of complement and which can be measured spectrophotometrically. Since the amount of leftover complement will depend on the amount used in the first antigen–antibody reaction, hemolysis gives an indirect measure of that reaction.

Microcomplement fixation is best applied to single proteins, since separate titration curves will apply for each protein in a complex mixture. A similar technique (in that it is indirect) which enables one to compare mixtures is *hemagglutination inhibition.* In this procedure, sheep RBC are treated with dilute tannic acid, which renders them capable of being coated with serum. They are then coated with the reference antigens (the stimulating serum) and added to a solution in which the appropriate antiserum has been absorbed with a cross-reacting antigen. To the extent that the cross-reacting serum is similar to the referent antigens, the agglutination of the RBC (which would ordinarily result from mixture of sensitized cells with antiserum) is inhibited. The degree of inhibition obtained with different absorptions is a measure of similarity of the antigens. This test is sensitive to differences even between congeneric species (Kirsch and Murray, 1969), when appropriate antisera are used.

Although most numerical taxonomic computer problems are not entirely suitable for analyzing serological data (Kirsch, 1969), when they are applied with a discriminating consideration of the nature of the data they can provide additional insight into relationships. In this chapter I have given the results of applying the "classical NT-SYS" programs developed at the University of Kansas to two sets of data involving immunoelectrophoretic tests on representatives of the major groups of marsupials. Correlation matrices were computed from the data and a variety of clustering algorithms applied; principal components analysis was also performed on each correlation matrix and provided the basis for three-dimensional representations of affinities.

## V. RESULTS

### A. Cytology

The range of chromosome numbers in Marsupialia is not great. The most variable family is Macropodidae, with from $2n = 10$ to 32 chromosomes, only the phalanger *Pseudocheirus cupreus* having as few as $2n = 10$ and no other marsupial having more than $2n = 24$ (*Tarsipes spencerae*). Thus, as compared with other mammalian groups, marsupials are fairly uniform and low in number; among vertebrates only the Indian muntjac (*Muntiacus muntjak*, $2n = 6$ in the female and 7 in the male) has fewer chromosomes (Wurster and Benirschke, 1970).

The sex-determining mechanism in marsupials is normally XX/XY, with the male always heterogametic, but some kangaroos (*Potorous tridactylus* and *Wallabia bicolor*) and the rabbit-eared bandicoot (*Macrotis lagotis*) have trivalent $XX/XY_1Y_2$ systems, a result of incorporation of autosomal material in the sex chromosomes. Another kangaroo, *Lagorchestes conspicillatus*, is possibly polymorphic for the autosomes and in addition has an $X_1X_1X_2X_2/X_1X_2Y$ sex-determining mechanism. Cooper *et al.* (1971) suggest that X-inactivation at least in kangaroos is not random, for the evidence is that it is always the paternal X which is inactivated; bandicoots other than *Macrotis* normally eliminate one X (in females) or the Y (in males) from most somatic tissue cells, which may represent an extreme example of the Lyon hypothesis. Two bandicoots, *Echymipera kalubu* (Hayman *et al.*, 1969) and *E. rufescens* (Sharman, 1973b), and the greater gliding phalanger *Schoinobates volans* (Hayman and Martin, 1965), have supernumerary chromosomes. In *E. kalubu* the supernumeraries are lost along with one of the sex chromosomes in somatic cells; accordingly, Hayman *et al.* (1969) suggest that such extra chromosomes may derive from the same genetic material as the sex chromosomes.

Further aspects of these unusual cytological phenomena are summarized in Schneider (Chapter 2); here we are most concerned with the numbers and distributions of "normal" karyotypes. Figure 14 is a histogram of the frequencies of occurrence of the diploid numbers of chromosomes found in marsupials, and representative karyotypes are shown in Figs. 15, 16, and 17.

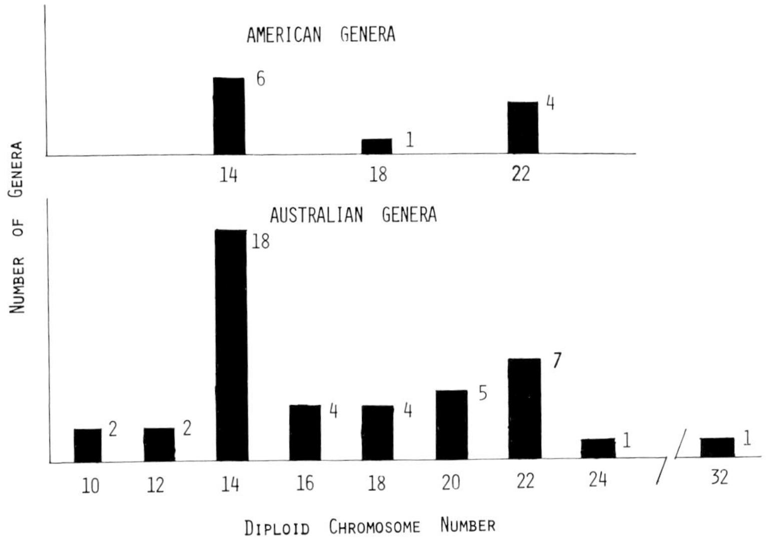


Fig. 14. Histogram showing the frequencies of known chromosome numbers of marsupials. [Redrawn after Hayman (1976); reprinted by permission of Macmillan Press, London.]

As Sharman (1961) pointed out, there are modes at $2n = 14$ and 22 among the Australian marsupials, and subsequent work has shown that both modes occur among American marsupials as well. Interestingly, few species possess intermediate numbers.

Table I gives the ranges of numbers found in each extant family, and as Fig. 15 shows, in most of the forms with $2n = 14$ chromosomes those chromosomes have strikingly similar morphologies. The autosomes consist of three pairs of large metacentrics, one pair of medium metacentrics (or submetacentrics), and two pairs of short, usually metacentric, chromosomes. The sex chromosomes are variable in size. On the other hand, there is less consistency among the morphologies of the $2n = 22$ karyotypes: Fig. 17 indicates that the autosomes are usually all acrocentric in American marsupials with $2n = 22$, but they may be different in Australian species (Fig. 16). Hayman and Martin [(1974); but earlier in Martin and Hayman (1967)] therefore argue that, on frequency of occurrence, similarity of chromosome form, and breadth of taxonomic distribution, the $2n = 14$ pattern should be considered the basic one from which other karyotypes have been derived; this point, and its phylogenetic implications, are taken up in Section VI.

Martin and Hayman (1967; Hayman and Martin, 1974) showed that not only do the 14-chromosome patterns look much alike, but their DNA

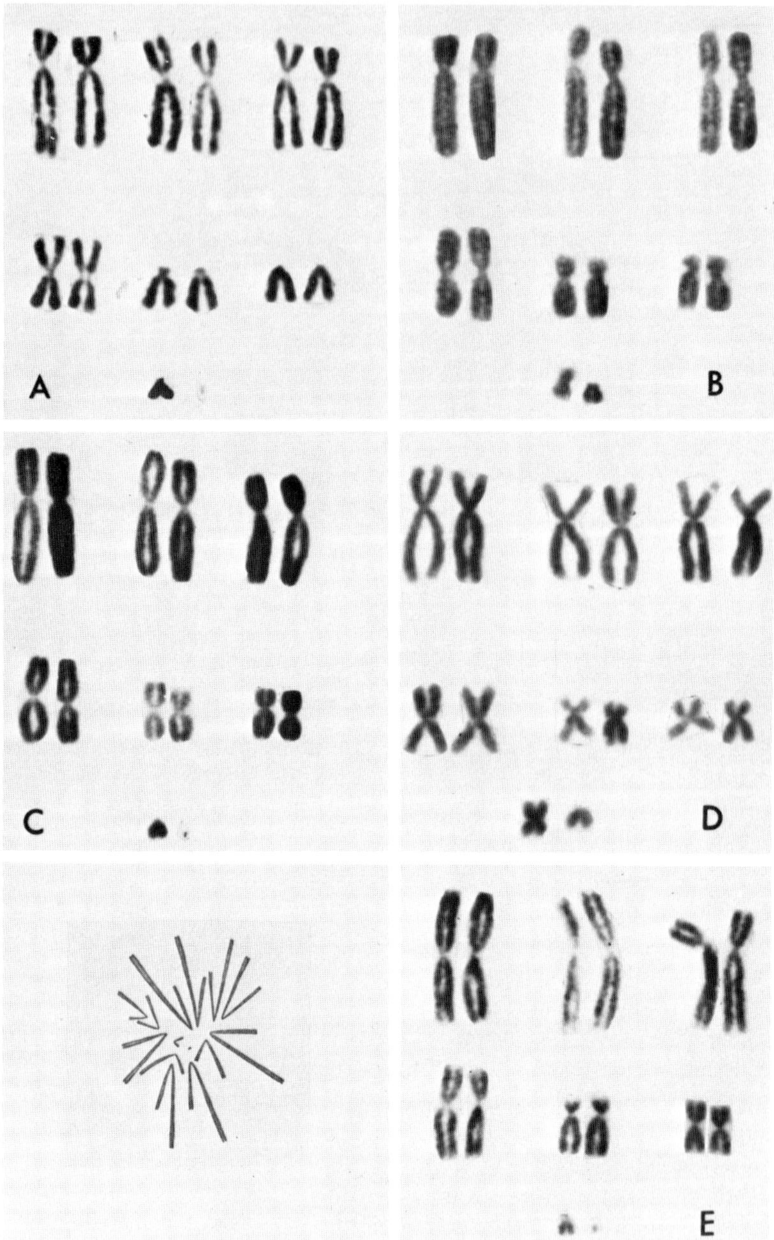


Fig. 15. Comparison of the karyotypes of marsupials with the basic form of the complement $2n = 14$, and Hayman and Martin's reconstruction of the karyotype of a Cretaceous marsupial (lower left). (A) *Acrobates pygmaeus*, a phalangeroid; (B) *Echymipera kalubu*, a perameloid; (C) *Caenolestes obscurus*, a caenolestoid; (D) *Marmosa* (*Marmosa*) *fuscata*, a didelphoid; (E) *Dasyurus geoffroii*, a dasyuroid. [After Plate 1 and Fig. 22 in Hayman and Martin (1974).]

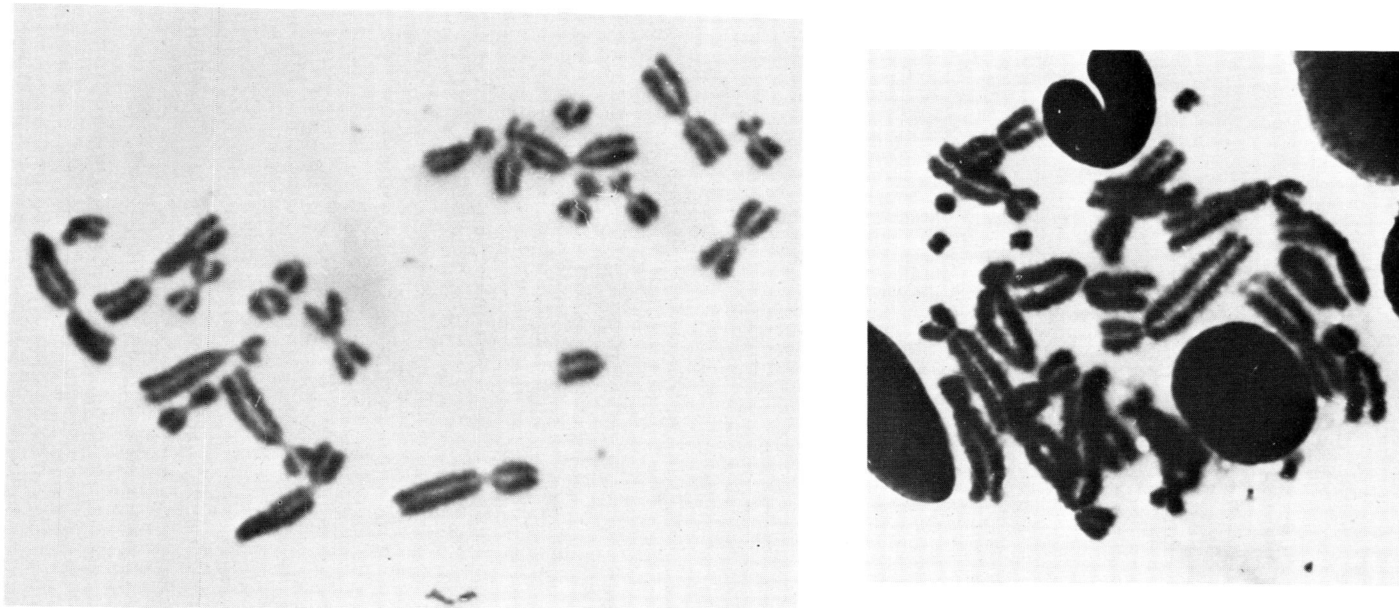

A B

Fig. 16. The chromosomes of two phalangeroid marsupials with $2n = 22$, illustrating the differences in form between these complements and those of didelphids with the same number (cf. Fig. 17). (A) *Bettongia penicillata*, a macropodid; (B) *Schoinobates volans*, a petaurid which possesses a variable number of supernumerary chromosomes. (Photographs by D. L. Hayman.)

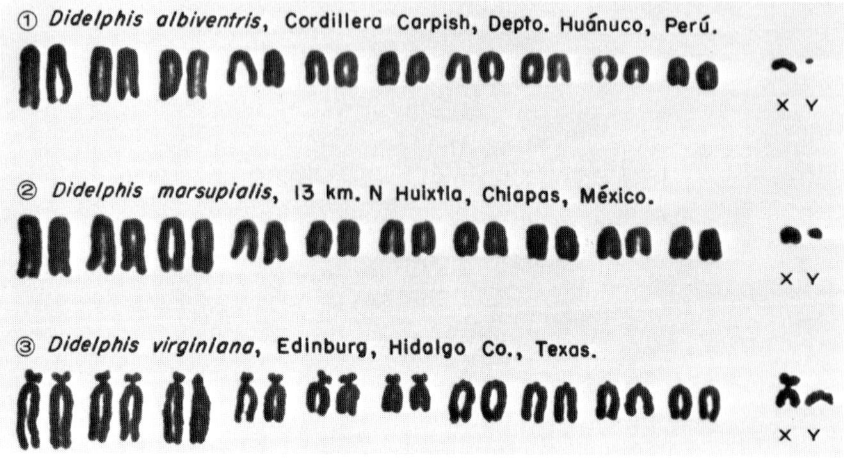


Fig. 17. Comparison of the karyotypes of the three species of *Didelphis*. [From Gardner (1973); reprinted by permission of Texas Tech Press.]

content is frequently similar and is similarly distributed among individual arms; variation is least among the karyotypically uniform Dasyuridae, and thus it was possible for Martin and Hayman to establish a "standard dasyurid" against which other marsupials might be compared after arm lengths had been standardized for DNA content (see Section IV,A). They measured 182 arms in 13 species, finding that the most common DNA-corrected length was 4.0 to 4.5% of the total arm lengths, and that this length occurred 23 times. On the basis of this distribution, the probability of two species sharing a chromosome arm of this (corrected) length was calculated and provided a basis for interspecific comparisons. Sharing of two two-armed chromosomes was considered minimal evidence of relationship.

The comparisons showed that the pigmy phalangers *Cercartetus*, which have the same apparent karyotype as the dasyurids but with slightly different arm lengths, share two chromosomes with the koala *Phascolarctos cinereus* and the wombat *Vombatus ursinus*; the other wombat tested, *Lasiorhinus latifrons*, shares two with *Phascolarctos* as well. In the Peramelidae, two species of *Isoodon* share two chromosomes; *Perameles nasuta* shares two each with *Isoodon macrourus* and with *Lasiorhinus*. No sharings were evident between dasyurids and bandicoots.

So far as these comparisons go, then, they do indicate some interrelationships between and within superfamilies, but they would seem to be equally significant in suggesting that within a basic form of karyotype considerable rearrangement of genetic material can take place. On the other hand, in an earlier paper involving kangaroos Martin and Hayman (1965) demonstrated sharings among four species, though the diploid chromosome numbers were quite different—two species had 16, one 10, and one 20 chromosomes.

No DNA measurements have yet been made on Didelphidae, but Table II shows that this is a family with little numerical diversity. Most species have either $2n = 14$ or 22 chromosomes, but the species of *Monodelphis* examined interestingly enough have 18. Other than in *Monodelphis*, this number appears only in the female *Macrotis lagotis* and in four genera of phalangeroids, all Australian marsupials. There is some variation in the appearance of certain of the autosomes (pairs 5 and 6) in different species

**TABLE II**

Chromosome Numbers of American Marsupials[a]

| Species | 2*n* | Reference |
|---|---|---|
| Family Didelphidae | | |
| *Didelphis albiventris* | 22 | Sharman (1961) |
| *Didelphis marsupialis* | 22 | Sharman (1961) |
| *Didelphis virginiana* | 22 | Sharman (1961) |
| *Philander opossum* | 22 | Biggers *et al.* (1965) |
| *Lutreolina crassicaudata* | 22 | Sharman (1961) |
| *Chironectes minimus* | 22 | Hayman and Martin (1974) |
| *Marmosa cinerea* | 14 | Hayman and Martin (1974) |
| *Marmosa elegans*[b] | 14 | Reig *et al.* (1972) |
| *Marmosa fuscata* | 14 | Reig and Sonnenschein (1970) |
| *Marmosa mexicana* | 14 | Biggers *et al.* (1965) |
| *Marmosa murina* | 14 | Hayman and Martin (1974) |
| *Marmosa robinsoni* | 14 | Reig (1968) |
| *Metachirus nudicaudatus* | 14 | Hayman and Martin (1974) |
| *Monodelphis brevicaudata* | 18 | Reig and Bianchi (1969) |
| *Monodelphis dimidiata* | 18 | Reig and Bianchi (1969) |
| *Caluromys derbianus* | 14 | Biggers *et al.* (1965) |
| *Caluromys lanatus* | 14 | Hayman and Martin (1974) |
| *Caluromys philander* | 14 | Hayman and Martin (1974) |
| *Dromiciops australis* | 14 | Reig *et al.* (1972) |
| Family Caenolestidae | | |
| *Caenolestes convelatus*[c] | 14 | Hayman *et al.* (1971) |
| *Caenolestes fuliginosus* | 14 | Hayman *et al.* (1971) |
| *Caenolestes obscurus* | 14 | Hayman *et al.* (1971) |
| *Lestoros inca* | 14 | Hayman *et al.* (1971) |

[a] All known species have an XX/XY sex-determining mechanism. After Table I in Hayman (1976).

[b] This is the sole species of the subgenus *Thylamys* studied.

[c] Included in *C. fuliginosus* in Hayman *et al.* (1971).

of *Marmosa,* and the Virginia opossum, *Didelphis virginiana* (which had been considered conspecific with *D. marsupialis*), has some submetacentric autosomes. In this respect, *D. virginiana* differs from the other two species of the genus (Fig. 17) and from other didelphids with $2n = 22$, and such a difference would appear to constitute an effective breeding barrier. In his revision of *Didelphis,* Gardner (1973) showed that the North American species is sympatric with the mainly South American *D. marsupialis* at least from southern Mexico to Costa Rica. Besides the chromosome differences, the two opossums are distinct in pelage and in some details of osteology. Gardner also found behavioral differences, such as a tendency of the southern form not to "play 'possum." Gardner found no evidence of hybridization, and there can be little doubt that the Virginia opossum is a distinct species.

## B. Serology

Nuttall's early studies (1904) on a few marsupials demonstrated their distinctness from other mammals and the homogeneity of the families represented. Weymss' later work (1953) indicated that the Virginia opossum is very different from the Australian marsupials, and that marsupial families seem to be more distinct from each other than are the families of placental orders.

The present author's work has confirmed these general points, and has demonstrated a number of subgroupings of the taxa included in Marsupialia. Some of these results, and the inferences drawn from them, have been summarized in previous papers (Kirsch, 1968a,b, 1969, 1970, 1973; Hayman *et al.,* 1971), and the major part of the data on which they are based is in preparation. Some results which pertain to the major classification of marsupials, and to the generic groupings of didelphids, will be described briefly here.

The stereogram shown in Fig. 18 was derived by NT-SYS analysis of immunoelectrophoretic data on the 14 species included in the figure. The tests employed nine antisera against nine species. There is one artifact in the figure: the slight separation of *Perameles nasuta* from the other bandicoots does not correspond to the indications of the tabulated data. Otherwise, the homogeneity of some families (Dasyuridae, Peramelidae, Macropodidae) is as expected from classical taxonomy, and there is an indication of a special association of the dasyurids and bandicoots. The phalangeroid species (Simpson's sense) group together, but *Phascolarctos* and *Vombatus* stand apart from the rest of that group and are close to each other.

Additional experiments with phalangeroids, encompassing sera from a

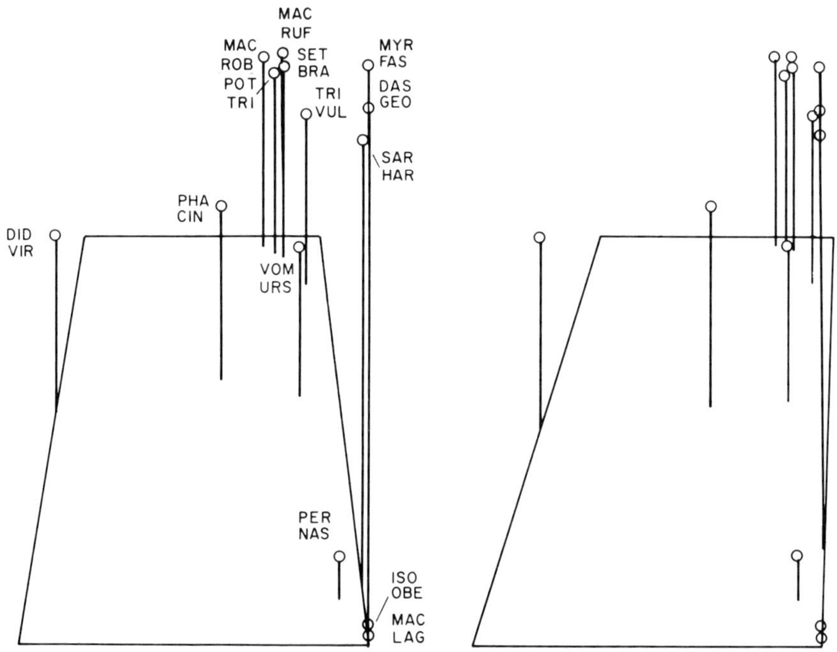


Fig. 18. Three-dimensional representation of the affinities of 14 species of mostly Australian marsupials, derived by NT-SYS analysis of immunoelectrophoretic data. The "balls" represent species whose locations in the cube of space above the plane are indicated by the "wires" to which they are affixed. Abbreviations: DID VIR, *Didelphis virginiana*; DAS GEO, *Dasyurus geoffroii*; SAR HAR, *Sarcophilus harrisii*; MYR FAS, *Myrmecobius fasciatus*; PER NAS, *Perameles nasuta*; ISO OBE, *Isoodon obesulus*; MAC LAG, *Macrotis lagotis*; PHA CIN, *Phascolarctos cinereus*; VOM URS, *Vombatus ursinus*; TRI VUL, *Trichosurus vulpecula*; POT TRI, *Potorous tridactylus*; SET BRA, *Setonix brachyurus*; MAC ROB, *Macropus robustus*; MAC RUF, *Macropus rufus*.

wider range of species, show that the phalangers comprise several groups as distinct from each other as they are from the kangaroos. Most relevant to this discussion, the small phalangers (*Burramys, Cercartetus, Acrobates*), all with $2n = 14$ chromosomes, form a definite group. The morphologically aberrant *Tarsipes* is also distinctive serologically and, on that basis, seems to represent a third major group of Australian marsupials.

Similarly, *Notoryctes* serum shows no special affinities with any group of Australian species, but more similarity to Australian marsupials as a whole than to several American marsupials with which *Notoryctes* was compared.

Figure 19 is a perspective drawing of the affinities of another group of 14 species, mainly American marsupials, derived from immunoelectrophoretic tests employing eight antisera. Since the experimental coverage of

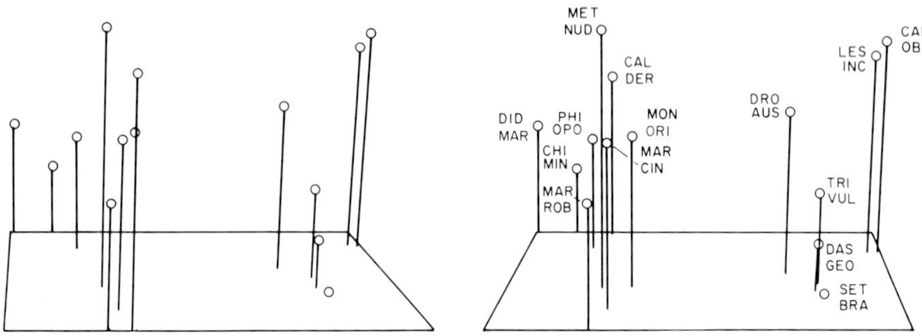


Fig. 19. Three-dimensional representation of the affinities of 14 species of mostly American marsupials, derived by NT-SYS analysis of immunoelectrophoretic data. The "balls" represent species whose locations in the cube of space above the plane are indicated by the "wires" to which they are affixed. Abbreviations: DID MAR, *Didelphis marsupialis*; PHI OPO, *Philander opossum*; CHI MIN, *Chironectes minimus*; MET NUD, *Metachirus nudicaudatus*; MAR ROB, *Marmosa* (*Marmosa*) *robinsoni*; MAR CIN, *Marmosa* (*Marmosa*) *cinerea*; MON ORI, *Monodelphis orinoci*; CAL DER, *Caluromys derbianus*; DRO AUS, *Dromiciops australis*; DAS GEO, *Dasyurus geoffroii*; TRI VUL, *Trichosurus vulpecula*; SET BRA, *Setonix brachyurus*; CAE OBS, *Caenolestes obscurus*; LES INC, *Lestoros inca*.

didelphids was greatest, the NT-SYS analysis tends to group marsupials of other families closer to each other than the data in fact indicate. Those data also show more clearly than the drawing that *Caluromys* stands apart from the other didelphids (except *Dromiciops*; see later in this paragraph). The affinities of *Marmosa* and *Metachirus* to each other are not closer than to other genera, and *Marmosa cinerea* and *M. fuscata* are no closer to each other than to *Monodelphis*, which, however, is closer to *Marmosa* than to any other genus. Significantly, *Dromiciops* is very different from all other didelphids; of American marsupials only the caenolestids are more distinct, and these are at least as different from the didelphids as are the Australian marsupials. Finally, caenolestids and Australian marsupials show no special affinities with each other, although, again, the projection tends to suggest that this is the case because of limitations in the number of antisera involving these taxa.

Many of these affinities have been checked by other methods, and in experiments employing antisera evoked in marsupials against the proteins of other marsupials; double-diffusion, hemagglutination inhibition, and turbidimetric tests are all consistent with the statements just made.

Thus, a composite phenogram of the serological affinities of the major groups of marsupials and most of the genera of didelphids can be constructed with some confidence (Fig. 20). In summary, it shows that there are three major groups of marsupials, represented by the didelphids, the caenolestids, and the Australian marsupials. Within the last there are four

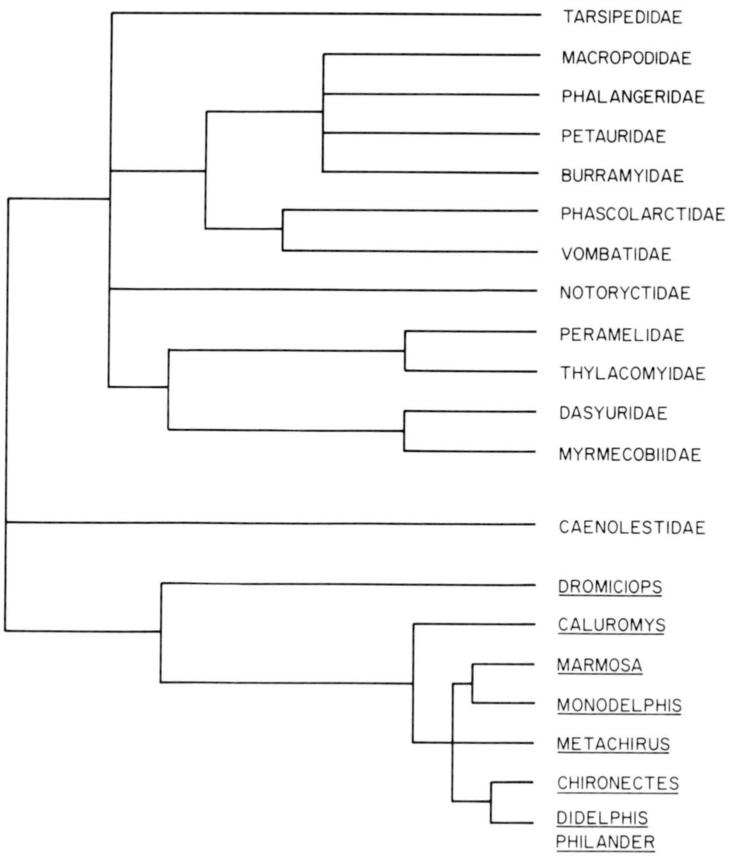


Fig. 20. Phenogram showing the serological affinities of most of the extant marsupial families, with additional detail for the genera of didelphoid marsupials, collated from several sets of experiments and NT-SYS analysis of them. Only the horizontal axis has dimension, with linkages of groups being successively less close from right to left.

subgroups. In general the results agree with anatomy, although the morphological evidence is that the marsupial mole (*Notoryctes*) is probably a derivative of the dasyurid–peramelid association, and that *Tarsipes* is surely part of the phalangeroid group.

## VI. DISCUSSION

This discussion will be aimed at showing how the evidence from karyotypes of and serological tests on living marsupials can be used together to indicate the probable phylogenetic relationships of marsupials, and to rationalize a classification of marsupials which incorporates these data and the phylogenetic insights they afford. The process of classification proceeds in three stages, of which the construction of phenetic groupings is the first, the inference of phylogeny is the second, and the crafting of a formal classification which takes into account both phenetic and phylogenetic data is the third. The results (Section V) establish the phenetic groupings, and we are thus ready to consider the phylogenetic inferences which may be drawn from them.

### A. Phylogeny

To begin with, what do we mean by phylogeny? Although the term is most commonly taken to mean the sequence of branchings of lineages in a group's genealogical tree, this, strictly speaking, is *cladistics*. Other and equally valid aspects of phylogeny are the directions and amounts of evolution in those lineages, including the sequences of changes in particular characters from "primitive" or ancestral to "advanced" or descendent. Knowledge of such sequences enables us to arrange organisms in series from the least altered to most modified, but it must be emphasized that such arrangements can only be made in terms of characters: there are no primitive organisms, only primitive character states. A final aspect of phylogeny concerns the absolute dates of divergence of the various lineages.

Cytological and serological approaches to classification are applied in studies of living organisms. This means that the delineation of phylogenetic relationships must be made on the basis of extrapolation backward in time to the probable characteristics of the ancestral species, and such extrapolation in turn depends upon assumptions or judgments as to how these characters evolved. Because chromosomes and antigens probably evolved differently, their meanings for phylogeny are bound to be different, but, as will be shown, complementary.

Karyotypes give two basic pieces of information: the numbers of chromosomes and their shapes. On these bases, groupings of species with similar patterns may be established. However, groupings formed by allying organisms with the same or similar karyotypes and excluding those with different karyotypes do not always result in arrangements which correspond to taxa solidly based on other characters. For example, it would clearly be absurd to set *Didelphis virginiana,* which has some submetacentric autosomes, apart from the other *Didelphis* species and from *Philander, Chironectes,* and *Lutreolina,* all of which have only acrocentric autosomes in their 22-chromosome diploid complements. On the other hand, it would be difficult to defend as real a grouping of all marsupials with the "standard dasyurid" pattern of $2n = 14$, since that group would include organisms as morphologically distinct as *Marmosa* and *Burramys*, but exclude *Didelphis*.

Evidently, in each example what we are dealing with is a situation in which one pattern is more common and has given rise to another (or others). Thus it seems reasonable to infer that *D. virginiana* has a derived pattern, its submetacentrics having arisen by modification from the pattern of all acrocentrics which characterizes the other members of its genus and three other genera of didelphids as well. Similarly, the 22-chromosome pattern seems likely to have been derived from the more common marsupial karyotype of $2n = 14$.

A signal feature of comparative karyology is therefore that it permits us to infer primitive and advanced stages in karyotypes. Operationally, the inference of primitiveness is made on the basis of one or more criteria. The most important of these is the frequency of occurrence and taxonomic distribution of the patterns: when one form of chromosome complement is common to a great number of taxonomically distinct taxa, it is assumed to be the ancestral pattern for the group. When two or more patterns occur with about the same frequency and distribution, a decision between them may be made by reference to the karyotypes found in groups known to be closely related to the one under scrutiny—this is the "outside primitive" criterion of Uzzell and Spolsky (1974). Finally, or alternatively, a decision may be made on the basis of what is known about the possible kinds of changes that may take place in chromosomes. For example, there is a general reluctance to admit evolution that involves "fissioning" of chromosomes (leading to an increase in number), because no mechanism has been unequivocally demonstrated for a rise in number.

Applying this reasoning to the Didelphidae, $2n = 14$ is evidently the primitive number for the family, both because it occurs in many distinct genera and because a complement of similar form is present in other marsupial families.

A diploid number of 14 is probably also primitive for Marsupialia as a whole, since $2n = 14$ chromosomes of similar morphologies occurs in most didelphids and all dasyurids and caenolestids, and some phalangeroids and peramelids. The alternate choice of $2n = 22$ scarcely seems tenable, since the various groups with 22 usually display different chromosome morphologies. Nevertheless, Sharman (1973a,b) argues the possibility that 22 is the basic number largely because that would permit marsupial chromosome evolution to proceed mostly by fusion, but he overlooks the variation in the 22-chromosome complement morphologies. The only argument acceptable for the higher number, often hinted at but never clearly articulated, is that higher chromosome numbers are more typical of placentals, monotremes, and some reptiles; this is the "outside primitive" argument applied to mammals as a whole, and suggests that the very earliest metatherian may well have had a higher number.

Nevertheless, the evidence seems compelling that all the *living* marsupials (and probably most of the known fossil ones as well) had a community of ancestry in a form with 14 chromosomes of the pattern shown in Fig. 15.

Given the wide taxonomic distribution of this karyotype, it is evidently the differences, and not the similarities, in karyotypes which are most significant to understanding the interrelationships of marsupial taxa; cytology really indicates very little about the relationships of species with the same karyotypes—such as which of the opossums with 14 chromosomes is most primitive for Didelphidae, or what might be the sequence of derivation of the 14-chromosome marsupial families from each other. Nor is it always clear just what the steps are in an apparently related sequence involving different karyotypes. For example, *Monodelphis* is unique among didelphids in having 18 chromosomes of morphologies which could as easily have been derived from a form with $2n = 14$ as by reduction from 22 (Sharman, 1973a). The cytological data of themselves do not permit choosing between these two of the several possible alternatives. Reig and Bianchi (1969) favor derivation of *Monodelphis* from a 14-chromosome form, while Hayman and Martin (1974) favor reduction from 22.

Resolving these sorts of ambiguities is a matter of establishing cladistics, as will be demonstrated, but to infer the cladistics of Didelphidae or of living marsupials in general we need, ideally, a set of characters which have evolved regularly through time. For such characters, branching sequences could easily be worked out, because similarity between any two taxa would be directly proportional to the time elapsed since their most recent common ancestor.

An enormous advantage of comparative serology and similar biochemical techniques is that they permit direct comparisons of organisms over a very wide taxonomic range on a continuous quantitative scale. It is tempting to infer from such facile comparisons that the evolution of antigens has been regularly divergent through time. Without wishing to reopen the polemic (see, for example, Kirsch, 1969; Sarich, 1969) as to whether the evolution of antigens is constant or time dependent, it does seem important to indicate the sort of evidence that has been adduced to support the notion of regularity in antigen evolution. Comparative serology lacks an internal criterion of primitiveness, but if, for example, among the Australian marsupials any lineage changed much less than any other, we would expect it to be serologically more similar than other Australian marsupials to an organism taxonomically much removed from the Australian forms, e.g., *Didelphis*. However, the serological results indicate that this is not the case: all Australian marsupials are approximately "equidistant" from *Didelphis*, including the morphologically similar dasyurids. Equidistance is also found to be the case with respect to caenolestids. Both examples suggest that antigen evolution has proceeded roughly at the same rate in each lineage of marsupials.

But what is true in general, over long periods of time, may not be true in detail, and it is the details of classification and phylogeny which are frequently of greatest interest. Ultimately the only way to test regularity in such instances is against a good fossil record; where this is lacking, as it mostly is in the case of the marsupials, the indications of another set of characters can provide a critical test. To the extent that the indications of serological and other characters are consistent we can accept the assumption of regularity as valid.

As we have seen, the chromosome data permit some inferences as to the directions of change in Marsupialia. Taking again the example of Didelphidae, we can be reasonably sure that $2n = 14$ is the primitive number for this family, and that both 18 and 22 are derived. However, we do not know if it is permissible to group all the 14-chromosome didelphids or not. If we did, it would invalidate Reig's assignment of *Caluromys* and *Dromiciops* to the Microbiotheriinae, as Biggers *et al.* (1965) pointed out. Serology, however, indicates that the several genera with 14 chromosomes represent distinct lines, with *Dromiciops* particularly different from the other didelphids. *Marmosa* and *Metachirus* are no closer to each other than to the genera with $2n = 22$, but *Monodelphis* is probably allied with *Marmosa*. The 22-chromosome species are, however, very similar serologically.

Thus the serological and cytological evidence can be combined in a phylogeny which indicates both cladistics and directions of change. The serological data allow us to state that the 18-chromosome pattern is a separate derivative of 14 by way of *Marmosa*-like forms, since the close cladistic relationship of *Marmosa* and *Monodelphis* precludes the latter having had a more recent common ancestry with one of the extant 22-chromosome species.

What has been done for these didelphids may similarly be done for marsupials in general. Figure 21 is a phylogeny of marsupials incorporating the serological results and including the cytological and some morphological data. Again, either approach is insufficient when used alone; serology by itself cannot indicate whether, for example, the bandicoots or the dasyurids are more primitive. The chromosome data unequivocally indicate that in cytological terms it is the dasyurids, since in the $2n = 14$ bandicoots the sex chromosomes are larger than those of all other 14-chromosome forms and the DNA values are higher than those of dasyurids. (The unique karyotype of *Macrotis* provides additional evidence that Perameloidea is a derived group.)

Clearly, this is not the whole story: neither serology nor cytology can

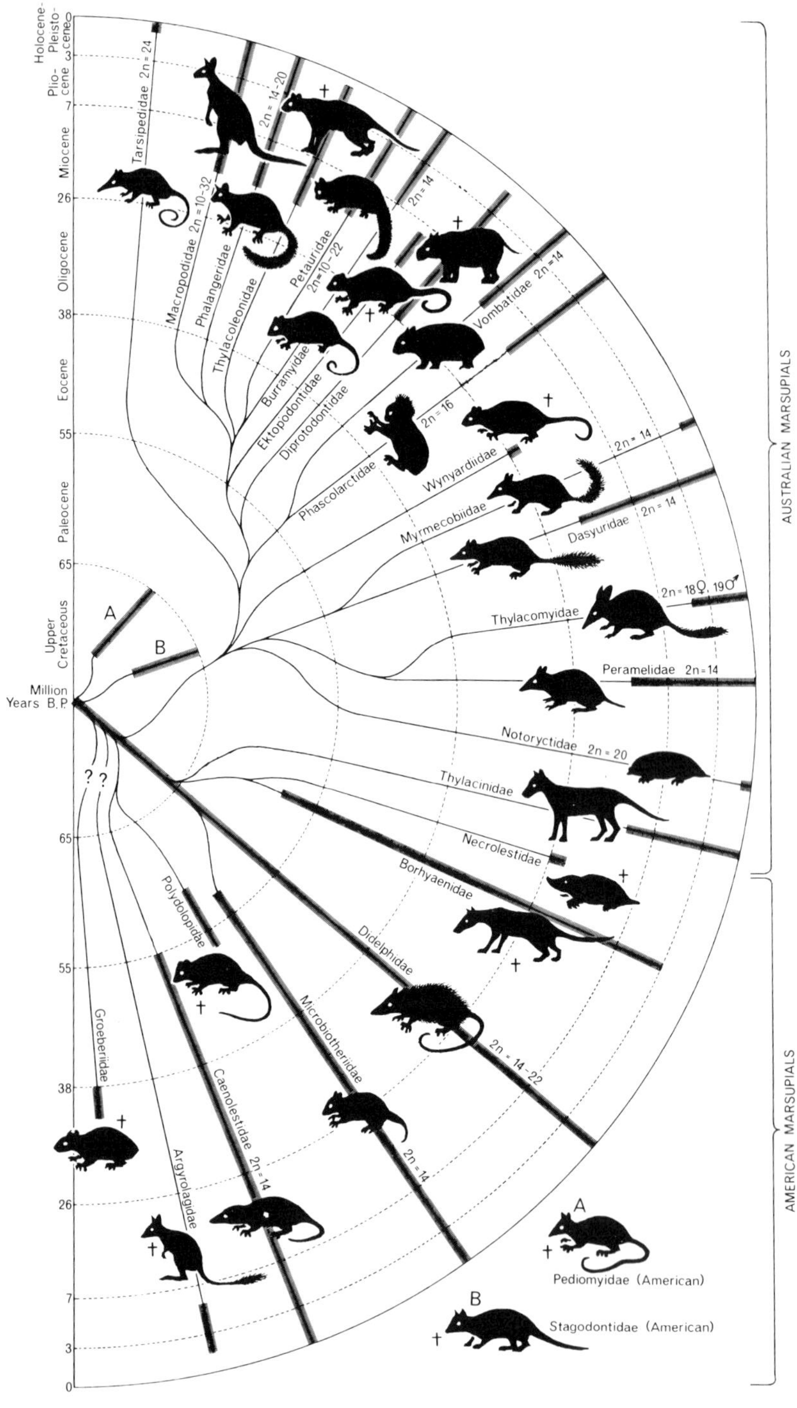
Holocene-Pleistocene
Pliocene
Miocene
Oligocene
Eocene
Paleocene
Upper Cretaceous
Million Years B.P.
0
3
7
26
38
55
65
Tarsipedidae 2n=24
Macropodidae 2n=10-32
2n=14-20
Phalangeridae
Thylacoleonidae
Petauridae 2n=10-22
2n=14
Burramyidae
Ektopodontidae
Diprotodontidae
Vombatidae 2n=14
Phascolarctidae 2n=16
Wynyardiidae
Myrmecobiidae
2n=14
Dasyuridae 2n=14
Thylacomyidae 2n=18♀, 19♂
Peramelidae 2n=14
Notoryctidae 2n=20
Thylacinidae
Necrolestidae
Borhyaenidae
Didelphidae 2n=14-22
Polydolopidae
Microbiotheriidae 2n=14
Groeberiidae
Caenolestidae 2n=14
Argyrolagidae
A
B
?
?
A
Pediomyidae (American)
B
Stagodontidae (American)
AUSTRALIAN MARSUPIALS
AMERICAN MARSUPIALS

indicate which of the 14-chromosome groups most nearly represents the common ancestor of all marsupials. Similarly, within the Australian group, neither cytology nor serology indicates whether the dasyurids or the 14-chromosome phalangers are most primitive.

To answer these questions reference must be made to morphological data. Applying the same criteria for deducing primitiveness of karyotypes to dentition, it is not unreasonable to conclude that polyprotodonty is primitive with respect to diprotodonty. Thus, dasyurids are primitive for Australian marsupials and didelphids are primitive with respect to the American radiation, since serology indicates that diprotodonty has evolved twice within the living marsupials. The higher incisor number in didelphids further indicates that they are primitive with respect to dasyurids.

We might, finally, wish to place a time scale on this phylogeny, and doing this surely depends on well-dated fossils. In this respect, the results on *Dromiciops* are especially significant. The distinctness of *Dromiciops* from other didelphids certainly supports Reig's contention that *Dromiciops* is a member of Ameghino's Microbiotheriidae. We are far from knowing the precise time of origin of that family, but it may have become distinct as early as the Paleocene (Tedford, 1974), not much later than the caenolestoids did. We thus have a tentative calibration point for dating the marsupial phylogeny and, incidentally, for testing the time-dependent assumption. In general it appears that the fossil and serological data are in accord: *Dromiciops* seems little more like the didelphids than are the caenolestids or the Australian marsupials, and divergences within the latter seem to have occurred relatively more recently.

## B. Classification

Many classifications of marsupials have been suggested, among the oldest being the typological ones of de Blainville and Owen into Polyprotodontia

←

Fig. 21. The complete phylogeny of Marsupialia, incorporating serological, karyological, morphological, and paleontological data, as discussed in the text. In general, relative amounts of divergence from a radius indicate the degree of departure from primitive characters where this can be assessed. Wholly extinct families are indicated by a dagger (†), and the period for which fossils are known is shown by the thickened portion of each lineage. The range of chromosome numbers for each family is indicated, when known. All families listed as "American" are South American, except for †Stagodontidae, which is only known from North America; †Pediomyidae, which is North and possibly South American; and Didelphidae, which occurs in North America and was also present in Europe from Eocene to Miocene. Fossil data collated from several sources, but especially Simpson (1948, 1970a), Patterson and Pascual (1972), Tedford (1974), and Woodburne *et al.* (1977, quoted by permission of Michael Archer). The time intervals are not shown in exact proportion in order to preserve clarity in the branching pattern, nor are the silhouettes drawn to scale. (Drawing by Frank Knight.)

vs Diprodotontia or Syndactyla vs Didactyla [see Gregory (1910) for a summary history]. With the discovery of the caenolestoids, it became clear that contradictions were inherent in these arrangements, and among others Gregory (1910) recognized the possibility that diprotodonty at least may have arisen twice; accordingly, he considered the caenolestoids a third, phylogenetically separate group.

In Simpson's standard classification (1945), he recognized six natural groups of marsupials but explicitly stated that these superfamilies did *not* have any special relationships with each other, and he therefore used no subordinal groupings.

If there is any significance to the results reported here, it is that there *are* meaningful similarities and thus phylogenetic relationships among at least some of the superfamilies, and it seems incumbent upon us to find some way of expressing these in a classification.

Of the available alternatives to Simpson's classification, one of the most interesting is that of Ride (1964), who argued that the antiquity, diversity, and breadth of Marsupialia indicated that marsupials should be considered several orders. In his arrangement the polyprotodonts, excluding the bandicoots, were grouped as the order Marsupicarnivora. The bandicoots represent his order Peramelina, and the phalangeroids and caenolestoids the Diprotodonta and Paucituberculata, respectively. Because Ride's system has won some acceptance, the present author, believing that it is basically sound, has chosen (Kirsch, 1968a) to build upon Ride's scheme rather than attempt an entirely new arrangement. This classification is shown in Table III; it is simplified from Kirsch (1968a), with modifications resulting from additional information on the American marsupials and on *Notoryctes*.

This author's major point of difference with Ride involves the assignment of the bandicoots. Since there is evidence for grouping peramelids with dasyurids, the earlier idea of Polyprotodonta was employed. The constitution of this order is, in fact, the most controversial and possibly misleading aspect of this and Ride's classifications. Were cladistics and temporal relations the only aspects of phylogeny worth expressing in a classification, zoogeographical and serological evidence would make it more logical to split the living Marsupialia into two subgroups, Australian vs American [as Simpson (1970a) has humorously suggested] or, more exactly, into three: Australian, didelphoid, and caenolestoid. But, as indicated in the foregoing pages, phylogeny involves more than branching sequences, and a proper consideration of the data allows us to adumbrate the directions of change as well as cladistics. Thus, from the karyotypes it is clear that all marsupials have a community of ancestry in a 14-chromosome form. Among the living marsupials the dasyurids, caenolestids, some phalangeroids, and some didelphoids all retain this karyotype. Moreover, dasyurids and didelphoids at

TABLE III

A Classification of Marsupials[a]

| | |
|---|---|
| Order Polyprotodonta | Order Paucituberculata |
| Suborder Didelphimorphia | Superfamily Caenolestoidea |
| Superfamily Didelphoidea | Family Caenolestidae |
| Family Didelphidae | †Family Polydolopidae |
| †Family Pediomyidae | ?Paucituberculata |
| †Family Stagodontidae | †Family Groeberiidae |
| Family Microbiotheriidae | Superfamily Argyrolagoidea |
| †Family Borhyaenidae | †Family Argyrolagidae |
| †Family Necrolestidae | Order Diprotodonta |
| Family Thylacinidae | Superfamily Phalangeroidea |
| Suborder Dasyuromorphia | Family Phalangeridae |
| Superfamily Dasyuroidea | Family Macropodidae |
| Family Dasyuridae | Family Petauridae |
| Family Myrmecobiidae | Family Burramyidae |
| Suborder Peramelemorphia | †Family Thylacoleonidae |
| Superfamily Perameloidea | †Family Ektopodontidae |
| Family Peramelidae | †Family Diprotodontidae |
| Family Thylacomyidae | Superfamily Vombatoidea |
| Suborder Notoryctemorphia | Family Vombatidae |
| Superfamily Notoryctoidea | Family Phascolarctidae |
| Family Notoryctidae | Superfamily Tarsipedoidea |
| | Family Tarsipedidae |
| | Superfamily Wynyardioidea |
| | †Family Wynyardiidae |

[a] (From Kirsch, 1968a, and in preparation). Wholly extinct families are indicated by a dagger (†).

least are morphologically primitive in dentition, pes, and general structure. Didelphoids have a long and nearly continuous fossil record as well, and there are only trivial morphological reasons for separating the members of Didelphoidea and Dasyuroidea into unrelated major taxa. For all of these reasons, it seems justifiable to group the dasyuroids and didelphoids and to place both in an order with the fossil forms to which they are manifestly allied, even though in cladistic and temporal terms it may be argued that dasyurids are closer to the other Australian marsupials than to didelphoids.

On the other hand, there is no serological, zoogeographical, or fossil evidence of an unequivocal sort to indicate relationships between the caenolestids and phalangeroids, and these *are* placed in separate orders. While the Diprotodonta are serologically closer to Australian polyprotodonts than are any of the American marsupials, Diprotodonta does represent a major group of Australian marsupials, distinct in several important morphological features. Moreover, in spite of the inclusion of the cytologically primitive burramyid phalangers (as this author considers them), Diprotodonta shows more diversity in karyotypes than any other subgroup of Marsupialia.

Several subgroupings within the Diprotodonta have also been suggested (see Table III), but since the details of the supporting studies have not been given here it does not seem appropriate to discuss the arrangement in detail.

Of special interest is the classification of the didelphoids. Since Reig's 1955 paper it has seemed appropriate that we recognize two subordinate taxa among living Didelphoidea. Reig originally proposed that *Caluromys*, *Caluromysiops*, and *Glironia* as well as *Dromiciops* should be considered living microbiotheres, but he has subsequently and independently concluded (O. A. Reig, personal communication) that these three genera are only convergently similar to microbiotheres, though undoubtedly distinct from typical didelphines. Segall's work (1969a,b) confirms both conclusions from study of morphology of the middle ear region in didelphoids.

In view of its "living fossil" status, *Dromiciops* may well prove to be one of the most interesting didelphoids for biological study, and its taxonomic distance from other opossums should be borne in mind when planning comparative investigations.

Gardner's demonstration that the karyotypic distinctness of *D. virginiana* is biologically meaningful will also be of significance in more general studies; moreover, the large geographical overlap of *D. virginiana* and *D. marsupialis* indicates the need for care in determining the origin of experimental animals and in keeping voucher specimens for identification.

As yet, too few species of *Marmosa* have been examined either cytologically or serologically, but the karyotypes do differ among species and thus have potential for taxonomy within the genus; there is also some indication that *M. cinerea* may be as distinct serologically from *M. robinsoni* as is *Monodelphis*. Thus, easy acceptance of the biological interchangeability of *Marmosa* species may not be justified.

While it is fairly easy to accommodate extinct didelphoids in the classification indicated by the relationships of living marsupials, this is not true for the American fossil diprotodonts. Simpson (1970a,b) has criticized both this author's and Ride's arrangements as difficult to reconcile with the new evidence from paleontology, and we should next consider Simpson's point.

Both the Argyrolagidae and Groeberiidae are typologically diprotodont, but are otherwise unlike any other marsupial group, and Simpson rightly questions whether either could be accommodated in a multiordinal scheme without adding two more major taxa. As he points out, adding these taxa would destroy the simplicity that both Ride and the present author were aiming for in erecting higher categories for the grouping of more than one superfamily.

This is not an entirely convincing argument, however, for two reasons: first, Simpson himself, explicitly in the case of the Argyrolagidae and implicitly in the case of *Groeberia,* proposes two new superfamilies to add to the five he now uses for the rest of Marsupialia, so that his arrangement is hardly parsimonious of major taxa; and second, it still seems valid to try at least to express the undoubted phylogenetic relationships of the Polyprotodonta—the only aim of a hierarchical classification is not, after all, to distribute diversity among the fewest possible groups [Van Valen (1971) went as far as possible in this direction by placing all the South American diprotodonts as well as the perameloids in the order Polyprotodonta].

But our disagreement does indicate that, for all their comprehensiveness, the serological and cytological data may in fact have limited relevance to the classification of marsupials as a whole. This is perhaps an unduly pessimistic conclusion. As we continue to learn more of the paleontology of any group, work on the previously known forms comes to seem more and more inadequate to understanding the classification of the entire group: the ability to classify, in the end, depends to a large extent on gaps in knowledge of extinct organisms. If the inadequacy of the present classification seems especially clear, it is perhaps because we are experiencing an expansion of knowledge of marsupials, but it is fair to say also that serology and cytology have each played a part in interpreting the new evidence from fossil studies.

In any case, we must have provisional classifications, and we will always benefit from their heuristic value; it is with the hope that this author's classification may contribute in this way that it is offered.

## ACKNOWLEDGMENTS

The original work reported here was done over a period of several years while I held, in succession, pre- and postdoctoral fellowships from the Australian–American Educational Foundation, the University of Western Australia, the National Science Foundation, and the Australian Research Grants Committee. The research was carried out at the universities of Western Australia, Kansas, and Adelaide, and at Yale University, and was supported by grants from those institutions, the National Science Foundation, and the Australian Research Grants Committee. The latter was especially generous in underwriting an expedition to South America for joint work by David Hayman, Peter Martin, and myself. I am grateful to all who helped me collect blood or animals. David Hayman and Rod Wells provided photographs, as indicated in the captions to the appropriate figures. The manuscript was read, at various stages, by Michael Archer, Pamela Parker, David Hayman, and Robert Henzell, all of whom provided helpful criticism; none of them should be blamed for the perversity with which I clung to many of my ideas, however. The award of a Yale College Junior Faculty Fellowship for 1974–1975 enabled me to complete the writing of this chapter.

**REFERENCES**


Abbie, A. A. (1937). *J. Anat.* **71,** 429.

Abbie, A. A. (1941). *Aust. J. Sci.* **4,** 77.

Archer, M. A. (1976). *Aust. J. Zool., Suppl. Ser.* **39,** 1.

Archer, M. A., and Kirsch, J. A. W. (1977). *Proc. Linn. Soc. N. S. W.* (in press).

Bensley, B. A. (1903). *Trans. Linn. Soc. London, Zool.* [2] **9,** 83.

Biggers, J. D., and Delamater, E. D. (1965). *Nature (London)* **208,** 402.

Biggers, J. D., Fritz, H. I., Hare, W. C. D., and McFeely, R. A. (1965). *Science* **148,** 1602.

Blish, J. (1958). "A Case of Conscience." Ballantine, New York.

Boyden, A. A. (1942). *Physiol. Zool.* **15,** 109.

Broom, R. (1898). *Proc. Linn. Soc. N. S. W.* **23,** 57.

Butler, P. M., and Kielan-Jaworowska, Z. (1973). *Nature (London)* **245,** 105.

Cabrera, A. (1919). "Genera Mammalium. Monotremata Marsupialia." Museo Nacional de Ciencias Naturales, Madrid.

Cabrera, A. (1957). "Catalogo de los Mamiferos de America del Sur." Museo Argentino de Ciencias Naturales "Bernardino Rivadavia," Buenos Aires.

Calaby, J. H., Corbett, L. K., Sharman, G. B., and Johnston, P. G. (1974). *Aust. J. Biol. Sci.* **27,** 529.

Clemens, W. A. (1966). *Univ. Calif., Berkeley, Publ. Geol. Sci.* **62,** 1.

Clemens, W. A. (1968). *Evolution* **22,** 1.

Collins, L. R. (1973). "Monotremes and Marsupials a Reference for Zoological Institutions." Random House (Smithsonian Inst. Press), New York.

Cooper, D. W., Vandeberg, J. L., Sharman, G. B., and Poole, W. E. (1971). *Nature (London), New Biol.* **230,** 155.

Flower, W. H. (1868). *Q. J. Geol. Soc. London* **24,** 307.

Gardner, A. L. (1973). *Spec. Publ., Mus. Tex. Tech. Univ.* **4,** 1.

Gardner, A. L., and Patton, J. L. (1972). *Occas. Pap. Mus. Zool., La. State Univ.* **43,** 1.

Goodman, M. (1961). *Hum. Biol.* **33,** 131.

Goodman, M. (1962). *Hum. Biol.* **34,** 377.

Gregory, W. K. (1910). *Bull. Am. Mus. Nat. Hist.* **27,** 1.

Gregory, W. K. (1951). "Evolution Emerging," Vol. II. Macmillan, New York.

Hall, E. R., and Kelson, K. R. (1959). "The Mammals of North America," Vol. I. Ronald Press, New York.

Hayman, D. L. (1976). *In* "The Biology of Marsupials" (B. Stonehouse, ed.), Macmillan, London (in press).

Hayman, D. L., and Martin, P. G. (1965). *Aust. Biol. Sci.* **18,** 1081.

Hayman, D. L., and Martin, P. G. (1974). "Animal Cytogenetics, 4: Chordata 4. Mammalia I.: Monotremata and Marsupialia." Borntraeger, Berlin.

Hayman, D. L., Martin, P. G., and Waller, P. F. (1969). *Chromosoma* **27,** 371.

Hayman, D. L., Kirsch, J. A. W., Martin, P. G., and Waller, P. F. (1971). *Nature (London)* **231,** 194.

Hofer, H. (1952). *Zool. Jahrb., Alet. Anat. Ontog. Tiere* **72,** 365.

Johnstone, J. (1901). *Trans. Liverpool Biol. Soc.* **15,** 354.

Jones, F. W. (1923). "The Mammals of South Australia," Part I. Government Printer, Adelaide.

Kirkpatrick, T. H. (1969). Ph.D. Thesis, University of Queensland (unpublished).

Kirsch, J. A. W. (1968a). *Nature (London)* **215,** 418.

Kirsch, J. A. W. (1968b). *Aust. J. Sci.* **31,** 43.

Kirsch, J. A. W. (1969). *Syst. Zool.* **18,** 296.

Kirsch, J. A. W. (1970). *Congr. Zool. Latin., 4th, 1968* Vol. I, p. 117.
Kirsch, J. A. W. (1973). *Aust. Nat. Hist.* **17,** 301.
Kirsch, J. A. W., and Calaby, J. H. (1976). *In* "The Biology of Marsupials" (B. Stonehouse, ed.). Macmillan, London (in press).
Kirsch, J. A. W., and Murray, M. D. (1969). *Aust. J. Zool.* **17,** 799.
Lillegraven, J. A. (1969). *Univ. Kans. Paleontol. Contrib.* Art. 5 (Vert. 12), 1.
Lillegraven, J. A. (1974). *Annu. Rev. Ecol. Syst.* **5,** 263.
Marshall, L. G. (1972). *Proc. R. Soc. Victoria* **85,** 51.
Martin, P. G., and Hayman, D. L. (1965). *Evolution* **19,** 156.
Martin, P. G., and Hayman, D. L. (1967). *Chromosoma* **20,** 290.
Merrilees, D. (1968). *J. R. Soc. West. Aust.* **51,** 1.
Nuttall, G. H. F. (1904). "Blood Immunity and Blood Relationship." Cambridge Univ. Press, London and New York.
Osgood, W. H. (1921). *Publ. Field Mus. Nat. Hist., Zool. Ser.* **14,** 1.
Pascual, R., and Herrera, H. E. (1973). *Ameghiniana* **10,** 36.
Patterson, B., and Pascual, R. (1972). *In* "Evolution, Mammals, and Southern Continents." (A. Keast, F. C. Erk, and B. Glass, eds.), pp. 247–309. State University of New York Press, Albany.
Pine, R. H. (1973). *Proc. Biol. Soc. Wash.* **86,** 391.
Reig, O. A. (1955). *Invest. Zool. Chil.* **2,** 121.
Reig, O. A. (1964). *Publ. Mus. Mun. Cienc. Nat. Mar del Plata* **1,** 203.
Reig, O. A. (1968). *Experientia* **24,** 185.
Reig, O. A., and Bianchi, N. O. (1969). *Experientia* **25,** 1210.
Reig, O. A., and Simpson, G. G. (1972). *J. Zool.* **167,** 511.
Reig, O. A., and Sonnenschein, C. (1970). *Experientia* **26,** 199.
Reig, O. A., Fernandez, R., and Spotorno, A. (1972). *Z. Saeugetierkd.* **37,** 37.
Reig, O. A., Gardner, A. L., Patton, J. L., and Bianchi, N. O. (1977). In preparation.
Ride, W. D. L. (1962). *In* "The Evolution of Living Organisms" (G. W. Leeper, ed.), pp. 281–306. Melbourne Univ. Press, Melbourne.
Ride, W. D. L. (1964). *J. Proc. R. Soc. West. Aust.* **47,** 97.
Ride, W. D. L. (1970). "A Guide to the Native Mammals of Australia." Oxford Univ. Press, Melbourne.
Riggs, E. S. (1934). *Trans. Am. Philos. Soc.* [N.S.] **24,** 1.
Sahni, A. (1972). *Bull. Am. Mus. Nat. Hist.* **147,** Art. 6, 321.
Sarich, V. M. (1969). *Syst. Zool.* **18,** 286.
Sarich, V. M., and Wilson, A. (1967). *Proc. Natl. Acad. Sci. U.S.A.* **58,** 142.
Scott, W. B. (1913). "A History of Land Mammals in the Western Hemisphere." Macmillan, New York.
Segall, W. (1969a). *Acta Anat.* **72,** 489.
Segall, W. (1969b). *Acta Anat.* **73,** 176.
Sharman, G. B. (1961). *Aust. J. Zool.* **9,** 38.
Sharman, G. B. (1973a). *Aust. Mammal.* **1,** 137.
Sharman, G. B. (1973b). *In* "Cytotaxonomy and Vertebrate Evolution" (A. B. Chiarelli and E. Capanna, eds.), pp. 485–530. Academic Press, New York.
Sigé, B. (1968). *C. R. Hebd. Seances Acad. Sci. Ses Fil.* **267,** 1495.
Sigé, B. (1972). *Bull. Mus. Natl. Hist. Nat., Sci. Terre* **19,** 375.
Simpson, G. G. (1930). "Fossilium Catalogus, 1: Animalia. Post Mesozoic Marsupialia." Junk, Berlin.
Simpson, G. G. (1941). *Am. Mus. Novit.* **1118,** 1.
Simpson, G. G. (1945). *Bull. Am. Mus. Nat. Hist.* **85,** 1.

Simpson, G. G. (1948). *Bull. Am. Mus. Nat. Hist.* **91,** 1.
Simpson, G. G. (1970a). *Bull. Mus. Comp. Zool.* **139,** 1.
Simpson, G. G. (1970b). *Breviora* **362,** 1.
Sinclair, W. J. (1906). *Rep. Princeton Exped. Patagonia* **4, pt. 3,** 333.
Slaughter, B. H. (1968a). *Science* **162,** 254.
Slaughter, B. H. (1968b). *Tex. J. Sci.* **20,** 3.
Slaughter, B. H. (1971). *In* "Early Mammals" (D. M. Kermack and K. A. Kermack, eds.), pp. 131–143. Academic Press, New York.
Stockard, C. R. (1941). *Am. Anat. Mem.* **19,** 1.
Symington, J. (1898). *J. Anat. Physiol.* **32,** 278.
Tate, G. H. H. (1933). *Bull. Am. Mus. Nat. Hist.* **66,** 1.
Tedford, R. H. (1974). *Tulsa Soc. Econ. Paleontol. Mineral., Spec. Publ.* **21,** 109.
Tedford, R. H., Banks, M. R., Kemp, N. R., McDougall, I., and Sutherland, F. L. (1975). *Nature* (*London*) **255,** 141.
Thomas, O. (1888). "Catalogue of the Marsupialia and Monotremata in the Collection of the British Museum (Natural History)." British Museum, London.
Thomas, O. (1921). *Ann. Mag. Nat. Hist.* [9] **8,** 137.
Turnbull, W. D. (1971). *In* "Dental Morphology and Evolution" (A. A. Dahlberg, ed.), pp. 151–179. Univ. of Chicago Press, Chicago, Illinois.
Tyndale-Biscoe, H. (1973). "Life of Marsupials." Am. Elsevier, New York.
Uzzell, T., and Spolsky, C. (1974). *Am. Sci.* **62,** 334.
Van Valen, L. (1971). *Evolution* **25,** 420.
Weymss, C. T. (1953). *Zoologica* **38,** 173.
Wilson, A. C., and Sarich, V. M. (1969). *Proc. Natl. Acad. Sci. U.S.A.* **63,** 1088.
Woodburne, M. O., Tedford, R. H., Plane, M., Turnbull, W. D., Archer, M. A., and Lundelius, E. L. (1977). *Proc. Geol. Soc. Am.* (in press).
Wurster, D. H., and Benirschke, K. (1970). *Science* **168,** 1364.
Yadav, M. (1973). *Aust. J. Zool.* **21,** 285.

# 2

# Marsupial Chromosomes, Cell Cycles, and Cytogenetics

LAWRENCE K. SCHNEIDER

## 1. INTRODUCTION

Marsupials possess a variety of unusual biological characteristics which, combined with the relative ease of procurement and maintenance in the laboratory, make them extremely valuable animals for scientific research. Nowhere, perhaps, are these differences more pronounced, and the scientific value greater, than in their chromosomes. In general, marsupials possess low diploid chromosome numbers ($2n = 10$ to 32), large chromosomes, and perhaps most importantly, morphologically recognizable sex chromosomes (and in many cases, distinctive autosomes); these conditions, for the most part, do not exist in placental mammals, where $2n = 17$ to 84. These attributes have resulted in a great deal of research on marsupial chromosomes of somatic or germ line cells taken directly from the animal or from a variety of cells in tissue culture. As a prerequisite to cytogenetic studies, it has been necessary to obtain a knowledge of the kinetics of the various cell types either *in vivo* or *in vitro*.

Numerous investigators have utilized the advantages offered by marsupial chromosomes to study such problems as subchromatid organization at the electron microscopic level (Brinkley and Humphrey, 1969); chromosome morphology after prolonged incubation *in vitro* (Bick and Brown, 1969b); various aspects of cytogenetics (Shaw and Krooth, 1964; Hayman and Martin, 1965a; Moore and Uren, 1965; Walen, 1965; Schneider and Rieke, 1967; Sinha, 1967a; Reiss *et al.*, 1970; Schneider, 1970, 1971); and the effects of various agents, for example, heliotrine (a pyrrolizidine alkaloid with radiomimetic and mutagenic properties), X-rays, and LSD (Bick and Jackson, 1968; Bick and Brown, 1969a; Bick, 1970). Many others have cited the value of marsupial chromosomes for similar types of research (Walen and Brown, 1962; Legator *et al.*, 1966; Moore and Uren, 1966; Goh, 1967; Jackson and Ellem, 1968; Moore and Radley, 1968; Gibson, 1970; Heneen, 1970; Jakob and Ruiz, 1970; Sharman *et al.*, 1970; Thrasher, 1970).

The purpose of this chapter is to integrate and discuss many of the recent findings on marsupial chromosomes, cell cycles, and cytogenetics as well as certain of the more classic studies. It is not, however, intended to comprise an exhaustive review of all the known chromosome complements in Marsupialia. Several excellent tables have been published which list chromosome numbers, location of animal(s), sex chromosome mechanisms, and references to these observations. References to these tables will be cited in this chapter, but in the interest of space, they will not be duplicated or collated here. Rather, emphasis will be placed on those chromosome complements which lend themselves particularly well to expanded or additional research in the fields of cellular kinetics, nucleic acid and protein synthesis, sex chromatin, and chromosomal evolution.

## II. CHROMOSOMES OF AUSTRALIAN MARSUPIALS

In the chromosome descriptions which follow, the nomenclature employed will, with two exceptions, be that proposed in the Denver Report (1960) in which the longer chromosome is arbitrarily assigned the lower number. The exceptions occur in the numbering of the multiple Y chromosomes of the wallaby (*Wallabia bicolor*) and the Tasmanian rat kangaroo (*Potorous tridactylus*). In these examples, the presumed ancestral Y chromosome (see Sections II,A and B) is assigned the lower number regardless of size. The chromosomes of about half of the estimated 160 species of Australasian marsupials have been investigated (Hayman and Martin, 1969). References to these observations may be found in a number of sources (Sharman and Barber, 1952; McIntosh and Sharman, 1953; Sharman, 1961; Hayman and Martin, 1969; Thrasher, 1972). Emphasis in this section will be placed on the chromosomes of *Wallabia bicolor, Potorous tridactylus,* and certain other species of exceptional value.

### A. *Wallabia bicolor* (Wallaby or Black-Tailed Wallaby)

*Wallabia bicolor* possesses one of the lowest diploid chromosome numbers in Mammalia, where $2n = 10$ in the female and 11 in the male (Fig. 1). The sex chromosomes consist of XX in the female and $XY_1Y_2$ in the male. Moore and Gregory (1963) reported that each pair of autosomes and the sex chromosomes are distinguishable from one another. Autosome pair number 1 is the largest and possesses submetacentric centromeres. Pair number 2 is about the same size as 1, but the centromeres are subtelocentric. Autosome pairs 3 and 4 are about one-half the size of pairs 1 and 2; pair number 3 possesses submetacentric centromeres while those of pair 4 are subtelocentric. The X chromosome is approximately the same size and configuration as autosomal pair 2, but may be identified by an achromatic region at the base of the short arm. $Y_1$ is a very small telocentric, and $Y_2$ is a large telocentric chromosome. $Y_1$ is frequently not seen at metaphase, perhaps due to loss during spread preparation or to masking by the larger chromosomes (Radley, 1966).

It is believed that in this species and in *Potorous tridactylus* (see Section II,B) the multiple sex chromosome mechanisms ($XX/XY_1Y_2$) originated from a sex chromosome/autosome translocation (Hayman and Martin, 1965a; Moore and Uren, 1965). In each species, the two X chromosomes of females possess a short arm and a long arm. The short arms are believed to be the original (ancestral) X chromosomes which are translocated onto autosomes (long arms). In males, the single X consists of the ancestral X translocated onto an autosome. $Y_1$ is the ancestral Y, and $Y_2$ is the remain-

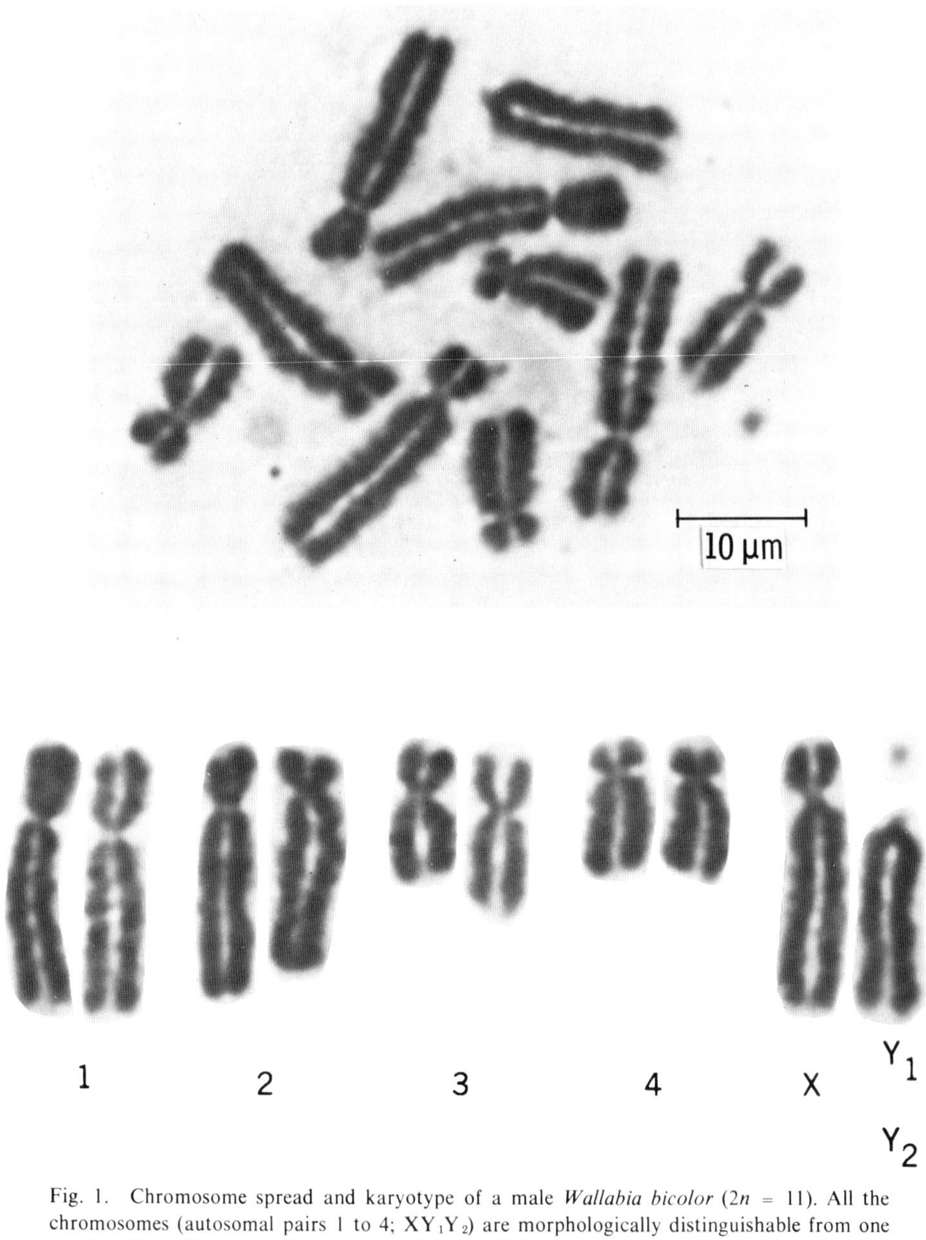


Fig. 1. Chromosome spread and karyotype of a male *Wallabia bicolor* ($2n = 11$). All the chromosomes (autosomal pairs 1 to 4; $XY_1Y_2$) are morphologically distinguishable from one another. From Moore and Uren, *Exp. Cell Res.*, 1965.

ing autosome (with no ancestral X attached). Deviation from the Denver Report nomenclature is considered justified in naming the Y chromosomes of these two species, since $Y_1$ should connote the "true" Y chromosome, not simply the longest (Thrasher, 1972). Moore and Uren (1965) have demonstrated that the long arm of the X is significantly longer than the $Y_2$, and, therefore, the ancestral X may not consist of only the short arm of the present X, but a small portion of the long arm also. Radioautographic evidence which supports the concept of sex chromosome/autosome translocation will be presented in Section VI.

### B. *Potorous tridactylus* (Tasmanian Rat Kangaroo or Potoroo)

*Potorous tridactylus* is another marsupial in which all chromosomes are morphologically distinct; the diploid number ($2n = 12$ in the female and 13 in the male) is only slightly larger than that of *Wallabia bicolor.* This species also possesses an $XX/XY_1Y_2$ sex chromosome mechanism. The chromosomes of the rat kangaroo (Fig. 2) are described in detail by Shaw and Krooth (1964). Autosome pair number 1 is the longest and possesses subtelocentric centromeres. Pair number 2 is almost as long as 1, but in this case the centromeres are submetacentric. Pair number 3 is also submetacentric, but is shorter than pair 2. Pairs 4 and 5 are metacentric; number 4 is 1.6 times as long as number 5. The X chromosomes are comparable in length and centromere position to autosomal pair number 3, but they may be recognized by a secondary constriction in the long arm near the centromere. Sharman *et al.* (1950) identified the $Y_1$ as the smallest of the dual Y chromosomes; both $Y_1$ and $Y_2$ are telocentric in form.

Sharman and Barber (1952) reported that the short arm of the X chromosome and the proximal portion of its long arm together with the $Y_1$ in *P. tridactylus* demonstrated the "typical behaviour of marsupial sex chromosomes." They observed that most of the long arm of the X and the $Y_2$ chromosomes behaved like the autosomes, and that the $Y_1$ paired with a terminal segment of the X which did not include the centromere. Hayman and Martin (1965a) believe that the multiple sex chromosome mechanism in *P. tridactylus* originated in the same fashion as was described for the wallaby in the preceding section. In addition, they feel that the rat kangaroo must have derived from a $2n = 14$ species as a result of the sex chromosome/autosome translocation (Hayman and Martin, 1969). This type of translocation, accompanied by the loss of a centromere, had been postulated by Sharman *et al.* (1950) to explain the "aberrant mechanism" observed in *P. tridactylus.*

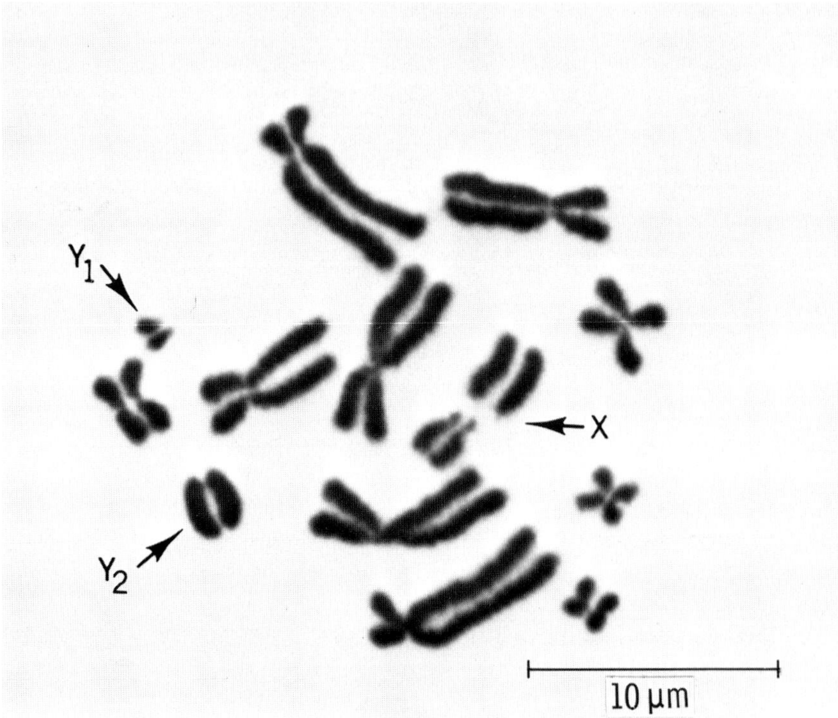


Fig. 2. Chromosome spread of a male *Potorous tridactylus* ($2n = 13$). All the chromosomes of this species are also morphologically distinguishable from one another. From Bick and Brown, *Cytobios,* 1969b.

## C. Other Species

There are a number of additional species of Australian marsupials whose chromosomal constitutions, although not as simple and distinct as those of *Wallabia bicolor* or *Potorous tridactylus,* are nevertheless worthy of discussion. For example, Hayman and Martin (1965b) found that in somatic cells of *Perameles nasuta* (the Australian long-nosed bandicoot), *Isoodon obeselus,* and *Isoodon macrourus,* the diploid number is 13, and the sex chromosome constitution is XO. Hayman *et al.* (1969) found that *Echymipera kalabu* does not contain all of the chromosome complement in each of its somatic cells; one X is missing in females and the Y in males. There is a similar elimination of supernumerary chromosomes which appear in this species. The full chromosome complement does exist, however, in the corneal epithelium and cells of the reproductive tissues. Supernumerary chromosomes were also found in *Schoinobates volans*, an animal which demonstrates a wide variety of chromosome numbers (Hayman and Martin, 1965c).

The chromosomes of *Perameles nasuta* were investigated more recently by Jackson and Ellem (1968), who did not confirm the 13 XO finding of Hayman and Martin (1965b). Rather, they found a diploid number of 14 with an XX/XY sex chromosome constitution. They were unable to offer a satisfactory explanation for the variation in these observations other than a suspected mosaicism in this animal. The chromosomes are distinguishable from one another only with careful observation and measurement. Autosome pair number 1 is the largest and possesses submetacentric centromeres. Pair number 2 is also submetacentric, but shorter than the former. Number 3 is about the same size as numbers 1 and 2, but nearly metacentric in form. Autosomal pair 4 is metacentric, and 5, which is smaller than 4, is submetacentric; both are considerably smaller than pairs 1 to 3. Pair number 6 is satellited, giving it a characteristic morphology. The X chromosomes are identifiable metacentrics, and the Y is the only telocentric chromosome.

Another marsupial with unusual chromosome numbers and with a complex sex chromosome system is the spectacled hare wallaby (*Lagorchestes conspicillatus*) studied by Martin and Hayman (1966). They found chromosome numbers to be 15 in males and 16 in females. The sex chromosomes are $X_1X_1X_2X_2$ in the female and $X_1X_2Y$ in the male. The X chromosomes are acrocentric, and the Y is metacentric.

Moore and Uren (1966) found the diploid number of the marsupial mouse (*Antechinus swainsonii*) to be 14, with the more typical XX/XY sex chromosome constitution. The autosomes are distinguishable from each other upon analysis of individual lengths and/or arm ratios. The X appears to be a very small submetacentric chromosome, and the Y (which cannot be seen in approximately 20% of otherwise normal spreads) is visible only as a small dot at metaphase.

Marshall Graves (1967) studied the chromosomes of the grey kangaroo (*Macropus giganteus*) and the tammar or Kangaroo Island wallaby (*Macropus eugenii*), and found each animal to have seven pairs of autosomes, all of which are distinguishable except pairs 4 and 5. The X is clearly identifiable, but of special interest is a morphological difference between the two X chromosomes in females of both species. One, which is designated $X_1$, is less condensed than its homologue $X_2$. More specifically, $X_1$, contains a prominent secondary constriction in its short arm which is absent or less prominent in $X_2$. Male cells contain the $X_1$ chromosome (Fig. 3).

Finally, Sharman *et al.* (1970) compared normal karyotypes of the tammar wallaby (*Macropus eugenii*), the euro (*Macropus robustus*), and the brush possum (*Trichosurus vulpecula*) with chromosome constitutions of intersexual animals of the same species. The authors noted that whereas cases of intersexual marsupials were known to exist, few had been recorded,

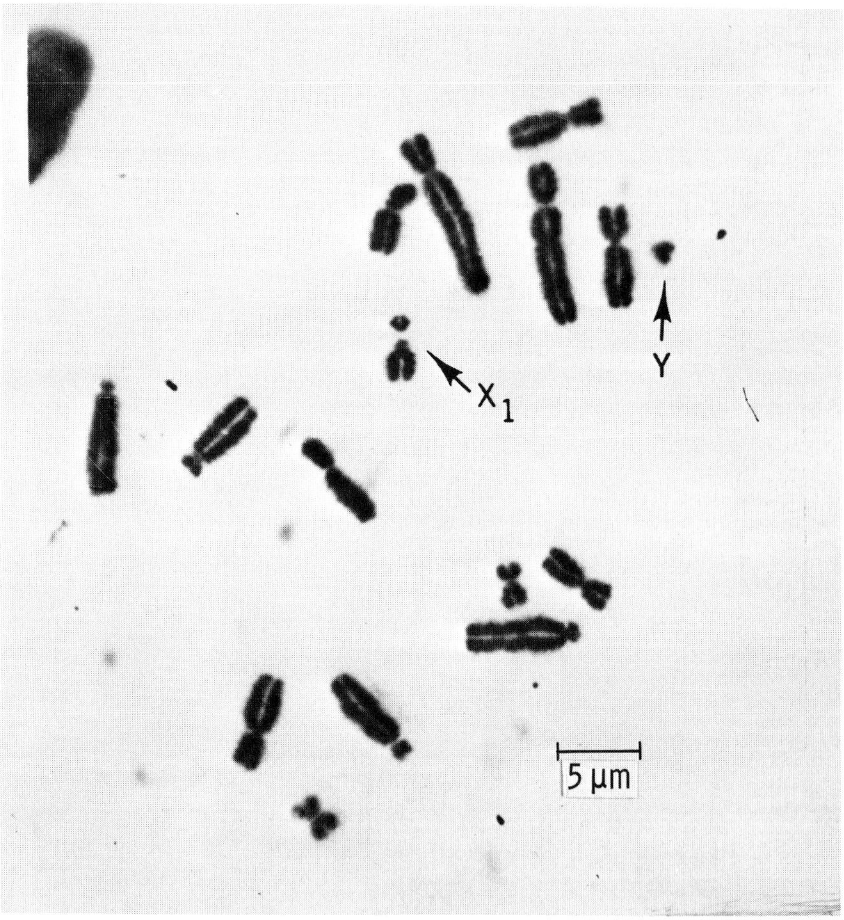


Fig. 3. Chromosome spread of a male *Macropus giganteus* ($2n = 16$). The male X chromosome demonstrates a prominent secondary constriction. In females, one X is identical to this one, while the other is more condensed and lacks the secondary constriction. From Marshall Graves, *Exp. Cell Res.*, 1967; photograph courtesy of D. L. Hayman.

and they concluded from their study that the Y chromosome in marsupials is strongly male determining.

## III. CHROMOSOMES OF AMERICAN MARSUPIALS

A relatively small number of the estimated 69 species of American marsupials (Cabrera, 1957) have been investigated for chromosomal constitutions, although recent emphasis is increasing in this area. Those species which have been investigated, however, have appeared in the literature in considerable detail (especially *Didelphis virginiana,* the common opossum). Chromosome numbers and sex chromosomal mechanisms of most of the American marsupials which have been investigated to date are listed by Sharman (1961), Hayman and Martin (1969), and discussed recently by Thrasher (1972). In addition to possessing fewer and larger chromosomes, marsupials also differ from eutherian mammals in that the former possess a bimodal, rather than a unimodal distribution of chromosome numbers (Sharman, 1961). These peaks of 14 and 22 chromosomes become readily apparent when analyzing the opossums of the Western Hemisphere.

### A. Opossums with 14 Chromosomes

Biggers *et al.* (1965) studied the chromosomes of the woolly opossum (*Caluromys derbianus*) and a murine opossum (*Marmosa mexicana*), both from Nicaragua. The karyotypes of these two species were found to be similar. Both possess three pairs of large submetacentric autosomes (group A), one pair of medium metacentric autosomes (group B), and two pairs of medium subtelocentric autosomes (group C). The X chromosome in each case is a small acrocentric, and the Y is a very small telocentric. The description of the chromosomes of *Caluromys derbianus* reported by Legator *et al.* (1966) differs slightly from that just cited. They separated autosomal pairs 5 and 6 into groups C and D, respectively, on the basis of a slightly different centromeric position. All sex chromosomes were classified as telocentrics, and they noted that the Y is one-half the length of the X, and that the X is less than one-half the length of the smallest autosome.

The karyotypes of the pouchless opossum *Marmosa robinsoni* (= *Marmosa mitis*) (Fig. 4) and *Marmosa fuscata* are also remarkably similar to that of *Caluromys derbianus* despite wide variations in morphology and evolutionary history between the two genera (Reig, 1968; Reig and Sonnenschein, 1970). Reig and Sonnenschein believe that the chromosomes of

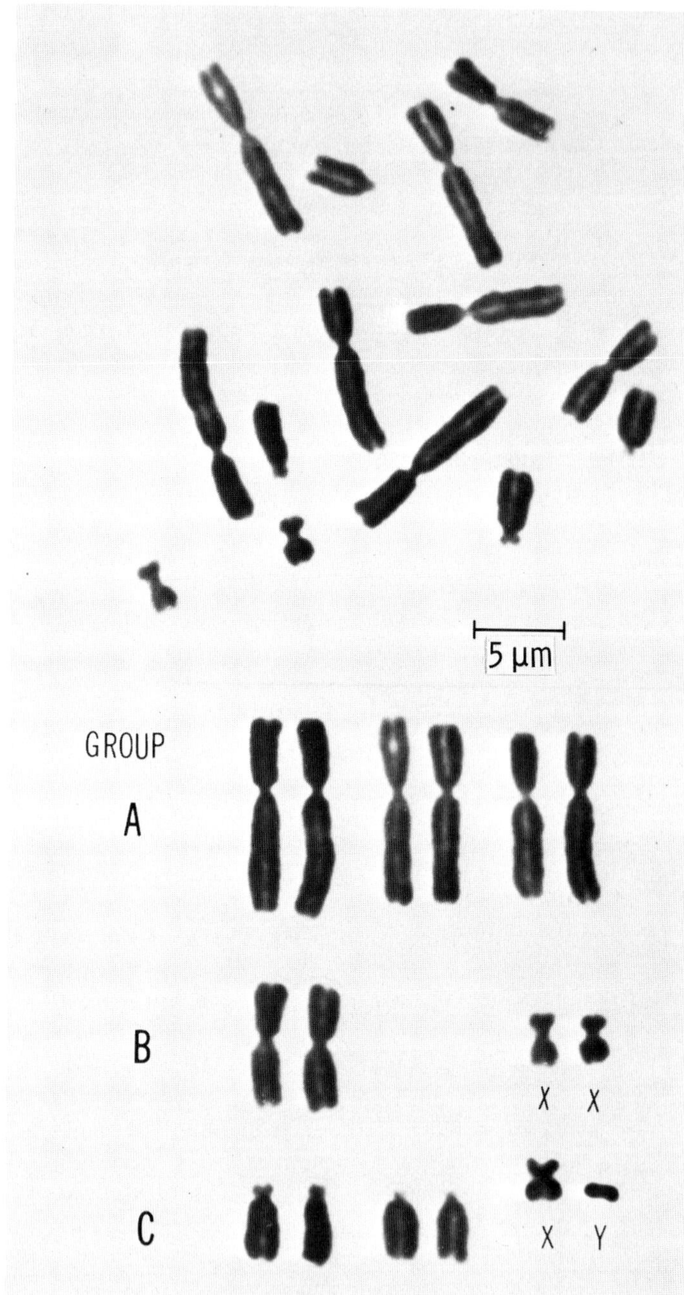
5 µm
GROUP
A
B
X X
C
X Y

groups A and B are probably the same in all these animals. Differences occur in group C, however (no mention was made of a group D): group C autosomes in *M. robinsoni* and *C. derbianus* are subterminal, whereas those of *M. fuscata* are metacentric (all are of the same relative size). The X is metacentric in *M. robinsoni* and *M. fuscata,* but acrocentric in *C. derbianus*; the Y chromosomes are similar in all three animals (telocentric). Reig and Sonnenschein concluded, therefore, that *M. robinsoni* and *M. fuscata* differ possibly only by two pericentric inversions in group C autosomes, and that these two species differ more than *M. robinsoni* and *C. derbianus,* the latter two demonstrating variation only in X chromosome structure.

Before discussing opossums with 22 chromosomes, it is pertinent to mention the work of Reig and Bianchi (1969), who investigated the chromosomes of two species of the genus *Monodelphis* which contain a diploid number of 18. Prior to this paper, a $2n = 18$ had not been reported in marsupials. They studied the chromosomes of *Monodelphis brevicaudata palliolata* from Venezuela and *Monodelphis dimidiata* from Argentina. The authors suggested possible interrelationships between these animals and *Marmosa* and *Caluromys* which will be discussed in Section VI,H.

### B. Opossums with 22 Chromosomes

To date, four Western Hemisphere opossums which possess 22 chromosomes have been examined cytologically and reported in the literature. These are *Didelphis marsupialis, Didelphis albiventris,* and *Philander opossum,* all from Central and/or South America, and *Didelphis virginiana,* which is also found in North America. The latter, perhaps owing to its relative ubiquity, has been employed extensively in chromosomal and cytogenetic investigations, and, therefore, will be given initial attention.

Jordan (1911) observed 17 chromosomes in dividing spermatogonia and 9 in primary spermatocytes of *Didelphis virginiana*; he concluded that the diploid number was 17 and that the male possessed an XO sex chromosome constitution. Hartman (1919) found 10 to 12 chromosomes in the polar body spindles of ova, and since 12 was most frequently represented, he concluded that this was the haploid number, and, therefore, $2n$ equaled 24. The first accurate assessment was made by Painter (1922), who observed 22

←

Fig. 4. Chromosome spread and karyotype of a female *Marmosa robinsoni* ($2n = 14$). The sex chromosomes of a male are included in the karyotype. The autosomes are subdivided into groups A (submetacentrics), B (metacentrics), and C (subtelocentrics), and arranged in order of decreasing size. From Wolf *et al.*, *Cytobios*, 1970.

chromosomes in spermatogonia, including the two smallest which constituted a typical XY sex chromosome mechanism. Painter believed that Jordan's error resulted from faulty fixation, and Hartman's from precocious splitting of the X chromosome tetrad in the polar spindles. In equatorial plates of somatic cells, Hoy and George (1929) observed an autosomal ring surrounding the sex chromosomes which were morphologically distinguishable due to the relatively small size of the Y. They identified 6 "macrochromosomes," 14 "mesochromosomes," and the smaller XX/XY, depending on sex. One pair of macrochromosomes in both males and females appeared larger than the other two, an observation which has not been substantiated in the recent literature. The mesochromosomes demonstrated a general decrease in size.

Prior to the advent of modern cytological techniques (metaphase arrest, hypotonic spread preparation), it was believed that all the chromosomes of *D. virginiana* were rod shaped, and, therefore, acrocentric or telocentric in form (Painter, 1922; Dreyfus and Campos, 1941). With the advantages of tissue culture and a metaphase spreading technique, Tjio and Puck (1958) were able to distinguish clearly the X and Y chromosomes, but they did not construct an idiogram of the autosomes. Ohno *et al.* (1960) used metaphase figures which had undergone prolonged hypotonic treatment to analyze chromosomes from cultured cells of *D. virginiana.* Using degree of heteropycnosis and relative size as guidelines, they incorrectly assumed, however, that the X chromosome was acrocentric in form.

Shaver (1962) performed a detailed study of the form and relative size of the chromosomes of *D. virginiana.* The autosomes were divided into two groups: group A (6 pairs) are subtelocentric, and group B (4 pairs) are telocentric in form. These observations were confirmed by Biggers *et al.* (1965). Schneider and Rieke (1967) further subdivided the chromosomes of group A into 3 pairs of large subtelocentrics and 3 pairs of small subtelocentrics; the X chromosome is a considerably smaller subtelocentric, and the Y a very small telocentric chromosome (Fig. 5). Thus, Sharman's (1961) statement that all didelphids possess 20 acrocentric autosomes and a metacentric X does not hold true for *D. virginiana.* It similarly does not apply to *D. albiventris* or *D. marsupialis* in which the sex chromosomes, as well as the autosomes, are telocentric in form. The approximate mean content of chromatin in each autosome was presented by Shaver (1962): each large subtelocentric contains 6.4% of the total chromatin, each small subtelocentric 4.4%, and each telocentric 3.7%. Each X contains 2.7% of the total chromatin, and the Y, 1.4%. These findings have recently been confirmed with slight variations by L. K. Schneider (unpublished observations), who noted that each X averages 66% the length of the smallest subtelocentric autosome, while the Y measures only 38% the length of the

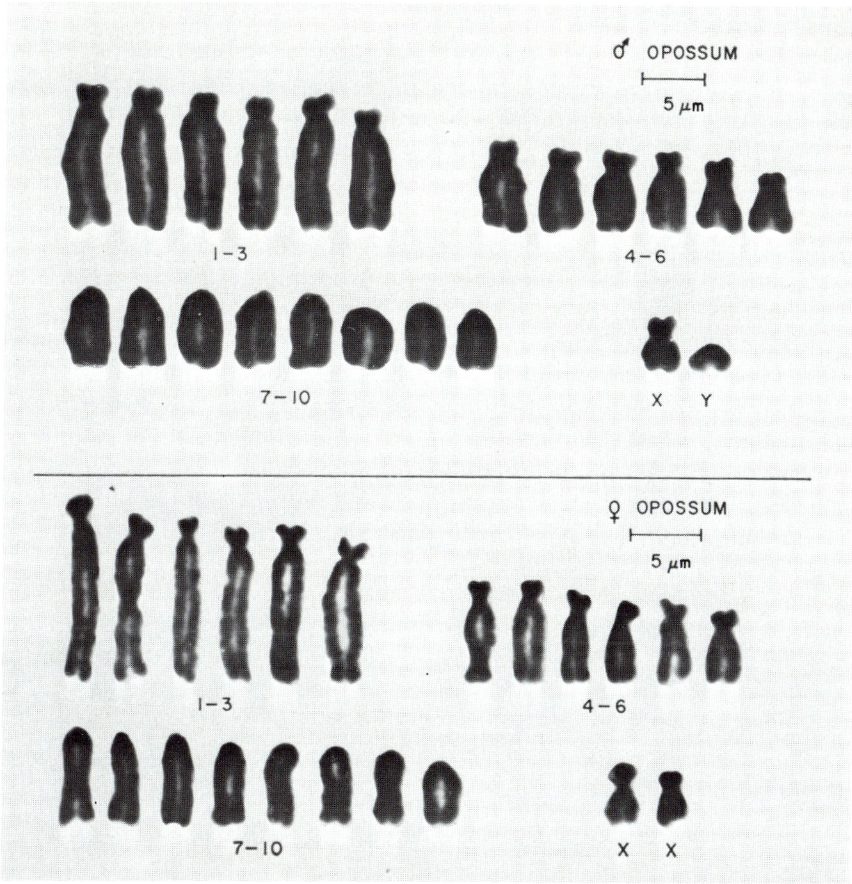


Fig. 5. Karyotypes of a male and female *Didelphis virginiana* ($2n = 22$). The autosomes are subdivided into large subtelocentrics (pairs 1 to 3), small subtelocentrics (pairs 4 to 6), and telocentrics (pairs 7 to 10). The sex chromosomes are clearly distinguishable by size. From Schneider and Rieke, *J. Cell Biol.*, 1967.

smallest telocentric autosome. Thus, the sex chromosomes are easily detectable by sight in hypotonic spreads of cells in prophase as well as in metaphase.

The remaining opossums with $2n = 22$ which have been investigated possess chromosome complements in which all the autosomes are telocentric in form. Since analysis of the chromosomes of *Didelphis marsupialis aurita* was performed without the advantage of hypotonic spreading (Dreyfus and Campos, 1941), it is difficult to ascertain their accurate morphology. The investigators reported that "All of the autosomes are more or less rod-shaped with telomitic insertion . . . ." The X was said to be composed of "two elements" whereas the Y was "constituted by one round-shaped element." It is apparent that more work is necessary in order to clarify the chromosomal constitution of this animal.

*Didelphis albiventris* was said to possess chromosomes with the shape of a telomitic rod; the complement was divided into 3 large pairs, 7 median pairs, and the XY (Saez, 1931). More recently, Perondini and Perondini (1965) published the karyotype of *D. albiventris* which shows that all autosomes and the sex chromosomes are telocentric. The Y is one-half the size of the X, both being considerably smaller than the smallest autosome.

Biggers *et al.* (1965) showed that *Philander opossum* (four-eyed opossum) also possesses three pairs of large telocentric autosomes (group A), seven pairs of medium telocentrics (group B), a small telocentric X, and a very small telocentric Y chromosome. This was confirmed by Perondini and Perondini (1966) who reported that the Y chromosome, which is the smallest in the complement, is about one-half the length of the X, the latter being shorter than the smallest autosome (Figs. 6 and 7).

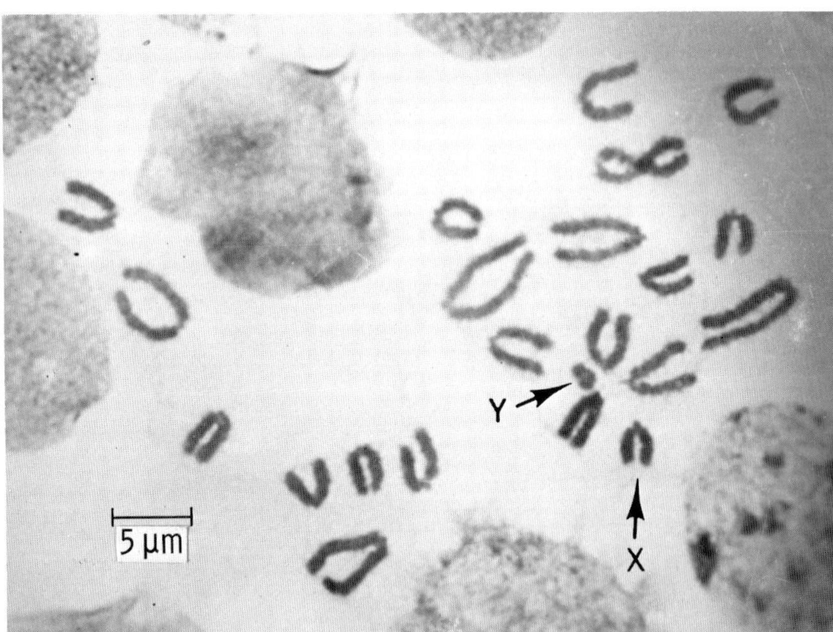


Fig. 6. Chromosome spread of a male *Philander opossum* ($2n = 22$). All the chromosomes are telocentric in form. The sex chromosomes are the smallest of the complement. From Perondini and Perondini, *Cytogenetics*, 1966.

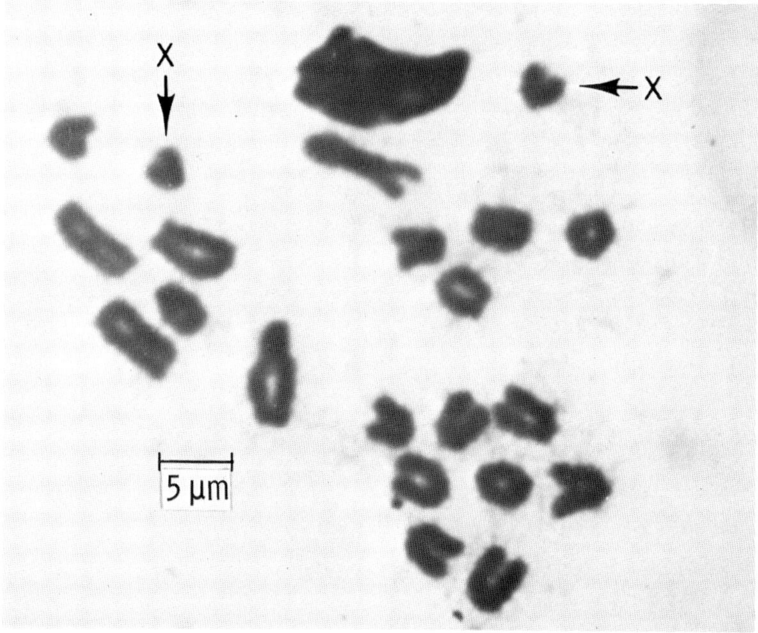


Fig. 7. Chromosome spread of a female *Philander opossum*. From Perondini and Perondini, *Cytogenetics*, 1966.

The karyotype of the Brazilian opossum, *Didelphis marsupialis* (Cestari and Perondini, 1964), appears essentially comparable to those of *Didelphis albiventris* and *Philander opossum* (Fig. 8). It is probable that *D. marsupialis aurita* is also very similar, making all four species closely related on a chromosomal basis. *Didelphis marsupialis* possesses 11 pairs of chromosomes which are classified as large (pairs 1 to 3), medium (pairs 4 to 10), and short (the sex chromosomes). All are telocentric (sometimes referred to as acrocentric) in form. The investigators noted that there is a "sharp difference between the length of the X chromosome and the smallest autosome," and that the Y chromosome measured about one-half the length of the X. If the autosomes are arbitrarily divided into the same categories as employed for *D. virginiana,* i.e., pairs 1 to 3, 4 to 6, and 7 to 10, it is found that each chromosome of pair 1 to 3 contains an average of 6.9% of the total chromatin, while pairs 4 to 6 contain 4.4%, and pairs 7 to 10, 3.7%. Each X chromosome contains 1.7% of the total chromatin, and the Y, 1.1%. Thus, autosomal pairs 1 to 3 are larger in *D. marsupialis* than in *D. virginiana,* whereas the opposite is true for the sex chromosomes. The

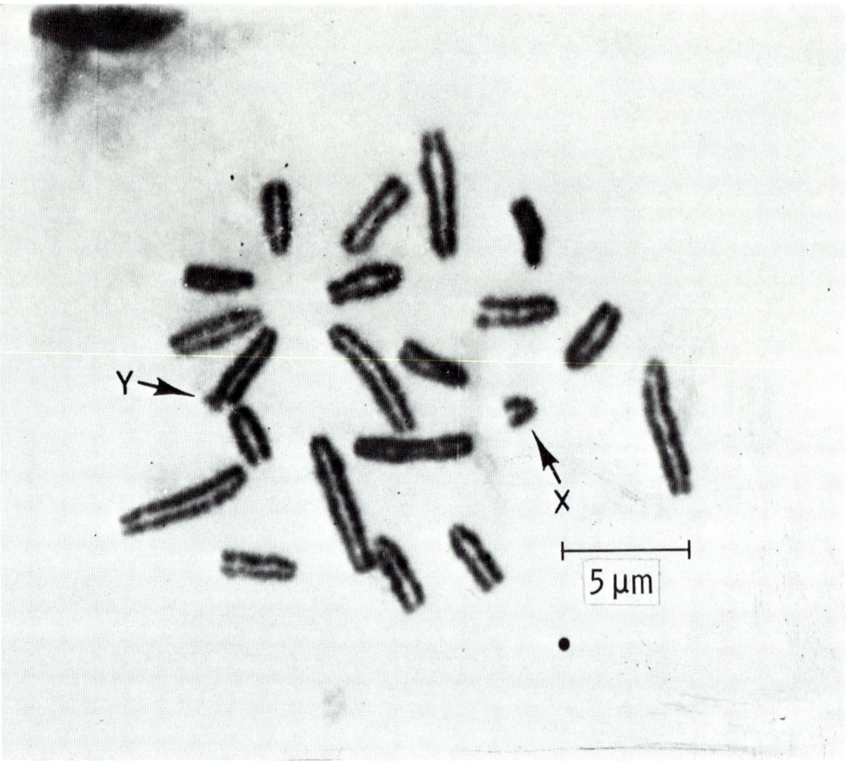


Fig. 8. Chromosome spread of a male *Didelphis marsupialis* ($2n = 22$). All the chromosomes of this species are also telocentric in form. The sex chromosomes are the smallest of the complement; the Y is very closely associated with an autosome in this photograph. From Perondini and Perondini, *Rev. Bras. Biol.*, 1965.

percentages of total chromatin contained in the remaining autosomes are equivalent in the two species.

The absolute sizes of marsupial chromosomes are large. Reig (1968) reported that the largest autosomes of *Marmosa robinsoni* and *Caluromys derbianus* measure about 11 $\mu$m in length. The largest chromosome pair in *Didelphis virginiana* measures 13–14 $\mu$m in length which is three times or more the size of the largest human autosome (Schneider and Rieke, 1968).

## IV. TISSUE CULTURE OF MARSUPIAL CELLS

Modern cytological techniques have advanced to a level where chromosome preparations may be obtained from nearly any warm-blooded animal with rapidity, accuracy, and reproducibility. This ability has developed with the advent of the following basic techniques: hypotonic solutions which swell the cell and allow for chromosomal separation (Hsu and Pomerat, 1953), mitotic inhibitors such as colchicine and vinblastine which arrest cells in metaphase, and tissue culture. Especially significant has been the development of the short-term lymphocyte culture technique, which has greatly simplified the procedure of obtaining chromosome spreads compared to classic tissue culture techniques. The latter system allows investigation of cells other than lymphocytes, however, and offers the advantage of prolonged incubation *in vitro*.

### A. Lymphocytes

Prior to the discovery by Nowell (1960) that phytohemagglutinin (PHA), an extract from the red kidney bean (*Phaseolus vulgaris*), is mitogenic for leukocytes *in vitro*, it had been employed as a red cell agglutinin in serum production. Immediately following this discovery, Moorhead *et al.* (1960) published a method for chromosome preparations from PHA-stimulated human leukocytes which has provided the common basis for almost all subsequent methods of human and animal leukocyte culture. It is now known that it is the small lymphocyte which responds to PHA by enlargement, nucleic acid and protein synthesis, and ultimately by division (for a review, see Robbins, 1964).

In the majority of instances in which animal lymphocytes have been used, one or more modifications of the original technique have been necessary. A reliable method for the short-term culture of lymphocytes from *Didelphis virginiana* was established after considerable experimentation by Schneider and Rieke (1968). Since this method may prove effective for the majority of marsupials, the rationale behind the various approaches will be discussed briefly, followed by a summary of the culture technique. The key to successful culturing, and, thus, to the preparation of high quality metaphase plates, of *D. virginiana* lymphocytes lies in the elimination of as many red blood cells from the lymphocyte fraction as is possible. This may best be accomplished by the collection of thoracic duct lymph, but this is an often difficult and time consuming procedure which necessitates sacrificing the animal. Peripheral blood obtained by cardiac puncture comprises the most readily accessible source of lymphocytes, but additional steps must be taken in their purification. This is possible by employing $NH_4Cl$ and a cotton-packed tube (Agostoni and Idéo, 1965) which lyses the RBC's (red blood cells) and removes the majority of polymorphonuclear leukocytes, respectively, but this technique often results in vastly reduced cell yields. Another approach is to agglutinate and sediment the RBC's by treating whole blood with PHA-M (Difco) for 24 hr at 4°C. This results in approximately a 40% supernatant of leukocyte-rich plasma.

The nature of the difficulty caused by RBC's in the final culture is unknown, but it was noted that when large numbers were present, they had lysed by 24 hrs of incubation, and lymphocyte growth was invariably poor. It was found that the type of serum used is important in cellular response (Table I). Twenty percent human, horse, or autologous serum either does not support growth *in vitro,* or supports it poorly. Bick and Jackson (1968) noted that concentrations of horse serum greater than 12% sometimes proved toxic to *P. tridactylus* lymphocytes in culture. Burton (1968b) reported that the serum of the grey kangaroo (*Macropus giganteus*) had to undergo defibrination before it was effective in cultures of lymphocytes from these animals. Rat serum which has been heated at 56°C for 30 min to inactivate complement (Carpenter, 1965) is extremely effective in supporting growth of *D. virginiana* lymphocytes *in vitro.* Since lysis of red blood cells does not occur in these cultures, it is possible that the lysis, or perhaps an antibody which causes it, is responsible for the depressed growth response (Schneider and Rieke, 1968). Heat inactivation of the serum complement eliminates this problem.

The culture technique may be summarized as follows: 3 to 5 $\times$ $10^6$ cells/ml of Eagle's minimal essential medium which is supplemented with 20% heat-inactivated Sprague–Dawley or Lewis rat serum, 40 $\mu$g/ml PHA-P (Difco), and 100 units penicillin/ml are incubated in small Erlenmeyer

**TABLE I**

Effects of Various Sera on Opossum (*Didelphis virginiana*) Lymphocyte Response as Determined by Cellular Transformation and $^3$H-Thymidine Labeling[a]

| Serum used | Cell differential at 45 hr incubation: Small(%) | Medium(%) | Large(%) | Percentage of labeled cells (all sizes) at 45 hr |
|---|---|---|---|---|
| Human | Did not support growth in culture | | | |
| Horse | 31 | 50 | 19 | 0.4 |
| Opossum | 16 | 52 | 32 | 0–7 |
| Rat | 14 | 22 | 64 | 0–31 |
| Heat-inactivated rat on TDL lymphocytes | 8 | 25 | 67 | 12–31 |
| Heat-inactivated rat on blood lymphocytes[b] | 5 | 32 | 63 | 14 |
| Control(TDL) | 94 | 2[c] | 4[c] | 3–12 |

[a] From Schneider and Rieke (1968).
[b] Incubation 55 hr.
[c] Occasional animals showed as many as 33% medium and large cells.

flasks at 37°C for 36–48 hr. Utilization of a technique developed by Evans (1965) allows easy localization of large numbers of well-spread metaphase plates. Use of MacNeal's (1922) tetrachrome stain provides a rapid and simple staining technique.

With various modifications of the technique of Moorhead *et al.* (1960), successful short-term cultures have been established with lymphocytes from the following marsupials: *Potorous tridactylus* (Shaw and Krooth, 1964; Hayman and Martin, 1965a; Moore, 1965; Goh, 1967; Bick and Jackson, 1968), *Wallabia bicolor* (Moore and Gregory, 1963; Hayman and Martin, 1965a; Moore and Uren, 1965; Radley, 1966), *Macropus eugenii* (Marshall Graves, 1967), *Macropus giganteus* (Marshall Graves, 1967; Burton, 1968a,b), *Macropus rufus* (Moore, 1965), *Lagorchestes conspicillatus* (Martin and Hayman, 1966), *Marmosa robinsoni* (Wolf *et al.*, 1970), and *Didelphis virginiana* (Biggers *et al.*, 1965; Schneider and Rieke, 1967, 1968; Sinha, 1967a,b; Reiss *et al.*, 1970). Hayman and Martin (1965c) reported that with one exception, all attempts to culture leukocytes from *Schoinobates volans* were unsuccessful.

### B. Other Cell Lines

Walen and Brown (1962) established tissue cultures of brain, eye, lung, liver, kidney, and testis of a male *Potorous tridactylus*. Only kidney and testis cells multiplied reasonably well; cultures had been maintained with transfers for 6 months at the time of writing. The most satisfactory medium for these cells was 90% LEY (2% lactalbumin enzymatic hydrolysate and 2% yeast hydrolysate in Earles salt solution) and 10% fetal calf serum. Walen (1965) was also able to culture *P. tridactylus* cells in Eagle's medium containing 4% fetal calf serum. A kidney cell line ($HPK_1$) from this animal was established and maintained for 4 years and 4 months with 420 passages *in vitro* (Bick and Brown, 1969b); the cells (500,000/ml) were cultured at 37°C in medium 199 with 12% horse serum. Cultures of a *Potorous tridactylus* cell line which are commercially available have also been employed in chromosomal research (Jakob and Ruiz, 1970).

Cultures of lung, heart, and kidney cells from *Didelphis virginiana* were established by Shaver (1962) in 70% medium 1066 (Connaught), 20% calf serum which had been preheated to 60°C for 30 min, and 10% of a 0.5% lactalbumin hydrolysate solution. Penicillin and streptomycin (100 units/ml of medium) were added. The cultures were incubated at 37°C for 3 to 8 days in screw cap and Leighton tubes which contained coverslips. The pH was adjusted to 7.2. Cestari and Perondini (1964) cultured kidney cells of *Didelphis marsupialis* by the plasma clot method. The cells were washed in

Hanks' salt solution with antibiotics and clotted with chicken plasma and chicken embryo extract on coverslips in Leighton tubes. One ml of fresh fibroblast medium, supplemented with 15% calf serum, was added to each tube. The cultures were grown at 37°C for 1 to 2 weeks at a $CO_2$-adjusted pH of 7.2. Lung, kidney, skin, and peritoneal cells from *Caluromys derbianus, Marmosa mexicana,* and *Philander opossum* were cultured in Eagle's medium with 15% fetal calf serum or in medium 1066 with 10% lactalbumin hydrolysate (0.5%) and 20% inactivated calf serum (Biggers *et al.*, 1965).

A pseudodiploid cell line (JU56) from explants of buccal mucosa of a male *Wallabia bicolor* was established by Uren *et al.* (1966). The explants were embedded in plasma clots in Eagle's minimal essential medium (MEM) modified by the addition of pyruvate and serine, and enriched with 25% fetal calf serum. The outgrowth of fibroblasts was removed and grown as monolayers in 4 oz McCartney bottles in the same medium supplemented with 2% lyophilized beef embryo extract. Cells from subcutaneous tissue of *Antechinus swainsonii* have been cultured under similar conditions (Eagle's MEM with 25% fetal calf serum in 5% $CO_2$ in air) by Moore and Uren (1966), as have cells of skin and various organs of *Perameles nasuta* (Eagle's MEM, 15% calf serum, and 5% fetal calf serum) by Jackson and Ellem (1968).

## V. MARSUPIAL CELL CYCLES

Knowledge of cell cycles, with their component parts, is prerequisite to understanding the time sequences of nucleic acid and protein synthesis as well as the kinetics of proliferation within a given cell population. Chromosomes are visible only during mitosis (M); the cellular events which culminate in division take place largely in interphase, which is subdivided into $G_1$ (presynthetic gap or phase), S (DNA synthetic phase), and $G_2$ (postsynthetic gap or phase). Information on the duration of these phases in various cells has been obtained largely by radioautographic techniques (for a review, see Baserga and Malamud, 1969).

### A. *In Vivo* Cycles

Thrasher (1969) determined the durations of various phases of the cell cycle for the body and pyloric regions of the stomach of *Didelphis virginiana* by constructing curves of percent labeled mitoses (PLM) after injec-

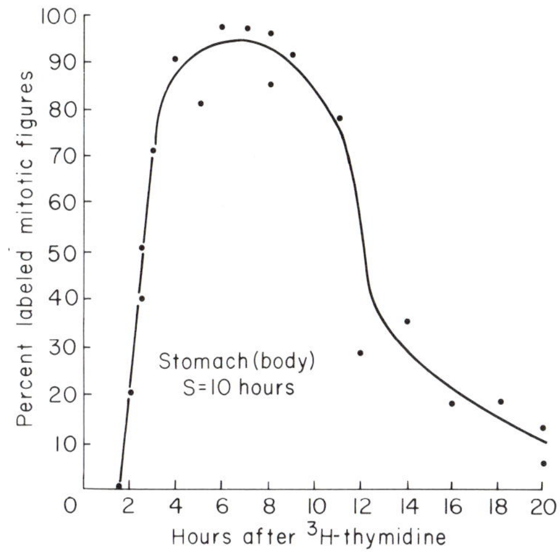


Fig. 9. Percent labeled mitoses (PLM) curve of the body portion of the stomach mucosa of *Didelphis virginiana*. Each dot represents at least one animal. Redrawn slightly from Thrasher, *Exp. Cell Res.*, 1969.

tion of $^3$H-thymidine (Figs. 9, 10). $G_2$, M, and S phases were found to be 1.5, 2.5, and 10.0 hr, respectively. These times are longer than those reported for intestinal epithelia of eutherian mammals, a condition which Thrasher believes may be due to the inherent lower body temperature of didelphids (approximately 34°C). Cell cycle and S phase durations of small intestinal epithelia of *Marmosa robinsoni* are also longer than corresponding times reported for placentals. In addition, the average duration of S within different examples of renewing epithelia of *M. robinsoni* (Thrasher *et al.*, 1971) was highly variable; for instance, in the stomach it was greater than 14 hr, whereas in the duodenum it was only 8.5 hr. In contrast, they found that the mean duration of $G_2$ plus one-half of M was fairly constant in these tissues.

### B. *In Vitro* Cycles and Kinetics

The PLM curve for *Didelphis virginiana* lymphocytes extending from 1 to 20 hr is presented in Fig. 11 (L. K. Schneider, unpublished observations).* Work is currently in progress to extend the curve beyond 20 hr, but the first wave is sufficient to establish that the average $G_2$ is 2 hr and the average S is 9 hr in these cells (Schneider and Rieke, 1967). If the second peak of the

* See addendum.

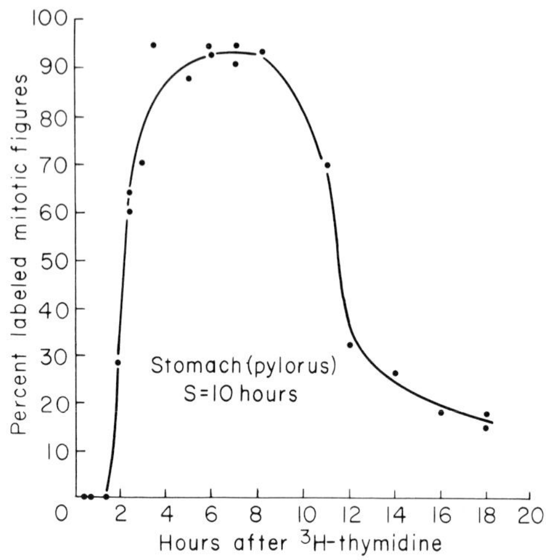


Fig. 10. PLM curve of the pyloric portion of the stomach mucosa of *Didelphis virginiana*. Each dot represents at least one animal. Redrawn from Thrasher, *Exp. Cell Res.*, 1969.

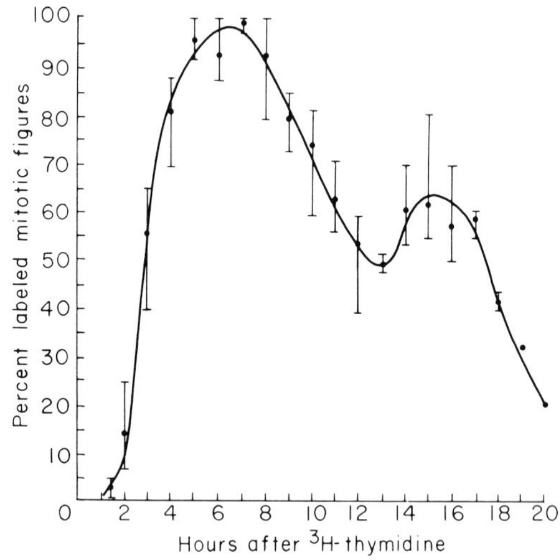


Fig. 11. PLM curve of cultured lymphocytes of *Didelphis virginiana*. Each dot represents from two to five cultures; ranges of labeled mitoses observed at individual times are indicated by the vertical lines. The initial wave is very similar to those reported for stomach mucosa of *D. virginiana* (see Figs. 9 and 10).

curve is accepted as an additional cycle, the generation time for these cells is just slightly more than 11 hr. Allowing approximately ½ hr for mitosis, this would indicate a nonexistent $G_1$ phase which, in turn, would indicate a rapid recycling of the cells in culture. This concept must await extension of the curve for verification, however, due to the diminutive size of the second peak. Reiss *et al.* (1970) estimated the $G_2$ period for bone marrow leukocytes of *D. virginiana* to be 3 to 4 hr. Cultured lymphocytes from *Wallabia bicolor* demonstrated what Moore and Uren (1965) considered to be a variable $G_2$ phase.

Working with the $PTK_1$ cell line form *Potorous tridactylus*, Levan (1970) found that the $G_2$ plus prophase period always exceeded 2 hr, was over 4 hr in 85.5% of the cells analyzed, and exceeded 6 hr in 12.6%. This variability in the length of $G_2$ is apparent in most PLM curves, since the ascending limb of the first peak is usually skewed to the right. The time between the first appearance of labeled mitotic figures and when the peak reaches 100% comprises the total extent of $G_2$; the average $G_2$ lies somewhere between these two points, depending on the angle of the ascending limb. Levan estimated the average length of $G_2$ plus prophase in these cells to be about 5 hr.

Cultured lymphocytes of *Didelphis virginiana* demonstrated an unusual occurrence of DNA synthetic peaks (Fig. 12) over the incubation period

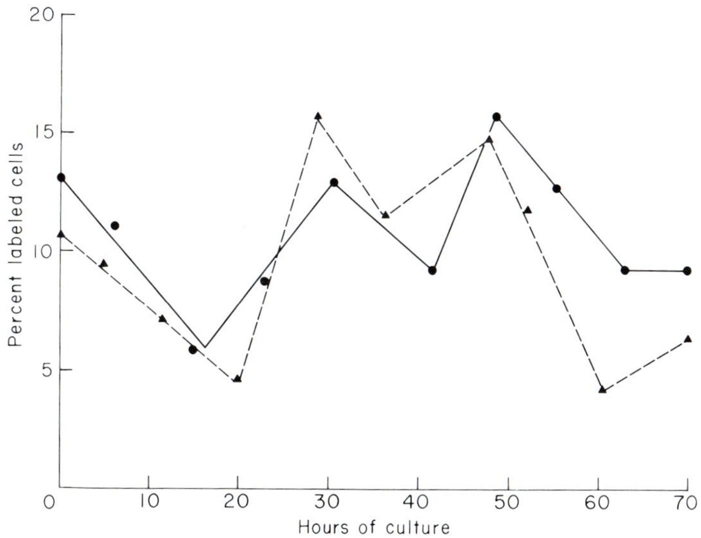


Fig. 12. $^3$H-thymidine labeling percentages in cultures of lymphocytes from *Didelphis virginiana* (two experiments). Each point represents the percentage of DNA synthesizing cells when individual cultures were labeled terminally with the isotope. Redrawn from Schneider and Rieke, *Cytogenetics*, 1968.

(Schneider and Rieke, 1968) compared to other reports in the literature. Whereas 11–13% of the lymphocytes from this animal demonstrated incorporation of $^3$H-thymidine at the initiation of incubation, only 0.06% of human lymphocytes demonstrated this phenomenon (Bond *et al.*, 1958). Cooper *et al.* (1961) and Michalowski (1963) reported a 24-hr lag in DNA synthesis in human lymphocytes *in vitro*, with a single peak at 70 hr of incubation. Lymphocytes from the opossum showed two equivalent peaks at approximately 30 and 48 hr of culture. In addition, these investigators observed initial mitotic activity in human cultures between 42 and 48 hr with a peak between 65 and 77 hr of incubation. Opossum lymphocytes, on the other hand, began to divide at 20 hr after initiation of incubation, and demonstrated a peak at 46 hr (Schneider and Rieke, 1968). Burton (1968b) reported that the onset of DNA synthesis and initial increase in RNA synthesis were much more rapid in lymphocytes of *Macropus giganteus* than in human lymphocytes *in vitro*. These rapid bursts of nuclear activity have not been observed in lymphocyte cultures from most other animals. Whether this is the result of a higher incubation temperature than body temperature of some marsupials awaits further investigation.

The mitotic index (percent of cells in mitosis during various intervals of incubation) provides one means of assessing cellular response *in vitro*. Shaw and Krooth (1964) reported the rather low mitotic index (MI) of 0.38% in peripheral blood cultures of *Potorous tridactylus*. That reported by Legator *et al.* (1966) for incubated lymphocytes from *Caluromys derbianus* was considerably higher (3.3%). Wolf *et al.* (1970) obtained an extraordinarily high MI of 15.7 to 22.0% in cultures of lymphocytes form *Marmosa robinsoni*. These cultures contained 5% PHA, however, which is four times that employed in human lymphocyte cultures, and they were treated with colchicine for 9 hr before harvest.

## VI. MARSUPIAL CYTOGENETICS

In addition to the important advances mentioned earlier (mitotic arrest, hypotonic chromosome spreading, and lymphocyte culture), the field of cytogenetics has benefited greatly from three other major discoveries. These are (1) the observation that mammalian somatic cells present a morphologically recognizable sexual dimorphism with regard to the presence or absence of sex chromatin (Barr and Bertram, 1949); (2) the realization of the specific incorporation of $^3$H-thymidine into newly replicated DNA of plants and animals, and of the semiconservative mode of DNA replication (Taylor *et al.*, 1957); and (3) the finding by Hayhoe and Quaglino (1965) that the isotope [5-$^3$H]uridine (in which the tritium molecule in located on the 5 position of the pyrimidine ring) is specifically incorporated into newly synthesized RNA. Marsupials, by virtue of their exceptional chromosomal characteristics, have been very useful in the application of these discoveries, leading to a wealth of new (and often unusual) cytogenetic information.

### A. DNA Synthesis in *Wallabia bicolor*

Moore and Uren (1965) performed a detailed analysis of chromosomal DNA synthetic patterns in this species. They reported that the autosomes as a whole completed DNA synthesis at about the same time, although different regions of the chromosomes terminated at different times. An underlying pattern could be detected, but it was characterized by considerable variation. The investigators postulated that DNA synthesis is initiated at a small number of fixed sites and that there is a characteristic distribution of these sites in each chromosome. The small $Y_1$ chromosome completed DNA synthesis early. This chromosome demonstrates positive heteropycnosis in prophase and is represented as a small dot at metaphase. Moore and Uren believe that early completion in the $Y_1$ may be due to its small size or that it may be causally related to its positive heteropycnosis. The latter belief is contradictory to the concept that heteropycnotic chromosomes or segments are late in DNA replication (Schmid, 1963; Lima-de-Faria *et al.*, 1965). The short arm of the X synthesized DNA late into the S phase, and the authors stated that this portion of the X "is subject to some form of control different from that of the autosomes." Late DNA replication in the short arm of one X was also observed by Hayman and Martin (1965a). They reported that the long arm of the X chromosomes and homologous autosomes replicated synchronously, but that some pairs of nonhomologous autosomes synthesized DNA asynchronously.

### B. DNA Synthesis in *Potorous tridactylus*

Walen (1965), studying *P. tridactylus*, confirmed the findings of other workers that there are multiple sites for DNA synthesis along the chromosomes, and demonstrated that segments or whole chromosomes (whether or not homologous) may replicate asynchronously. The latter observation was most obvious in the sex chromosomes. Evidence was also presented for a nonrandom distribution of old and new DNA along the chromosomes. Hayman and Martin (1965a) observed the same form of DNA replication in the sex chromosomes of this species as they did in *Wallabia bicolor*, that is, the short arm of one X replicated asynchronously while the long arms replicated concurrently with the autosomes. No sex chromatin was observed in

500 buccal smear cells from a male *P. tridactylus* (Shaw and Krooth, 1964); this was taken to mean that none of the three sex chromosomes replicates late in the male. Levan (1970) investigated DNA synthesis in the chromosomes of the $PTK_1$ line of *P. tridactylus* after over 100 passages in tissue culture. The results indicated that the cell line chromosomes derived from the homologous pairs of this species: members of homologous autosomal pairs demonstrated similar labeling patterns, whereas differences between the active and inactive X chromosomes were apparent.

### C. DNA Synthesis in *Didelphis virginiana*

Schneider and Rieke (1967) studied DNA synthetic patterns in the autosomes and sex chromosomes of cultured lymphocytes from *Didelphis virginiana* throughout the S phase. In general, the individual groups of autosomes (pairs 1–3, 4–6, and 7–10) began replication synchronously and terminated asynchronously (Fig. 13). This observation confirms the earlier findings of German (1961) and Painter (1961), who reported that human

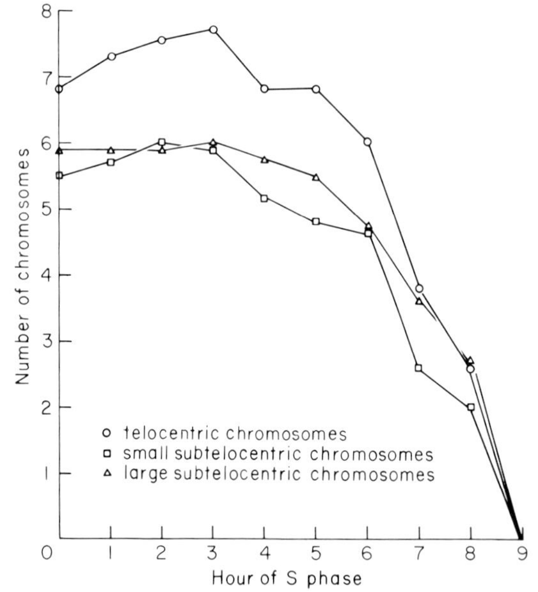


Fig. 13. DNA synthetic patterns in the autosomes of cultured lymphocytes from *Didelphis virginiana*. The graph demonstrates that initiation of replication was generally synchronous, but termination was asynchronous in these chromosomes. Redrawn from Schneider and Rieke, *J. Cell Biol.*, 1967.

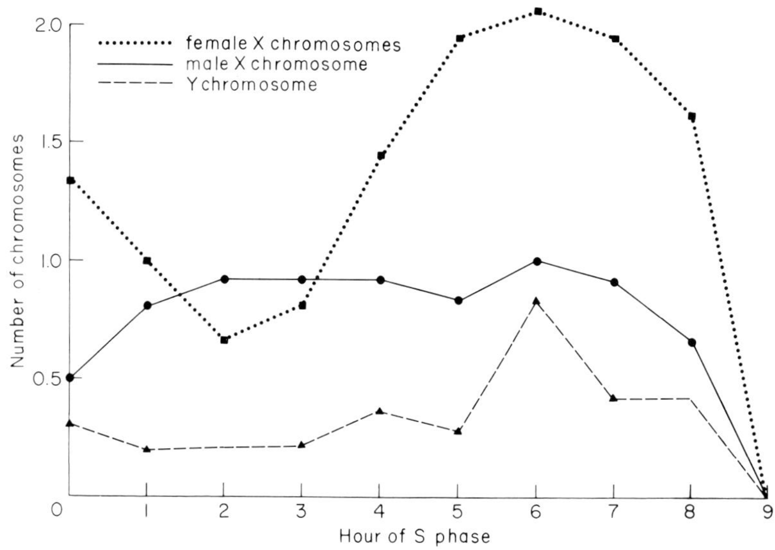


Fig. 14. DNA synthetic patterns in the sex chromosomes of cultured lymphocytes from *Didelphis virginiana*. The graph demonstrates that one X chromosome of female cells replicated DNA throughout the S phase, while its homologue began synthesis about 3.5 hr later. The male X synthesized throughout S, while the Y showed a late peak of replicative activity. Termination of replication was abrupt and synchronous in the sex chromosomes. Redrawn from Schneider and Rieke, *J. Cell Biol.*, 1967.

chromosomes terminated replication asynchronously. Also in agreement with human chromosome studies (German, 1964; Kikuchi and Sandberg, 1964) was the observation that the long autosomes began DNA synthesis earlier than the short autosomes, but in contrast to their findings, the long autosomes in *D. virginiana* did not replicate later than the short ones. Labeling intensity decreased in all the autosome groups as the S phase progressed toward completion in opossum lymphocytes.

Of particular interest are the labeling patterns in the sex chromosomes of *D. virginiana* (Fig. 14). One X in the female synthesized DNA throughout the S phase, while the other began DNA synthesis approximately 3.5 hr later. The two terminated replication synchronously, just slightly later than the autosomes. This type of X chromosome replication, where one X begins late but finishes at the same time as its homologue, is apparently unique to *Didelphis virginiana*. In all other species investigated to date, one X chromosome of somatic cells terminates DNA synthesis later than the other X, whether it began later, as in the Chinese hamster (Taylor, 1960; Hsu, 1964), or at the same time, as in the cow (Gartler and Burt, 1964). Thus, if one X chromosome of the opossum is considered to undergo late replication, this is manifest only by late *initiation*. In male cells, the Y chromosome replicated primarily during the latter half of the S phase, and, as was the case in the female, terminated DNA synthesis synchronously with the other sex chromosome (Schneider and Rieke, 1967). Grain patterns in the sex chromosomes of *D. virginiana* lymphocytes near the termination of the S phase are demonstrated in Figs. 15 and 16.

Assuming that no gaps exist in the DNA synthetic sequence in sex chromosomes throughout the S phase, it is logical to assume that if one X began late and terminated late with regard to its homologue, grain counts in both would be approximately equal. In addition, it must also be assumed that both X chromosomes contain the same amount of DNA. Thus, if both X's begin DNA synthesis synchronously, but one continues later (as in the cow), one might expect to observe generally reduced grain counts in the latter, since it is synthesizing DNA at a slower rate than its homologue. On the other hand, in the case of *D. virginiana*, where one X begins late but both

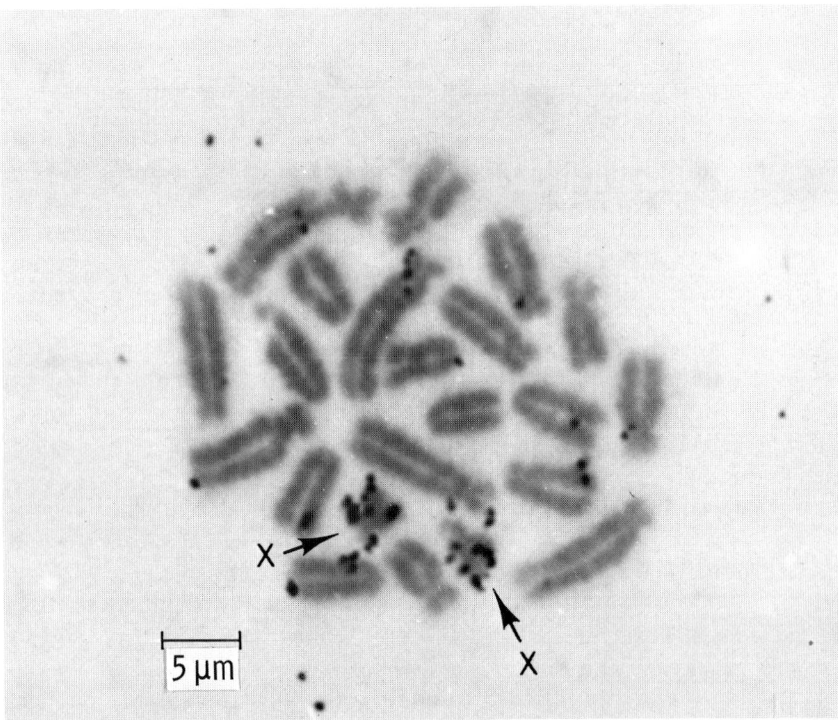


Fig. 15. Chromosome spread from a lymphocyte culture of a female *Didelphis virginiana*. The culture was labeled with $^{3}$H-thymidine near the termination of the S phase. Both X chromosomes demonstrate continued incorporation of $^{3}$H-thymidine after DNA replication had terminated in the autosomes. From Schneider and Rieke, *J. Cell Biol.*, 1967.

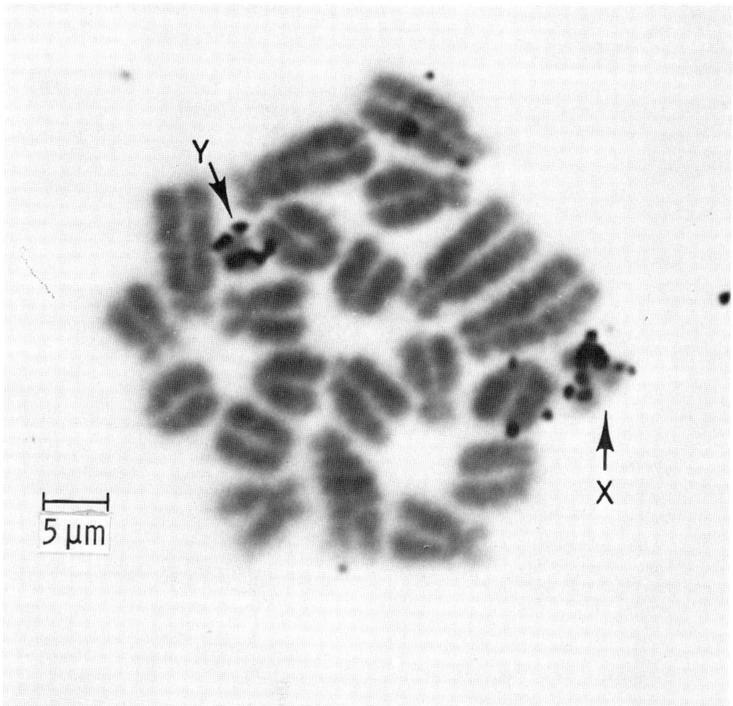


Fig. 16. Chromosome spread treated identically to that in Fig. 15, only from a lymphocyte culture of a male *D. virginiana.* In this case both X and Y chromosomes have continued DNA replication after the autosomes have ceased. From Schneider and Rieke, *J. Cell Biol.*, 1967.

terminate synchronously, one might expect a heavier grain count in the late starter, since it must synthesize its DNA at a faster rate than its homologue. The latter situation was frequently observed in the opossum, where one X chromosome contained considerably more grains than its homologue toward the end of the S phase (Fig. 17).

Sinha (1967a) also observed unusual DNA synthetic patterns in the chromosomes of cultured lymphocytes from *Didelphis virginiana.* Certain aspects of these observations confirm those reported above; others are in conflict. One female X chromosome and a large portion of its homologue plus a similar portion of the male X and entire Y chromosomes were "remarkably allocyclic in behavior." Grain counts on the female X chromosomes were similar in this study, and Sinha accredited this to approximately equivalent amounts of heterochromatin in each. Terminal patterns of DNA replication did demonstrate asynchrony in the overall complement; however, in this case, certain autosomes continued synthesis

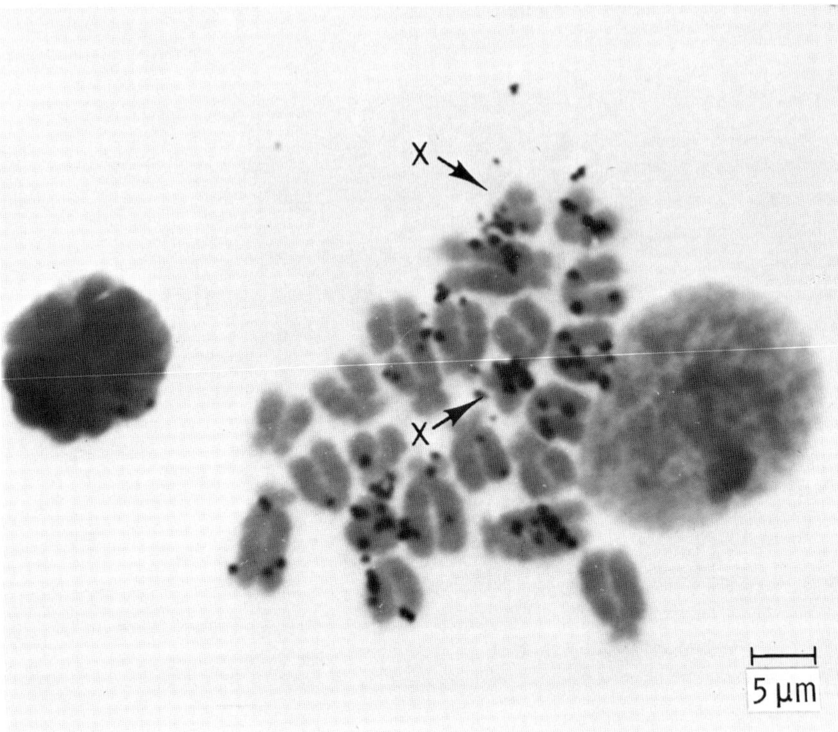


Fig. 17. Chromosome spread (from a female opossum) treated the same as that in Fig. 15, only labeled slightly earlier in the S phase (some of the autosomes have not yet completed replication). Both X chromosomes are labeled, but the one in the center of the spread more intensely. From Schneider and Rieke, *J. Cell Biol.*, 1967.

later than the sex chromosomes. The Y chromosome in the male did not complete replication later than the X, and the female X chromosomes, while displaying asynchronous termination of synthesis among themselves, were early replicators. Sinha believes there is more heterochromatin in certain autosomes than in the sex chromosomes, and this may explain this apparent reversal of "normal" DNA synthetic patterns.

Considerable variation is evident in the DNA replication patterns of *D. virginiana* bone marrow leukocyte chromosomes (Reiss *et al.*, 1970). In females, 27.8% of labeled metaphases demonstrated a single late replicating X chromosome; 3% showed one heavily labeled X and one lightly labeled X. In males, 28.6% demonstrated late labeling sex chromosomes. In 15.7%, both X and Y were late replicating, in 10% the Y alone was late, and in 2.9% the X was singularly late labeling. The investigators feel that the differences observed between their study and those of Schneider and Rieke

(1967) and Sinha (1967a) are due to the use of PHA by the latter investigators. They similarly believe that their results are more closely related to the *in vivo* situation, although the cells were incubated at 37°C for 6 hr in medium TC 199 with 15% fetal calf serum. At best, the sex chromosomes in less than 30% of female cells, and in only 10% of male cells, demonstrated a response similar to the majority of mammalian somatic cells; no mention was made of how the remaining cells responded. Thus, it appears that there is no typical late replicating X or Y chromosome in somatic cells of *D. virginiana*.

### D. DNA Synthesis in Other Marsupials

Marshall Graves (1967) investigated DNA replication in *Macropus giganteus* and *Macropus eugenii*, each of which demonstrates dimorphic X chromosomes. The more condensed $X_2$ chromosome was late in initiating DNA synthesis, and was the last to finish replication in each species. Corresponding segments or arms of homologous autosomes demonstrated no significant differences in amount of DNA synthesis. A recent study (Sharman, 1971) of hybrid female kangaroos, in which the sex chromosomes of the male and female parents are readily distinguishable by size, demonstrated late replication in the paternally derived X chromosome only. The author suggested "The mode of dosage compensation in kangaroos, and perhaps in all marsupials, thus appears to be paternal X inactivation in contrast to the random X inactivation characteristic of eutherian mammals" (see Section VI,E). It should be kept in mind, however, that these experiments were performed on hybrid chromosome complements in which the normal interrelationships and controlling mechanisms of the sex chromosomes may not necessarily apply.

### E. Sex Chromatin in Marsupial Cells

Lyon (1961) hypothesized that one X chromosome (of either maternal or paternal derivation) in female mammalian somatic cells is genetically inactive, a condition which would provide dosage compensation between male and female cells. It was also speculated that the inactive X chromosome is the one which forms the heterochromatic sex chromatin body of female cells at interphase. Evidence for this hypothesis derived from the observations that XO female mice are phenotypically normal and are fertile, and that female mice which are heterozygous for some sex-linked mutants show a mosaic phenotype. Since heterochromatin is believed to demonstrate late

DNA replication (see Section VI,A), it is now generally accepted that the sex chromatin body, the late labeling X chromosome, and the genetically inactive X chromosome are one and the same structure. Certain marsupials (see below) possess unusual sex chromatin patterns which are relevant to an understanding of the relationships between DNA synthesis, RNA synthesis, and genetic inactivation in these species (for a review of sex chromatin patterns in various animals, see Moore, 1966).

Of the marsupials which have been examined cytologically to date, three possess sex chromatin in somatic cells from *both* sexes, a condition which has not been reported in other mammals. A chromocenter was observed in intermitotic nuclei of various tissues from both sexes of *Didelphis virginiana,* although it was larger in female cells (Barr, 1959; Graham and Barr, 1959). This sex chromatin body could frequently be resolved into a bipartite structure in both males and females, and was believed to originate from heterochromatic regions of both X chromosomes in the female, and from the X and Y chromosomes in male cells. This finding is in accord with the previously mentioned observation that the heterochromatic content of the two X chromosomes is about the same in this animal (Sinha, 1967a).

Perondini and Perondini (1966) also observed a conspicuous chromocenter in interphase nuclei of somatic cells of both sexes of *Philander opossum.* In males, it is about one-half the size that it is in females. Prophase, prometaphase, and metaphase cells from both sexes demonstrated a small, heteropycnotic chromosome which is also about one-half as big in males as it is in females. In contrast to the conclusion reached by Graham and Barr for *D. virginiana,* however, Perondini and Perondini stated that in *Philander opossum* ". . . the sex chromatin body is formed by one condensed X chromosome in females and by the condensed Y chromosome in males."

Hayman and Martin (1965b) and Hayman *et al.* (1969) observed that in certain somatic cells of *Perameles nasuta* and *Echymipera kalabu,* one X was missing in females, and the Y was missing in males. They believe that this may be explained on the basis of dosage compensation in which the inactive sex chromosome in each sex is lost. Walton (1969), however, studying corneal epithelium of *P. nasuta* (the cells of which do contain the full chromosome complement), observed sex chromatin of equivalent size in both males and females. The presence of a sex chromatin body in male cells could not be explained by the currently accepted concept that in females it represents an X chromosome. Walton observed, "If one accepts the interpretation that the male sex chromatin in the marsupials studied so far is the Y chromosome, then in *P. nasuta* the extent of heterochromatization of the X and Y chromosomes in interphase nuclei is equivalent." Additional research on the sex chromosomes and sex chromatin patterns of these and other marsupial species should prove enlightening.

### F. RNA Synthesis in *Didelphis virginiana*

The majority of cytogenetic evidence supports the Lyon hypothesis that one X chromosome in female somatic cells is genetically inactive (see Schneider, 1970). Certain other observations, however, suggest that both X chromosomes in certain organisms may be *active* in genetic transcription during brief portions of the cell cycle. For example, substantial percentages of given populations of female somatic cells routinely do not demonstrate sex chromatin at interphase (Klinger *et al.*, 1968). Chromosomal RNA synthesis has been observed in both X chromosomes of human lymphocytes (Back and Dörmer, 1967) and of bovine lymphocytes (Mann and Mukherjee, 1970). *Didelphis virginiana* provides an exceptional opportunity to analyze sex chromosomal RNA synthesis due to characteristics previously described (morphologically recognizable sex chromosomes, sex chromatin in both sexes, approximately equal amounts of heterochromatin in both X chromosomes, and the lack of a typically late replicating X or Y chromosome).

Schneider (1970, 1971) reported that both X chromosomes in lymphocytes from female opossums and the X and Y from males synthesized RNA when incubated briefly with [5-$^3$H]uridine immediately prior to mitosis (Tables II and III). Grain counts on the X chromosomes in female cells revealed an average of 3.4% of the total grains per "active" X and 1.9% per "inactive" X. Since each X chromosome comprises 2.7% of the area of the total chromosome complement, the "active" X synthesized RNA at a rate slightly greater than that of the autosomes. The "inactive" X was found to be 55% as active in RNA synthesis as its homologue. Figure 18 demonstrates [5-$^3$H]uridine labeling in both X chromosomes of a prophase figure from a culture of *D. virginiana* lymphocytes; one X (the so-called "active" X chromosome) is more heavily labeled than its homologue. The X chromosome of male cells contained the same average percentage of grains as the "active" X of female cells (3.4%), and the Y possessed a mean count of 1.1%. Since the Y chromosome represents 1.4% of the total area of chromatin, it synthesized RNA at a rate slightly lower than the autosomes (Fig. 19).

**TABLE II**

RNA Synthesis in the Sex Chromosomes of the Female Opossum (*Didelphis virginiana*)

| No. of cells | Mean No. of grains (total complement) | Mean No. of grains per active X | Mean No. of grains per inactive X |
|---|---|---|---|
| 89 | 90.1 | 3.1 (3.4%) | 1.7 (1.9%) |

**TABLE III**

RNA Synthesis in the Sex Chromosomes of the Male Opossum (*Didelphis virginiana*)

| No. of cells | Mean No. of grains (total complement) | Mean No. of grains per X | Mean No. of grains per Y |
|---|---|---|---|
| 89 | 39.84 | 1.36 (3.4%) | 0.44 (1.1%) |

These results indicate that both sex chromosomes in male and female cultured lymphocytes from *D. virginiana* are "active" in RNA transcription, although one to a lesser degree than the other in each sex. Such findings are not in conflict with earlier observations made by other investigators. For example, Hsu and Schmid (1963) noted that whereas heterochromatin was originally believed to be totally inert, ". . . results have shown that some genetic activity may be detected in heterochromatin." They further stated "It is obvious that neither X chromosome of a female is *genetically* inert (italics theirs). The inertness, therefore, must be a metabolic expression. Thus, a chromosome or part of a chromosome showing heteropycnosis is not necessarily devoid of genes." Back (1967) feels that the terms "inactive" and "heterochromatic" should not be used to describe the late replicating X chromosome since neither term fully applies to the circumstances. Hamerton (1968) believes that the heterochromatic sex chromosome in both sexes may

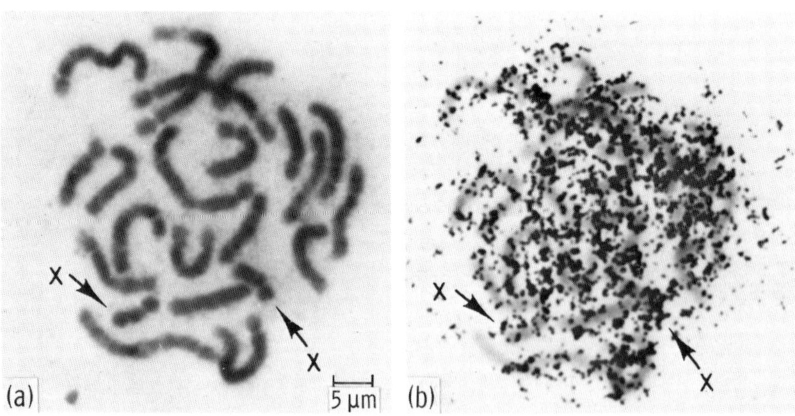


Fig. 18. (a) Prophase chromosome spread from a culture of female *D. virginiana* lymphocytes incubated terminally with [5-$^3$H] uridine; and (b) the same spread after radioautography. Both X chromosomes demonstrate RNA synthesis, although one to a greater degree than the other. From Schneider, *Experientia*, 1970.

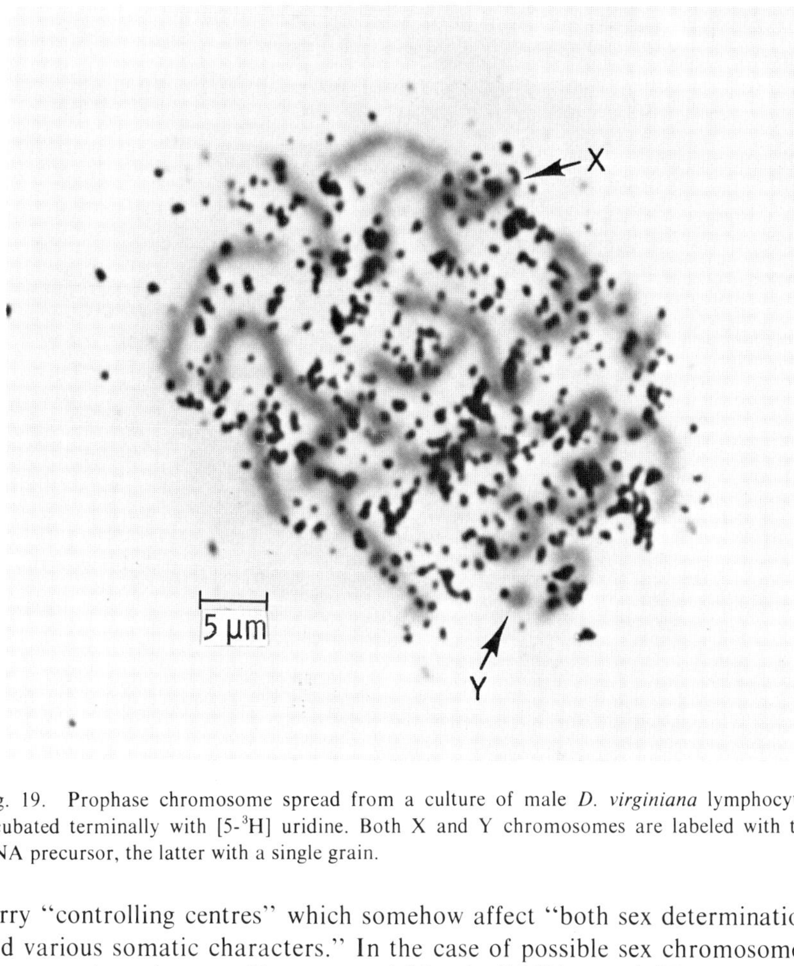


Fig. 19. Prophase chromosome spread from a culture of male *D. virginiana* lymphocytes incubated terminally with [5-$^3$H] uridine. Both X and Y chromosomes are labeled with the RNA precursor, the latter with a single grain.

carry "controlling centres" which somehow affect "both sex determination and various somatic characters." In the case of possible sex chromosome/autosome interchanges, Hayman and Martin (1969) stated "Dosage compensation requires inactivation of only that portion of the X-chromosome which is unequally represented in both males and females."

### G. Protein Synthesis in Marsupial Cells

Chromosomal protein synthesis was investigated in lymphocytes from *Didelphis virginiana* following incorporation of $^3$H-arginine *in vitro* (Schneider and Rieke, 1967). By analyzing the percentage of labeled mitotic figures which appeared with time after introduction of the isotope, it was shown that chromosomal protein synthesis gradually increased throughout the S and into the $G_2$ phase (Fig. 20). These results confirmed those of Cave (1966), who investigated the incorporation of $^3$H-lysine into human chromosomes during interphase. Cave reported a lack of functional dependence between protein synthesis and DNA synthesis, noting that chromosomal protein synthesis occurred throughout interphase, while DNA synthesis was limited to the S phase.

Burton (1968a,b) studied the effects of PHA on cellular kinetics in lymphocyte cultures from *Macropus giganteus*. Increases in protein synthesis, cellular dry mass, and cell volume were observed soon after the addition of PHA. Burton suggested that the early commencement of DNA synthesis in these cells (12 hr earlier than in human lymphocytes *in vitro*) was the result of the earlier attainment of the necessary critical cell mass. Concomitant with the increase in cytoplasmic dry mass (and total cell mass) was a 20–25% decrease in stainability with alkaline fast green and an initial decrease in nuclear dry mass. Burton believes that these findings argue in favor of the hypothesis that PHA induces blastogenesis in lymphocytes by the removal of histones from the deoxyribonucleoprotein complex.

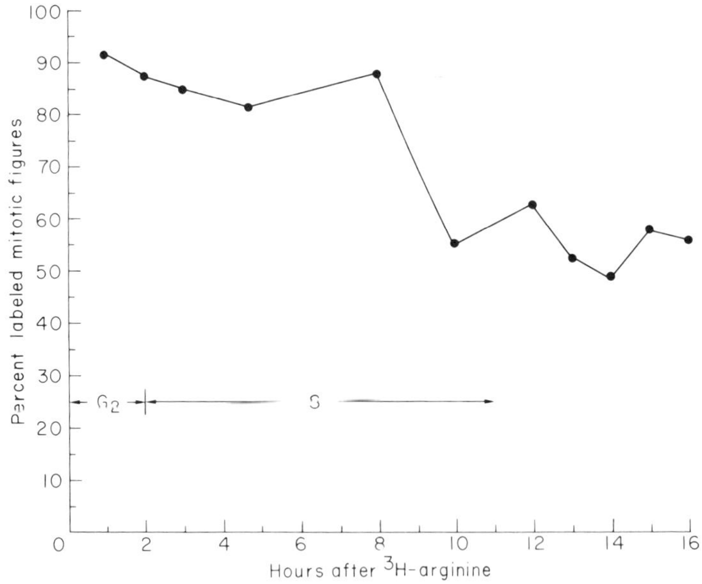


Fig. 20. Chromosomal protein synthesis in *Didelphis virginiana* lymphocytes *in vitro*. The curve demonstrates an increase in $^3$H-arginine incorporation throughout S and into $G_2$ phase, and, therefore, a lack of functional dependence between chromosomal protein synthesis and DNA synthesis. Redrawn from Schneider and Rieke, *J. Cell Biol.*, 1967.

### H. Chromosomal Evolution in Marsupials

McIntosh and Sharman (1953) suggested that chromosome evolution in Australian marsupials has led to reduced diploid numbers and a greater percentage of "mediocentric" chromosomes, probably as a result of translocations. Martin and Hayman (1966) presented evidence that *Lagorchestes conspicillatus* ($2n = 15$ in males and 16 in females) evolved from a form with 22 chromosomes which included a small X chromosome and a very small Y. They observed late DNA replication in the proximal one-fifth of the current $X_1$ chromosome. They believe that this small segment comprises the ancestral X chromosome which has been translocated onto an autosome. The fate of the original Y was uncertain, but they suggested that it forms the centromeric portion of the present Y in this species.

With regard to *Potorous tridactylus* and other species with large X chromosomes, Hayman and Martin (1969) expressed difficulty in explaining the origin of the additional chromosomal material. They noted "If the material was autosomal in origin then, unless this same material was present on the Y chromosome, there would be a genetic imbalance which would be lethal." Their data suggest that the X and Y chromosome possess a special capacity to increase in size which is not shared by the autosomes. They believe that *P. tridactylus* derived from a $2n = 14$ form, and that the latter could also evolve into a species with $2n = 20$ or $2n = 22$ with the necessary addition of centromeres and telomeres.

Regarding chromosomal evolution in American marsupials, Reig and Bianchi (1969) compared the karyotypes of *Monodelphis* with those of *Caluromys* and *Marmosa*, and concluded that the chromosomes of groups B and C are similar in each case and are probably comprised of the same elements. They believe that these chromosomes in *Monodelphis* may have derived from *Marmosa* and *Caluromys* by centromeric fission of two pairs of large submetacentric chromosomes into four pairs of telocentric chromosomes. The possibility of a reverse process ($2n = 18$ evolving to $2n = 14$) was mentioned, but they believe *Monodelphis* to be more primitive than *Marmosa* or *Caluromys*. The investigators also considered the possibility that didelphids ($2n = 22$) may have derived from the $2n = 14$ pattern "by full transformation of 4 pairs of biarmed chromosomes into 8 pairs of telocentric ones."

The chromosomal differences between *Didelphis virginiana* and *Didelphis marsupialis* (autosome pairs 1 to 3 longer, and sex chromosomes shorter in the latter species in addition to variations in form) have been discussed (see Section III,B; Cestari and Perondini, 1964). Gardner (1970; and personal communication) has suggested a possible evolutionary sequence between these two species which would involve pericentric inversions in autosomal pairs 1 to 6 of *D. marsupialis,* plus translocation of small segments of autosomes 1 to 3 to the sex chromosomes of this animal. Although this non-Robertsonian type of chromosomal evolution is admittedly rare, the possibility of its occurrence nevertheless exists (Ohno, 1969). The presence of autosomal material on the sex chromosomes of *D. virginiana* might explain the unusual DNA synthetic patterns (Schneider and Rieke, 1967; Sinha, 1967a), heterochromatin content (Sinha, 1967a), RNA synthetic activity (Schneider, 1970, 1971), and sex chromatin patterns (Barr, 1959; Graham and Barr, 1959) observed in this animal.

Galton (1966) hypothesized that sex chromosomes in mammals arose from a homologous autosomal pair by some translocation which led to the formation of a larger metacentric X and a smaller Y chromosome. He stated that the X chromosome is nearly always metacentric in the species studied so far. However, this statement does not, in general, hold true for marsupials, in which the X chromosome is not metacentric, at least in the majority of species discussed in this chapter. Thus, if Galton's hypothesis is correct, it is possible that marsupials have followed a path of sex chromosome evolution which is different from the remainder of Mammalia.

## VII. CONCLUSIONS AND SUMMARY

Evidence has been presented which confirms the often stated observation that marsupial cells are valuable tools in chromosomal and cytogenetic investigations. This definition of worth derives mainly from the basic simplicity of the karyotypes and from the multitude of unusual characteristics displayed by these cells. Thus, the following generalities may be made:

1. Marsupials possess low diploid chromosome numbers, large chromosomes, recognizable sex chromosomes, and often distinctive autosomes.
2. Marsupial cells may be incubated *in vitro* by simple, reliable, techniques; large numbers of well-spread mitotic figures may be rapidly obtained from a variety of species. Marsupial lymphocytes respond to phytohemagglutinin with rapid bursts of DNA, RNA, and protein synthesis.
3. Marsupials present cell cycles which demonstrate variations within the animal as well as differences from eutherian mammals, conditions which may be related to body temperature.
4. Certain marsupials demonstrate unusual DNA replication patterns and RNA synthetic activities, especially in the sex chromosomes. Sex chromatin has been observed in both sexes of three species investigated so far.

5. The nature of marsupial chromosomes facilitates evolutionary investigations; pathways for sex chromosome evolution may differ from those of placental mammals.

## VIII. ADDENDUM

A number of publications have appeared in the literature which shed additional light on some of the problems and questions discussed in the text. Cell lines from *Potorous tridactylus* have been employed extensively in the analysis of various aspects of chromosome structure. Grewal *et al.* (1971) studied the quinacrine fluorescent patterns in mitotic chromosomes of this animal and reported distinctive banding in both autosomes and sex chromosomes. Bick and Brown (1972) continued their work on chromosomal sensitivity of cells from *P. tridactylus* to X-rays and heliotrine. Chromosome measurements and DNA synthetic patterns were analyzed for cultures of the $PTK_2$ cell line from this animal by Lorenz and Ainsworth (1972). It is interesting to note that chromosomal measurements obtained by these investigators differed from data for primary bone marrow cells of this species reported by Shaw and Krooth (1964). Roos (1973) performed an extensive light and electron microscopic study of the formation and breakdown of the mitotic apparatus in cells from *P. tridactylus*.

Numerous reports have also appeared regarding research performed on the cells and chromosomes of *Didelphis virginiana*. Schneider (1972, 1973a) published additional observations and interpretations of RNA synthesis in the sex chromosomes and in connectives between chromosomes from lymphocytes of this animal. With regard to the latter, he suggested that as a result of their apparent RNA synthetic ability, surface extensions may represent the active chromatin fraction of *D. virginiana* chromosomes during prophase and metaphase. Sinha *et al.* (1972) published the interesting observation that with utilization of the cytological DNA denaturation–reassociation technique, C bands (centromeric heterochromatin) were found only in the sex chromosomes of *D. virginiana*. The fact that the C-band pattern of the two X chromosomes was nearly identical lends additional support to the suggestion made in the text that the X chromosomes in this animal are more similar than dissimilar. These similarities apply to both structure and function.

Schneider (1973b) extended the generation time curve for *D. virginiana* lymphocytes *in vitro*, and noted a general trailing off of the curve to 28 hr. Of interest is the observation that the sum of $G_2$, S, and M as defined from the curve was greater than the total generation time. This was attributed to variation, primarily in the duration of $G_1$, of perhaps two or more functionally diverse populations of lymphocytes. In a publication by Schneider and Goldman (1974), the hypothesis of Thrasher (1969) that longer $G_2$ and S phases for *D. virginiana* may be related to the lower body temperature of this animal (see text, Section V,A) was tested experimentally. Cardiac blood lymphocytes from five male opossums were cultured with phytohemagglutinin at 34° and 37°C, all other conditions being identical. The investigators observed that at the higher temperature the $G_2$ phase was shortened, whereas the S phase was lengthened. The results were discussed with regard to different temperature controlled rate-limiting reactions for the different phases.

## ACKNOWLEDGMENTS

The author expresses appreciation to the following individuals who have assisted in the preparation of this chapter: Sharron L. Pierce for technical assistance, Karen G. Schneider and Dr. Philip H. Krutzsch for editorial suggestions, and Patricia A. Broome for typing the final manuscript. In addition, he is grateful to Dr. Y. A. E. Bick, Dr. D. L. Hayman, Dr. Ruth C. Moore, Dr. A. L. P. Perondini, Dr. Jack D. Thrasher, and Henry G. Wolf for their willing contributions to the photographs and illustrations employed herein.

## REFERENCES

Agostoni, A., and Idéo, G. (1965). *Experientia* **21,** 82.
Back, F. (1967). *Lancet* **2,** 258.
Back, F., and Dörmer, P. (1967). *Lancet* **1,** 385.
Barr, M. L. (1959). *Science* **130,** 679.
Barr, M. L., and Bertram, E. G. (1949). *Nature (London)* **163,** 676.
Baserga, R., and Malamud, D. (1969). "Modern Methods in Experimental Pathology. Autoradiography: Techniques and Application." Harper (Hoeber), New York.
Bick, Y. A. E. (1970). *Nature (London)* **226,** 1165
Bick, Y. A. E., and Brown, J. K. (1969a). *Mutat. Res.* **8,** 613.
Bick, Y. A. E., and Brown, J. K. (1969b). *Cytobios* **2,** 123.
Bick, Y. A. E., and Brown, J. K. (1972). *Cytobios* **5,** 189.
Bick, Y. A. E., and Jackson, W. D. (1968). *Aust. J. Biol. Sci.* **21,** 469.
Biggers, J. D., Fritz, H. I., Hare, W. C. D., and McFeely, R. A. (1965). *Science* **148,** 1602.
Bond, V. P., Cronkite, E. P., Fliedner, T. M., and Schork, P. (1958). *Science* **128,** 202.
Brinkley, B. R., and Humphrey, R. M. (1969). *J. Cell Biol.* **42,** 202.
Burton, D. W. (1968a). *Exp. Cell Res.* **49,** 300.
Burton, D. W. (1968b). *Exp. Cell Res.* **53,** 329.
Cabrera, A. (1957). *Rev. Mus. Argent. Cienc. Nat. "Bernadino Rivadavia" Inst. Nacl. Invest. Cienc. Nat., Cienc. Geol.* **4,** Part 1.

Carpenter, P. L. (1965). "Immunology and Serology." Saunders, Philadelphia, Pennsylvania.
Cave, M. D. (1966). *J. Cell Biol.* **29,** 209.
Cestari, A. N., and Perondini, A. L. P. (1964). *In* "Mammalian Cytogenetics and Related Problems in Radiobiology" (C. Pavan *et al.*, eds.), pp. 157–162. Pergamon, Oxford.
Cooper, E. H., Barkhan, P., and Hale, A. J. (1961). *Lancet* **2,** 210.
Denver Report (1960). *Lancet* **1,** 1063.
Dreyfus, A., and Campos, J. E. (1941). *Univ. Sao Paulo, Fac. Filos. Cienc. Let.* **17,** 3.
Evans, H. J. (1965). *Exp. Cell Res.* **38,** 511.
Galton, M. (1966). *Lancet* **2,** 1397.
Gardner, A. L. (1970). Abstract of paper presented at the 50th Annual Meeting of the American Association of Mammologists, College Station, Texas.
Gartler, S. M., and Burt, B. (1964). *Cytogenetics* **3,** 135.
German, J. L. (1961). *J. Clin. Invest.* **40,** 1041.
German, J. L. (1964). *J. Cell Biol.* **20,** 37.
Gibson, D. A. (1970). *Nature* (*London*) **227,** 164.
Goh, K. (1967). *Cytologia* **32,** 416.
Graham, M. A., and Barr, M. L. (1959). *Arch. Anat. Microsc. Morphol. Exp.* **48,** 111.
Grewal, M. S., Dev, V. G., Miller, D. A., and Miller, O. J. (1971). *Exp. Cell Res.* **69,** 241.
Hamerton, J. L. (1968). *Nature* (*London*) **219,** 910.
Hartman, C. (1919). *J. Morphol.* **32,** 1.
Hayhoe, F. G. J., and Quaglino, D. (1965). *Nature* (*London*) **205,** 151.
Hayman, D. L., and Martin, P. G. (1965a). *Cytogenetics* **4,** 209.
Hayman, D. L., and Martin, P. G. (1965b). *Genetics* **52,** 1201.
Hayman, D. L., and Martin, P. G. (1965c). *Aust. J. Biol. Sci.* **18,** 1081.
Hayman, D. L., and Martin, P. G. (1969). *In* "Comparative Mammalian Cytogenetics" (K. Benirschke, ed.), pp. 191–217. Springer-Verlag, Berlin and New York.
Hayman, D. L., Martin, P. G., and Waller, P. F. (1969). *Chromosoma* **27,** 371.
Heneen, W. K. (1970). *Chromosoma* **29,** 88.
Hoy, W. E., Jr., and George, W. C. (1929). *J. Morphol. Physiol.* **47,** 201.
Hsu, T. C. (1964). *J. Cell Biol.* **23,** 53.
Hsu, T. C., and Pomerat, C. M. (1953). *J. Hered.* **44,** 23.
Hsu, T. C., and Schmid, W. (1963). Research Report, pp. 75–77. University of Texas, M. D. Anderson Hospital and Tumor Institute, Houston.
Jackson, L. G., and Ellem, K. A. O. (1968). *Cytogenetics* **7,** 183.
Jakob, H., and Ruiz, F. (1970). *Exp. Cell Res.* **62,** 310.
Jordan, H. E. (1911). *Arch. Zellforsch.* **7,** 41.
Kikuchi, Y., and Sandberg, A. A. (1964). *J. Natl. Cancer Inst.* **32,** 1109.
Klinger, H. P., Davis, J., Goldhuber, P., and Ditta, T. (1968). *Cytogenetics* **7,** 39.
Legator, M., Jacobson, C., Perry, M., and Dolimpio, D. (1966). *Life Sci.* **5,** 397.
Levan, G. (1970). *Hereditas* **64,** 85.
Lima-de-Faria, A., Reitalu, J., and O'Sullivan, M. A. (1965). *Chromosoma* **16,** 152.
Lorenz, P. R., and Ainsworth, J. W. (1972). *Chromosoma* **38,** 431.
Lyon, M. F. (1961). *Nature* (*London*) **190,** 372.
McIntosh, A. J., and Sharman, G. B. (1953). *J. Morphol.* **93,** 509.
MacNeal, W. J. (1922). *J. Am. Med. Assoc.* **78,** 1122.
Mann, K., and Mukherjee, B. B. (1970). *Can. J. Genet. Cytol.* **12,** 44.
Marshall Graves, J. A. (1967). *Exp. Cell Res.* **46,** 37.
Martin, P. G., and Hayman, D. L. (1966). *Chromosoma* **19,** 159.
Michalowski, A. (1963). *Exp. Cell Res.* **32,** 609.

Moore, K. L. (1966). *In* "The Sex Chromatin" (K. L. Moore, ed.), pp. 16–58. Saunders, Philadelphia, Pennsylvania.
Moore, R. (1965). *Cytogenetics* **4**, 145.
Moore, R., and Uren, J. (1965). *Exp. Cell Res.* **38**, 341.
Moore, R., and Uren, J. (1966). *Exp. Cell Res.* **44**, 273.
Moore, R., and Radley, J. M. (1968). *Exp. Cell Res.* **49**, 638.
Moore, R. C., and Gregory, G. (1963). *Nature (London)* **200**, 234.
Moorhead, P. S., Nowell, P. C., Mellman, W. J., Battips, D. M., and Hungerford, D. A. (1960). *Exp. Cell Res.* **20**, 613.
Nowell, P. C. (1960). *Cancer Res.* **20**, 462.
Ohno, S. (1969). *Annu. Rev. Genet.* **3**, 495.
Ohno, S., Kaplan, W. D., and Kinosita, R. (1960). *Exp. Cell Res.* **19**, 417.
Painter, R. B. (1961). *J. Biophys. Biochem. Cytol.* **11**, 485.
Painter, T. S. (1922). *J. Exp. Zool.* **35**, 13.
Perondini, A. L. P., and Perondini, D. R. (1965). *Rev. Bras. Biol.* **25**, 377.
Perondini, A. L. P., and Perondini, D. R. (1966). *Cytogenetics* **5**, 28.
Radley, J. M. (1966). *Exp. Cell Res.* **41**, 217.
Reig, O. A. (1968). *Experientia* **24**, 185.
Reig, O. A., and Bianchi, N. O. (1969). *Experientia* **25**, 1210.
Reig, O. A., and Sonnenschein, C. (1970). *Experientia* **26**, 199.
Reiss, J. A., Neu, R. L., Kajii, T., and Gardner, L. I. (1970). *Experientia* **26**, 917.
Robbins, J. H. (1964). *Science* **146**, 1648.
Roos, U. P. (1973). *Chromosoma* **40**, 43.
Saez, F. A. (1931). *Am. Nat.* **65**, 287.
Schmid, W. (1963). *Cytogenetics* **2**, 175.
Schneider, L. K. (1970). *Experientia* **26**, 914.
Schneider, L. K. (1971). *Anat. Rec.* **169**, 420.
Schneider, L. K. (1972). *Can. J. Genet. Cytol.* **14**, 905.
Schneider, L. K. (1973a). *J. Cell Biol.* **56**, 608.
Schneider, L. K. (1973b). *Experientia* **29**, 468.
Schneider, L. K., and Goldman, H. D. (1974). *Experientia* **30**, 1192.
Schneider, L. K., and Rieke, W. O. (1967). *J. Cell Biol.* **33**, 497.
Schneider, L. K., and Rieke, W. O. (1968). *Cytogenetics* **7**, 1.
Sharman, G. B. (1961). *Aust. J. Zool.* **9**, 38.
Sharman, G. B. (1971). *Nature (London)* **230**, 231.
Sharman, G. B., and Barber, H. N. (1952). *Heredity* **6**, 345.
Sharman, G. B., McIntosh, A. J., and Barber, H. N. (1950). *Nature (London)* **166**, 996.
Sharman, G. B., Robinson, E. S., Walton, S. M., and Berger, P. J. (1970). *J. Reprod. Fertil.* **21**, 57.
Shaver, E. L. (1962). *Can. J. Genet. Cytol.* **4**, 62.
Shaw, M. W., and Krooth, R. S. (1964). *Cytogenetics* **3**, 19.
Sinha, A. K. (1967a). *Experientia* **23**, 889.
Sinha, A. K. (1967b). *Exp. Cell Res.* **47**, 443.
Sinha, A. K., Kakati, S., and Pathak, S. (1972). *Exp. Cell Res.* **75**, 265.
Taylor, J. H. (1960). *J. Biophys. Biochem. Cytol.* **7**, 455.
Taylor, J. H., Woods, P. S., and Hughes, W. L. (1957). *Proc. Natl. Acad. Sci. U.S.A.* **43**, 122.
Thrasher, J. D. (1969). *Exp. Cell Res.* **57**, 441.
Thrasher, J. D. (1970). *Experientia* **26**, 871.
Thrasher, J. D. (1972). *In* "Methods in Cell Physiology" (D. M. Prescott, ed.), Vol. 5, pp. 127–166. Academic Press, New York.

Thrasher, J. D., Berg, N. B., and Hauber, E. (1971). *Cell Tissue Kinet.* **4,** 185.
Tjio, J. H., and Puck, T. T. (1958). *J. Exp. Med.* **108,** 259.
Uren, J., Moore, R., and van den Brenk, H. A. S. (1966). *Exp. Cell Res.* **43,** 677.
Walen, K. H. (1965). *Genetics* **51,** 915.
Walen, K. H., and Brown, S. W. (1962). *Nature (London)* **194,** 406.
Walton, S. M. (1969). *Experientia* **25,** 535.
Wolf, H. G., Klein, A. K., and Foin, A. T. (1970). *Cytobios* **6,** 97.

# 3

# Ecology of New World Marsupials

DON HUNSAKER II



## I. INTRODUCTION

The ecology of most species of New World marsupials is virtually unknown. Many species are known only from museum records and sketchy reports based on a few observations. The Virginia opossum, *Didelphis virginiana,* has been extensively studied by a variety of scientists; the murine opossum, *Marmosa,* has minimal information available; and the four-eyed *Philander* and woolly opossum *Caluromys* are only casually known from field studies.

The following material is an attempt to bring together as much information as possible from the published literature, colleagues, and observations by the author. Some preliminary deductions can be made on ecological strategies and adaptations of the family, but many conclusions on any species other than *Didelphis virginiana* will have to await future research.

## II. GENERAL ECOLOGY

### A. Origin

Marsupials are believed to have originated in North America. According to Clemens (1968) there is no evidence of the presence of marsupials on any other continent prior to the latest Cretaceous epoch. The oldest fossil teeth of mammals that can be considered to be a common ancestor of marsupials and placentals are from *Aegialodon,* recovered from early Cretaceous deposits in England (Kermack *et al.,* 1965). Slaughter (1968) described the teeth from the earliest known marsupial, *Holoclemensia,* from slightly younger Cretaceous beds in Texas. Slaughter considered that the marsupial–eutherian separation was distinct by this period, but Lillegraven (1974) considers that more work in early Cretaceous sediments needs to be done to accurately confirm the date of dichotomy. The first major radiation of the group probably occurred in the early part of the late Cretaceous, about 94 million years before present. Kirsch (Chapter 1) discusses the evolution of the group in greater detail. Elements of the early didelphids invaded Central and South America during this time. During the Tertiary in South America, some differentiated into the Caenolestidae, others remained didelphid, and some evolved into a variety of other families including the large carnivorous borhyaenids. The latter family became extinct as did *Didelphis* and other genera in North America until the Pleistocene (Hershkovitz, 1969). *Didelphis* and other members of the family are considered to be fairly recent migrants into North America (Clemens, 1968). Hibbard *et al.* (1965) reported *Didelphis* remains in material deposited during the Sangamon interglacial stage of the Pleistocene. Subsequently, the genus is well represented in post-Wisconsin deposits. Members of this family are now found from southern Canada to Tierra del Fuego. The caenolestids are represented by several species limited to South America.

The genus *Didelphis* now ranges from southern Ontario, Canada (Peterson and Downing, 1956) to as far south as Argentina (Hershkovitz, 1969; Gardner, 1973). The northern boundary of its distribution is presently considered to be a line extending from eastern New York through southern Wisconsin, and eastern Nebraska (Farris, 1950). Within this vast area,

*Didelphis* is mainly confined to the more temperate wooded areas, although in the past few decades it has been successfully introduced into areas as arid as Arizona and southern California. Introduced into California in 1890, it is well established throughout the state at the present time, and has extended its range north to British Columbia in western Canada. Figure 1 shows the present distribution of *Didelphis* in North and Central America.

None of the other genera of didelphids are as widespread as *Didelphis*. *Marmosa* is a tropical form distributed from about 25° N latitude (*M. canescens*) to 30° S latitude (*M. elegans*) and *Caluromys* is restricted to the equatorial regions.

The didelphids have radiated into almost all the habitats of the Western

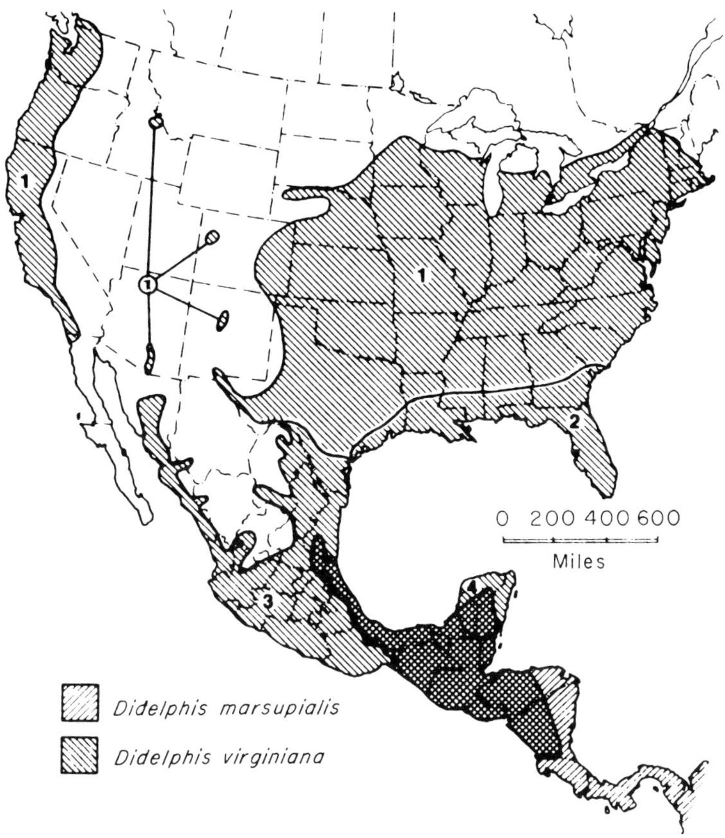


Fig. 1. Distribution of *Didelphis virginiana* and *Didelphis marsupialis* in North America. (1) *Didelphis virginiana virginiana*; (2) *Didelphis virginiana pigra*; (3) *Didelphis virginiana californica;* (4) *Didelphis virginiana yucatanensis*. From Gardner, 1973.

Hemisphere but do not dominate any single niche. Their success lies in an extremely adaptable behavior (McManus, 1970; Hunsaker and Shupe, Chapter 5) and a physiology which can support the species in a variety of habitats. *Didelphis* and *Marmosa* both have extended their ranges into mesic brush, grassland, and forest habitats and have been successful.

## B. Distribution

### 1. Distribution and Habitat of Genera

The following list of genera gives the range, physical description, habitat, and diet of representative species.

**a. Didelphidae.** i. *Didelphis.* Common opossum. With three species in the genus, *Didelphis virginiana* (Virginia opossum) (Fig. 2), *Didelphis marsupialis* (South and Central American or black-eared opossum), and *Didelphis albiventris* (white-eared opossum), *Didelphis* is one of the most widely distributed mammalian genera (except for man) in the Western Hemisphere, and is also the most widely distributed marsupial in the world. The Virginia opossum, *D. virginiana,* having the northernmost range of its

Fig. 2. Virginia opossum, *Didelphis virginiana.* A male from San Diego County, California. Photo by Hunsaker.

Fig. 3. Gray four-eyed opossum, *Philander opossum*. From Venezuela. Photo by P. F. Waller.

family, extends from southern Canada throughout Mexico to Central America at altitudes from 0 to 3600 m. *Didelphis marsupialis* is found throughout most of South America, east and west of the Andes below 1400 m. *Didelphis albiventris,* the white-eared opossum, is a temperate zone species and lives mostly at middle level altitudes above 1000 m. The species is found from Colombia, where it lives in a habitat very similar to that of *D. virginiana,* in temperate woodlands, to Argentina. All are omnivorous and are found in a variety of habitats that are principally wooded or brushy areas near watercourses. The species is mainly terrestrial, but also is adept at climbing. *Didelphis* varies in total length from 645 to 1017 mm, 255 to 535 mm tail length, hind foot 48 to 80 mm, and weighs approximately 2 to 5.5 kg. A well-developed marsupium is present in the female. Both light gray and dark color phases are present, even in the same litter. *Didelphis virginiana* and *D. marsupialis* have black ears; *D. albiventris* has pinkish-white ears.

ii. *Philander.* Gray and black four-eyed opossums. Pine (1973) stated positively that *Philander* is incorrect and that *Metachirops* in the preferred and legal name of this genus. However, Hershkovitz (1976) disagrees. This genus is comprised of two species, *Philander opossum* (Fig. 3) and the most recently described *Philander mcilhennyi. Philander opossum* ranges from

Tamaulipas, Mexico to Brazil and northeastern Argentina. *Philander mcilhennyi* is known only from the type locality in Balta, Rio Curanja, Amazonian Peru. *Philander* live in tropical forests up to 1200 m, are common in second growth and plantations, and are often found near streams, rivers, and swamps where they lead a carnivorous life feeding on small mammals, birds, eggs, carrion, and fruit. They are primarily terrestrial, but are good climbers and swimmers. Nothing is known of the habits or natural history of *P. mcilhennyi,* but it is assumed to be similar to *P. opossum.* This genus is aggressive and possesses a very irritable manner. *Philander opossum* is a rather slim animal and ranges in size from 250 to 400 g, measuring 530 to 610 mm total length, 380 mm tail length, with a 29- to 50-mm hind foot. *Philander mcilhennyi* measurements are 551 to 610 mm total length and 276 to 329 mm tail length. Female *Philander* possess distinct pouches. Coloration is brown, gray, or dark; the tail is haired at the basal third. There are distinct white spots above the eyes.

iii. *Metachirus.* Brown four-eyed opossum or brown-masked opossum (Fig. 4). The only species, *Metachirus nudicaudatus,* is found from Nicaragua to northern Argentina at altitudes below 700 m. It is mostly arboreal and terrestrial, living in the lowlands, in heavy forests, or thickets in open brushy country, is often found on the ground, and is strictly

Fig. 4. Brown four-eyed opossum, *Metachirus nudicaudatus.* A female from Venezuela with young. Photo by Kirsch.

Fig. 5. Water opossum, *Chironectes minimus*. A male from Venezuela. Photo by Kirsch.

nocturnal. It nests in tree branches or occasionally under rocks and logs. Total length is 570 to 630 mm, and the tail is 305 to 330 mm. Weight is approximately 800 g. Females have lateral folds of skin on the lower abdomen instead of pouches. Predominantly frugivorous, they do eat some insects, birds' eggs, and small vertebrates. Color is brown with two distinct light spots above the eyes. The tail is almost entirely bare.

iv. *Chironectes.* Water opossum or yapock. The single species, *Chironectes minimus* (Fig. 5), ranges from Oaxaca, Mexico to northeastern Argentina. This is the only opossum completely adapted to an aquatic life. It lives in wet, gallery, and deciduous forests close to water and prefers small streams and ponds. It has been found at elevations up to 1500 m. It utilizes subterranean dens which it enters from holes in stream banks just above water level. They appear to be basically crepuscular, and are often seen during the mid-morning hours. Size varies from 460 to 710 mm total length, 355 to 395 mm tail length, and 70 to 72 mm hind foot length. Both sexes possess a well-developed marsupium. The male is said to utilize his as a water-tight pouch into which he can draw his scrotum. *Chironectes* is predominantly carnivorous with a diet consisting of insects, shrimp, fish, crayfish, and frogs. It may also eat some aquatic vegetation and fruit. Body color is dark gray or black with a pattern of light gray bands across the sides. The tail is laterally compressed.

v. *Lutreolina.* Lutrine opossum. The single species, *Lutreolina crassicaudata,* has a discontinuous range. The northern population ranges from the savannas of eastern Colombia to Venezuela and Guianas. The southern group ranges from south of the Amazon River basin in southern Brazil to Buenos Aires in Argentina. Members of this genus are nocturnal, fair swimmers, and occupy many habitats near water including parkland forests, marshes, streams, lake shores, and open dry grasslands. Basically they occupy open savanna–pampa lowland areas. The species does not appear to be exceedingly rare, but is seldom observed. Some live near cities and are fairly tolerant of humans, especially those who have chicken houses in which to feed. They are terrestrial, but also climb well, although they do not have a prehensile tail. Measurements are 250 to 400 mm for the head and body length, 216 to 318 mm for the tail length, and adults weigh 200 to 540 g. Like most of the opossums, it is omnivorous, including in its diet small rodents, marsupials, fishes, insects, plants, and seeds.

vi. *Marmosa.* Murine or mouse opossum. A large genus consisting of some 50 or more named species, *Marmosa* range through Mexico and all of South America, except the high puna and paramo zones, the arid Chilean deserts, and Patagonia. Some of the species will probably be combined when the genus is further studied, and it could be split into two genera, but at the present time it is a confusing problem. They live in a variety of habitats ranging from warm rain forests in the humid tropics to thorn scrub, pampa, desert area, and up to 3400 m in the low Andean paramo. This genus is primarily arboreal and nocturnal. *Marmosa robinsoni* (= *mitis*) (Fig. 6) does not use a permanent nest, but will sleep in whatever suitable shelter is available at daybreak. Different species of *Marmosa* vary in size from 75 to 215 mm head and body length, and 89 to 280 mm tail length. *Marmosa cinerea* is one of the largest species, *Marmosa elegans* among the smallest. This genus is pouchless and has a prehensile tail. *Marmosa* is omnivorous, eating insects, fruits, small rodents, lizards, and birds' eggs. Species inhabiting the colder regions tend to be more carnivorous.

vii. *Monodelphis.* Short-tailed opossum. The genus *Monodelphis* encompasses 11 named species. They range from eastern Panama to central Argentina. Though terrestrial, they can climb fairly well. *Monodelphis* are often found near human habitations, preferring to live near water. The species *Monodelphis brevicaudata* is very common in Venezuela but scarce in Colombia. Individuals vary in size from 160 to 205 mm total length, and 45 to 64 mm tail length. The pouch in the female is almost nonexistent. *Monodelphis* feed on small rodents, insects, fruits, seeds, and carrion. Pelage is basically brown, with the back darker and underparts light gray.

Fig. 6. Murine opossum, *Marmosa robinsoni*. A male from Colombia. Photo by Hunsaker.

The short tail is typical of the genus and is not prehensile. They breed throughout the year and produce litters with from 8 to 14 young. The mammae number up to 27, the largest number of any mammal.

viii. *Lestodelphys*. Patagonian opossum. One species, *Lestodelphys halli*, is known only from the coast of Patagonia near the Gulf of San Jorge. Thought to be nocturnal, it is more terrestrial than arboreal and also probably inhabits the adjacent pampas. The area in which it is collected is a low scrub vegetation with very few trees. Little is known about its habits except that it is cold adapted, having the most southerly distribution of any marsupial. It is probably more carnivorous than other small mammals of the family due to the structure of the teeth and skull and the fact that fruits and insects are scarce that far south. Measurements of the type specimen (male) are 114 mm for the head and body length, and 93 mm for the tail length. The tail is heavily furred for about 20% of the base and well haired for the remainder.

ix. *Caluromys*. Woolly opossum. The three species, *Caluromys lanatus* (Ecuadorean woolly opossum), *Caluromys philander* (bare-tailed woolly opossum, the type species), and *Caluromys derbianus* (Derby's woolly opossum) range from southern Mexico to central South America, and the island of Trinidad. *Caluromys derbianus* (Fig. 7) range from

Fig. 7. Derby's opossum, *Caluromys derbianus*. A female (with young) from Nicaragua. Photo by Kirsch.

Veracruz, Mexico to the Cauca Valley of Colombia, and to western Ecuador on the slopes of the Andes. *Caluromys lanatus* range throughout the Amazon and Orinoco basins to southern Brazil (Sao Paulo). *Caluromys philander* range from the Guianas to Venezuela, and northeastern and south central Brazil.

*Caluromys* are relatively common and are found in deciduous and rain forests from the lowlands to the mountains. Very agile for their size, they are more arboreal than other large didelphids. Measurements for the genus range from 450 to 800 mm total length, 270 to 490 mm tail length (which is always longer than the combined measurements of the head and body), and 45 to 47 mm hind foot length. *Caluromys philander* is considerably smaller than *derbianus*. The marsupium in the female of this species is rudimentary in some individuals but is developed in some older females. It generally assumes the form of a lateral fold along the ventral side. *Caluromys* is basically frugivorous but will eat a wide variety of foods. Body color is reddish or dark brown with a dark stripe on the face. The tail is well furred for almost half the length and is very prehensile.

x. *Caluromysiops*. Black-shouldered opossum. This genus is very closely related to *Caluromys*, but treated separately by most authorities. The single species, *Caluromysiops irrupta* (Fig. 8), is known from Marcapata Valley, southeast Peru, and the Rio Guamues in southern Colombia. It is adapted to tropical, very wet forests and is probably arboreal. To date, few specimens have been collected. Body color is light gray on sides with a well-defined black shoulder region and light face. Hair extends almost the total length of the tail, almost white on the last 25%. Some investigators consider *Caluromysiops* as a species of *Caluromys*. The genera are quite distinct since they have both dentition and coloration differences. Measurement of the adult animal are 430 to 460 mm total length; tail length is 220 to 240 mm. *Caluromysiops* is both nocturnal and arboreal in its habits. It is an active climber, hardy, and a good forager. Food habits are probably similar to *Caluromys*.

xi. *Glironia*. Bushy-tailed opossum. *Glironia venusta* lives in north Ecuador, east of the Andes and north of Bolivia. The genus extends to Peru where the population is referred to as *Glironia criniger*. There are only two species presently recognized which might be conspecific. These opossums

Fig. 8. Black-shouldered opossum, *Caluromysiops irrupta*. New York Zoological Society photo.

live in dense, humid forests and, although their habits are unknown, are presumed to be arboreal and nocturnal. They range in size from 165 to 203 mm for the head and body length, and 195 to 230 mm for the tail length. Diet is unknown but is expected to be similar to that of *Marmosa*—insects, fruits, and seeds. *Glironia* is very rare in collections and almost nothing is known about the ecology and behavior of the genus.

xii. *Dromiciops*. Monitos del Monte. *Dromiciops australis* (Fig. 9) occurs in south central Chile, including Chiloe Island, and slightly east into the mountains of Argentina. The island population is morphologically distinct from the mainland and is given the subspecific designation *Dromiciops australis gliroides*. The species is probably a relict population of a once widespread species. Arboreal, it inhabits dense, humid forests, particularly cane or Chilean bamboo (*Chusquea*) thickets, and in far southern latitudes, araucaria pine forests. It has a prehensile tail. A small opossum, it measures about 125 mm for the head and body length, and 90 to 100 mm tail length. Females have a small pouch with up to five mammae and breed in the spring. Litter size is from two to five. Diet consists mainly of insects, especially Coleoptera and other invertebrates. Gray patches of color on the darker brown fur give it a blotched pattern. Kirsch (Chapter 1) gives a discussion of Reig's (1955) paper in which he suggested *Dromiciops* is closely related to the extinct *Microbiotherium* and therefore should be included in

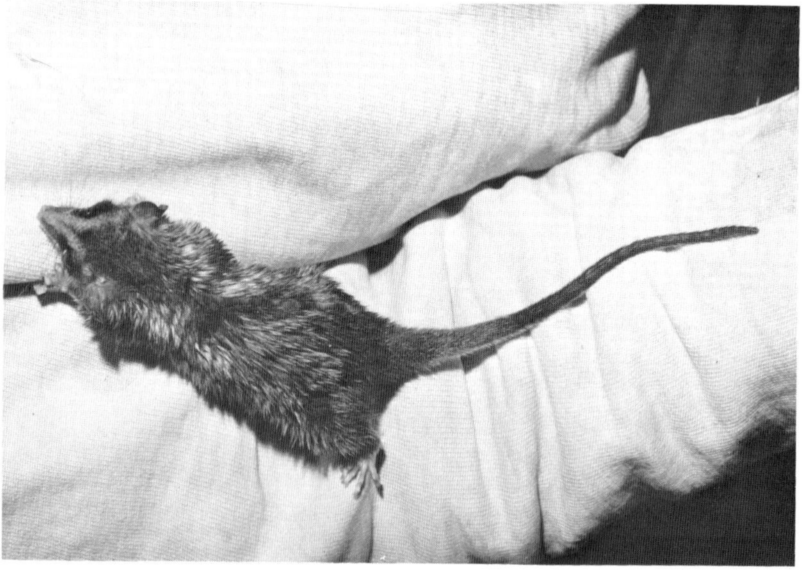

Fig. 9. Monito del Monte, *Dromiciops australis*. A female from Chile. Photo by Kirsch.

Fig. 10. Colombian caenolestid, *Caenolestes obscurus*, from Colombia. Photo by Kirsch.

the subfamily Microbiotheriinae. This discussion of dividing the family into subfamilies is generally considered to group *Dromiciops, Caluromys, Caluromysiops,* and *Glironia* into the Microbiotheriinae. The rest of the family would therefore be in Didelphinae.

**b. Caenolestidae.** *Caenolestids.* The caenolestids generally include three genera and several species limited to the Andean zone of western South America. The general appearance is similar to a shrew, with well-developed ears and tail, small in size with reduced eyes and an elongated rostrum. It is one of the least-known families of mammals.

i. *Caenolestes.* There are five named species, *Caenolestes caniventer* (grey-bellied caenolestid), *Caenolestes convelatus* (blackish caenolestid), *Caenolestes obscurus* (Colombian caenolestid) (Fig. 10), *Caenolestes tatei* (Tate's caenolestid), and *Caenolestes fuliginosus* (type species). They are distributed in the high intermontane valleys, cloud forests, and low paramo zones of the Andes of Colombia and the Paramo Tama of Venezuela, and in Ecuador. They are terrestrial, using ground level trails which are very well marked on the surface of the ground. In thick grass, the runways will actually be in the form of a tunnel. *Caenolestes convelatus, C. obscurus,* and *C. fuliginosus* are crepuscular and nocturnal, living in the damp, cool, high altitude zones where they feed primarily on insects and other invertebrates. *Caenolestes caniventer* and *C. tatei* are lowland species and probably have similar habits, but are adapted to higher temperatures. Trapping records suggest that in certain areas *Caenolestes* are very common and can be collected in relatively large numbers. Either at other times of the season or in other areas, they do not appear frequently in trap lines. They are small animals, with total length ranges from 190 to 267 mm, and tail lengths from 95 to 127 mm. The caenolestids have a very reduced pouch that is completely absent in some individuals. The females have four mammae. The coloration is uniformly dark on the back, ranging from dark black to sooty brown. The ventral area tends to be lighter in some species than others. Eyes are very reduced with ears protruding well above the fur. They eat caterpillars, beetles, ants, centipedes, and spiders. In two species, *C. obscurus* and *C. fuliginosus*, it is reported by Kirsch that the tails are prehensile but not to the point of being able to support the hanging animal. This is true of *Lestoros* as well.

ii. *Lestoros*. Peruvian caenolestid. *Lestoros inca* is the single species referred to in this genus. It is very similar to *Caenolestes* with an elongated snout and small eyes. The majority of specimens have been collected above 2800 m in the highest Andean slopes and valleys of southern Peru. The size ranges from 192 to 267 mm total length, body length varies from 90 to 127 mm, and tail length is between 102 and 140 mm. Trapping data suggest that it is terrestrial and nocturnal; stomach contents include insects and other small invertebrates. Other information on the basic biology of the species, including reproduction and behavior, is generally unknown. This genus is synonymous with *Orolestes* and *Cryptolestes*.

iii. *Rhyncholestes*. Chilean caenolestid. *Rhyncholestes raphanurus*, the only species of this genus, is represented by very few specimens collected in southern Chile from sea level up to 1000 m. The type specimen was collected by Osgood in 1923 on Chiloe Island. A pair was collected from Chiloe Island and a mainland specimen was also collected that same year. Subsequent specimens have been extremely rare in collections. Its general appearance is that of a typical caenolestid with an elongated head possessing reduced eyes and small size. The tail is somewhat shorter than other members of the family, varying between 65 and 89 mm, with total length ranging from 179 to 217 mm. The specimens in the Chicago Natural History Museum are dark brown with no distinctive coloration; the tail is dark and appears to be thickened at the base. It is probably seasonally incrassated. The specimens were collected during intensive trapping so it is considered to be uncommon, terrestrial, and nocturnal in habit. The stomach contents were small invertebrates, and nothing is known of the reproductive physiology or behavior. The female specimen was lacking a pouch and had five mammae.

## 2. Distribution of *Didelphis virginiana*

The geographic range of the Virginia opossum has been considerably expanded during historic times. One would think, if the distribution were entirely limited by topographic or climatic factors, that the species would have expanded to the limits of its environment before this century (especially since we know it has been in the United States for at least the last 50,000 years). The more recent range expansions are probably related to the impact of civilization on the habitat, which provide the species with inhabitable locations near cities and farms where cultivated food supplies are more readily available now than they were before the advent of modern man. Certainly in the region west of the Rocky Mountains man has been responsible for the recent establishment of the population in that area. In a rather interesting analysis of archaeological sites in North America, Guilday (1958) found Virginia opossum bones associated with numerous prehistoric Indian sites. He pointed out that the presence or absence of bones in a series of sites enables early distribution (within rough limits) to be determined by this method. The sites he studied varied in age from 3000 or 4000 years BP to 1400–1600 AD, well before the area was occupied by Europeans. The evidence he presented strongly suggests that the opossum was a common food item of the Indians of Indiana, Ohio, and Kentucky and had not yet extended its range to the Appalachian Plateau of northern West Virginia, western Pennsylvania, and southern New York State. Figure 11 shows the distribution of *Didelphis virginiana* at various periods in North America.

Hamilton (1933) noted an interesting abundance of opossum pelts in the collections of trappers in New York State during the 1920's and early 1930's, and cited a considerable number of people who were under the impression that the opossum populations were undergoing a rapid expansion in that part of New York during those decades. The annual reports of the New York State Conservation Commission point out that in 1920, 375 opossum pelts were listed and in 1929, 1538 were reported, presumably with the same trapping pressure during both years.

R. L. Peterson, Curator of Mammals at the Royal Ontario Museum, has stated (personal communication) that he is under the impression that opossums occasionally move into central Ontario and become established during mild winters. The populations are subsequently decimated during

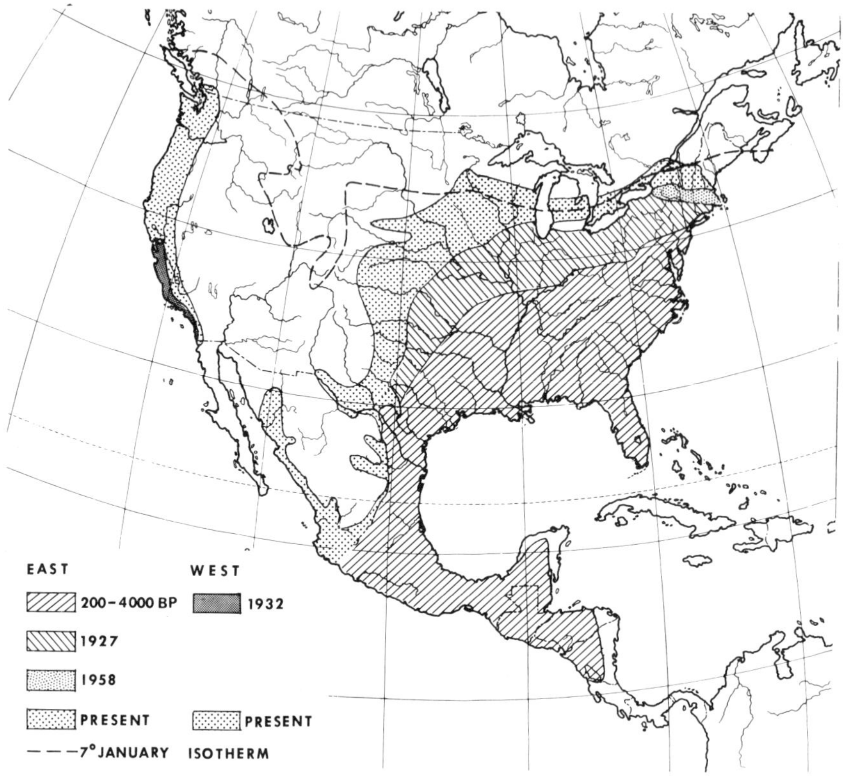


Fig. 11. A representation of approximate boundaries of *Didelphis virginiana* in North America, based on several authors' reports. 200–400 BP (before present) line from Guilday (1958) and −7° January isotherm from Tyndale-Biscoe (1973).

severe winters, but eventually reappear when climatic conditions are not so harsh. *Didelphis* was first collected in Ontario in 1858 (Anonymous, 1858). There were no details given but it does suggest that the opossum was spreading north at that time. Peterson and Downing (1956) stated that most opossums probably entered Ontario during the winter months by crossing ice or on ice rafts; the most northerly location in western Ontario is in Goderich Township, Huron County. Fleming (1913), Cross and Dymond (1929), Saunders (1932), and Anderson (1947) have reported on the occurrence of *Didelphis* in Ontario. Figure 12 shows the years that opossums were captured in Ontario plotted with the mean January temperatures for Detroit, Michigan. Detroit is adjacent to the area of Ontario where *Didelphis* have been collected (collection data from Peterson and Downing, 1956). These data show that the animals are present during the warmer years, when the mean January temperature of the area is above −2°C. During particularly cold periods of time, i.e., 1935–1945, *Didelphis* were not recorded in Ontario. This supports Peterson's hypothesis that the Virginia opossum populations suffer from particularly severe winter conditions and that subsequent warm periods enable the species to reinvade the region. Blair (1936) pointed out that opossums die from cold and hunger during periods of cold weather in Michigan after he observed one that was found dead in a snow drift after 6 or 8 weeks of cold weather. Tyndale-Biscoe (1973) considered that the −7°C January isothermal line in the northern limit of the species range is due to climatic limitations. From the present author's data and his, it appears that the −7°C line would be an absolute limit, and that the populations would be stressed and partially limited at the −2°C January climate.

The record of *Didelphis* populations in the western United States is reasonably well documented. In 1890, the species was first introduced into southern California near Los Angeles. By 1906, museum specimens were collected in other parts of Los Angeles County and the population was well established and expanding by 1915 (Grinnell, 1915). By 1924 it had expanded into adjacent Ventura county. J. R. Kocher imported seven *Didelphis* from Tennessee in 1910 and maintained them at his residence in San Jose in central California. Five subsequently escaped and formed the nucleus of the northern population of California opossums.

Grinnell (1915) was not sure that California should consider itself lucky in having added the species to its native fauna, but was convinced that the population was destined to be spread from one end of the state to the other.

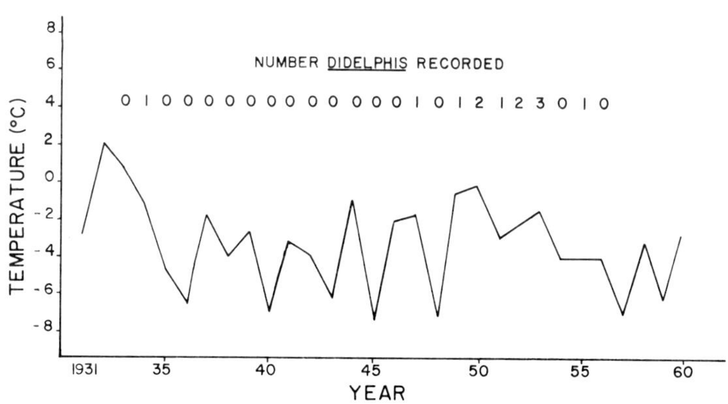


Fig. 12. Mean January temperature for Detroit, Michigan and number of *Didelphis* recorded in Canada from 1931 to 1960. The majority were observed during years of relatively warm periods of temperature above −2°C.

The species began appearing in collections in San Diego County, at the southern end of the state, in 1927 (Bryant, 1927) and 1928 (von Bloeker, 1928). By 1932, it occupied the state from San Francisco and Sacramento south to the Mexican border and by 1958 had expanded its range to all the area of California below 1500 m altitude that was not desert. Although not well represented by museum specimens, the species does occur in northern Baja California at the present time.

At the time the species was expanding rapidly in California, circa 1910, it was introduced into Oregon and Washington. The Oregon population was apparently introduced in Umatilla County by escaped pets, or liberated by some well-meaning opossum lover, between 1910 and 1921. Jewett and Dobyns (1929) stated that a Mr. Wade liberated a pair of opossums on McKay Creek, Umatilla County, and that these were considered as the ancestral stock of the population in that area. These were liberated about 1914, and by 1923 trappers were taking opossums in set traps. At least 50 opossums were taken along McKay Creek within a distance of 22 miles for a period of 6 or 7 years. In the trapping season of 1927–1928, 12 opossums were taken by local trappers within a 5-mile area along the creek. From these early beginnings, the Oregon population is now very well established and the species can be found in a variety of habitats throughout the state.

The Washington state population tells the same story again: a series of escapes and liberated pets during the 1910–1920 period were responsible for the rapid expansion and establishment of the species in that area (Scheffer, 1943). By 1958, it had entered British Columbia and can be considered a permanent resident. It is rather curious that an animal lover would select an opossum as one of the favorite exotic species to import into a new area. Probably the greatest incentive would come from an interest in eating the meat or from childhood memories of boy and dog running through the fields of Tennessee and Missouri and fondly remembering these experiences associated with hunting opossums. Such apparently was the case of S. M. White, who very systematically began liberating opossums in Arizona. He liberated a female with 13 young on the Santa Cruz River north of Tucson in the summer of 1927, and several opossums were recovered during ensuing years from the area. He had imported them from Arkansas and was still in the process of liberating opossums in 1950. Hock (1952) traced the evolution of the Arizona population in great detail, and was another of the scientific community who was not too sure that the opossum should be considered an asset to the fauna of Arizona. While at the University of Arizona, he had noted that in 1949 and 1950 reproducing animals were established in the Tucson area because young animals were collected. If the species is permanently established, it only occurs in specific areas and is not well represented in scientific collections. Cockrum (1964) stated that opossums have been introduced in southern and central Arizona, but were not well established anywhere in the state at that time. Although it occurs naturally in eastern New Mexico (Bailey, 1931) and in the state of Sonora, northern Mexico (Bailey, 1933), the intervening region between Sonora, New Mexico and Tucson is a very harsh desert which is an effective barrier to the species expanding from either of those two populations. *Didelphis,* when well established in Arizona, will be restricted to riparian and mesic habitats associated with irrigated farms and cities. Once established west of the Rocky Mountains, *Didelphis* had access to a variety of suitable habitats located in the western states. For better or for worse, the species is a permanent resident in the west and can be expected to continue increasing its population until natural ecological factors begin to suppress its population expansion.

## C. Habitat Selection

All of the neotropical marsupials are primarily adapted to a forest habitat. These sylvan adaptations are reflected in numerous arboreal species that abound in the tropical and temperate forests of the Western Hemisphere. Many genera are terrestrial like *Lutreolina* and *Didelphis,* while the water opossum (*Chironectes*) is very specialized for an aquatic existence. Figure 13 shows the varied habitats occupied by the didelphid marsupials.

The Virginia opossum (*Didelphis virginiana*) primarily is associated with woodland areas that contain small streams. Lay (1942) found that in eastern Texas there were almost four times as many opossums in the wooded areas as there were on the sandy prairie. Fitch and Sandidge (1953), working in northeastern Kansas, found indications of greater concentrations of opossums in densely wooded areas than in surrounding areas that were more sparsely wooded. In Maryland, Llewellyn and Dale (1964) found that opossums preferred low dense woodland near water to dry upland woods or open areas. Stout and Sonenshine (1974) stated that opossums in Virginia definitely preferred forested areas. These findings are consistent with what would be expected from the opossum's needs for the protective cover of the forest and the streams for foraging. In all probability, the northern distribution is dependent upon wooded areas with suitable dens for protection from cold weather.

*Didelphis* is basically a terrestrial species which spends most of the time on the ground. Although they can climb, they are somewhat clumsy above ground (McManus, 1970; Hunsaker and Shupe, Chapter 5) and use the arboreal habitat only occasionally. The majority of climbing is done primarily in search of food.

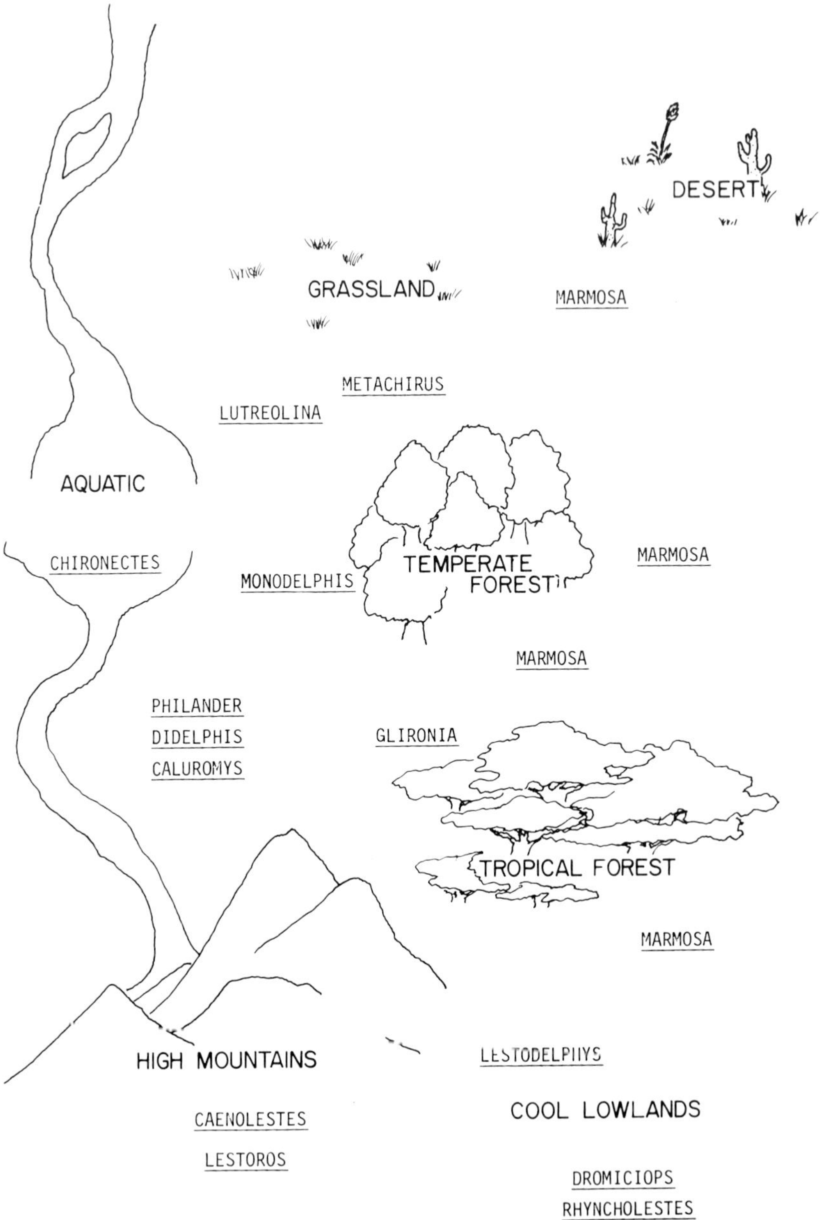


Fig. 13. Habitat selections of Didelphidae and Caenolestidae.

The opossum is very versatile in selection of dens. Lay (1942), in eastern Texas, found that opossums preferred dens with woody cover. They often used armadillo burrows under the roots of bushes as dens. He found the typical den on the prairie was a simple burrow about 6 feet long. Wiseman and Hendrickson (1950), in southeastern Iowa, found that opossums extensively used burrows of other animals as dens. They also discovered the opossums in the area using hay stacks, hollow logs, and holes under barns as dens. In northeastern Kansas, Fitch and Sandidge (1953) found opossums often used rock ledges on hillsides as dens.

*Marmosa* is the most varied genus in the family Didelphidae, with over 50 named species (Tate, 1933). Some species have secondarily become adapted to a terrestrial habitat, especially *Marmosa elegans*. These species have shorter tails, small feet with reduced claws, and short vibrissae. *Marmosa microtarsus, Marmosa fuscata,* and related species are also terrestrial.

*Marmosa* also inhabits arid semidesert areas as well as the wet tropical forests. *Marmosa canescens* of southern Mexico, *Marmosa robinsoni* of Ecuador and Peru and *Marmosa quichua* of eastern Peru are all adapted to dry sections. *Marmosa robinsoni* can be collected near *Opuntia* cactus near Santa Marta, Colombia but is generally a mesic habitat species. *Marmosa elegans* inhabits the drier areas of the chaco and pampas in Argentina. The altitudinal range is extreme, from sea level to 3400 m, where *Marmosa aceramarcae* has been collected. The genus primarily occupies the regions below 2500 m. *Marmosa pusilla, M. robinsoni, M. fuscata, M. cinerea,* and *M. impavida* are all species which have a wide range of altitudinal tolerance, all of which have more than 2000 m vertical variation in their ranges. Other species are much more restricted to a single habitat type.

*Marmosa robinsoni* appears to prefer secondary forests and farmland to unmodified areas. The population can be expected to be significantly larger in disturbed areas. The present author and Fleming (1972) both have observed this to be reflected in trap line results and other observational data. They are especially common in the vines and bushes adjacent to cleared areas.

The most specialized adaptation to a particular habitat is found in the water opossum *Chironectes minimus*. This species is almost entirely aquatic, has webbed feet, and is an excellent swimmer. It is found in forested areas that have streams and rivers where it hunts for food. *Lutreolina* is another specialized genus that is remarkably similar to the mustelids in appearance. It prefers brushy and grassland habitats to heavily forested areas. Although not aquatic, there are numerous records of them near watercourses. They have an abnormally thick basal section of the tail which is distinctive. Cabrera and Yepes (1940) state that *Lutreolina* are good swimmers, but Davis (1966) and Collins (1973) reported that a specimen in captivity appeared to be a poor swimmer. Most reports the present author has heard indicate that they are good swimmers but are not basically an aquatic animal.

The caenolestids, which are restricted to the colder regions, represent what appear to be three relict populations at the generic level. *Caenolestes* and *Lestoros* are found in the high, cold areas of the Andes and *Rhyncholestes* is found in the colder southern regions of Chile. The habitat of *Caenolestes* includes the paramo, a very damp and cold area that is characteristic of the high mountain valleys above 3000 m. The weather is cloudy and foggy most of the time and very damp (Osgood, 1943). *Caenolestes* thrives in these conditions and frequently has a surprisingly high buildup of population numbers in a good habitat.

### D. Food Preference and Nutrition

The Virginia opossum is probably one of the most omnivorous vertebrates known. Food consumption apparently is based on whatever food is most readily available; because of this, there is a great variation in the diet reported in the literature. Studies conducted in different areas at different times of the year have produced data which suggest *Didelphis virginiana* eats the food which is seasonally abundant. One study by Lay (1942), in eastern Texas, found the opossum in that area to be more carnivorous than herbivorous, while another study by Fitch and Sandidge (1953), in northeastern Kansas, found just the opposite. Although they do eat a great variety of material, personal observation and some casual experiments show that they do show a preference for meat and insects when they are available.

The eastern Texas study by Lay (1942) was done by means of stomach analysis. The foods found and the proportions which they comprised were as follows: insects and worms, 45%; fruit, 11.8%; green leaves, 11%; trash (leaf and log litter), 10.6%; mammals, 7%; acorns, 4.7%; birds, 4.3%; crayfish, 3.3%; snails, 0.75%; grass seeds and cellophane traces, 1.5%. As the results indicated, about 60% of the Virginia opossum's diet was of a carnivorous and insectivorous nature.

In northeastern Kansas, by examining the feces or "scats," Fitch and Sandidge (1953) found that during the fall and winter fruit made up the majority of the opossum's diet. Of the fruits eaten by the opossum, the ones found most often were grapes, hackberry fruit, wild plum, and wild crabapple. Other food items by volume were crayfish, insects, corn, rabbit carcasses, young snakes, snails, frogs, and lizards, the importance of which varied by season. As in the Lay study, the opossum's diet consisted of a wide variety of foods. In contrast, however, the diet was found to be more herbivorous than in the Lay study. Even with only a limited knowledge of the two areas in which the studies were conducted, it would be safe to assume that the differences could easily be accounted for by the availability of the different foods rather than preference.

Wood (1954) studied a series of 25 digestive tracts and 23 fecal droppings of the Virginia opossum from east Texas. Thirty-nine different foods were found in the analysis. The digestive tracts contained 36 of the total of 39 food items recorded, 26 more than were identified in the scats. This suggests the digestive tract, especially the stomach, is a more reliable source of data than droppings. In the process of digestion, many food objects are rendered unidentifiable when they appear in the fecal material.

Young opossums made up 1.6% of the annual volume of food. There is a cannibalistic trait in the opossum, but since carrion is known to be eaten it is possible all of these records represent carrion, but probably not. Wilson (1970) observed a *Didelphis marsupialis* attack and eat a *Philander opossum* in Panama, and there are numerous records of cannibalism in captivity. Wood found the following percentages of material in opossums: mammal, 14.9%; bird, 3.8%; cold-blooded vertebrates, 7.4%; insects, 25.0%; other vertebrates, trace; plants, 44.8%; unidentified, 3.2%.

Opossums show some preference for a variety of foods offered them, probably mediated through both olfaction and taste. Behaviorally, they appear to show little taste sensitivity unless the food has a particularly strong taste. They do, however, respond in an overt manner to high concentrations of sweetened or very bitter food. They appear to have gustatory responses more similar to the carnivores than to rodents. *Didelphis virginiana* are unique in that they are the only animal tested that shows a preference for the bitter taste of quinine over water.

Pressman and Doolittle (1966) tested the opossums' taste preferences and compared them to values for other animals (Carpenter, 1956). Opossums showed a distinct preference for quinine hydrochloride at 0.000004 to 0.000008 *M* over distilled water. Carpenter's data (1956) showed that cats avoided quinine hydrochloride at those concentrations. Rabbits and hamsters did not discriminate it from tap water at that level. The opossums tested preferred sucrose at the 0.0512 *M* level, which is similar in concentration to sucrose preference levels in rabbits. Hamsters preferred higher levels and cats did not prefer sucrose to tap water at any level. Salt (NaCl) was preferred at the 0.4 *M* level and avoided at 0.64 *M* and above. The reaction to NaCl was more similar to that of cats and hamsters than to rabbits.

Tamar (1961) monitored the chorda tympani nerve to test taste sensitivity of the opossum and also found that the species has low taste sensitivity. They are especially sensitive to $NH_4Cl$ and they react about double to KCl as to NaCl. LiCl also elicits a greater response than NaCl. Maximum response to the most stimulating salt is reached at concentrations of over 2 moles. The anterior half of the tongue is most sensitive to the sour stimulus and the bitter modality is detected on the posterior part of the tongue.

In comparison to data collected by Beidler, Fishman, and Hardiman (1955) on a variety of vertebrate species, Tamar (1961) concluded that the opossum resembles the Carnivora more than the Rodentia. He also made the observation that if the opossum did in fact represent the primitive condition of taste sensitivity, then more evolutionary change has occurred in rodents than in carnivores.

The Virginia opossum is almost completely nocturnal in its search for a wide variety of foods. In all of the recent studies, opossums were seldom seen out during the day, unless of course they had been trapped or the weather was particularly cold at night. Wiseman and Hendrickson (1950), in a study of the Virginia opossum in southwestern Iowa, found by tracking, that in the course of an evening opossums forage for 1.6 km to 2.4 km (1 to 1½ miles) and return to their dens each morning. They would stay in their dens at night only in the winter when the temperature was near or below 0°C. Fitch and Sandidge (1953), in northeastern Kansas, reported similar results. They also found, however, that if opossums had been forced by the cold to stay in for several nights, they would occasionally forage for food during the day when it was warmer.

In a study in Michigan, the energy requirements of the opossum were estimated as a function of oxygen consumption by Brocke (1970). He plotted the oxygen consumption of these winter acclimated animals over an ambient temperature range from 0° to 25°C.* Brocke calculated an opossum would ingest 1380 kJ (kilojoules) of energy on a typical winter night's foraging. He calculated an energy consumption of 264 kJ/day at an ambient temperature of 25°C. This was found to be below average compared to other marsupials and considerably lower than similar sized eutherian species. With a drop in temperature, the metabolic rate ($O_2$ consumption) of the opossum increases. An ambient temperature of 0°C boosts the energy consumption to 840 kJ/day, 0.5 ml $O_2$/g/hr. On this basis, in areas where temperatures commonly drop to 0°C and below, the opossum is forced to partake in two foragings over a period of 3 days to meet its energy requirements.

During the 1.6 to 2.4 km nightly forage for food, the opossum has been found to follow a very "circuitous and erratic" route which usually changes from night to night and is seldom associated with trails. Being mainly ter-

* They have a very low basal metabolic rate, 0.15 ml $0_2$/g body weight per hour (for a 3.5 kg animal).

restrial, the opossum probes along the ground for food with his snout and is seemingly guided mostly by olfactory and tactile stimulation. Observational evidence and signs indicate that the opossum scratches at the base of rocks and turns them over in search of food (Fitch and Sandidge, 1953; Lay, 1942; Llewellyn and Dale, 1964; Wiseman and Hendrickson, 1950).

Food availability probably is an important factor in survival of individuals. In captivity they will continue to eat as long as food is available and can become obese. Ad lib food in the lab is not conducive to the opossum's good health, especially the female's. Jurgelski (1974) reported female *Didelphis virginiana* with proptosis and strabismus of the eyes as a result of great amounts of orbital fat storage. He recommends only 108 g of wet, supplemented diet daily during the nonbreeding season and twice this amount for the nursing females.

The mouse opossum, *Marmosa,* and woolly opossum, *Caluromys,* are also nocturnal but much more arboreal than *Didelphis*. Observations by the author and other researchers indicate *Marmosa* are mainly insectivorous, but will also eat fruit, snails, carrion, and other small animals. *Caluromys* is basically frugivorous but will also eat a variety of other foods.

Enders (1935) found the *Marmosa* of Panama to lean strongly toward insectivorous habits. They ate bananas, peaches, grapes, figs, bat carcasses, viscera of mice and birds, earthworms, moths, and grasshoppers, but preferred insects to flesh. Captive animals were seen to die when fed only fruit. Upon the addition of grasshoppers and other insects to their diet, the animals remained in much better condition. The figs on the island he studied were badly infected with larvae of various insects. *Marmosa* were partial to those ripe figs and ate their inner pulp and seeds, larvae and all. Weak *Marmosa* were attacked and partially eaten by their conspecifics, which thus exhibited cannabalistic tendencies.

In the laboratory, *Marmosa* will live and gain weight on commercial dog or cat food preparations. A diet of this nature is inadequate for reproduction and supplements of fresh fruit, meat, insects, minerals, and vitamins are necessary for a healthy colony. *Lutreolina* is carnivorous, feeding on small rodents, marsupials, and aquatic animals.

### E. Home Range

Opossums are nomadic, solitary animals that do not consistently restrict their activities to any particular area. Definite home ranges are difficult to determine because they travel over long distances and remain in one area for relatively short periods of time. They require considerable effort to trap and males appear to be somewhat more trap shy than females. Over a 24-

month study period in eastern Texas, Lay (1942) reported that 49% were taken only once. In an analysis of his data on 29 individuals trapped three times or more, the mean distance traveled was 445 m (1460 feet). Other investigators have reported similar movements; Llewellyn and Dale (1964), who trapped *Didelphis* for 8 consecutive years in Maryland, reported 50% of a series of 224 opossums were trapped only once and only 3% (6 individuals) were recorded for periods of a year or more. Their longest record was a male who was captured seven times in 29 months and had wandered more than 1½ miles from his original home range. Wiseman and Hendrickson (1950) trapped and tracked opossums in the snow of Iowa and observed tracks that passed within a meter or so of baited traps, but the opossums would not enter the trap. In view of their good olfactory ability, one would tend to suspect the animals were basically wary of traps and hesitated to enter them, even for food. Previous experience could also account for their avoidance of the traps, but in any case, these studies point out the difficulty of the determination of an accurate home range for the species from trap records.

Home range is the area commonly frequented by a particular animal and in which the animal spends most of its time. It differs from a territory in that it may overlap with home ranges of several other conspecifics and it is not defended. Orientation within a home range is aided by visual landmarks such as trees and rocks. Sign posts of urine, dung, and glandular secretions indicate the present use of a given area to other animals. Most importantly, the animal's familiarity with its home range more easily enables it to detect predators and find cover with a minimum of delay.

Hunsaker and Shupe (Chapter 5) stated that the Virginia opossum is not territorial in the strict definition of the word, but represents a solitary species which excludes other individuals because it is very aggressive to other opossums when they meet. Most of the large sizes of home ranges represent a wide ranging, nomadic way of life for these animals. The males advertise their presence in a particular area by marking sign posts and do a characteristic fighting dance (Reynolds, 1952; McManus, 1970; Hunsaker and Shupe, Chapter 5) when they encounter the scents of other male opossums. The sedentary and fixed area of habitation which characterizes a territorial species is not found in *Didelphis*.

The only accurate method of determining the exact home range of a species is to keep individuals under continuous observation. This is particularly difficult for small and nocturnal animals and a variety of techniques have been utilized. Tracks, colored droppings, trapping, and automatic monitoring devices have all been used. Some animals can be studied by remote census techniques i.e., radio tracking, as Shirer and Fitch (1970) did with opossums, but the majority of home range data for vertebrates is derived from trapping records.

Several methods of analysis of trapping data are used to determine home range, all with greater or lesser error factors inherent in the technique. Grid systems, random placement, and selected trap sites have all been used. Critical analyses of these systems have been made by Davis (1953), Hayne (1949, 1950), Mohr (1947), Sanderson (1966), and others. The basic problem is that animals do not lay out their home ranges in exact squares or circles. A capture in a line of traps probably does not represent the actual location of the limit or edge of the movements of an animal. Drawing a line halfway between the capture site and the adjacent trap will compensate for this error, but is still not absolutely accurate.

Distance between captures has also been used to determine home ranges, the maximum distance (Evans and Holdenreid, 1943) as well as the distance between successive captures (Davis *et al.*, 1948; Davis, 1953). The best approach appears to utilize a system to determine the center of activity (Hayne, 1949) of an individual and to record the distance between successive recaptures to determine the area used by the animal. Basically, the questions we really want to answer are what specific types of habitat the animals lives in, how various species distribute themselves in a habitat, and how many animals a particular habitat will support. The distribution, both spatially and temporally, is particularly difficult to determine from a trap grid. By using as standardized procedures as possible, comparative studies can be made, but they are difficult when we have to use the variety of data analysis systems utilized by different investigators.

Lay (1942) reported that the average minimum area between stations of 29 recapture sites of *Didelphis virginiana* was 4.7 hectares (11.5 acres). The mean greatest distance traveled between stations was 445 m (1460 feet), which would form a theoretical circle of 15.5 hectares (38.4 acres). The home range of these animals would appear to be approximately 10 hectares or 25 acres, only about half the area of minimum and maximum movements. During 2 years of study, Lay could recapture only 29 opossums three times or more.

Fitch and Sandidge (1953), in northeastern Kansas, encountered similar problems in determining the opossum's home range. As in other studies of opossums they found no evidence of territoriality. The home ranges were reported to overlap and shift frequently. They were described as relatively small areas familiar to the animal and capable of providing for all of its ecological needs. There was no significant difference found between the home ranges of the males and females. For the 37 opossums in this study that were caught two, three, or four times, the average distance between trapping stations was found to be 249 m (817 feet). They calculated that this would indicate an average home range of about 19.5 hectares or 48 acres.

In Maryland, Llewellyn and Dale (1964) also found that the home ranges of opossums do not appear to be as highly developed as in some other mammals. They further suggested that their home ranges appear to be less circular and more elongated than had been suggested. This is probably true, since the elongations of the home range could be a function of the prime habitat along stream courses. Stumpf and Mohr (1962) also pointed out that most mammals tend to have elongated home ranges. Circular home ranges are easy to calculate in the laboratory but are probably rare in nature, especially for a stream bed inhabiting species. Based on the 50% of the 541 trappings of 224 opossums which had been trapped more than once, Llewellyn and Dale found the average length of range to be 981 m or 0.61 miles. The average distance between successive captures was 463 m or 1520 feet. This study was an 8-year effort and the opossums in this area preferred low, damp woods rather than upland woods. They also appeared to be more sedentary than in other studies. The opossums appeared to stay in certain areas, since one female was trapped at the same trap location for 6 of 13 recaptures, although the maximum distance between successive captures was 1670 m (5478 feet).

Verts (1963) studied an opossum population in a cultivated area in Illinois and found that they were far from sedentary. He calculated the home range of *Didelphis* in that habitat to be 54 (134 acres) to 82 hectares (203 acres), on the basis of a circular area. Verts compared the length and breadth of the home ranges of his animals and those of Fitch and Sandidge (1953) and found the opossums in his study had a home range 2.9 times longer than broad and those in Kansas (Fitch and Sandidge, 1953) to have a home range 2.7 times longer than broad. Stumpf and Mohr (1962) calcu-

**TABLE I**

Movements of Virginia Opossums in Kansas, from Their Dens, According to Age and Sex[a]

| Number of individuals | Age or sex group | Average distance in meters between den and field point[b] | Average movement in meters from den to den |
|---|---|---|---|
| 2 | Newly independent young (June) | | 11 (3 to 25) in 3 |
| 4 | Young in third month of independence (August) | 23 (9 to 49) in 5 | 89 (9 to 245) in 10 |
| 5 | Young in fifth month of independence (October) | 158 (62 to 367) in 10 | 140 (128 to 147) in 5 |
| 3 | Young in sixth month of independence (November) | 196 (16 to 429) in 18 | 204 (83 to 429) in 13 |
| 8 | Adult male | 229 (31 to 588) in 40 | 305 (25 to 734) in 130 |
| 7 | Adult female | 155 (9 to 620) in 38 | 299 (31 to 636) in 93 |

[a] From Fitch and Shirer (1970).
[b] Range in parentheses.

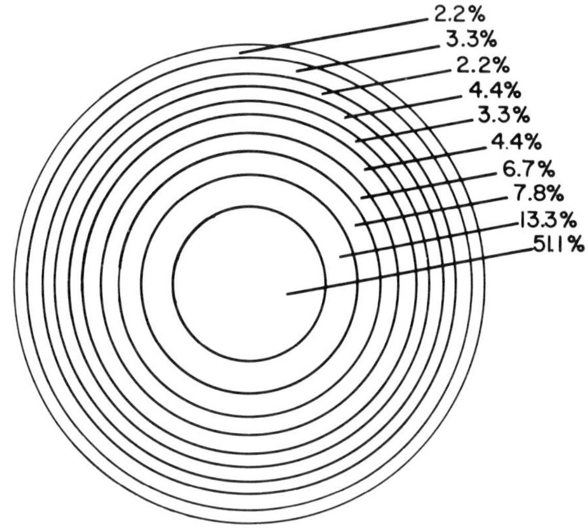


Fig. 14. A representation of the hypothetical circular home range of an opossum divided into 10 equal areas. The numbers show the relative amount of activity in each of the areas. From Fitch and Shirer (1970).

lated the opossum ranges in Lay's study in Texas (1942) to be 2.7 times longer than broad. Verts stated his calculations of home ranges based on these figures were 13.3 hectares (33 acres) in Kansas, 12.5 hectares (31 acres) in Texas, and 38.8 hectares (96 acres) in Illinois.

Fitch and Shirer (1970) utilized radio transmitters to effectively locate and track opossums in Kansas. They monitored the movements of 27 opossums during 1965–1966. The transmitters were crystal controlled from 26.5 to 26.65 MHz and placed on the animals by a collar which also served as an antenna (Shirer and Downhower, 1968). This technique was very successful. Ninety locality records were obtained and 88 dens located during the study. Fitch and Shirer stated that the home ranges appeared to be circular, with the animals utilizing the various parts of the habitat equally. The foraging activity, ranging out from the den in a random fashion, produced circular home ranges that had an average radius of 620 m, approximately 120 hectares (see Table I). Rather than moving throughout a particular home range, opossums ranged out from whatever den site they were using at that particular time. As one would expect, the central area around the den was used the most, approximately 51.1% of the activity being within a circle comprising 10% of their hypothetical home range (see Figure 14). The authors calculated that 24.2% of the activity was spent in the central one-fiftieth of the total home range.

The data on the same opossums were analyzed in conjunction with similar material on raccoons (*Procyon lotor*) and striped skunks (*Mephitis mephitis*) in the study area by Shirer and Fitch in 1970. They found that opossums utilized a particular den only for an average of 2.2 days and shifted an average of 990 feet between dens. On one occasion an opossum and skunk were observed to den together.

Information on the home range of other genera of didelphids is very scarce. Fleming (1972), Enders (1935), and Davis (1945a,b) have trapped in Central and South America and have reported on *Marmosa robinsoni, Philander opossum, Caluromys philander,* and *Didelphis marsupialis.* Fleming's 1972 work included information on all three species. He trapped a 5.1 hectare grid in the Panama Canal Zone, using National live traps in a 10-row grid of traps with 10 trap stations per row, each 25 m apart. Additional traps were placed in trees 1 or 2 m above ground level. In general, his data indicated that all the species were nomadic and stayed in the trapping area for only 2 or 3 months.

*Didelphis marsupialis* had an average distance between successive trap sites of 69 to 81 m. The longest distance between successive capture sites was 122 to 170 m. These data were calculated on 14 animals trapped six times or more. Using the average distance between successive captures, 74 m, as the diameter of a circle, the present author estimates a circular home range of 0.43 hectares or 1.06 acres.

Davis (1945a,b) reported *D. marsupialis* to have similar habits of wandering and that it did not remain in one locality for any length of time. Males especially were seldom recaptured but one female stayed in a 5000 $m^2$ (approximately) area for 11 months. Although his traps were 100 m apart, they followed a winding trail and the traps the female was caught in formed a rough square. She was recaptured 19 times from August 1942 to June 1943. (See Table II.)

*Marmosa robinsoni* is the only species of murine opossum that has been extensively trapped in live traps and recaptured. No actual home ranges have been calculated but some conclusions can be drawn from the data available. The species appears to be basically nomadic and solitary. Much like *Didelphis,* the females are more sedentary and tend to exclude other females from their area of occupancy. None of the reports mention the occurrence of two adults in the same nest, and the author has not observed more than one adult per nest in several *Marmosa* nests. Enders (1935) trapped *Marmosa robinsoni* on Barro Colorado Island, Panama, and also recovered them from nesting boxes he erected in trees. The *Marmosa* would occupy a nest box for a few days, then leave, to return at a later date or never to be seen again. He observed a female in a nest box about 2 m above the ground. The box had a nest of leaves when the female was first discovered. She vacated the nest box and did not return for 25 days. However,

**TABLE II**

A Record of Captures and Recaptures of *Didelphis marsupialis* in Brazil during 1942 and 1943[a]

| | | Location of capture | | | | | | | | | | | |
|---|---|---|---|---|---|---|---|---|---|---|---|---|---|
| | | 1942 | | | | | 1943 | | | | | | |
| Age | Sex | Aug | Sept | Oct | Nov | Dec | Jan | Feb | Mar | Apr | May | June | July |
| Adult | ♀ | 1.5 | 2.8 | 1.5 | 3.5 | 1.5 | 1.5 | 1.5 | | 1.5 | 1.5 | 1.5 | |
| | | 2.2 | 3.2 | 4.8 | 1.5 | 2.5 | | | | | | | |
| | | 2.8 | 1.5 | 4.8 | | | | | | | | | |
| | | | | 2.2 | | | | | | | | | |
| Adult | ♀ | 5.6 | | X | | | | | | | | | |
| | | 5.1 | | 5.3 | | | 5.4 | | | | | | |
| | | | | 9.5 | | | | | | | | | |
| Adult | ♂ | | | 9.0 | | | | | | | | | |
| Adult | ♂ | | | | 9.5 | | | | | | | | |
| Adult | ♂ | | | | 5.3 | | | | | | | | |
| Adult | ♂ | | | | | | 7.7 | X | | | | | |
| Immature | ♀ | | | | | | 9.4 | X | | | 7.6 | 8.2 | |
| | | | | | | | 9.4 | | | | 9.2 | | |
| Young of | ♀ | | | | | | 5.4 | | 4.9 | 2.5 | 1.6 | 1.6 | |
| M 14669 | | | | | | | | | | 2.2 | 3.8 | | |
| | | | | | | | | | | | 4.2 | | |
| Adult | ♂ | | | | | | | | | 7.8 | 6.0 | 9.0 | |
| Immature | ♂ | | | | | | | | | | 6.0 | | |
| Adult | ♂ | | | | | | | | | | | 1.0 | |
| Adult | ♂ | | | | | | | | | | | 4.3 | |
| Adult | ♀ | | | | | | | | | | | | 2.9 |
| Adult | ♂ | | | | | | | | | | | | 2.7 |

[a] From Davis (1945b). Numbers indicate stations 100 m apart; 9.2 = a capture site 20 m from station 9, toward station 10.

a series of three other *Marmosa* were observed in the box during her absence.

Fleming's (1972) work on *M. robinsoni* in the Panama Canal Zone indicated the males were more nomadic than females and remained in an area only half as long as females. The females' longest period of average residence was 135 days, a little more than 4 months. The longest period between first and last capture was 210 days for two individuals.

As one would expect, the home ranges and distances of mouse opossums are reasonably small; the greatest distance mentioned was a female that moved 300 m in 2 days. The males had an average distance between successive capture sites of 58.5 m, and for females the distance was 46.6 m; the longest distance between successive capture sites was 130.4 m for males and 81.9 m for females. The home ranges of males overlapped extensively but the females appeared to be territorial and excluded conspecific females. Based on the average distance between successive captures of 53 m, which would suggest a home range with a 26.5-m radius, a circular area would be 0.22 hectares (0.54 acres).

*Caluromys, Philander,* and the other genera of didelphids appear to be generally nomadic and have varied home ranges. *Caluromys* is considerably more arboreal than *Didelphis* and can be captured in live traps set in trees. They readily investigate platforms built in trees and are not too trap shy. This author has observed two *Caluromys lanatus* hunting together in Leticia, Colombia, but it was not known if they were a mated pair or siblings. Davis (1945a,b) reported that *Caluromys philander* was more sedentary than *Didelphis* in his Brazil study area near Rio de Janeiro. Like other species, the females of *Caluromys* moved less than males. One male moved an estimated 300 m during 1 month and then was recaptured in the first trap 3 months later (see Table III). Females tended to restrict their radius of activity to areas of 100 m or less.

Fleming (1972) stated that *Philander opossum* had a home range about the same size as *Didelphis* in the Canal Zone. Males and females were

**TABLE III**

A Record of Captures and Recaptures of *Caluromys philander* in Brazil during 1942 and 1943[a]

| | | Location of capture | | | | | | | | | | | | |
|---|---|---|---|---|---|---|---|---|---|---|---|---|---|---|
| | | 1942 | | | | | 1943 | | | | | | | |
| Age | Sex | Aug | Sept | Oct | Nov | Dec | Jan | Feb | Mar | Apr | May | June | July | Sept |
| Adult | ♀ | 9.2 | | | | | | | | 9.0 | | | | 9.0 |
| | | 9.0 | | | | | | | | | | | | |
| | | 9.0 | | | | | | | | | | | | |
| Adult | ♀ | 8.1 | 9.2 | 9.0 | 9.0 | 9.2 | 9.0 | 9.2 | 9.0 | | 9.2 | | 8.1 | |
| | | | 9.0 | 9.0 | 9.0 | 9.0 | | 8.1 | | | | | | |
| | | | | 9.2 | 9.2 | | | | | | | | | |
| | | | | 9.0 | 9.2 | | | | | | | | | |
| Adult | ♂ | | | 4.9 | | | 4.9 | | | | | | | |
| | | | | 9.0 | | | | | | | | | | |
| Adult | ♀ | | | | | | 4.9 | | | 4.9 | 4.9 | | | |
| Immature | ♀ | | | | | | | | 9.0 | | | | 6.0 | |
| | | | | | | | | | 8.1 | | | | | |
| Immature | ♀ | | | | | | | | | 6.0 | | | | |
| Adult | ♂ | | | | | | | | | 9.0 | 9.2 | 9.0 | | |
| Adult | ♂ | | | | | | | | | 6.0 | | | | |

[a] From Davis (1945b). Numbers indicate stations 100 m apart; 9.2 = a capture site 20 m from station 9, toward station 10.

**TABLE IV**

Home Ranges of *Didelphis*, *Marmosa*, and *Philander*

| Species<br>Reference | Shape | Area | |
|---|---|---|---|
| | | Hectare | Acre |
| *Didelphis virginiana* | | | |
| Fitch and Sandidge (1953) | Circular | 19.5 | 48.1 |
| | Ellipse[a] | 13.2 | 33.0 |
| Lay (1942) | Circular | 4.7–15.5 | 11.5–38.4 |
| | Ellipse[a] | 12.4 | 31.0 |
| Verts (1963) | Circular | 54.2–82.2 | 134–203 |
| | Ellipse | 38.9 | 96.0 |
| Fitch and Shirer (1970) | Circular | 254 | 627.4 |
| Holmes and Sanderson (1965)[b] | Circular | 14.2 | 35.0 |
| Llewellyn and Dale (1964)[b] | Circular | 16.7 | 41.7 |
| *Didelphis marsupialis* | | | |
| Fleming (1972)[b] | Circular | 0.43 | 1.06 |
| *Marmosa robinsoni* | | | |
| Fleming (1972) | Circular | 0.22 | 0.54 |
| *Philander opossum* | | | |
| Fleming (1972) | Circular | 0.34 | 0.84 |

[a] Calculated from authors' data by Verts (1963).
[b] Calculated from authors' data (average distance between captures = diameter).

observed to travel an average of 60 to 73 m between successive capture sites, and the longest recorded movements were 108 to 125 m. An average distance between successive captures is 66 m, which would be a 33-m radius of a circular home range of 0.34 hectares. Although *Philander* are about the same size as *Caluromys*, the larger range of activity of the former could be a reflection of its more carnivorous diet. They probably have to search a wider area to capture prey than the more frugivorous woolly opossums. This author has live trapped *Metachirus nudicaudatus* in Colombia and noted that a marked individual appeared to move about 50 m a night in a series of three recaptures during a 7-day trapping period. This author knows of no extensive field studies of this genus. Table IV presents a compilation of the various home range sizes of didelphids from data collected by several authors.

### F. Population Densities

The density of a vertebrate population is in direct proportion to the productivity of the habitat and competition with other species. There are many complex interactions, including the behavioral capacities to tolerate other individuals in their home ranges, seasonal climate variations, local disruptive occurrences such as fire and floods, and a variety of other factors. In the higher latitudes, the most dramatic shift in population densities occurs during the warm weather of spring and summer when the peak production of juveniles occurs. Conversely, exceptionally cold and inclimate weather in the winter will reduce populations down to their lowest ebb, especially in the peripheral areas of the range.

Several field studies to determine the population densities of the Virginia opossum have been conducted. In eastern Texas, Lay (1942) found an average of one opossum to 0.8 hectares (2 acres) in wooded areas. On the sandy coastal prairie he found the ratio to be one in 0.04–5.9 hectares or one to 14.6 acres. Wiseman and Hendrickson (1950) in southeastern Iowa found six opossums per 259 hectares (1 square mile). By using the Lincoln Index method, Fitch and Sandidge (1953) in northeastern Kansas found one opossum to 8.1 hectares or 20 acres in the fall and half that number in the spring. Verts (1963) studied the density of opossums in a 6-square-mile study area of northwestern Illinois. Land was 95% under cultivation with corn, hay, and oats. He found that opossums in that particular habitat ranged farther than in a noncultivated natural habitat. He came to the conclusion that there were probably less than 10 opossums per 259 hectares (square mile), which he calculated to be one-third of the population densities reported by Fitch and Sandidge (1953) in Kansas, one-sixteenth of the population density reported by Lay (1942) in east Texas, and one-twenty-fifth of the population density reported by Sanderson (1961) in a wooded area in Illinois. Table V shows a summary of the population densities of *Didelphis, Marmosa,* and *Philander* as calculated by various authors. *Didelphis virginiana* appears to have the most variable densities, as we would expect from the extremely wide range and variety of habitats that this species inhabits. Throughout its total range, the Virginia opossum appears to have a density of approximately one opossum per 4 hectares in suitable habitat. The type habitat which has the highest population is a temperate woodland with numerous streams.

Fleming (1972) found *Didelphis marsupialis* in Panama to have a density of 0.09 to 1.32 per hectare, which roughly approximates that of the Virginia opossum in the north. Density, based on either production of young or number of animals captured, varied, with the greatest density occurring at the height of the rainy season and the beginning of the dry season, which in Panama is October through January. Based on total number of animals captured, *Didelphis marsupialis* appeared to have a denser population in the mature, moist tropical forest rather than the dry tropical forest with considerable secondary growth.

**TABLE V**

Population Density of *Didelphis*, *Marmosa*, and *Philander*

| Species<br>Reference | Study location | *N*/Hectare | *N*/Acre |
|---|---|---|---|
| *Didelphis virginiana* | | | |
| Fitch and Sandidge (1953) | N.E. Kansas | 0.12 | 0.05 |
| | | 0.06 | 0.025 |
| Wiseman and Hendrickson (1950) | S.E. Iowa | 0.02 | 0.009 |
| Lay (1942) | East Texas | 0.62 | 0.25 |
| Verts (1963) | N.W. Illinois | 0.04 | 0.016 |
| Holmes and Sanderson (1965) | East-Central Illinois | 1.16 | 0.47 |
| Stout and Sonenshine (1974) | Virginia | 0.05 | 0.02 |
| Hamilton, Jr. (1958) | New York | 0.07 | 0.03 |
| *Didelphis marsupialis* | | | |
| Fleming (1972) | Panama Canal Zone | 0.09–1.32 | 0.036–0.53 |
| *Marmosa robinsoni* | | | |
| Fleming (1972) | Panama Canal Zone | 0.31–2.25 | 0.12–0.90 |
| *Philander opossum* | | | |
| Fleming (1972) | Panama Canal Zone | 0.55–0.65 | 0.22–0.26 |

*Marmosa robinsoni* at Fleming's study sites were considerably more plentiful in the secondary growth than in the mature forest. With population levels highest during the dry season and at the end of the rainy season (November through April), Fleming estimated minimum density of this species to be from 0.31 to 2.25 per hectare. *Philander opossum* were present in the dry secondary forest at a density of 0.65 per hectare and in the mature forest at 0.55 per hectare, with a peak early in the dry season of January and February. The peak densities, based on trapping records of *Philander* in the mature forest, were February and in the secondary forest January. *Philander* does not appear to achieve the density of populations that *Caluromys* and *Didelphis* have. This is probably due to the more carnivorous habits and larger home range required by *Philander* than *Caluromys*.

Population density estimates based on trapping records must be carefully considered. The increase in the number of animals in a trap can indicate an actual increase in the population numbers, or just as easily, it can indicate a decrease in available food with the animals coming to the traps for the bait. Fluctuations in numbers of animals trapped also can indicate increased movements during the reproductive season when the males and females are breeding, or a decrease in activity when the females with pouch young are more sedentary. However, these data give us an accurate representation of the comparative densities of these species in the study areas mentioned and provide us with insight into the carrying capacities and adaptations to a variety of habitats that the didelphids have.

## III. PHYSIOLOGICAL ECOLOGY

### A. Thermoregulation

Marsupial body temperatures are usually higher than those of monotremes and lower than those of placentals. Traditionally, in part because they are primitive vertebrates, marsupials have not been considered to be as efficient temperature regulators as placentals. However, recent investigators have reported that in many respects marsupials are as efficient as the Eutheria but that they have variable body temperatures and lower temperatures associated with a reduced metabolic rate. Hoversland *et al.* (1975) stated that their data do not support the hypothesis that there are marked differences between the metabolic activity of marsupials and placentals, as suggested by Kinnear and Brown (1967).

The mean body temperature ($T_B$) of *Didelphis virginiana* is approximately 35.2°C. The Virginia opossum has a variable temperature depending upon the animals state of excitement, movements, and wakefulness. This accounts for the differences in the mean body temperatures estimated by various authors (Higginbotham and Koon, 1955; Morrison and Petajan, 1962; McManus, 1969; Dills and Manganiello, 1973; and others). Recorded $T_B$ of *Didelphis* ranges from 17.0° to 37.4°C under a variety of conditions.

Diel temperature variation in the body temperature of the Virginia opossum, *Marmosa robinsoni,* and several Australian species have been established. Dills and Manganiello (1973) recorded a 3.0°C average change in the body temperature of the Virginia opossum. They utilized a blocking ocillator transmitter planted in the body cavity to monitor the temperature of five male opossums. These animals showed the highest $T_B$ during their active periods at night; the maximum recorded was 38.9°C and the lowest $T_B$ during inactive periods throughout the day was the minimum recorded of 32.8°C.

Figure 15 shows a $T_B$ recording of female *Marmosa robinsoni* in Hunsaker's laboratory utilizing an implanted Yellowsprings recording thermometer probe. These *Marmosa* had a variable temperature with a mean $T_B$ of 31.9°C and a range of 28° to 35°C. The $T_B$ of resting or sleeping animals was 30.7° and 33.2°C in active animals. In a series of several animals measured, the fastest rate of temperature change was the increase of $T_B$

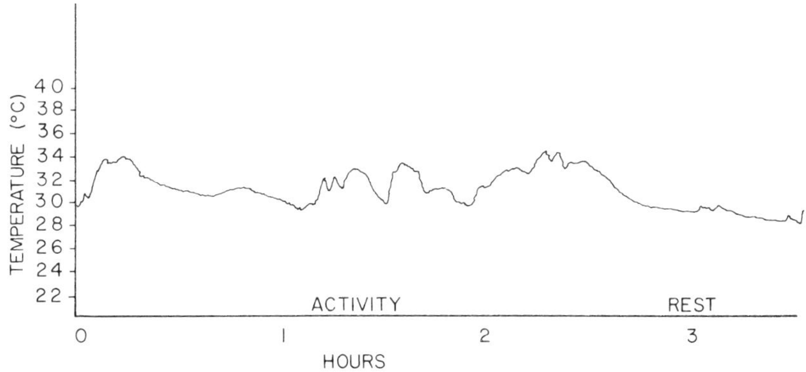


Fig. 15. Body temperature of *Marmosa robinsoni* in laboratory during periods of activity and rest. $T_A$ = 24°C.

when the animal was excited and moving. The maximum rate of change was 0.7°C per minute, which is close to the 0.5° rate of change Morrison (1965) recorded in Australian marsupials.

The extremely wide distribution of *Didelphis* in the Western Hemisphere attests to the fact that it is an effective homeotherm. Its nocturnal behavior protects it from excessive daytime heat, and it appears to be limited in its northern and southern distribution primarily by the very low temperatures in the high northern and southern latitudes. The Virginia opossum can efficiently control body temperature at ambient temperatures ($T_A$) ranging from 0° to 37°C. The Virginia opossum cannot maintain stable $T_B$ at $T_A$ of −7°C and rapidly loses heat at −10°C. Since they do not hibernate, this limits their foraging and survival abilities at low $T_A$. There have been several reports on temperature regulation of *Didelphis, Marmosa,* and *Metachirus* (Enders and Davis, 1936; Brown, 1909; Wislocki, 1933; Scott, 1938; Johnston, 1931; Britton and Atkisson, 1938; Britton and Klein, 1939; Morrison, 1946; Nardone *et al.*, 1955; Petajan *et al.*, 1962; Morrison and Petajan, 1962; Petajan and Morrison, 1962; Morrison and McNab, 1962).

The didelphids employ a variety of responses in order to effect thermoregulation. Among these responses are behavioral avoidance, evaporative cooling, vasoregulation, shivering, piloerection, and torpor.

McManus (1969) did extensive work on temperature regulation in *Didelphis* and found that there appeared to be no heat stress at $T_A$ 25°–35°C. Above this latter temperature he observed increased activity from $T_A$ 35° to 37°C, at which point the animals began to lick themselves and to salivate extensively. They licked their forefeet, snouts, and hind feet first, and as the ambient temperature increased, the abdomen, hind limbs, and tail were covered with saliva. At $T_A$ 40°C, McManus observed the highest level of saliva spreading, especially on the posterior region of the body. Urination and defecation occurred at these higher temperatures, and thermoregulatory activities were almost continuous. At and above $T_A$ 40°C, the heat generated by overt body activities was apparently greater than that lost due to the evaporation of saliva, and the licking activities were interrupted by periods during which the animals rested by lying on their sides. Although breathing was rapid and deep, the animals did not pant in the strict sense of the word. In the process of salivating and licking the pouch area, females with pouch young would cover their babies with saliva, but this did not appear to be so much a result of maternal care as of the position of the babies in the pouch. Severe temperature stress appeared above $T_A$ 41°C, with the animals spending the majority of their time lying on their sides breathing rapidly. When licking occurred, it was very rapid, almost frenzied, and was accompanied by extensive salivation.

Although the utilization of the evaporative heat loss via licking and salivation appears to be the most important modality in the effective control of high $T_B$, nevertheless the importance of behavioral avoidance cannot be discounted. The Virginia opossum is protected from excessive daytime heat simply by confining its foraging activities to the cool nighttime hours. In more than 1000 days of field work, Fitch and Sandidge (1953) reported seeing opossums abroad in daylight on only four occasions. In laboratory situations, opossums will utilize posturing responses in order to radiate as much heat as possible. This is especially noticeable when the animals are asleep. It is not uncommon to see opossums sleeping with mouth open, all four legs extended, and the tail directed straight out, thus providing for maximum air circulation around the body.

*Didelphis* and other marsupial genera (*Philander, Caluromys, Chironectes, Philander,* and *Marmosa*) have a well-pigmented scrotum which apparently serves to radiate heat. During high body temperatures, the cremaster muscle is relaxed and the scrotum becomes extended. Several authors (vanden Broek, 1910; Finkel, 1945; Biggers, 1966) have noted that *Didelphis virginiana* has a well-pigmented and pendulous scrotum. In all the New World and most of the Australian marsupials, the testes are permanently scrotal in the adults. The wombat has scrotal testes only in the breeding season and the marsupial mole (*Notoryctes typhlops*) has permanent inguinal testis (Stirling, 1891). Biggers (1966) suggested the melanin deposits of the tunica vaginalis could serve to enhance heat exchange between the testes and the environment. This probably does occur, especially to radiate heat from the animal to the environment. Heat conservation of the testes is accomplished by maintaining the scrotum in close association with the abdominal wall.

*Marmosa robinsoni* utilizes the same thermoregulatory responses at higher $T_B$ as does *Didelphis*. Body temperature of 35.2°C appears to be the upper limit of thermoneutrality for this species. Above $T_B$ 35.2°C *Marmosa* in this author's laboratory relied primarily upon active body movements in an attempt to escape from thermal stress, although considerable saliva spreading in the inguinal area was noted. The scrotum was noticeably extended from the body and there was no evidence of sweating. At $T_B$ 36°C, escape behavior consisted primarily of erratic leaping and crawling under pieces of paper or litter on the floor of the experimental chamber. At $T_B$ 36.5°C, the body movements consisted of circling, with no leaping or frantic movements. Above $T_B$ 37°C, frantic escape movements had entirely ceased, breathing was irregular with much rapid panting, muscular coordination and tone decreased, and the animal approached a lethargic state. *Marmosa* showed significant heat stress at $T_A$ 41.5°C, and 42°C is considered the effective behavioral lethal temperature, at which point the animal is unable to control its body temperature. We estimate that $T_A$ 43°–44°C is lethal, although this author never took his animals to the actual point of death.

Robinson and Morrison (1957) found that many Australian marsupials utilize saliva spreading to reduce $T_B$, and Morrison (1961), in a study of some Australian marsupials, reported that the short-nosed bandicoot (*Isoodon obesulus*) regulates heat by evaporative cooling. Bartholomew and Hudson (1962), in studies of the pygmy opossum (*Cercartetus nanus*), reported an increase in breathing rate at $T_A$ 38°C which resulted in a significant increase in evaporative water loss. Tyndale-Biscoe (1973) reports that the brush opossum (*Trichosurus vulpecula*) can maintain a $T_B$ of 39°C against $T_A$ 30°–44°C provided that humidity is low and that there is available drinking water. Dawson (1969) found that preventing *Trichosurus vulpecula* from licking did not affect its internal temperature. He also found that water loss through the skin made up only 10% of the animal's total water loss, while water loss through panting made up 90%. He concluded that the 400% increase in respiratory rate, which he noted occurring between $T_A$ 30° and 40°C, is the primary means of cooling.

Higginbotham and Koon (1955) reported panting, profuse salivation, and licking by the Virginia opossum at a rectal temperature of 38°C. They tested the effectiveness of saliva spreading by anesthetizing an experimental animal, thus preventing it from spreading saliva. The animal died at $T_A$ 44°C with a $T_B$ of 43°C, while control animals not under anesthesia survived the same $T_A$. Higginbotham and Koon concluded from their experiment that evaporative heat loss through the utilization of saliva spreading was the most important mechanism in the effective control of high $T_B$.

Tyndale-Biscoe (1973), however, considered Higginbotham and Koon's experiment to be an inadequate demonstration of the importance of saliva spreading, and pointed out that quokkas (*Setonix brachyurus*) and brush opossums (*Trichosurus vulpecula*), as studied by Bentley (1960) and Dawson (1969), did not show that salivation was an important part of their thermoregulatory ability. In observations on *Didelphis* and *Marmosa* under temperature stress in the present author's laboratory, it appears to the author that salivation and saliva spreading do represent important aspects of thermoregulatory behavior in the didelphids. The author agrees with Tyndale-Biscoe to the extent that additional research on the subject would be most welcome, and would represent an important contribution to our knowledge of thermoregulation of the didelphids.

Several workers have studied the responses of some didelphids to low ambient temperatures. McManus (1969) found differences between deep body and tail surface temperatures of *Didelphis virginiana* at low $T_A$. Tail surface temperatures varied widely, however, and appeared to be primarily dependent upon how much of the tail was beneath the animal's body. McManus concluded that the most important response of the Virginia opossum to low $T_A$ was vasoconstriction and reduced blood flow to the peripheral areas of the body, as evidenced by skin pallor. He considered behavioral avoidance by denning to be the next most important means of thermoregulation at low $T_A$, followed by postural responses such as tail curling and piloerection.

Fitch and Sandidge (1953) found that, in northwestern Kansas, extremely cold weather prevented the Virginia opossum from leaving its den to forage for food. Noting an increased mortality rate among live-trapped animals in early spring, they concluded that the animals' generally weakened condition due to reduced winter food supply had lessened their resistance to cold. Furthermore, second litter young, lacking the heavy layer of body fat found in healthy adults, were found to suffer an increased percentage of trap mortality on unusually cold autumn nights. However, Higginbotham and Koon (1955) found that a single 4-month-old animal maintained a constant $T_B$ of approximately 34.6°C against a $T_A$ of 10°C for a period of 1 hr as efficiently as adult animals. *Didelphis* does not forage at $T_A$ below −7°C and if accidentally exposed to temperatures lower than that will surely die after a few hours.

In southwestern Iowa, Wiseman and Hendrickson (1950) found that cold weather was one of the main mortality factors for the Virginia opossum in that area. They reported that one-third of the animals had frost-damaged ears, and several had frost-damaged tails.

Morrison and Petajan (1962) found the earliest appearance of thermoregulation in 60-day-old Virginia opossums (*Didelphis virginiana*) at $T_A$ 30°C. Thermoregulatory ability increased with age and development to the point that 90-day-old animals (about weaning age) could maintain a $T_B$ of approximately 34.5°C against a $T_A$ of 16°C. They noted considerable differences in individual response to cold stress, even among litter mates. These responses included body curling, tail curling, shivering, struggling, and even licking of the extremities. All of these responses, as well as the lack of overt physical response, may cause marked differences in body temperature. The author thinks that these factors should be considered when examining data based upon small sample sizes or limited repetitions in thermoregulation studies.

Nardone *et al.* (1955) recorded electrocardiograms and colonic temperatures in eight specimens of the Virginia opossum (*Didelphis virginiana*) which had been anesthetized with Nembutal and then placed in a refrigerator at $T_A$ to $-10$°C. They found that the animals could not maintain normal $T_B$ for more than 20 min, possibly because metabolic rate (as determined by heart rate) did not increase. The electrocardiogram indicated distinct irregularities from the normal pattern at 21°C. This feature of the Virginia opossum characterizes it as a nonhibernator.

Reynolds (1952) studied the beginning of effective temperature control in his physiological survey of the young. Direct measurement of the body temperature in response to a low ambient temperature, and measurement of the metabolic rate (as determined by oxygen consumption) at five different ambient temperatures, were used to quantify the study. In considering his data, Reynolds concluded that thermoregulation begins between the seventy-fifth and eighty-fifth days, but it is not completely effective during long periods of low ambient temperatures until the young opossums reach 90 to 95 days of age. Of the older animals that were able to control their body temperature, a minimum metabolic rate was observed at 30°C (roughly the lower limit of the zone of thermal neutrality). In interpreting Reynolds' graph of these 90- to 95-day-old animals, one can see that below 30°C, the amount of oxygen consumed is inversely proportional to the ambient temperature. Above 30°C, oxygen consumption increased slightly with an increase of ambient temperature.

Lustick and Lustick (1972) also investigated metabolism as a function of ambient temperature for a clue to thermoregulation in the Virginia opossum. Oxygen consumption was measured over an ambient temperature range of 2.5° to 37°C to determine the animals' metabolic rate. Their graph of oxygen consumption in response to different environmental temperatures resembled that of the oldest group of Reynolds' animals, except for a definite zone of thermal neutrality noted in the more recent study. Lustick and Lustick contend that the ability to increase heat production (metabolism) and not heat conservation (peripheral heterothermy) at low ambient temperatures, especially when the high thermal conductivity of the animals is considered, is the major factor in the opossum's homeothermic mechanism. They pointed out that the zone of thermal neutrality (29°–35°C) is high for a mammal of its size, but that it agrees more closely with that of similar sized tropical animals and that this can be expected because of the opossum's tropical origin.

For a tropical animal, this high thermal neutrality is advantageous as it minimizes thermal stress at the high environmental temperatures of the tropics. The thermal conductance of the opossum was found to be twice the expected value for a mammal of its size, another adaptation to the warm ambient temperatures of tropical nature. All factors considered, the physiological and distributional evidence suggest that *Didelphis virginiana* evolved from a tropical ancestor that has opportunistically spread from the warm environment and moved poleward to the limits of its thermoregulatory ability. It can be seen that outside of the zone of thermal neutrality the effort to maintain body temperature is energetically expensive under low ambient temperatures and hydrologically costly in conditions of high ambient temperatures.

Morrison (1946) reported that the Eten opossum (*Didelphis marsupialis caveae*) had an average $T_B$ of 35.5°C and could maintain a $T_B$ of approximately 34°C against a $T_A$ of 10°C. The brown opossum (*Metachirus nudicaudatus dentaneus*) was able to maintain a $T_B$ of 32°–34°C against a $T_A$ of 10°C. Morrison reported *Metachirus* had an average $T_B$ of 33.8°C during the day and 35.7° at night. Enders and Davis (1936) reported a $T_B$ of 33.9°C for the same subspecies of *Metachirus*. Enders and Davis (1936) also reported $T_B$ of 33.7°C for a female *Philander* (*Caluromys ?*) sp., 34.7°C for *Chironectes minimus*, and 35.4°C for *Philander opossum*.

*Marmosa cinerea* has been reported by Eisentraut (1960) to very precisely maintain an average temperature of 34.7°C during the day. In cases of great excitement however, extremes may go from 29.3° to 37.8°C or higher. Our studies of the mouse opossum (*Marmosa robinsoni*) showed that body temperatures will vary 5°C within a 3-min period. This species will also go into a torpid condition for a few hours at low $T_A$, and will lose all ability to control its body temperature.

Morrison and McNab (1962), studying daily torpor in a single specimen tentatively identified as *Marmosa microtarsus microtarsus*, found that the animal went dormant in response to cold stress. They concluded that this species might have adapted its estivation response, suitable to withstanding high temperatures in its normally tropical habitat, to the physiologically similar response of hibernation when its range extended into a cooler climate.

*Dromiciops* becomes torpid when the body temperature is lowered to 4.5°C (Mann, 1955; Greer, 1965). This is a lower temperature than that required for *Marmosa* to enter torpor, and could be an adaptation to the colder climates that *Dromiciops* lives in.

Temperature control mechanisms of the New World marsupials are similar to those in Australian species. Bartholomew and Hudson (1962) reported that, under laboratory conditions, the pygmy possum (*Cercartetus nanus*) would enter torpor merely upon being left by itself, regardless of lighting, $T_A$, noise, or activities in the laboratory. Morrison (1965), in a study of some Australian species, determined that the short-nosed bandicoot (*Isoodon obesulus*) regulates heat by evaporative cooling. The body temperature in this nocturnal animal reached a high of 37°C at midnight. This dropped to 36°C during the inactive period from 0600 to 1700 hours. The rabbit bandicoot (*Macrotis lagotis*) very precisely maintained a temperature of 37.1°C during its 9-hour active period from 2000 to 0500 hours. During the 15-hr inactive period, however, its body temperature was maintained much less precisely. It was also reported that during the inactive period, the rabbit bandicoot slept "profoundly" even when disturbed. If awakened and forced into activity during the day, its temperature would rise. At ambient temperatures between 30° and 40°C *Isoodon* showed effective $T_B$ regulation but *Macrotis* did not. A plausible explanation of this offered by Morrison (1965) was that in the desert environment in which the rabbit bandicoot lives it would be disadvantageous to dissipate a lot of water.

The Virginia opossum has reached the northern limit of distribution near the −2° to −7° January isothermal line which roughly approximates its range boundary in the northern and eastern parts of the United States. The restrictions to this and warmer temperature is simply one of thermoregulation and physiological adaptations to a tropical temperate rather than a colder climate. Since the Virginia opossum does not hibernate, the energy budget to maintain body temperature at 0°C ambient temperature requires a considerable amount of energy expenditure. The calories to replace this energy must be supplied from food which is foraged rather than stored. This input of food is just not possible during extended periods of cold weather and the individuals die.

The physiologically controlled limitations of water balance and temperature control appear to be the limiting factors of range expansion in *Didelphis virginiana* and probably other members of the family also. Because of the moderate renal concentrating ability and the lack of hibernation to carry them through the winter, the population of Virginia opossums appears to have reached the limits of its present distribution, except possibly in a few localities in Arizona and the Midwest.

### B. Water Balance

The Didelphidae have good control over their water balance and most of them live in habitats where water conservation is important but not critical.

The Virginia opossum is almost always within 300 m of surface water and drinks often. Captive opossums require a considerable amount of surface water to avoid dessication. In the laboratory, pans of water are always available for opossums since their intake is insufficient when they only have drip watering devices. This suggests that surface water is necessary for survival, and they cannot compensate for loss by intake of water with food or by metabolic manufacture.

Water within the body is a result of ingestion or a product of metabolism. It may be ingested as part of a food or as a free liquid. Water is lost through urination, defecation, evaporation from the cells of the skin and lungs, sweating, and related thermoregulatory mechanisms. Evaporative cooling through saliva spreading is very expensive since it significantly depletes water reserves which have to be replaced. It is especially costly where free water is not available. Since didelphids utilize saliva spreading when under heat stress, it indicates that the family probably evolved in a habitat that had adequate levels of surface water available. Behavioral avoidance of solar incidence can be seen as an attempt to conserve water in many species, especially the nocturnal ones, but the physiological adaptations also determine an animal's existence in a given area over extended periods of time. The adaptations and mechanisms controlling the delicate balance of the process in marsupials have not been completely researched and further studies are needed.

Water balance in the Virginia opossum has been researched by Plakke and Pfeiffer (1970). Opossums were fed diets of varying protein and electrolyte levels and later deprived of water. The authors found the opossum is able to raise its urine concentration capacity to compensate for reduced water intake. They can concentrate urea more effectively than electrolytes; this is a renal response also exhibited by a variety of species including man, dogs, kangaroo rats, rabbits, and muskrats. This is in contrast to the physiological abilities of the beaver and pig which employ a fixed urine osmotic ceiling. A high protein diet causes increased amounts of nitrogenous waste products in the form of urea in the urine. The Virginia opossum and other didelphids are able to raise their urine osmotic ceiling to further concentrate the urea and eliminate it with a relatively smaller quantity of water. This results in the conservation of water and enables some species to inhabit arid habitats. Plakke and Pfeiffer and others indicated that there is a good correlation between the relative medullary thickness of the kidney and the relative length of the loops of Henle with the ability to increase urine concentration. Animals with adaptations for more effective urine concentration characteristically develop their kidneys with inner and outer medullary zones comprised of grouped nephron segments. The inner zone and/or enlarged pyramids aid in increasing the urea concentration of papillary tissue by exchange of urea across the epithelial lining. Although the Virginia opossum has a relatively thick medullary area, it does not exhibit exceptional urine concentrating capacities. Zonation of the medulla is also evident in *Trichosurus vulpecula,* which possesses a kidney with a functional capacity remarkably similar to that of the opossum (Tyndale-Biscoe, 1973). Both *Trichosurus vulpecula* and *Didelphis virginiana* are mesically adapted species.

The ability to concentrate urine is undoubtedly an advantageous adaptation in desert inhabiting species and can limit the distribution of a species which is not as efficient. Compared to the xeric adapted macropods and other desert species, the opossum can be considered to only moderately concentrate urine to conserve water. This limitation on *Didelphis* is probably why they are restricted to riparian habitats and cannot exist in arid sections. *Marmosa* can maintain its water balance with reduced water intake to a greater extent and does occur in areas where *Didelphis* appears to be absent because of insufficient water availability.

As a part of his extensive research on *Didelphis virginiana,* Reynolds (1952) examined the dehydration rate of the young. The young were exposed to a rapid flow of dry air at a temperature comparable to the marsupium. The weight loss from the time of initial exposure until the animal became inanimate was the standard used to ascertain dehydration rate. Two individuals 6 and 16 days of age were used in the experiment. After roughly 8 hr of exposure and 30% loss of body weight, they became motionless. Upon moistening, respiratory movements of the 6-day-old young resumed. Reynolds believes that, under normal conditions, there is little danger of lethality due to dehydration of the pouch young since water loss (by dehydration) is more than compensated for by milk intake.

Kennedy and Heinsohn (1974) investigated water metabolism in the brush-tailed possum, *Trichosurus vulpecula,* and rock wallaby, *Petrogale penicillata,* under captive and wild conditions. They found the water turnover of these species, regardless of season, remained rather constant. Green grass remained on the study area and was available to the rock wallaby for consumption throughout the wet and dry seasons of the study period. Similarly, green leaves were attainable by the brush-tailed possum for the duration of the year. Both the rock wallabies and the possums also ingested moisture from dew on the leaves and grasses.

Since rock wallabies were never observed near the creek, Kennedy and Heinsohn concluded that these wild animals did not drink free water. They also noted that animals maintained on a diet of green grass (without drinking water) did not loose weight and turned over amounts of water similar to animals in the wild. Possums were found near the stream bed in the study area, but since the water turnover rates approximated those of the rock wallaby, Kennedy and Heinsohn considered that both the rock wallaby and the possum apparently can maintain a satisfactory water balance with availability of sufficient green feed and without free drinking water.

There is no doubt that the macropods can exist in more arid conditions than the didelphids, especially *Didelphis*. Natural selection has produced a much more efficient water balance system in *Trichosurus*, *Petrogale*, and a variety of other species which do well under deserticulous conditions. Some of the *Marmosa* species can approach this efficiency, but in general they are adapted to a reasonable amount of surface water.

*Didelphis* appears to be a lowland tropical genus that has invaded the temperate zones which have sufficient available water to support them. The deserts of northwestern Mexico and the west of the United States served as barriers to the western expansion of the range. It was not until man transported them across these barriers that the opossum rapidly filled the extensive habitat of California, Oregon, and Washington, all areas with surface water available. In all probability, it was water balance rather than thermal problems that restricted planted populations in Arizona from becoming well established. When *Didelphis* establishes itself in Arizona and other arid areas, it will probably be closely associated with riparian and irrigated habitats near the cities or farm areas.

## IV. REPRODUCTIVE CYCLES

Studies conducted in the United States have shown that the opossum (*Didelphis virginiana*) usually produces two litters of young per year (Lay, 1942; Reynolds, 1945, 1952; Fitch and Sandidge, 1953; Llewellyn and Dale, 1964). An earlier study, however (Hartman, 1928), reported that unusually fecund females in some southern states may produce a third litter. G. Cadzow (personal communication) and the author have both noted the possibility of an occasional third litter period in southern California, but it is unknown if this is the result of a female giving birth to three litters a year or if some of the population give birth later than others. The 9-month breeding season beginning in mid-January and ending in mid-October (the period from the first estrus to the time that the last young of the year are weaned) seems to be representative of the United States population in general, but it does vary with localities.

In southern California, Cadzow and the author have independently captured female opossums with pouch young as early as the second week of January. Embryons at such an early date would indicate that the onset of the first estrous cycle would be during the last week of December. Although more common in southern areas, an occasional early breeding female is observed in the colder climates. Grote and Dalby (1973) reported a female Virginia opossum with pouch young that they estimated were conceived in the first week of January.

Lay (1942), in eastern Texas, found young in the pouch from January to July. The first estrous cycle was reported to begin about the third week of January. Most of the first litter young are therefore born in February. Second litters were born from early April to as late as the third week in May.

In central Missouri, Reynolds (1945) reported that the first litters were born from February 12 to April 2. The second litters were born from May 16 to June 4. Eight of the 42 litters in this study or about one-fifth were born 6 to 9 weeks after the beginning of the breeding season. His explanation for the delay was that possibly these females had failed to mate during their first estrous cycle, but did mate some time after their regular diestrus of about 28 days.

Fitch and Sandidge (1953), in northeastern Kansas, found that although there was variation resulting from climatic differences and differences in individuals, most of the population, including less than full grown young of the preceding year, began breeding about the middle of February. These newly born young remained in the pouch for approximately 2 months. Shortly after the first litters had vacated the pouches in early June the second litters were born. March and June were also found to be the average times for the births of the first and second litters by Llewellyn and Dale (1964) in Maryland. Stout and Sonenshine (1974) reported pouch young in a population in Virginia from March to August.

In the Virginia opossum, there is a great difference in litter size at birth, pouch litter size, and the number of young that are weaned. The 0.75-mm ova are released during ovulation and are fertilized in the fallopian tubes. While in the tubes, albumin and shell membranes are added and the fertilized eggs reach the uterus within 24 hr. The first mortality occurs in the attachment process since there is an average of 13 teats, yet usually 21 embryons are born. Hartman (1952) reported as many as 56 embryons born to a single female. Not all the nipples are functional, and as a result the species sustains about a 35% mortality in the attachment process which results in a final mean litter size in the pouch of 7.5. In eastern Texas, on the basis of 65 litters, Lay (1942) found an average of 6.8 young per litter. In central Missouri, Reynolds (1945), on the basis of 42 litters, found the average litter size was 8.9 with a range of 5 to 9. The average litter sizes found by Wiseman and Hendrickson (1950) in southeastern Iowa, Fitch and Sandidge (1953) in northeastern Kansas, and Llewellyn and Dale (1964) in Maryland were, respectively, 9, 7.4, and 7.74.

There is a definite reduction in litter sizes from northern to southern latitudes. Burns and Burns (1957) and Fleming (1973) have referred to this. The reproductive potential of *Didelphis* is about the same regardless of the latitude, but there is a considerable difference in litter size and number of litters per year. In New York, litter size averages 8.66 (Hamilton, 1958), and the females have the major breeding period in the spring and a minor peak in summer. Further south, in Florida, litter size averages 6.26 (Burns and Burns, 1957), and there are two distinctive periods of reproduction. Fleming (1973) reported a 6.0 average litter size for *D. marsupialis* in Panama, and Tyndale-Biscoe and Mackenzie (1976) reported the smallest litter size for the genus, 4.2, for *Didelphis albiventris* in Colombia, but they produce three litters per year. Farris (1950) showed that photoperiod was a factor in estrus in the laboratory, so it probably functions in nature also. This, in addition to the rainy/dry seasonality and nutrition, probably plays an important part in the estrous cycles of didelphids.

**TABLE VI**

Tabulation of the Breeding Season and Reproduction in Some New World Marsupials

| Species | Reproductive period |
|---|---|
| *Caluromys derbianus* | In Central America commences during dry season (January to June) and into early parts of the season of heavy rains (July to December). Breeding all year in some areas. Average litter size is 3, range 1–6 |
| *Chironectes minimus* | Reproduction from November to February. December to January in Brazil. Average litter size of 3.5, ranges up to 5 |
| *Didelphis virginiana* | Begins in southern United States during the first week of the year and reaches peak in the third week. In central California breeding begins during third week of January and in February. In Missouri, begins during the fifth week of the year, with the greatest number during mid-February. Two litters per year in north of range, 3 in some areas of the southern part of the range. Gestation period of 13 days, average litter size of 7.5, range at birth is 8–18 |
| *Didelphis marsupialis* | From Panama to Colombia breeding begins with onset of the dry season in January. Second litters in April and May. Some third litters recorded in August from higher elevations in Colombia. In Brazil, some litter in August and October. Average litter size is 6.5 |
| *Didelphis albiventris* | Limited data indicates February to September breeding season, 4.2 average litter size |

**TABLE VI (Cont.)**

| Species | Reproductive period |
|---|---|
| *Dromiciops australis* | Breeds October through May, pouch young in November and December. Female with well-developed mammae in May. Male and female become sexually mature in second year. Litter size ranges from 2 to 5 |
| *Lutreolina crassicaudata* | Twice a year, once in the spring and again when the young of year become independent. Gestation is 14 days |
| *Marmosa* sp. | Breeds throughout the year in tropical regions, and diminishes as the range extends into cooler and temperate zones to 1 time per year. Gestation period is 14 days |
| *Marmosa robinsoni* | Two peaks during the year, February to June and July to December. Average litter is 8; gestation 14 days |
| *Metachirus nudicaudatus* | Breeds in November in Central America. Five mammae; litters of 1–3. Can raise single young. Litter size up to 9 |
| *Monodelphis* sp. | Breeds throughout the year. In tropical areas litter size, depending on species, varies from 8 to 14 individuals. A female *M. brevicaudata* had 5 uterine embryos on May 23 in Venezuela |
| *Philander opossum* | Reproductively active at all times of the year, but peaks during March to August. Females with pouch young from February to October. Litter size averages 5, ranges from 3 to 7 |

Fitch and Sandidge (1953), in northeastern Kansas, found the first litter to be the most important as far as the population success was concerned. They reported that 67% of the breeding females in the area had been marked as first litter pouch young the preceding year. Llewellyn and Dale (1964), in Maryland, also found the first litter to be the most productive; 63 of the 105 young that he found were from the first litter. A definite interpretation of these findings is lacking, but it has been suggested by Fitch and Sandidge (1953) that some females may be eliminated after the first litter and before the second. Others may be too physiologically exhausted from carrying the first litter to produce the second. Climate and nutrition are undoubtedly important factors in the production of multiple litters.

Table VI shows the reproductive period for the didelphids as reported by a variety of authors. Only sketchy data are available for most of the species in the New World from the few investigators who have studied or reported on reproduction in them (Cabrera and Yepes, 1940; Collins, 1973; Davis,

1945; Enders, 1935, 1966; Fleming, 1972; Hall and Dalquest, 1963; Phillips and Jones, 1968; Tate, 1933).

## V. GROWTH, LONGEVITY, AND SURVIVAL

The growth and maturation of *Didelphis* is more rapid than for many other omnivorous or carnivorous species of the same size and habits. Raccoons, skunks, and other carnivores require a few weeks longer from conception to independence than the 4-month period for opossums and reach sexual maturity later. However, the lower metabolism and other factors produce a different relationship with the smaller herbivores. Tyndale-Biscoe (1973) has pointed out that although a quokka and a rabbit reach adult size at the same time, the biotic potential of a rabbit is much greater. He calculated that a female rabbit can produce 35 young in 240 days, while the quokka can only produce one young that has less than one-third the weight of the rabbit progeny. The opossum then has a biotic potential much closer to the rabbit than the quokka.

At birth, the young of *Didelphis virginiana* are at a truly embryonic stage of development. They weigh about 0.13 g apiece and are only 10 mm in length. Compared to other mammals, this is a remarkably small size. Reynolds (1953) pointed out that if a human infant was born at a comparable stage it would weigh as much as a U.S. nickel. The opossum sex ratio at birth has an equal number of males to females. The embryons' development in the pouch proceeds at a rapid pace. At 20 days the young are 40 mm in length and weigh almost 2 g; by 40 days they are 50 mm long and weigh about 7 g; at 60 days they are up to 20 g and are 95 mm in length. By this age, they are actively crawling over the mother, can walk quite well, and can reattach themselves to the nipple with ease. Sexual maturity is reached within the first year after birth (Hartman, 1923a), and opossums can breed at 6 months when they weigh as little as 650 g (Hartman, 1923b). Maximum weight of females appears to be 3000 g and maximum body length is 760 mm. Males are larger, up to 5500 g and 850 mm total length. As can be found with other mammals, there is a large variation in the size and weight of the species depending on latitude. Specimens from Central America, Mexico, and southern United States are less than half the size of animals in southern Texas and adjacent Mexico. Two thousand grams appears to be maximum for *Didelphis marsupialis* in Colombia and South America. Figure 16 gives a representative growth curve for *Didelphis*.

During the period of suckling, the female produces milk that is higher in solids, fats, and proteins but lower in carbohydrates than typical eutherian

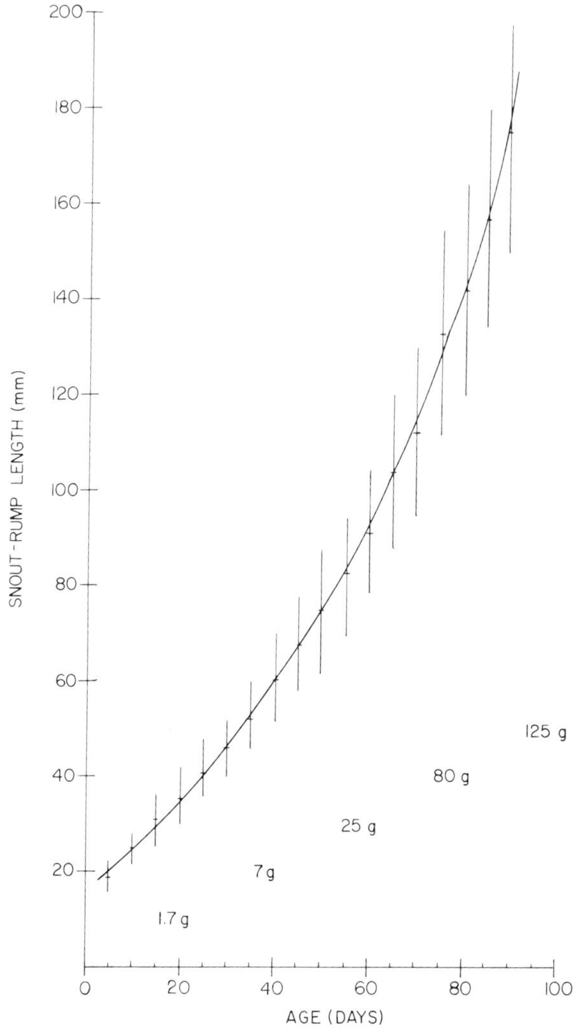


Fig. 16. The average growth of length and weight of *Didelphis virginiana*. The growth is a composite from the author's observations and published records. The variability is a reflection of geographic and individual differences in growth rate.

milk. Barker *et al.* (1967) and Bergman and Housley (1968) published analyses of opossum milk and stated that marsupial and eutherian milk also differ in the nature of their carbohydrate components. Galactose is a primary constituent of marsupial milk and lactose is the main constituent of eutherian milk (Jenness *et al.*, 1964; Barker *et al.*, 1967). This condition is also found in the following Australian species: *Macropus rufus* (Jenness *et al.*, 1964), *Trichosurus vulpecula* (Gross and Bolliger, 1958), *Setonix brachyurus* (Jenness *et al.*, 1964), and *Macropus robustus* (Bolliger and Pascoe, 1953). Barker *et al.* (1967) did not find lactose in *Didelphis* milk and considered the Jenness *et al.* (1964) report to be in error when it said lactose was present in *Didelphis*. Barker *et al.* (1967) also did not find free glucose but considered it to be bound and present with bound galactose in one or more of three oligosaccharide spots on his chromatogram. He reported the average ratio of bound galactose to bound glucose to be 5:1.

The life of the pouch young is a hazardous one, beginning with the long journey into the pouch on day 1. This blind and naked creature must grasp and crawl its way from the vagina to the marsupium, an average distance of 5 cm, with no help from the mother other than the prebirth preparation which consists of licking and cleaning the genital and pouch area and the removal of the amnionic membranes. The embryon has small, deciduous claws on the front digits which assist it to grasp the hairs of the mother's abdomen. These are shed within a few days after attachment to the nipples. The front limbs are much more well developed than the hind ones and locomotion is maintained only by the forelimbs. Once inside the pouch the embryon's chances of survival are even slimmer as it still must locate and enjoin itself to one of the 13 nipples. As opossums generally bear more than 13 embryons in one litter, nipples are at a premium and an embryo that does not immediately develop its grasp of the teat may be knocked or pushed off by a competitor. Once firmly attached, the embryon will remain so until about 55 to 60 days of age when it begins to develop an awareness of the outside environment.* During this period of pouch life, the average litter size of *D. virginiana* is 7.5. The young opossums do not even begin to lick solid food before the seventy-fifth day and are weaned at about 100 to 110 days of age. It is very unlikely that they ever leave the female before the ninety-fifth day and some will still be nursing after 110 days of age.

The female begins to cease any maternal care and the young disperse. In females that have produced a litter in the early spring, the second litter is born on the average of 110 days after the first, so it is important to the survival of the species that the first litter is weaned and dispersed before the second is born. This aids in reducing sibling competition as well as avoiding predation of the young by the adult males when they are seeking out the females in estrus.

*Didelphis virginiana* that survive through weaning probably live for about 2 years in the wild. After the second year of maturity, females apparently

* The pouch young are very tolerant of low $O_2$ levels and are highly resistant to asphyxia (Rink and Miller, 1967), yet they are also tolerant of high levels of $O_2$ (Parmelee *et al.*, 1960) without developmental problems.

cease to reproduce and a general degradation of vitality occurs in both sexes. A captive specimen was reported to live for 9 years at the New York Bronx Zoo (Jones, 1963), and several laboratories have maintained individuals for 3 to 5 years. Crandall (1964) noted that it was difficult to maintain accurate records of the numerous opossums in the zoo, so the 9-year longevity could be a mistake in the records. Hartman (1923b) suggested that they lived for at least 7 years in the wild and McCrady apparently thought they lived for this length of time also. However, these ages would be highly unlikely in the light of more recent investigations. Observations from studying both wild and captive populations by the author and others indicate that the longevity and reproductive activity of opossums is short compared to similar sized animals. These studies show the opossum is mature at 9 months and is reproductively fit for at least 1½ years and probably not for over 3 years. However, as noted by Reynolds (1952), Hartman (1928), and Jurgelski and Porter (1974), the reproductive ability of the opossum declines after the first reproductive year, as evidenced by atrophy of the female reproductive organs, decreased litter size, and nonproduction of young. Not only does the birth process become more of a hardship on the female, but maintaining the young becomes a major problem as she gets older.

Llewellyn and Dale (1964) conducted an 8-year study in Maryland, found the maximum trap life to be 29 months, and estimated the greatest age a wild opossum would attain would be 3 years. Very few of their animals survived more than 2 years. Although the nomadic activity pattern could account for a shorter period of time in the study area, the author considers that these data suggest the species has a very rapid turnover rate for individuals in the population.

Fitch and Sandidge (1953), in northeastern Kansas, found that there was an annual turnover rate of 95% of the breeding population. They estimated that 70% was due to predation, disease, or accidents and the remaining 25% to shifts in home ranges. In this study, 67% of the breeders examined were progeny of the previous year.

Petrides (1949) calculated the average age of an opossum in the wild to be 1.33 years, based on the number of young that lived to the end of the first year and the percentage of young present in the population, which agrees with Llewellyn's computations. Petrides calculated a turnover period of less than 1% of the original opossum population for 4.8 years.

Survival strategy of *Didelphis* is based on a high production of young to compensate for early mortality. The average number of eggs produced is 21, maximum 56 (Hartman, 1952), of which only 35% will survive until attached to a nipple and partially through pouch life. The average litter size in the pouch is a function of mortality that includes defective and unfertilized eggs, embryons that are lost in the birth migration, and those which attach to nonfunctional nipples.

Estimates of mortality while in the pouch vary from 23% by Fitch and Sandidge (1953) to 12% estimated by Sanderson (1961). Hamilton, Jr. (1958) claimed that loss was minimal while the young were in the pouch. Sequential counts of young in recaptured females do indicate that mortality is about 15 to 20% and probably can be higher or lower in some populations. Consideration also has to be given to pouch young mortality when a female dies with all her young or is eaten by a predator.

In all probability owls and man are the greatest predators of opossums. In the opinion of this author, the weather, nutrition, and disease probably account for the majority of natural deaths of adult opossums. Many authors and almost everyone that have trapped significant numbers of the animals have seen diseased, heavily parasitized opossums and individuals with wounds and infected areas. Coyotes probably do not account for heavy predation pressure on opossums. In an interesting experiment on coyote food preference, an investigator gave a series of coyotes a choice of food and recorded their preferences. They did not find opossums palatable and would not eat them when other food was available. With their anal secretions and large teeth, it is understandable that they would be less than the perfect prey species. Many authors have suggested that coyotes find opossums unpalatable, but some definitely do eat opossums. Owls are a significant predator and have been recorded as major predators by several authors. Fitch and Sandidge (1953) recorded an owl eating an opossum caught in the live trap.

Tropical species have a correspondingly greater degree of predation by carnivores and man and less death due to cold weather. *Caluromys* were extensively hunted for the fur trade in Central and South America. *Lutreolina* is undoubtedly an important predator of smaller marsupials and it in turn was heavily hunted for pelts in the past. This author has seen *Metachirus nudicaudatus* in the stomach of a Colombian fox, *Dusicyon thous*, *Caluromys derbianus* electrocuted in electric lines, and Wilson (1970) reported predation on *Metachirus* by *Didelphis* in Panama. Coatis, ocelots, and other carnivores also prey on the tropical species.

The population turnover is rapid and Petrides' (1949) estimate of an average age of an opossum in the wild of 1.33 years is probably accurate for northern areas with harsh winters. The more temperate south would afford a somewhat longer life span, possibly up to 2.5 years in southern climates. The success of the species is a result of a high biotic potential, an omnivorous food consumption, and a remarkably generalized and opportunistic life style.

Fleming (1972) reported on some aspects of the population dynamics of

*Didelphis, Philander,* and *Marmosa* in the Panama Canal Zone and found they have the same general pattern as *Didelphis virginiana.* In Panama, all the species produce young when food production is high and selection is strong to produce numerous offspring. There is no differentiation in reproductive season to reduce competition. This is probably due to the seasonal rainfall, food material, and general productivity of the Panama forests.

Growth and maturation data are not readily available on the other species. Some have been accumulated for *Marmosa robinsoni* in captive colonies (Barnes, 1968; Barnes and Barthold, 1969), but little or no field data have been collected. Adult males weigh about 75 g, females 50 g. The estrous cycle averages 23 days (ranges from 18 to 31 days); vaginal estrus is three days. There is a 14-day gestation period. Litter size in nature appears to be from seven to nine young. The females of *M. robinsoni* have 14 nipples. Litters are slightly smaller in captive animals, with an average of several laboratories of about 6.0. Males are larger than females and grow faster. In this author's colony at San Diego, young were 60–100 mg at birth. Males weighed 20 g, and females 17 g by 3 months (12 weeks) of age. Males doubled this weight by 6 months of age. Adult weight was gained at 11 months in males (75 g) and adult weight of females (50 g) gained at 1 year of age. The young are weaned at about 8 weeks (60 to 70 days, 10 g) and leave the female at 69 to 74 days of age. First estrus is at 265 to 275 days of age (40 g). Reproductive ability fades during the second year and life expectancy is probably less than 1 year in nature and considerably longer in captivity, about 3 years. The population turnover is probably very rapid and the species utilizes maximum high production of young to compensate for high mortality.

There have been few research projects involving the life span of the New World marsupials. Most of the data on longevity are from zoo records and other captive animals. Collins (1973) did an excellent job of summarizing the available information on captive monotremes and marsupials in zoological parks. In general, most of the didelphids appear to have a reasonably short life span.

Maximum life spans listed by Collins and other authors are as follows: *Caluromys,* 5 years, 2 months; *Caluromysiops,* 7 years, 10 months; *Marmosa,* 5 years (Enders, 1966), others average 1.5 years; *Philander,* 3 years, 6 months (Farris, 1950); *Metachirus,* 1 year, 2 months (Collins, 1973), but probably 3 to 4 years; *Lutreolina,* 3 years; and *Chironectes,* 2 years. Since most of the information suggests similar ecological adaptations as far as reproduction and survival are concerned, the survival strategy of *Didelphis* is probably representative of the family Didelphidae. *Didelphis* combines the protective value of large body size against the environmental influences of weather and predators with a remarkably fast generation time and high fecundity values. Most omnivores of this size range reproduce only once a year and produce smaller litters than the Virginia opossum. This combination of characteristics has proved to be very successful, and as a result *Didelphis* occupies a unique position among the vertebrates in that, for its size, it has a high biotic potential which offsets the greater levels of environmental resistance of the physical environment that temperate species encounter.

In variable habitats such as that of the temperate zone, survivorship appears to be controlled more by density independent (physical) factors of the environment than by density dependent (biotic) factors. This can be exemplified by the massive winter kills commonly seen in many animals. Invertebrates are particularly sensitive to these kills but birds and fish are commonly affected. The species involved have a high reproductive rate to fill the void in the population when conditions return to normal. Because of the lack of hardships imposed by competition and high densities, selection in individuals under these conditions favor high fecundity and rapid development, referred to as "b" selection by Hairston *et al.* (1970). Therefore these organisms typically have low competitive ability, are unspecialized, comparatively small and therefore energetically inexpensive to produce, and individually occupy broad niches. In the relatively constant conditions of the tropics, the biotic community imposes density dependent adaptations on the individual. The organism's energy is most economically used in producing fewer and larger progeny with high individual fitness (Dobzhansky, 1950). High densities and competition cause adaptation toward increasingly specialized stocks that replace lost individuals.

These two strongly opposed tactics are known, respectively, as $r$ and $K$ selection (MacArthur and Wilson, 1967). These strategies suggest that an $r$ (referring to $r_{max}$, the maximal intrinsic rate of natural increase) selected species operates more in a populational void and has low competition levels with conspecifics, and that a $K$ (referring to the carrying capacity) selected species attains large size, has a generation time of more than 1 year, and contends with higher levels of competition. Although it does hold true that many temperate species tend toward the $r$ end of the continuum and many tropical species tend toward the $K$ end, there are species that do not adhere to this pattern. As Pianka (1970, 1974) has pointed out, no animal is totally $r$ or $K$ selected, instead, the organism's strategy lies someplace along the $r$ and $K$ continuum relative to all other organisms and therefore may be better classified as a $r$ or $K$ strategist. There is a shift from $r$ to $K$ selection as the phylogenetic order progresses from primitive to advanced or from $r$ strategists, as generally represented by invertebrates, to $K$ strategists or vertebrates. Pianka has noted that this generalization does not hold true for aquatic organisms because they occur throughout the *r–K* selection continuum.

On the basis of the opossum's maximal rate of intrinsic increase ($r_{max}$) and generation time (good indicators of an organism's relative position along the *r–K* continuum), *Didelphis* appears as a strongly *r* selected strategist as far as mammals are concerned. The $r_{max}$ value of the opossum closely approximates those of *Rattus* and *Microtus*, which are smaller nonprimitive mammals and nonseasonal breeders. *Didelphis* employs both rapid development and high fecundity as selective features of its reproductive strategy. Intra- and interspecific competition are apparently minimal. Opossums occupy a wide food niche and are opportunistic omnivores that will eat almost anything. This behavior aids in reducing competition. As is indicative of *r* strategists, opossums exhibit an extremely steep juvenile mortality curve but have relatively high survivorship thereafter. Predation, disease, accidents, and unusually severe weather conditions are the common mortality factors for adult opossums.

*Didelphis* are not as phylogenetically primitive as some would have it. They are much closer to placentals than to the monotremes and, as we have shown previously, are very advanced in many characteristics. Neither *D. virginiana* nor *D. marsupialis* are *r* strategists because they are phylogenetically primitive. They have undergone genetic selection to adapt themselves for rapid dispersal, colonizing ability in marginal areas, to reduce competition, and probably a variety of other reasons which resulted in the reproductive strategy they have at the present time. There is also no reason to conclude that the same factors have operated equally in the two species of the genus to produce this type of life style. *Didelphis virginiana* is a temperate species which has been subjected to environmental extremes of temperature in the northern expansion of its range and the lack of water in the arid west. It is probably a very recently evolved species (from *D. marsupialis*), and dispersal and rapid colonization are important to the species during the periods of milder climate when they are able to cross an ecological barrier to a satisfactory habitat. There are few competing species in the temperate region where the Virginia opossum has been successful.

The species richness and environmental stability of the tropics present a completely different set of pressures for *Didelphis marsupialis*. The numerous competing species which are more specialized could well have provided the necessary selective pressure for evolution toward *r* strategy. John Kirsch (personal communication) also pointed out that if *Didelphis marsupialis* did in fact give rise to *virginiana*, then the Virginia opossum could have been a preadapted *r* type species when it arrived in the temperate zone.

In any case, we should consider the evolutionary ecology of these "mammalian weeds" as a product of selection due to environmental factors rather than a function of a primitive phylogenetic position, as some consider it.

## VI. SUMMARY

The family Didelphidae (including Microbiotheriinae) is represented by 12 living genera; there are three in the Caenolestidae. They are generally forest dwellers that have adapted to a variety of habitats, from deserts to high Andean paramos. *Chironectes* is a specialized aquatic animal and *Lutreolina* is a large, mustelidlike species inhabiting the grasslands. All the species appear to be omnivorous, with some species selecting more insects and animals in their diet while others tend to be more frugivorous. Little information is available on home range size and density of many of the genera. *Didelphis virginiana* has an average home range of about 20 hectares (50 acres) in a favorable habitat in the United States. Density of *Didelphis virginiana* averages 0.26 opossums per hectare (0.1 per acre) with a range from 0.02 to 1.16 per hectare (0.009–0.47 per acre).

Marsupial body temperatures are lower than placentals and are subject to considerable variation. *Didelphis* has an average body temperature of 35°C. *Didelphis* and *Marmosa* have a diel fluctuation in body temperature, higher during the night when they are active and lower during the day when they are inactive. The Virginia opossum can control its body temperature at ambient temperatures from 0° to 37°C and rapidly lose heat at ambient temperatures below −10°C. Others like *Marmosa* have more labile body temperatures and enter torpor at low ambient temperatures. Water balance of the Virginia opossum is maintained by high levels of surface water intake. Its water conserving ability is about what one would expect for a mesic, temperate-adapted species. It is seldom found beyond 300 m from surface water and is limited in its distribution by arid habitats. *Marmosa* is well adapted to more xeric habitats and can be found in cactus communities and wet tropical forests as well as the intermediate habitats.

Reproduction occurs all year round in the tropics and two or three times a year in temperate areas. *Didelphis virginiana* begins to breed in the latter part of January in the United States, earlier by 2 weeks in the south, and later in the northern states. Average litter size is 7.5. Apparently, there are larger litter sizes but fewer litters in the northern part of the range than in the southern part. Most of the tropical species, even the sympatric ones, begin to breed at the onset of the dry season, about February, and show two distinct peaks of reproductive activity. The largest peak is the first and the later one, from August to September, is somewhat lower. Some species breed all through the year in tropical areas.

Mortality is heavy during the neonatal period with a 30% estimated average loss of embryons during the birth process and an estimated 60% total loss from birth to weaning. Major mortality factors include old age, climate, diseases, and accidental death. Man and owls are major predators.

Annual turnover rate has been estimated as high as 95% of the breeding population per year. Average age of the Virginia opossum in the wild is probably 1.33 years.

## REFERENCES


Anderson, R. M. (1947). *Nat. Mus. Can., Bull.* **102,** Biol. Ser No. 31, 1–238.

Anonymous. (1858). "The Handbook of Toronto." Lovell & Gibson, Toronto.

Bailey, V. (1931). *North Am. Fauna* **53,** 7.

Bailey, V. (1933). *J. Mammal.* **14,** 241–243.

Barker, P. R., Bergman, H. C., Hunter, W. H., Lawrence, W. H., Steers, C. W., Wright, L. N., Willis, R. A., and Barrow, D. M. (1967). "Marsupial Biomodule Evaluation Study," Report no. SID66-1647. U.S. Defense Documentation Center, Cameron Station, Alexandria, Virginia.

Barnes, R. D. (1968). *Lab. Anim. Care* **18,** 251–257.

Barnes, R. D., and Barthold, S. (1969). *J. Reprod. Fertil.* **6,** 477–482.

Bartholomew, G. A., and Hudson, J. (1962). *Physiol. Zool.* **35,** 94–107.

Beidler, L. M., Fishman, L. Y., and Hardiman, C. W. (1953). *Am. J. Physiol.* **181,** 235–239.

Bentley, P. J. (1960). *Aust. J. Exp. Biol. Med. Sci.* **38,** 301–306.

Bergman, H. C., and Housley, C. (1968). *Comp. Biochem. Physiol.* **25,** 213–218.

Biggers, J. D. (1966). *Symp. Zool. Soc. London* **15,** 251–280.

Blair, W. F. (1936). *J. Mammal.* **17,** 410.

Bolliger, A., and Pascoe, J. V. (1953). *Aust. Sci.* **15,** 215–217.

Britton, S. W., and Atkinson, W. E. (1938). *J. Mammal.* **19,** 94.

Britton, S. W., and Klein, R. F. (1939). *Am. J. Physiol.* **125,** 4.

Brocke, R. H. (1970). Ph.D. Thesis, Michigan State University, East Lansing.

Brown, A. E. (1909). *Proc. Zool. Soc. London* **81.**

Bryant, H. C. (1927). *Calif. Fish Game* **13,** 146.

Burns, R. K., and Burns, L. M. (1957). *Rev. Suisse Zool.* **64,** 595–605.

Cabrera, A., and Yepes, J. (1940). "Mamiferos Sud-Americanos (vida, costumbres y descripcion)." Compania Argentina de Editores, Buenos Aires.

Cadzow, G. (1974). M.S. Thesis, Long Beach State College, Long Beach, California.

Carpenter, J. A. (1956). *J. Comp. Physiol. Psychol.* **49,** 139–144.

Clemens, W. A. (1968). *Evolution* **22,** 1–18.

Cockrum, E. L. (1964). "The Vertebrates of Arizona." pp. viii–276. Univ. of Arizona Press, Tucson.

Collins, L. (1973). *Smithson. Institution Publ.* **4888.**

Crandall, L. S. (1964). "The Management of Wild Animals in Captivity." Univ. of Chicago Press, Chicago, Illinois.

Cross, E. C., and Dymond, J. R. (1929). *R. Ont. Mus. Zool., Handb.* No. 1, p. 54.

Davis, D. E. (1945a). *Ecol. Monogr.* **15,** 243–295.

Davis, D. E. (1945b). *J. Mammal.* **26,** 119–127.

Davis, D. E. (1953). *J. Mammal.* **34,** 352–358.

Davis, D. E., Emlen, J. T., Jr., and Stokes, A. W. (1948). *J. Mammal.* **29,** 207–225.
Davis, J. A. (1966). *Anim. Kingdom* **69,** 112–117.
Dawson, T. J. (1969). *Comp. Biochem. Physiol.* **28,** 401–407.
Dills, G. C., and Manganiello, T. (1973). *J. Mammal.* **54,** 763–765.
Dobzhansky, T. (1950). *Am. Sci.* **38,** 209–221.
Eisentraut, M. (1960). *Bull. Mus. Comp. Zool.* **124,** 31–43.
Enders, R. K. (1935). *Bull. Mus. Comp. Zool.* **78,** 385–497.
Enders, R. K. (1966). *Symp. Zool. Soc. London* **15,** 195–203.
Enders, R. K., and Davis, D. E. (1936). *J. Mammal.* **17,** 165–166.
Evans, F. C., and Holdeinreid, R. (1943). *J. Mammal.* **24,** 231–260.
Farris, E. J. (1950). *In* "The Care and Breeding of Laboratory Animals" (E. J. Farris, ed.), pp. 210–267. Wiley, New York.
Finkel, M. P. (1945). *Am. J. Anat.* **76,** 93–152.
Fitch, H. S., and Sandidge, L. L. (1953). *Univ. Kans. Publ. Mus. Nat. Hist.* **7,** 305–338.
Fitch, H. S., and Shirer, H. W. (1970). *Am. Midl. Nat.* **84,** 170–186.
Fleming, J. H. (1913). *In* "The Natural History of the Toronto Region," pp. 206–211. Can. Inst., Ontario, Canada.
Fleming, T. H. (1972). *J. Mammal.* **53,** 619–623.
Gardner, D. L. (1973). *Spec. Publ., Mus., Tex. Tech. Univ.* **4,** 1–81.
Greer, J. K. (1965). *Publ. Mus., Mich. State Univ., Biol. Ser.* **3,** No. 2, 53–151.
Grinnell, J. (1915). *Calif. Fish Game* **1,** 114–116.
Gross, R., and Bolliger, A. (1958). *Aust. Sci.* **20,** 184–185.
Grote, J. C., and Dalby, P. L. (1973). *Ohio J. Sci.* **73,** 240–241.
Guilday, J. E. (1958). *J. Mammal.* **29,** 29–43.
Hairston, N. G., Tinkle, D. W., and Wilber, H. M. (1970). *J. Wildl. Manage.* **34,** 681–690.
Hall, E. R., and Dalquest, W. W. (1963). *Aust. J. Sci.* **30,** 105–106.
Hamilton, W. J. (1933). *J. Mammal.* **14,** 151–152.
Hamilton, W. J., Jr. (1958). *Cornell Agric. Exp. Sta.,* Mem. 354, 48 pp.
Hartman, C. G. (1923a). *Am. J. Anat.* **32,** 253–422.
Hartman, C. G. (1923b). *Smithson. Inst., Annu. Rep.* (1921), pp. 347–364.
Hartman, C. G. (1928). *J. Morphol. Physiol.* **26,** 143–216.
Hartman, C. G. (1952). "Possums." Univ. of Texas Press, Austin.
Hayne, D. (1949). *J. Mammal.* **30,** 1–18.
Hayne, D. (1950). *J. Mammal.* **31,** 26–39.
Hershkovitz, P. (1969). *Q. Rev. Biol.* **44,** 1–70.
Hershkovitz, P. (1976). *Proc. Biol. Soc. Wash.* (in press).
Hibbard, C. W., Ray, D. E., Savage, D. E., Taylor, D. W., and Guilday, J. E. (1965). *In* "The Quarternary of the United States" (H. E. Wright and D. G. Frey, eds.), pp. 509–525. Princeton Univ. Press, Princeton, New Jersey.
Higginbotham, A. C., and Koon, W. D. (1955). *Am. J. Physiol.* **181,** 69–71.
Hock, R. J. (1952). *J. Mammal.* **33,** 464–470.
Holmes, A. C., and Sanderson, G. C. (1965). *J. Wildl. Manage.* **29,** 287–295.
Hoversland, A. S., Murphy, W. S., Dhindsa, D. S., Parer, J. T., and Metcalfe, J. (1975). *Comp. Biochem. Physiol.* **50A,** 519–525.
Jenness, R., Regehr, E. A., and Sloan, R. E. (1964). *Comp. Biochem. Physiol.* **13,** 339–352.
Jewett, S. G., and Dobyns, H. W. (1929). *J. Mammal.* **10,** 351.
Jones, M. (1963). "Mammals in Captivity," rev. ed. (unpublished manuscript).
Johnston, G. E. (1931). *Q. Rev. Biol.* **6,** 430.
Jurgelski, W., Jr. (1974). *Lab. Anim. Sci.* **24,** 376–403.
Jurgelski, W., Jr., and Porter, M. (1974). *Lab. Anim. Sci.* **24,** 404–411.

Kennedy, P. M., and Heinsohn, G. E. (1974). *Comp. Biochem. Physiol. A* **47,** 829–834.
Kermack, K. A., Lees, P. M., and Mussett, F. (1965). *Proc. R. Soc. London, Ser. B* **162,** 535–554.
Kinnear, J. E., and Brown, G. D. (1967) *Nature* (*London*) **215,** 1501.
Lay, D. W. (1942). *J. Mammal.* **23,** 147–154.
Lillegraven, J. A. (1974). *Annu. Rev. Ecol. Syst.* **5,** 263–283.
Llewellyn, L. M., and Dale, E. H. (1964). *J. Mammal.* **45,** 113–122.
Lustick, S., and Lustick, D. D. (1972). *Comp. Biochem. Physiol. A* **43,** 643–647.
McArthur, R. H., and Wilson, E. O. (1967). "The Theory of Island Biogeography." Princeton Univ. Press, Princeton, New Jersey.
McManus, J. J. (1969). *J. Mammal.* **43,** 486–487.
McManus, J. J. (1970). *Am. Midl. Nat.* **84,** 144–169.
McManus, J. J. (1971). *J. Mammal.* **52,** 846–848.
Mann, G. F. (1955). *Invest. Zool. Chil.* **2,** 159–166.
Mohr, C. O. (1947). *Am. Midl. Nat.,* **37,** 223–249.
Morrison, P. R. (1946). *J. Cell. Comp. Physiol.* **27,** 125–137.
Morrison, P. R. (1965). *Aust. J. Zool.* **13,** 173–187.
Morrison, P. R., and McNab, B. K. (1962). *Comp. Biochem. Physiol.* **6,** 57–68.
Morrison, P. R., and Petajan, J. H. (1962). *J. Physiol. Zool.* **35,** 52–65.
Nardone, R. M., Wilber, C. G., and Musacchia, X. J. (1955). *Am. J. Physiol.* **181,** 352–356.
Osgood, W. H. (1943). *Field Mus. Nat. Hist., Zool. Ser.* **20,** 1–268.
Parmelee, A. H., Brooks, M. B., and Stone, R. S. (1960). *Anat. Rec.* **136,** 73–78.
Petajan, J. H., and Morrison, P. R. (1962). *J. Exp. Zool.* **149,** 45–57.
Petajan, J. H., Morrison, P. R., and Akert, K. (1962). *J. Exp. Zool.* **150,** 225–231.
Peterson, R., and Downing, S. C. (1956). *J. Mammal.* **37,** 431–434.
Petrides, G. A. (1949). *J. Mammal.* **30,** 364–378.
Phillips, C. J., and Jones, J. K. (1968). *J. Mammal.* **49,** 320–321.
Pianka, E. (1970). *Am. Nat.* **104,** 592–597.
Pianka, E. (1974). "Evolutionary Ecology." Harper, New York.
Pine, R. (1973). *Proc. Biol. Soc. Wash.* **86,** 391–402.
Plakke, R. K., and Pfeiffer, E. W. (1970). *Comp. Biochem. Physiol.* **34,** 325–332.
Pressman, T. G., and Doolittle, J. H. (1966). *Psychol. Rep.* **18,** 875–878.
Reig, O. A. (1955). *Invest. Zool. Chil.* **3,** 121.
Reynolds, H. C. (1945). *J. Mammal.* **26,** 361–379.
Reynolds, H. C. (1952). *Univ. Calif., Berkeley, Publ. Zool.* **53,** 113–275.
Reynolds, H. C. (1953). *Sci. Am.* **188,** 81–94.
Rink, R., and Miller, J. (1967). *Cryobiology* **4,** 24–29.
Robinson, K. W., and Morrison, P. R. (1957). *J. Cell. Comp. Physiol.* **49,** 455–478.
Sanderson, G. C. (1961). *J. Wildl. Manage.* **25,** 20–27.
Sanderson, G. C. (1966). *J. Wildl. Manage.* **30,** 215–235.
Saunders, W. E. (1932). *Trans. R. Can. Inst.* **47,** Part 2, 271–309.
Scheffer, V. B. (1943). *Murrelet* **24,** 27–28.
Scott, W. J. (1938). *J. Cell. Comp. Physiol.* **12,** 391–401.
Shirer, H. W., and Downhower, J. F. (1968). *Trans. Kans. Acad. Sci.* **71,** 463–479.
Shirer, H. W., and Fitch, H. S. (1970). *J. Mammal.* **51,** 491–503.
Slaughter, B. H. (1968). *Science* **162,** 254–255.
Stirling, E. C. (1891). *Trans. R. Soc. South Aust.* **14,** 154–187.
Stout, J., and Sonenshine, D. E. (1974). *Acta Theriol.* **19,** 235–245.
Stumpf, W. A., and Mohr, C. O. (1962). *J. Wildl. Manage.* **20,** 149–154.
Tamar, H. (1961). *Physiol. Zool.* **34,** 86–91.

Tate, G. H. H. (1933). *Bull. Am. Mus. Nat. Hist.* **66,** 6–236.
Tyndale-Biscoe, H. (1973). "Life of Marsupials." Am. Elsevier, New York.
vanden Broek, A. J. P. (1910). *Morphol. Jahrb.* **41,** 347–436.
Verts, B. J. (1963). *J. Wildl. Manage.* **27,** 127–129.
von Bloeker, J. C. (1928). *J. Mammal.* **9,** 62.
Wilson, D. (1970). *J. Mammal.* **51,** 386–387.
Wiseman, G. L., and Hendrickson, G. O. (1950). *J. Mammal.* **31,** 331–337.
Wislocki, G. B. (1933). *Q. Rev. Biol.* **8,** 385.
Wood, J. E. (1954). *J. Mammal.* **35,** 406–414.

# 4

# Central Nervous System of Marsupials

JOHN IRWIN JOHNSON, JR.

The purpose here is to review investigations into marsupial neurology, to sort out evidence of commonalities with eutherian nervous systems, to find distinctive features peculiar to the marsupial subclass, and to identify neural specializations of particular genera and species in this subclass. To accomplish this objective, the entire available literature for the years 1890–1970, and most of the publications from 1971–1975, dealing with marsupial central nervous systems, are reviewed. These sources are listed at the end of the chapter in a comprehensive bibliography on marsupial brains. In order to present this complete list there has been considerable sacrifice in reference to studies on nonmarsupials, these being restricted to a few contemporary secondary sources. All statements made were checked against original sources, however, and many of these can be found cited in the marsupial articles.

For an orderly approach, we will first consider the features common to all vertebrate nervous systems, and then take up individual subsystems to the extent that they can be dealt with separately. Each subsystem will be examined in detail for evidence of commonalities and differences in the placental and marsupial forms. Of course not all subdivisions of the nervous system have been studied to the same degree, and we shall encounter many instances where systems have never been studied in marsupials. Furthermore a discouragingly large proportion of investigations do not go beyond descriptive anatomy. Hopefully the account will encourage more investigations in the open fields of marsupial neural function.

## I. BASIC ORGANIZATION (Fig. 1)

### A. Neural Tube

The central nervous system of vertebrates can be considered as basically a neural tube running along the dorsal axis of the body. Communicating neural processes enter and leave the tube, carrying information to and from the various locations in the peripheral body.

#### 1. Afferent and Efferent Processes

Before entering the tube, processes carrying incoming (afferent) information are segregated from those carrying outgoing (efferent) messages. The afferent processes enter the tube on its dorsolateral aspect, in a series of dorsal roots (Fig. 2), and their cell bodies with few exceptions are located in ganglia near, but external to, the tube (see Fig. 3). Efferent processes, the motor fibers, leave the tube on the ventrolateral aspect, as a series of ventral roots, and their cell bodies are located ventrally within the tube (Fig. 1A). They terminate directly in striated muscle, or on other neurons which in turn activate smooth muscles or secretory glands.

#### 2. Sensory, Motor, and Internuncial Cell Columns (Fig. 1B)

The cell bodies of the motor processes form a column running the anteroposterior length of the tube, ventral and lateral to the lumen of the tube which is called the central canal or ventricle. There is a motor column on

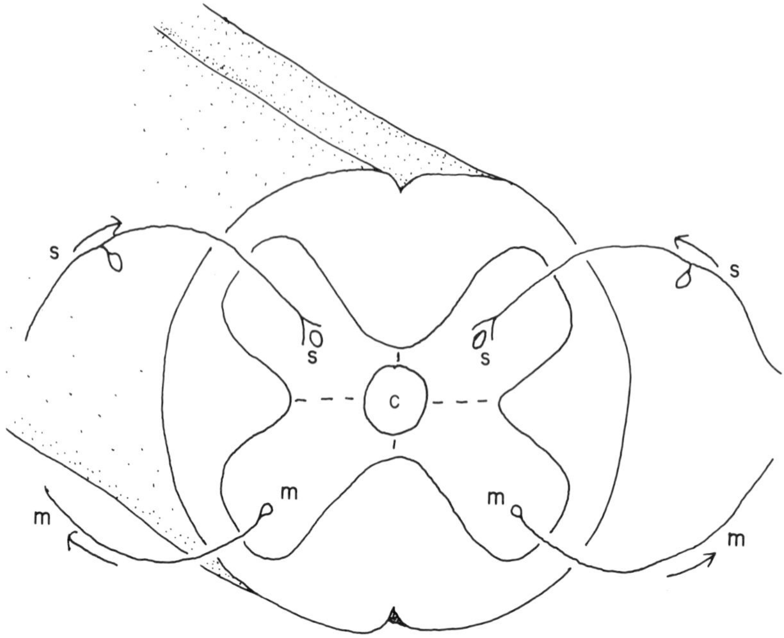


(A)

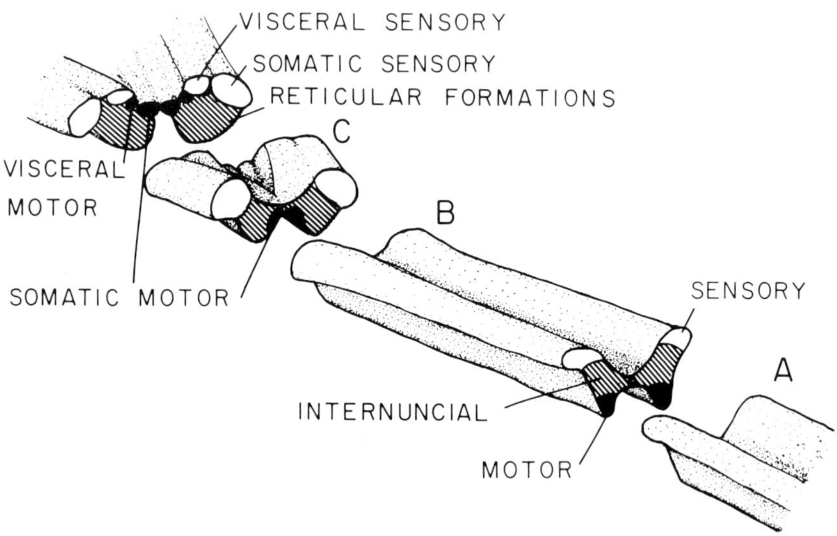


(B)

Fig. 1(A) and (B).

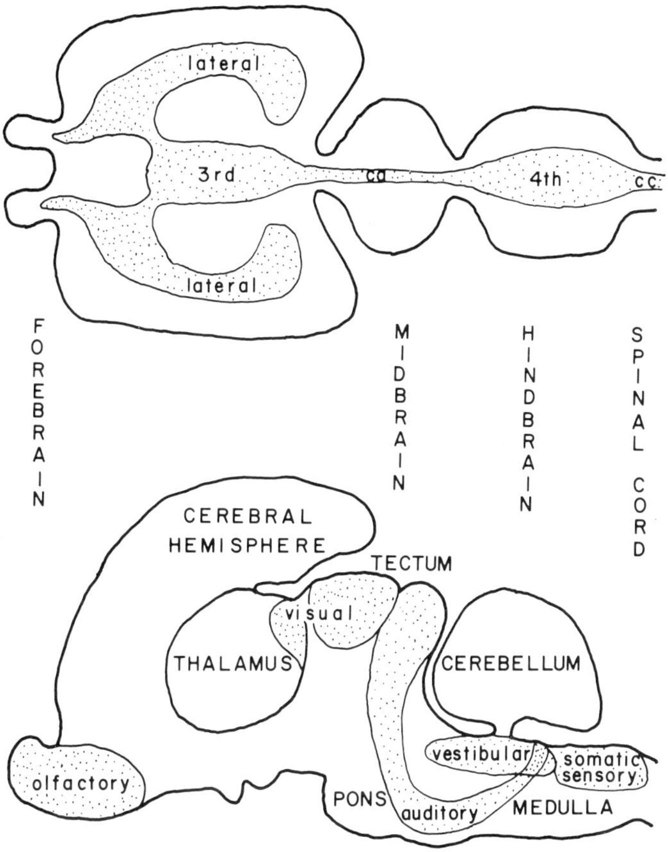


(C)

Fig. 1. (A) Basic neural tube formation seen in ontogeny of vertebrate nervous systems, shown schematically. c, central canal; m, motor process; s, sensory process. (B) Basic arrangement of cell columns of the spinal cord and brainstem of vertebrates, as found in adult marsupial opossums *Didelphis virginiana*. To reveal cross sections, the columns are interrupted at levels corresponding to sections seen in Fig. 2 (at A), Fig. 3 (at B), and Fig. 5 (at C). The central canal is relatively minute in adults. (C) Vesicular expansions of the front of the neural tube, seen as projected onto horizontal (upper) and sagittal (lower) planes. In the horizontal view, shaded areas represent the ventricles. ca, cerebral aqueduct; cc, central canal of the spinal cord. In the sagittal view, shaded areas represent some regions receiving input from sensory systems.

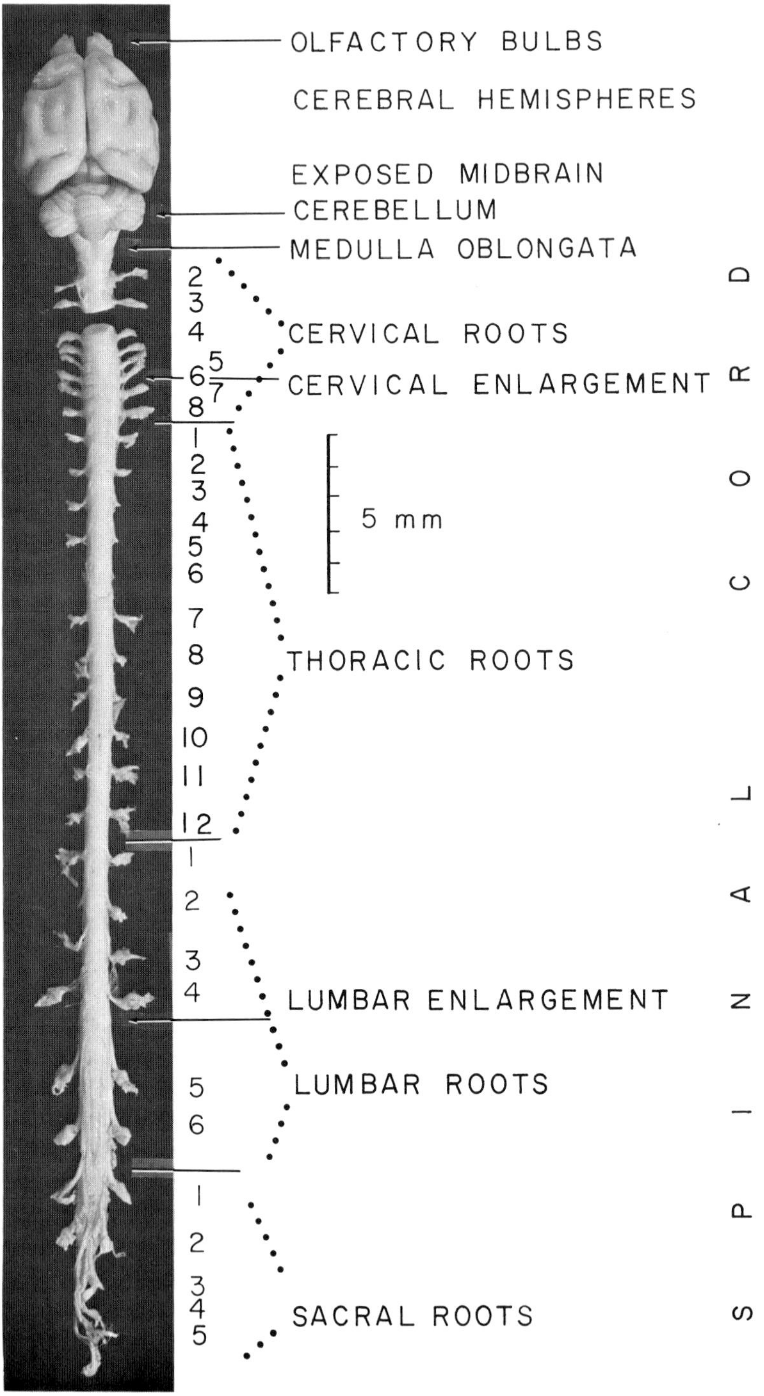
OLFACTORY BULBS
CEREBRAL HEMISPHERES
EXPOSED MIDBRAIN
CEREBELLUM
MEDULLA OBLONGATA
2
3
4
CERVICAL ROOTS
5
6
7
CERVICAL ENLARGEMENT
8
1
2
3
4
5
6
7
8
THORACIC ROOTS
9
10
11
12
1
2
3
4
LUMBAR ENLARGEMENT
LUMBAR ROOTS
5
6
1
2
3
4
5
SACRAL ROOTS
5 mm
SPINAL CORD

each side of the bilaterally symmetrical nervous system. Dorsolateral to the lumen, canal, or ventricle are corresponding sensory columns of cells, one on each side, where the incoming fibers terminate. Between the motor and sensory columns are cells which communicate between the columns as internuncials; their processes do not leave the tube. Thus the basic internal structure of the tube consists of a ventral pair of motor cell masses, a dorsal pair of sensory cell masses, and an internuncial cell mass connecting the other four columns.

Each of the sensory and motor columns has been segregated into "somatic" and "visceral" components (Ariëns Kappers *et al.*, 1936, Vol. 1, pp. 227–249*). The visceral columns send and receive communications to and from glands and smooth muscle, and they lie between the somatic columns. In marsupials visceral columns have been identified only in the medulla and brain stem of opossums, *Didelphis virginiana* (DuBois, 1929; Voris and Hoerr, 1932; Culberson and Kimmel, 1972).

Superimposed upon this basic plan of the central communication system are the various specialized elaborations in information gathering and processing systems that have developed in the course of evolution. The concentration of specialized sense organs at the anterior end of the body is related to the complicated enlargements of the anterior neural tube which constitute the brain (Figs. 1C, 2). The columnar structure is interrupted and distorted in many respects in the brainstem. The orderly pairs of dorsal and ventral spinal roots are replaced by the specialized cranial nerves, and the cell columns have become known as the nuclei of the cranial nerves (Figs. 4, 5).

The somatic sensory columns of the spinal cord are continuous with the sensory nuclear complex of the trigeminal (5th) nerve; Voris and Hoerr (1932), in the opossum, group the highly specialized auditory, vestibular, and somatic sensory nuclear complexes as elaborations of the sensory columns. The visceral afferent column is convincingly demonstrated only in the nucleus of the solitary tract (Fig. 5), which has been analyzed in *Didel-*

* Volume and page numbers cited throughout the text for this reference refer to the 1960 reprint.

Fig. 2. Central nervous system of wombat, *Vombatus ursinus* (no. 14), showing major morphological features in dorsal view. Most dorsal roots and their ganglia remain attached to the spinal cord. In marsupials and some placentals, there is no sensory root, only a ventral motor root, for the first cervical segment, defined as that segment whose fibers exit through a lumen in the atlas. Cervical and lumbar enlargements of the spinal cord occur at the levels of entry of roots innervating the fore- and hind limbs. The visible dorsal surface of the midbrain between the cerebellum and cerebral hemispheres is seen in several marsupials. Numbers of roots classified as "thoracic," etc., may vary due to variable number of ribs.

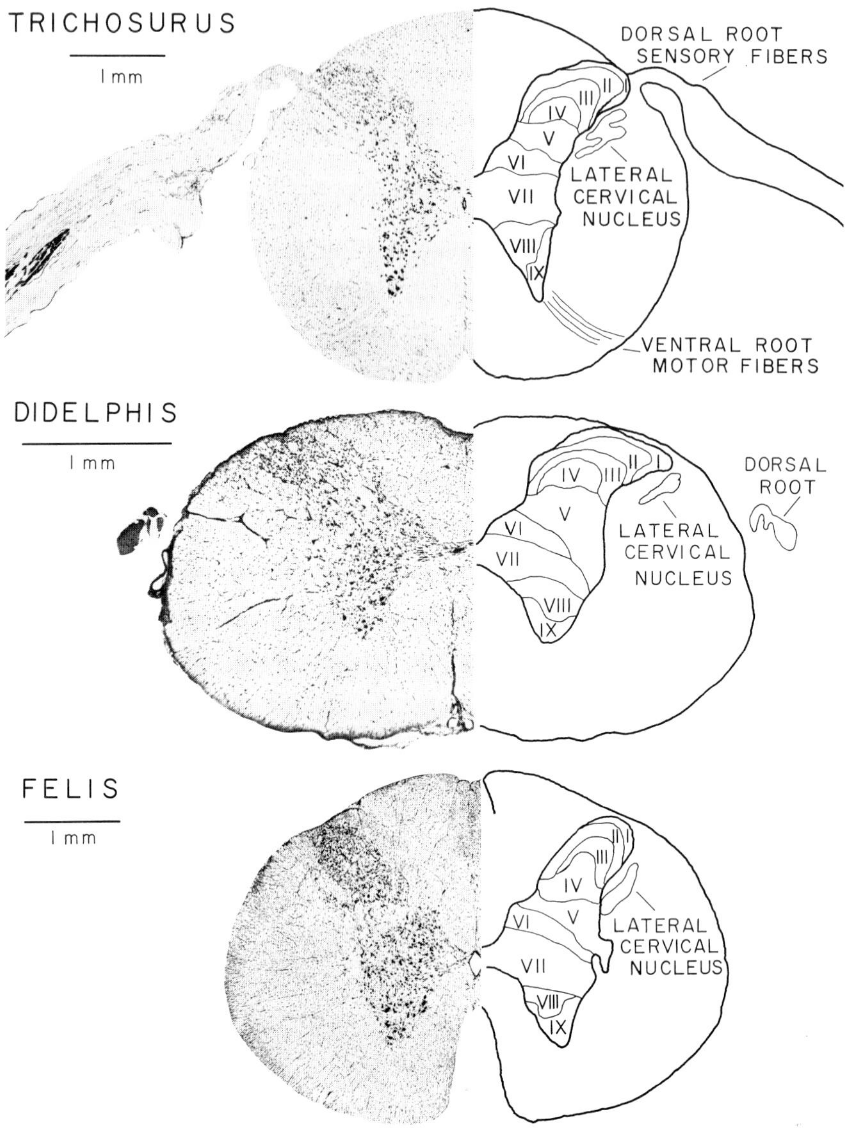


Fig. 3. Similarity of spinal cord architecture seen in transverse sections through the second cervical segment of two marsupials, *Trichosurus vulpecula* (no. 40, sec. 1504) and *Didelphis virginiana* (no. 501, sec. 418), and one placental, *Felis catus* (no. 324, sec. 1536). Specimen numbers in this and following figures refer to brains in the Michigan State University collection, for which data concerning brain and body measures, sex, and locality of origin are available. Roman numerals refer to cytoarchitectural laminae defined by Rexed (1952). All sections are stained with thionine. Notice the cells of the dorsal root ganglion of *Trichosurus* (second cervical nerve). The sensory column corresponds to laminae I–V, the internuncial column to laminae V–VII, and the motor column to laminae VIII and IX. Sparsity of cells in lamina V is due to the presence of a massive invasion of corticospinal fibers.

*phis virginiana* (DuBois, 1929; Culberson and Kimmel, 1972). Sensory components of the facial (7th), glossopharyngeal (9th), and vagus (10th) nerves make up the solitary tract and terminate in the surrounding small-celled nucleus. This nucleus is in close dorsal proximity to the visceral efferent nucleus, the dorsal motor nucleus of the vagus. This pair of nuclei thus constitute the first afferent and efferent stations for information to and from the major visceral organs innervated by the vagus nerve. The condition of these subdivisions of the columns in the brainstem in marsupials does not appear to be different from that in placental mammals.

### B. Cerebral Elaborations

As in most vertebrates, the front end of the neural tube in marsupials is elaborated into a set of enlarged vesicles surrounded by specialized masses of neural tissues (Figs. 1C, 2). The three major subdivisions constitute the forebrain, midbrain, and hindbrain. Extreme outgrowths occur on either side of the forebrain, becoming the cerebral hemispheres. A third major outgrowth, over the roof of the hindbrain, becomes the cerebellum. All of these elaborations are associated with the growth of large groups of sensory nuclei, which receive input from the specialized receptors concentrated in the head region. The anteriormost of these are olfactory bulbs, themselves outgrowths at the front of the forebrain. A primary visual input terminal occupies the anterior tectum on the roof of the midbrain. The posterior tectum and neighboring hindbrain regions receive nerves from auditory organs. Next to these are nuclei receiving input from vestibular and labryinthine sensors of gravity and motion. Next to these are somatic sensory nuclei receiving input from the body by way of the spinal cord and the somatic sensory columns of the brainstem. The cerebellum develops near the vestibular and somatic sensory nuclei and receives major inputs from them.

Most, and perhaps all, of these sensory nuclear regions send information to a large cell group at the center of the forebrain, the thalamus. The cerebral cortex has developed largely as a processing system handling this sensory input to thalamus.

In our detailed treatment, we will consider first the spinal cord, then the "general" somatic sensory systems, and then the cerebellum. Following this, for a functional approach we will consider each of the "special" sensory systems. After this are treatments of the integrative processes of the forebrain, including: the basal telencephalon, the great thalamocortical systems, the commissural connections, and the hemoneural interactions in

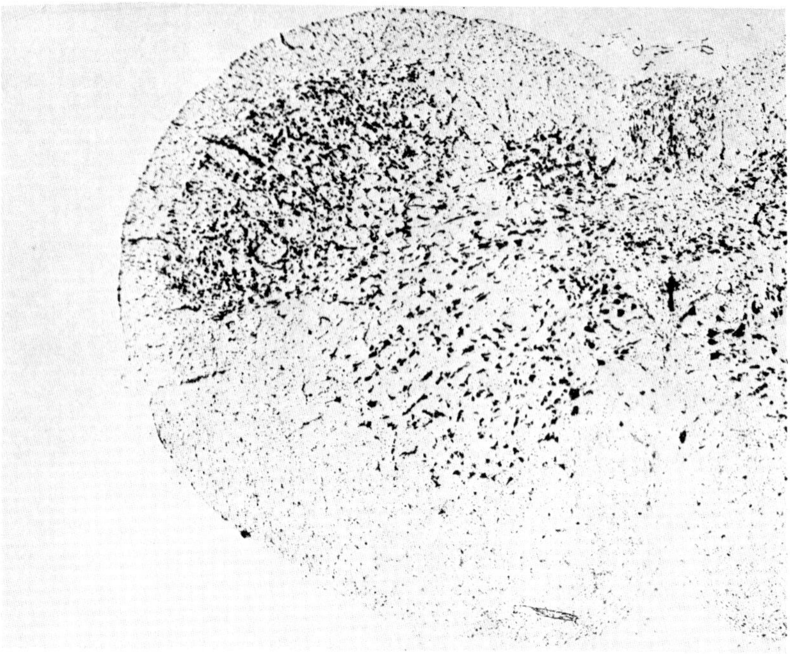

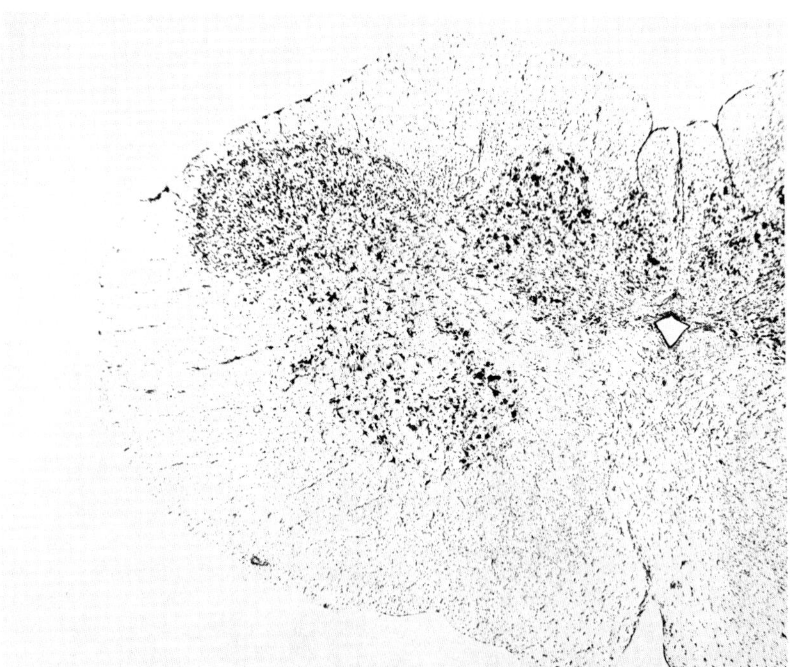

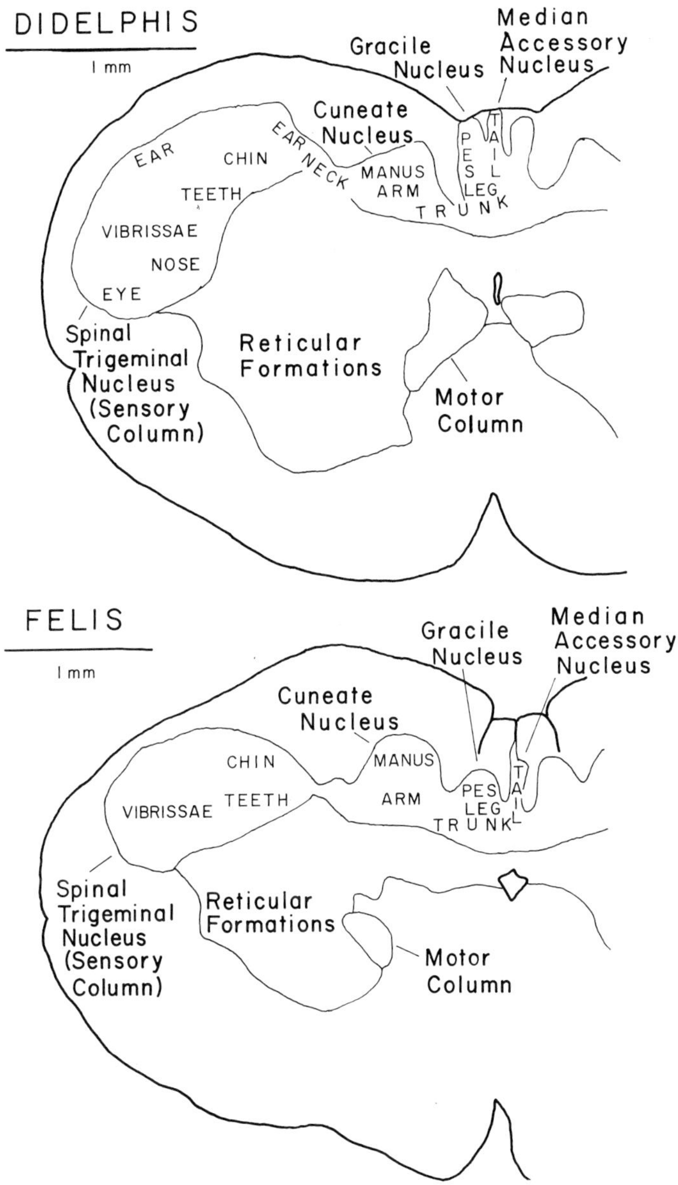


Fig. 4. Transverse sections stained with thionine through the caudal medulla of marsupial opossum *Didelphis virginiana* (no. 75, sec. 636) (top left) and placental cat *Felis catus* (no. 324, sec. 1356) (bottom left), showing the similarity of somatotopic projections to the cuneate–gracile nuclear complex and the spinal trigeminal complex. In the diagrams above, within the boundaries of these nuclear groups, are indicated the locations of cells which respond to mechanical stimulation of various parts of the body, as determined by electrophysiological recording in opossums (Hamilton and Johnson, 1973; Weller, 1971) and cats (Kruger *et al.*, 1961). At this level in *Didelphis* the motor column is represented by cells of the nucleus of the hypoglossal (12th) cranial nerve, and in both animals the sensory column (in the restricted sense) is the spinal nucleus of the trigeminal (5th) cranial nerve.

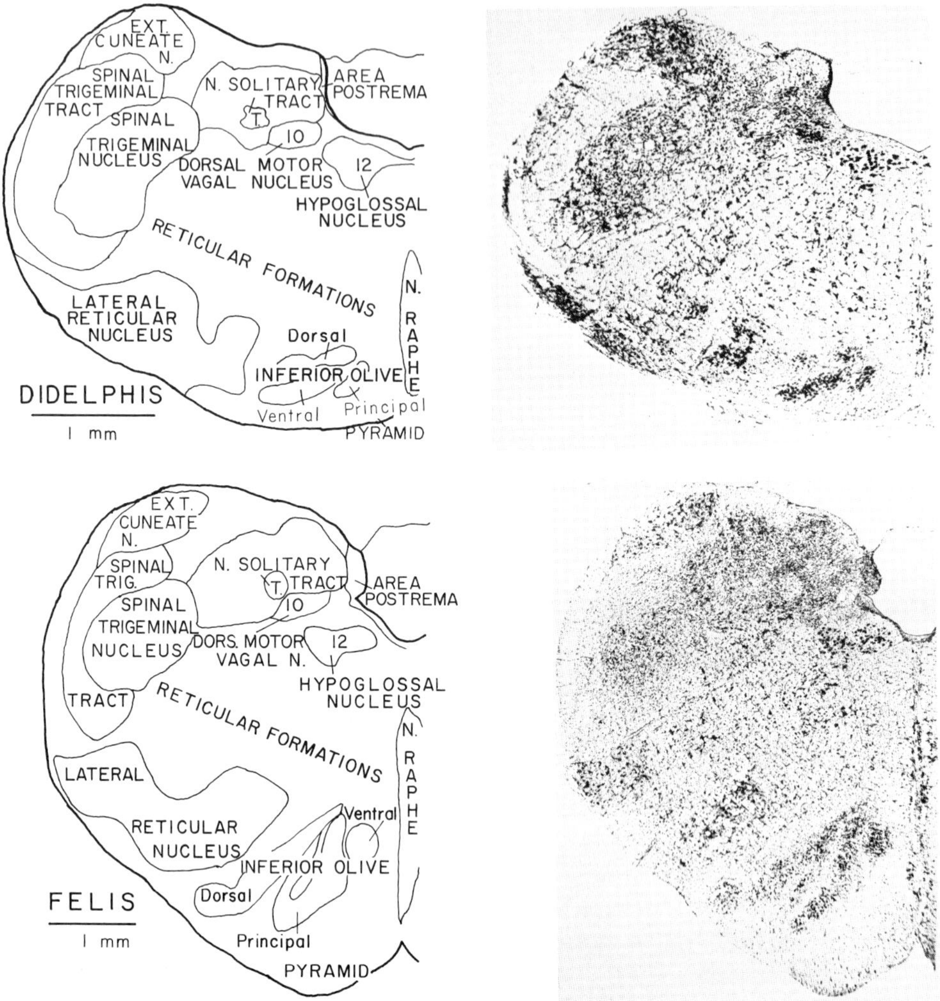


Fig. 5. Transverse sections, stained with thionine, through the medulla oblongata of marsupial opossum (top right) and placental cat (bottom right) (*Didelphis virginiana* no. 75, sec. 736; *Felis catus* no. 324, sec. 1232). The somatic sensory cell columns (in the restricted sense) are here represented by the spinal trigeminal nuclei. Embryologically related cell groups projecting to the cerebellum include the external cuneate nucleus, the lateral reticular nucleus, and the inferior olive. One interesting difference between marsupials and placentals is in the inferior olive, where the ventral subdivision lies lateral to the principal subdivision in marsupials, while their positions are reversed in placentals. The visceral sensory column here consists of the nucleus of the solitary tract, surrounding the solitary tract T. The visceral motor column forms the dorsal motor vagal nucleus 10 (of the tenth cranial nerve). The somatic motor column is the hypoglossal nucleus 12 (of the 12th cranial nerve). The reticular formations correspond to the internuncial column. The pyramid contains fibers from the forebrain on their way to the spinal cord.

the hypothalamus. Next descending systems are examined and, in conclusion, the general topics of sleep and circulation will be considered.

## II. THE SPINAL CORD

### A. Cell Columns and Lamination

The posterior portion of the neural tube, the spinal cord, retains the basic order of sensory, internuncial, and motor cell columns relatively undistorted. Figure 3 illustrates the structural plan with examples from the high cervical spinal cord of marsupials and placentals. The motor columns are represented by laminae VIII and IX, the sensory columns by I–IV at least. Laminae V, VI, and VII probably contain internuncials with connections between the other columns, although research into cellular function in these regions is greatly lacking in both placentals and marsupials.

The pattern of "lamination" (Rexed, 1952) of the cell columns in the spinal cord is remarkably similar in both placental and marsupial mammals. Two differences have been reported:

1. A special group of large cells at the medial base of the sensory column, where lamina VI meets VII, is present in thoracic and lumbar segments in placental mammals. These cells are known as the dorsal nucleus or Clarke's column; they receive input from muscle afferents from the hind limb, and project to the cerebellum. Voris (1928b) was unable to identify, by simple anatomical observation, such cells in the opossum spinal cord. Functional investigations are necessary, however, to document a real marsupial–placental divergence here, particularly since the forelimb homologue (the external cuneate nucleus) is obviously well developed in opossums, and responses meeting all the criteria of Clarke's column activity were recorded in spinal cords of phalangers (Australian brush-tailed possums *Trichosurus vulpecula*) (Magni and Oscarsson, 1962a,b).

2. Lamina V is massively invaded, in the cervical region, by fibers from the cerebral cortex, in both placentals and marsupials (Fig. 3). Most of these appear to terminate in lamina VI, though in some more dextrous species, including placental monkeys and raccoons and marsupial phalangers (*Trichosurus vulpecula*), terminals extend into more ventral laminae (Martin *et al.*, 1970; Rees and Hore, 1970). The terminal area and lamina VI itself were reported to be larger in placental cats and armadillos (*Dasypus novemcinctus*) than in the marsupial opossums and phalangers (Martin and Fisher, 1968; Fisher *et al.*, 1969; Rees and Hore, 1970). According to these authors, this reflects a greater degree of cortical involvement in spinal affairs in placentals when compared with marsupials.

### B. Spinal Roots and Enlargements (Fig. 2)

The fibers leaving the motor columns through the ventral spinal roots have similar distributions of diameters in placental cats and marsupial opossums (Rocha-Miranda and Oswaldo-Cruz, 1965). These two animals also show a similar arrangement of afferent fibers entering in dorsal roots (Oswaldo-Cruz *et al.*, 1965).

There are shifts and enlargements of the cord accommodating regions of greater input and output, in segments innervating the fore- and hind limbs, and they appear similar in marsupials and placentals.

## III. SOMATIC SENSORY SYSTEMS

Going rostrally, the first additions to the basic spinal organization are special centers subserving the processing and transmission of somatic sensory information to the anterior enlargements of the brain. Three main spinocerebral pathways have been identified in marsupials, and portions of what may be others. For all, similar findings were reported in placentals and marsupials, almost always by the same investigators.

### A. The Bulbothalamic Mechanosensory System

The first of these to be considered is the most prominent and most thoroughly investigated. Although a well-defined and recognized system, it still lacks a completely satisfactory name. It has been called a dorsal column system which ignores its trigeminal component. It has been called the medial lemniscal system, which allows for the trigeminal, but confuses the system with the spinothalamic systems which also travel in the medial lemniscus. We will call it the bulbothalamic mechanosensory system to distinguish it from the cervico- and spinothalamic paths. It can be regarded as the larger of two systems that are "pure mechanosensory" in that they are ordinarily relatively unresponsive to temperature and pain stimulation.

In this system, processes from mechanoreceptors in the body enter the spinal cord, and send branches rostrally in the dorsal funiculi, which terminate in nuclei between the dorsal sensory columns near the junction of the spinal cord and the medulla oblongata. In most mammals, marsupials included, there can be distinguished three of these cell masses (Fig. 4). At the midline is the median accessory nucleus, first identified by Bischoff (1899) in a kangaroo (species not given) as well as in placental rats, cats, and dogs. Ziehen (1899, 1901) pointed it out in the marsupial ringtail possum *Pseudocheirus peregrinus*. Just lateral to this is the nucleus of the gracile fasciculus, and lateral to this in turn is the nucleus of the cuneate fasciculus. These three make up the cuneate–gracile nuclear complex.

Lateral to the cuneate–gracile complex is a direct continuation of the dorsal sensory cell column of the spinal cord. But from the third cervical segment rostrally, many of the processes which terminate in the column are from receptors in the face, and their cell bodies lie in the semilunar ganglion of the trigeminal nerve, which serves as a massive "dorsal root" for somatic sensation from the front of the head. This portion of the column is thus known as the spinal trigeminal nucleus.

The central terminals from receptors in the head and body preserve the topological relationships that exist at the periphery, but with distortions related to the greater density of receptors in some body regions. This is an example of the "somatotopic organization" of sensory nuclei found in many places throughout vertebrate nervous systems. The size of nuclear regions receiving projections from receptor-dense regions is proportionally larger.

The four cell groups in this complex in the medulla, upon electrophysiological analysis, turn out to be such enlargements receiving projections from receptor-dense regions: the spinal trigeminal from the mouth, lips, vibrissae and nose; the cuneate from the forepaw; the gracile from the hind foot; and the median accessory from the tail (Fig. 4), in marsupial opossums as in placentals (Hamilton and Johnson, 1973; Magalhães-Castro *et al.*, 1971).

Cells of all four of these nuclei send processes to the ventrobasal nucleus of the thalamus; with a few exceptions in the regions receiving information from the nose and mouth, these processes cross the midline and terminate in the thalamus of the opposite side. Somatotopic organization, with tissue volume related to relative receptor density, has again been found in all mammals investigated, including opossums (Pubols and Pubols, 1966; Sousa *et al.*, 1971; Bombardieri *et al.*, 1975).

Cells in the ventrobasal thalamus project in turn upon sensory cerebral cortex, again in all mammals studied, including opossums (Pubols, 1968). These terminals again are somatotopically organized (Lende, 1963a).

Projections from the sensory cortex terminate in the ventrobasal thalamus in placental carnivores and primates (Petras, 1969) and marsupial opossums (Martin, 1968b); others terminate in the trigeminal and cuneate–gracile complexes in the medulla of placentals and marsupials *Didelphis* (Martin and West, 1967) and *Trichosurus* (Martin *et al.*, 1971). These cortical projections provide an architectural basis for feedback modulation, at two levels of mechanosensory input to the cortex.

### B. The Cervicothalamic Mechanosensory System

The second spinocerebral pathway to be considered is another "pure mechanosensory" system. Cells in the spinal cord which receive sensory input, including cells in lamina IV and in Clarke's column, send processes to the lateral cervical nucleus of the same side (Bowsher, 1965) in cats. The lateral cervical nucleus (see Fig. 3) occurs as a cell group lateral to the sensory column in the high cervical segments of the spinal cord. Cells in this nucleus project to the ventrobasal thalamus of the opposite side. They transmit information very much like that carried from the cuneate–gracile complex, to the thalamus which projects to the cortex. Somatotopically organized responses to mechanical stimulation are recorded in the absence of either the cuneate–gracile or the lateral cervical pathway, but interruption of both these pathways eliminates these responses in thalamus and cortex (Bowsher, 1965). Sousa *et al.* (1971) report all of these features in opossums.

### C. Spinothalamic Systems

A third spinocerebral pathway is known as the spinothalamic, or ventrolateral column, pathway. Transmission of activity induced by pain and temperature stimulation depends on the integrity of this path, and it carries mechanosensory information as well (Mountcastle, 1968, p. 1439). Fibers originating in cells of the sensory column travel rostrally through the ventrolateral sector of the spinal white matter, to the thalamus. These can be considered in two groups (Mehler, 1966). The *neospinothalamic* travels directly from the spinal cells through the ventrolateral columns and the medial lemniscus to the thalamus. There are very few of these direct fibers in opossums, compared with carnivores and primates where they form a substantial tract (Mehler, 1957). The *paleospinothalamic* group traverse a number of synapses on their way to the thalamus (Bowsher, 1965; Mehler, 1966).

In marsupials, only responses to mechanical and electrical stimulation have been studied in the spinothalamic system. Sousa *et al.* (1971) found responses, in ventrobasal thalamus, to mechanical and electrical stimulation of the body when all other thalamic input from the bulbo- and cervicothalamic paths had been eliminated. Direct fibers of the anterolateral columns were found to proceed no further rostral than the medulla in *Trichosurus vulpecula* (Clezy *et al.*, 1961) or in *Setonix brachyurus* (Blumer, 1963). Synaptic interruption of the pathway was assumed, since typical spinothalamic activity could be recorded in the upper brainstem.

Watson and Symons (1972), however, report a few spinothalamic terminals in *Trichosurus*.

Dennis and Kerr (1961a,b) reported somatic sensory responses in midbrain and posterior thalamic nuclei, in both *Trichosurus* and placental *Felis catus*. They also found responsive regions, in both animals, in the central tegmental fascicle leading to the parafascicular nucleus of the thalamus, and in the periventricular column of very small cells ("central gray") in the midbrain. The pathways of the inputs to these regions were not determined; they may arrive by way of internuncial column and reticular formation connections. Somatic sensory responses in posterior thalamic nuclei, and in the zona incerta of the ventral thalamus, were found in *Didelphis* (Erickson *et al.,* 1964; Pubols and Pubols, 1966).

## IV. CEREBELLUM

### A. Input

#### 1. Somatic Sensory

Still other somatic sensory systems carry mechanosensory information from receptors to the cerebellum. The dorsal spinal nucleus, or Clarke's column, has already been mentioned. These cells, in the thoracic and lumbar segments of the spinal cord, receive input from mechanoreceptors in the muscles and skin of the hind limb. Their axons travel rostrally in the ipsilateral spinal cord as the dorsal spinocerebellar tract and terminate in cerebellar cortex. Other cells, in laminae V, VI, and VII of lumbar segments of the lateral part of the internuncial column, receive information from muscle receptors in the hind limb; their axons ascend to the cerebellum on the contralateral side of the spinal cord as the ventral spinocerebellar tract. Both of these tracts were functionally identified and analyzed in *Trichosurus vulpecula*; they were not different from their counterparts in placental rabbits (Magni and Oscarsson, 1962a,b). Anatomically they followed the usual pattern in *Setonix* (Blumer, 1963) and *Trichosurus* (Watson and Symons, 1972). The mode of termination of the spinocerebellar fibers, in the anterior lobe of the cerebellum, in marsupials is different from that in placentals in one respect. In both *Didelphis* (Hazlett *et al.,* 1971) and *Trichosurus* (Watson, 1973) the tract is divided into five parasagittally oriented branches, while in placentals it forms from 7 to 11 branches. Analogues for information from the front limbs exist; cells of the external cuneate nucleus (Fig. 5) project to cerebellum as the cuneocerebellar tract and are analagous to the Clarke's column and dorsal spinocerebellar tract; a rostral spinocerebellar tract is the front limb analogue of the ventral spinocerebellar path. These front limb systems have not been analyzed in marsupials.

In Fig. 5, several other cell groups projecting to cerebellum can be seen. The lateral reticular nucleus receives information from more caudal spinal regions (Watson and Symons, 1972) and from cerebral cortex in *Trichosurus* (Martin *et al.*, 1971).

The cuneate–gracile complex apparently sends a few projections to cerebellum, as does the trigeminal nuclear complex. Other inputs come from the inferior and medial vestibular nuclei, as well as from direct fibers of the vestibular nerve. All of the projections thus far mentioned enter the cerebellum by means of the inferior cerebellar peduncle (the restiform body), and have been anatomically identified in opossums, though none has been functionally analyzed (Larsell, 1970, pp. 83–84).

## 2. Pontine

A large input to the cerebellum is provided by cells of the pontine nuclei, which receive input from cerebral cortex, and send their fibers to the cerebellar cortex by way of the middle cerebellar peduncle (the brachium pontis). Lesser inputs to the pontine nuclei derive from the cerebellum itself, the superior colliculus of the midbrain and neighboring pretectal nuclei and ventral lateral geniculate nuclei, and possibly the cuneate–gracile nuclear complex (Mihailoff and King, 1975).

The general morphology, connections, and ultrastructure of the pontine nuclei have been analyzed in a series of investigations using *Didelphis virginiana* (King *et al.*, 1968; Martin and King, 1968; Yuen *et al.*, 1974; Mihailoff and King, 1975). With one exception (the presence of an additional type of presynaptic terminal in opossums), all the features of the pons of *Didelphis* are reported to be similar to those seen in placentals.

The pontine nuclei contain large cells which project outside, many or all to cerebellum, and small interneurons. Both types receive four kinds of presynaptic terminals: large terminals with large round vesicles from cerebellum, small terminals with small round vesicles from cerebral cortex, small terminals with elliptical vesicles which are probably inhibitory contacts from the interneurons, and a few small terminals with large round vesicles of undetermined origin which may be unique to opossums or marsupials since they were not found in cats.

As was shown also in *Trichosurus* (Martin *et al.*, 1971), rostral regions of cerebral cortex project to more medial pontine nuclear regions; more caudal cortex projects to more lateral pontine regions. Cerebellar cells project to the same pontine region, as do cells of sensorimotor cortex; pontine regions receiving visual and auditory cortical projections are free of cerebellar input. A selective input from cerebellum is not unreasonable as a feedback modulating influence upon the connections to cerebellum from sensorimotor cortex, which is intimately involved in programming of muscular movements.

### 3. Inferior Olive

All mammals have a distinctive group of nuclei, collectively known as the inferior olivary nuclear complex. (The name derives from the olive shaped bulge seen in this region on the ventral aspect of the brainstem.) The small cell bodies of these nuclei are remarkably homogeneous in size and density, and they are believed to be the origin of the special climbing fiber input to the cerebellar cortex.

Three major nuclei, and an additional group of cells known as the cap of Kooy, are recognized in both placentals and marsupials and can be distinguished on the basis of their morphology and connections (Bowman and King, 1974; Watson and Herron, 1974; Martin *et al.*, 1975; King *et al.*, 1975).

1. A *dorsal nucleus* receives projections from the contralateral cerebellar nuclei, a few fibers from the midbrain tegmentum, and a strong input from the contralateral cuneate–gracile complex and the ipsilateral spinal cord. The spinal and cuneate–gracile inputs are distributed such that those activated by the contralateral upper limb lie medially, and those activated by the lower limb lie laterally, in the nucleus.
2. A large *principal nucleus* is folded into upper and lower lamellae. The upper receives strong projections from the ipsilateral red nucleus (completing a conceptual circuit, olive to cerebellum to red nucleus to olive). Here a difference was found between cats and opossums (Martin *et al.*, 1975); the "short circuit" additional projection seen in cats, from cerebellar nuclei to this region, was not evident in opossums. The lower lamella, in both cats and opossums, receives fiber from midbrain tegmentum and cerebellar nuclei. The principal nucleus varies in size with the relative development of neocortex and the size of the cerebellar hemisphere (Papez, 1929, p. 216).
3. The *ventral nucleus* receives spinal projections in its lateral region, fibers from sensorimotor cerebral cortex and cerebellum in its medial region, and processes from midbrain in all regions. The corresponding nucleus in

placentals is frequently designated the "medial accessory olivary nucleus." In all marsupials examined, however, it lies lateral, rather than medial, to the other olivary nuclei; thus it is good to adopt the term ventral nucleus for consistency among all mammals. This difference between marsupials and placentals may be related to different times of embryonic migration of cells forming the olivary nuclei (Watson and Herron, 1974); this region in marsupials can provide a system of some interest for study of factors governing cell migration and nuclear morphogenesis in the mammalian central nervous system.

The difference in position of the ventral nucleus is the major difference, and the absence of cerebellar projections to the lower lamella of the principal nucleus may constitute a minor difference between the inferior olivary complex of marsupials compared to that of cats and presumably other placentals. In other respects studied, the nuclear morphology and organization appear to be similar.

### B. Morphology

The cerebellum is a great appendage on the dorsal aspect of the rostral part of the neural tube (Figs. 1C, 2), possessed by almost all vertebrates. Its function is far from clear, but it does have some controlling influence on the programing of muscular contractions. According to the extensive analysis of the development of the cerebellum in pouch young opossums by Larsell (1970, pp. 71–77), it develops around two bundles of incoming sensory fibers that cross over the top of the tube between the sensory columns. The more rostral bundle are fibers of the trigeminal (5th) cranial nerve; the posterior bundle are processes from the vestibular portion of the acoustic (8th) cranial nerve. Cells aggregating around these commissural bundles give rise to an imposing structure of homogeneous yet complex architecture: a group of centrally located nuclear cell groups lies just above the central canal, which here is greatly enlarged into the fourth ventricle; and separated from these nuclei by a thick white "medulla" of communicating myelinated fibers, there is a vast expanse of laminated cortex, neatly folded into ridges and grooves to accommodate a large surface area into a compact volume. The innermost cortical layer contains small, closely packed granule cells; above this is a layer, one cell thick, of immense, geometrically arranged Purkinje cells; and the topmost layer consists chiefly of the extensive dendritic branches of Purkinje cells and the long branching axons of the granule cells (Fig. 6).

The cerebellum, while highly variable in shape from species to species, has the same basic organization in marsupials and placentals. In both, the

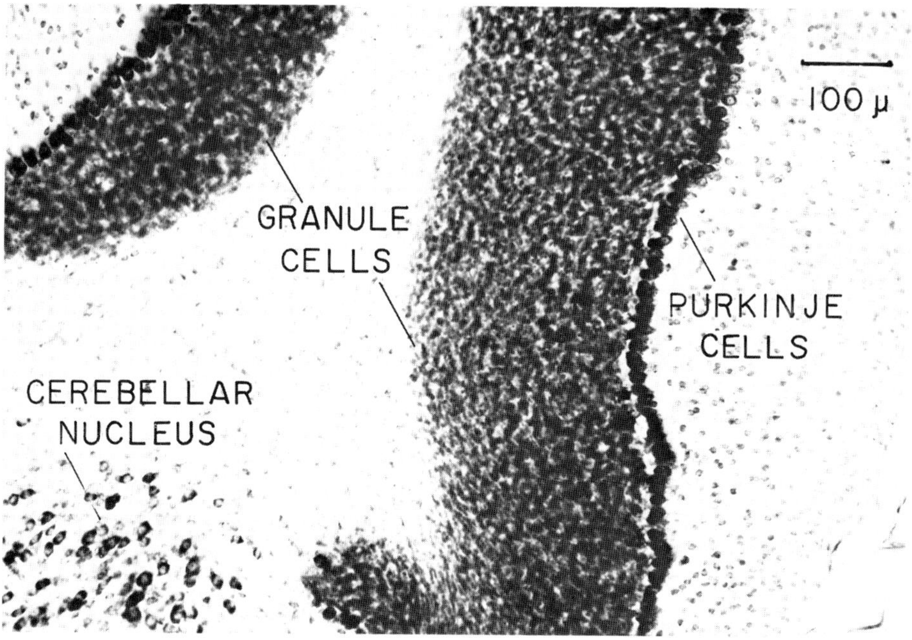


Fig. 6. Groupings of cell bodies in the cerebellum as seen in 25 $\mu$m sec. 392 from *Marmosa murina* no. 67. The large cells of the cerebellar nucleus lie dorsal to the fourth ventricle enlargement of the central canal. The white matter between the cerebellar nucleus and the layer of densely packed granule cells contains afferent fibers to the cerebellar cortex as well as axons of the Purkinje cells which terminate in the cerebellar nuclei. The cerebellar cortex includes the granule cell bodies, the single layer of Purkinje cell bodies, and outermost, the molecular layer. The molecular layer contains the extensive dendrites of the Purkinje cells and the long axons of the granule cells. The very small cell bodies seen in the white matter and in the molecular layer are those of neuroglial cells.

variations depend upon body size and probably on the complexity of muscular organization. A cylinder of folia at the midline, known as the vermis, has standard subdivisions. On either side of the vermis is a cerebellar hemisphere, and caudal and ventral to these are a flocculus and paraflocculus (Fig. 7).

Larsell (1970, pp. 77, 85–78, and 93) has pointed out the prominence of lobules I and II of the vermis in opossums, bandicoots (*Perameles*), and other marsupials and placentals with well-developed tail muscles, as an example of the relation between the complexity of cerebellar lobules and the size and functional importance of different locomotor organs. The cerebellum of the marsupial mole *Notoryctes typhlops* (Burkitt, 1938) seems to be the simplest mammalian cerebellum known; Larsell related this to the unusual mode of locomotion wherein the legs do not raise the weight of the

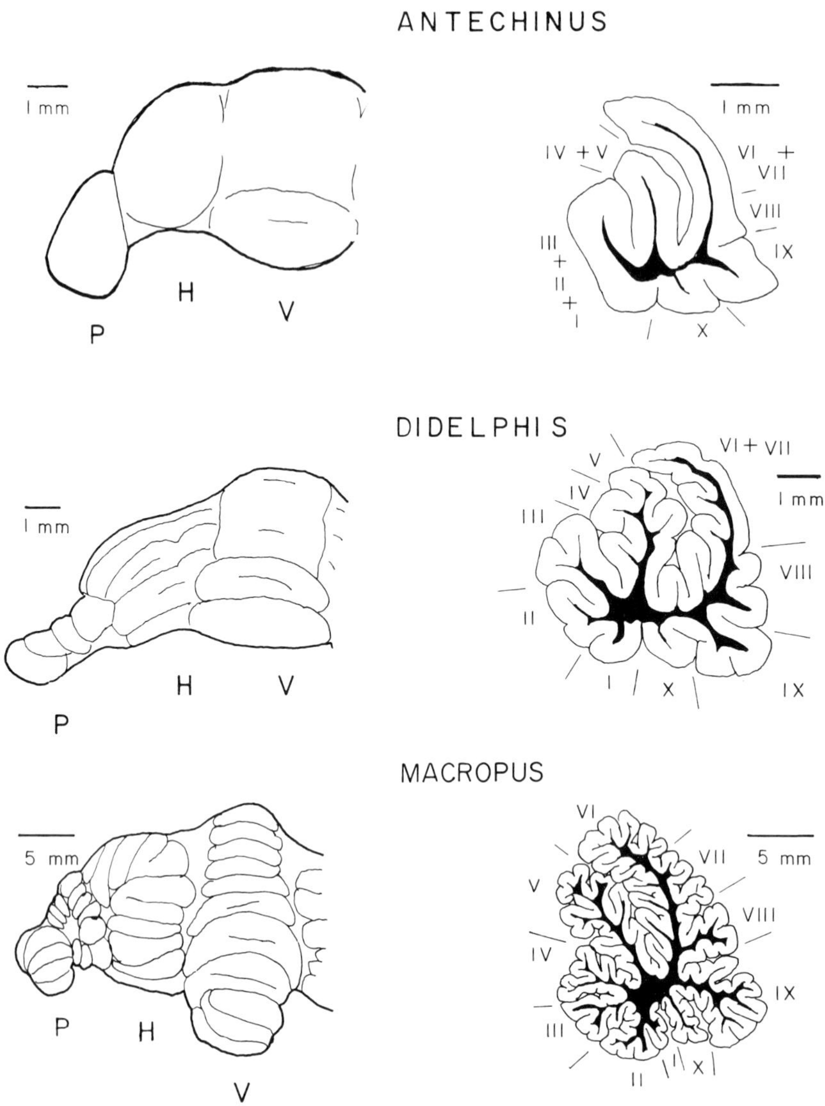


Fig. 7. Simple, intermediate, and complex degrees of cerebellar development in representative marsupials (left) and placentals (right). The outline drawings of the left sides of cerebellums viewed from their posterior aspect are traced from photographs of fixed brains. Under each, P indicates the parafloccular lobe, H the left hemisphere, and V the centrally located vermis. Next to each is a tracing of a median sagittal section through the vermis of the same brain. Around these the Roman numerals I through X indicate the homologous lobules of the vermis accord-

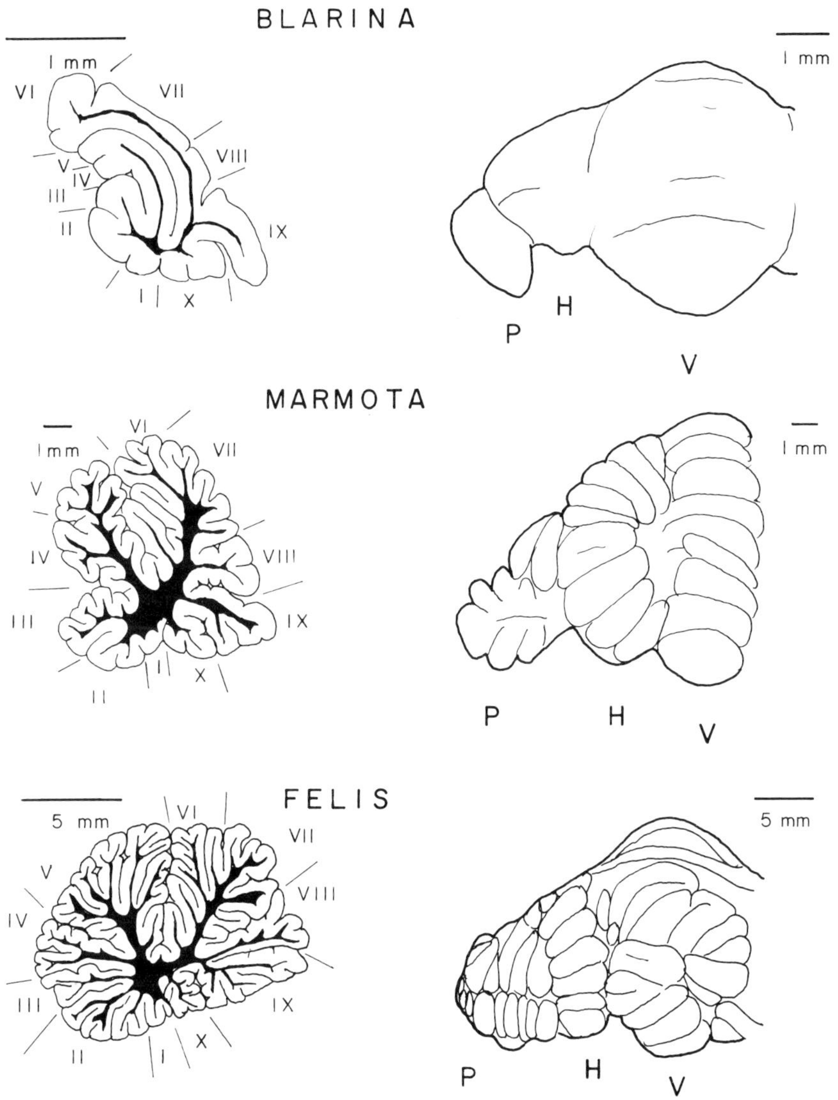


ing to Larsell (1970). For the sections the mm scales refer to the dehydrated, sectioned, and stained tissue which is considerably, and variably, shrunken from the normal size of the living specimen. Marsupials: Antechinus (an arboreal insectivorous dasyurid) *Antechinus flavipes* no. 46, sec. 167; American opossum *Didelphis virginiana* no. 73, sec. 274; red kangaroo *Macropus rufus* no. 21, sec. 632. Placentals: Short-tailed shrew *Blarina brevicauda* no 243, sec. 188; woodchuck *Marmota monax* no. 237, sec. 414; cat *Felis catus* no. 324, sec. 426.

body; vermis lobules I through V are represented by a simple anterior lobule. In other insectivorous forms which have relatively simple cerebella (these include placental moles *Talpa*, American marsupials *Caenolestes*, and Australian marsupials *Antechinus* and *Sminthopsis*) this anterior lobule is divided into distinct dorsal and ventral subdivisions (Fig. 7).

### C. Processing

The function of the various cellular components of the cerebellum have been extensively analyzed in placentals in recent years (Eccles *et al.*, 1967). The projections from the inferior olive directly stimulate the large Purkinje cells of the cerebellar cortex by means of extensive "climbing fiber" synaptic connections, where a single input fiber maintains contact with the Purkinje dendrite over a long extent of their surfaces. The pontine and the various spinocerebellar projections indirectly stimulate the Purkinje cells; they terminate as "mossy fibers" on granule cells, which in turn stimulate Purkinje cells by means of punctate synapses on the Purkinje dendrites, or inhibit them by activating intervening inhibitory basket cells. The Purkinje cells are the output channel for all cerebellar cortical processing; they terminate in inhibitory synapses upon cells of the cerebellar nuclei, which lie over the ventricle at the base of the cerebellum. Sreesai (1974) analyzed the pattern of projections from different regions of the cerebellar cortex to the separate cerebellar nuclei in *Didelphis*. The cortex of the vermis projected to the medialmost fastigial nucleus, the cortex of each of the remaining cerebellar regions projected widely to two or more of the cerebellar nuclei. The cells of the cerebellar nuclei are, in turn, the major output of the cerebellar system. In view of the remarkable similarities in gross morphology and fiber connections, it is reasonable to assume that marsupial cerebellar function is similar to that seen in placentals until it is demonstrated to be different. Rogers (1924) reported the effects of cerebellar lesions in *Didelphis* to closely resemble those seen in rabbits and dogs: rigidity, forced rolling movements, and lack of muscular control.

### D. Output

Axons from the cerebellar nuclei provide most of the output of the cerebellum to the rest of the nervous system, although some Purkinje cells terminate in vestibular nuclei in *Didelphis* (Sreesai, 1974), as in placentals. Fibers from the cerebellar nuclei leave the cerebellum by way of the superior cerebellar peduncle, the brachium conjunctivum. Their projections in opossums have been analyzed in great detail (Foltz and Matzke, 1960; Dom *et al.*, 1973; Martin *et al.*, 1974a). The principal projections were to the red nucleus in the midbrain tegmentum, the ventral nuclei of the dorsal thalamus, and the vestibular nuclei, in patterns very similar to those observed in placentals. Cerebellar nuclear projections were also found to the inferior olive (see Section V, A, 3) and a host of other regions in the brainstem and thalamus; their significance remains to be determined.

In *Didelphis*, as in placentals, cerebellar projections terminate in the anterior region of the ventral group of nuclei of the dorsal thalamus (Walsh and Ebner, 1973). In both cases, these thalamic regions send projections in turn to the "excitable motor" areas of cerebral neocortex, from which muscular movements can be induced by electrical stimulation.

Thus, the principal projections all appear to be directly related to the organization of muscular activity.

## V. SPECIAL SENSORY SYSTEMS

### A. Vestibular System

In marsupials as in all mammals, receptors sensitive to the forces of gravitation and acceleratory motion are found in the membranous labyrinth of the inner ear. These senses are called *vestibular* because this part of the labyrinth includes the vestibule and semicircular canals; the other part of the labyrinth is the cochlea which contains the auditory receptors. The course of development of the vestibular receptor organs was studied in detail in embryos and pouch young of *Didelphis* as an example of their ontogeny in mammals (Larsell *et al.*, 1935).

From the receptors the vestibular branch of the acoustic (8th) cranial nerve travels to the brainstem near the junction of the dorsal medulla and the cerebellum. This is one of the first nerves to become myelinated in opossums; this happens much later in the other, cochlear branch of the same nerve, for example. This precocity was presumed related to the role of vestibular sensation in basic postural and locomotor mechanisms in the developing young animal. Myelinated fibers were traced into the brainstem at 41 days of age, the time at which the first vestibular reflexes were observed (Langworthy, 1927; Larsell *et al.*, 1935).

In the brainstem, fibers of the vestibular nerve were found to terminate in the four vestibular nuclei (the lateral, medial, superior, and descending), and in the cerebellar nuclei. According to Langworthy (1927) this is the earliest and most prominent input to the cerebellum in the opossum pouch young 54 days of age. He also noted that in young of 46 days, fibers from the lateral vestibular nuclei could be traced into the spinal cord, while those from the superior nucleus went to the midline and possibly crossed to the other side.

All of these observations on young opossums are consistent with what is known of the connections of the vestibular nuclei in placentals. Anatomically, the vestibular nuclei in opossums (Stokes, 1912; Voris and Hoerr, 1932) appear similar to those of cats. There seem to be no studies of marsupial vestibular systems that reveal any features different from those seen in placentals.

Therefore, the following account of findings in cats (Brodal, 1969, pp. 382–383) probably apply in most respects to marsupials. The vestibular nuclei might well be termed additional cerebellar nuclei; their input from the cerebellar nuclei and the cortex of the vermis rivals that from the vestibular receptors [Watson and Symons (1972) also report an input to the vestibular nuclei from the spinal cord in *Trichosurus*]. Cells of the lateral vestibular nucleus project to the entire length of the ipsilateral spinal cord, in some cases their terminals directly contact motor neurons. In general, activation of these projections facilitates extensor reflexes. The medial and descending vestibular nuclei send processes to the cerebellar nuclei and the cortex of the caudal vermis and flocculus. All four nuclei project to midbrain and forebrain regions, and to reticular formations in the brainstem. Some of the projections have a controlling influence over nystagmic movements of ocular muscles, and some affect the general level of activation.

## B. Auditory System

### 1. Receptors in the Cochlea

The mammalian receptors for hearing lie in an extension of the vestibular labyrinth known as the cochlea. The cochlea is a set of membranous canals encased in bone; the receptors lie along one of the membranes separating the canals and are activated by sound pressure waves traveling in the fluid of the canals. In therapsid ancestral mammals the cochlea is straight; in monotremes it is curved; in marsupials and placentals it is spirally coiled, narrowing to an apex, like a snail shell (Fernandez and Schmidt, 1963). Using pouch young *Didelphis*, the development of the cochlea was correlated with the development of auditory sensitivity. Perception of high and low tonal frequencies and the maturation of the apical and basal ends of the cochlear coil occurred sometime after completion of the middle of the coil and the onset of discrimination of the middle range of tonal frequency

(McCrady *et al.*, 1937; Larsell *et al.*, 1944). This was a striking confirmation of the correspondence of differential frequency perception with the location of receptors within the cochlea. These studies constitute a major portion of the experimental findings in marsupial auditory systems, and their results fit well with what is known of placental systems.

### 2. Cochlear Nuclei

In placentals, the receptors are contacted by the distal processes from cells in the spiral ganglion, located in the axis of the cochlear coil. The other processes from these bipolar cells travel centrally as the auditory branch of the 8th cranial nerve, and terminate in the dorsal and ventral cochlear nuclei at the base of the inferior cerebellar peduncle, described in marsupials *Didelphis* (Stokes, 1912), *Macropus giganteus* (Holmes, 1903), and *Macropus rufus* (Cowley, 1973).

Cells of the cochlear nuclei project to the superior olivary complex and to the nuclei of the lateral lemniscus (Stokes, 1912; Whitfield, 1967; Cowley, 1973).

### 3. Superior Olivary Complex

The superior olivary complex is subdivided into a number of different structures; their number and degree of differentiation vary in both placental and marsupial species (Fig. 8). Their connections and function have been studied only in placentals.

Quantitative and qualitative analyses (Lyons, 1972) indicate that all the subdivisions are present in marsupials, although, as in placentals, they are much more distinct in some species than in others (Fig. 8). The medial superior olive is difficult to identify in *Didelphis virginiana*, but appears prominently in such diverse forms as the ringtail possum *Pseudocheirus peregrinus* (Petauridae) and the broad-nosed bandicoot *Isoodon obesulus* (Peramelidae). All marsupials examined, including representatives of five families, possess a large preolivary cell group, similar to that seen in the placental *Rattus*. In the marsupials the nucleus of the trapezoid body is separable into two cell groups: the lateral group of cells are smaller and more homogeneous in size than any others in the olivary complex; the cells of the medial group are distinguished by eccentric nuclei. Morest (1968) has analyzed, in pouch young opossums, the course of development of the distinctive synaptic endings of axons from the cochlear nucleus upon cells of the medial nucleus of the trapezoid body. These endings, the calyces of

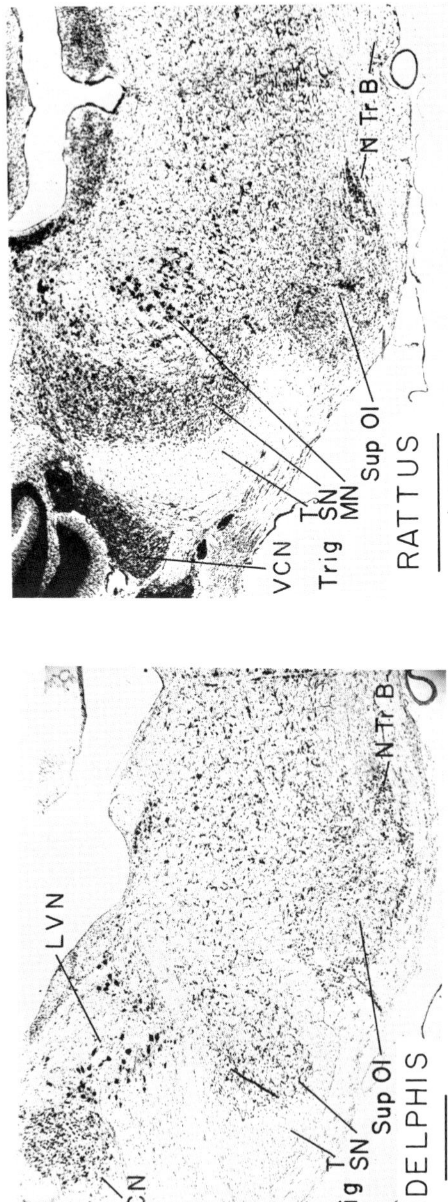
VCN
Trig
T
SN
MN
Sup Ol
N Tr B
RATTUS
1 mm
LVN
VCN
Trig
T
SN
Sup Ol
N Tr B
DIDELPHIS
1 mm

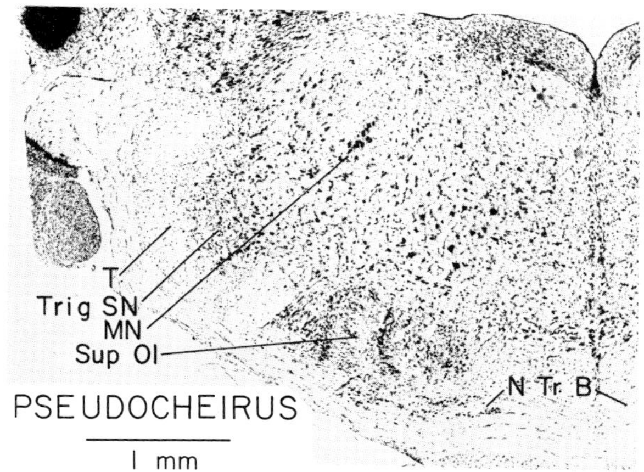


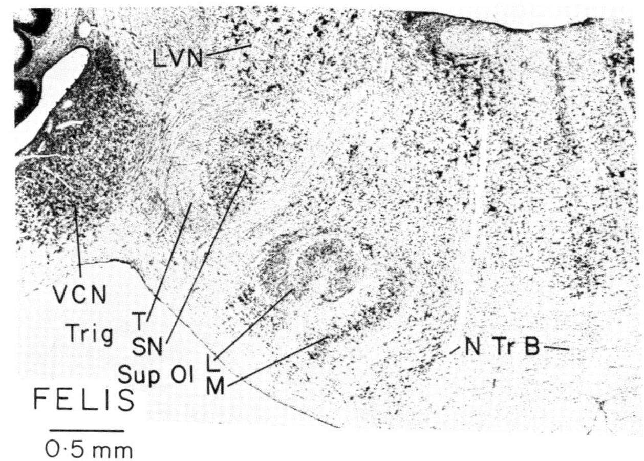


Fig. 8. Coronal (transverse) sections, stained with thionine for nerve cell bodies, through the widest part of the superior olivary complex, near the junction of the medulla and pons, beneath the cerebellum, from two marsupials (left) *Didelphis virginiana* no. 75, sec. 918 and *Pseudocheirus peregrinus* no. 38, sec. 874, and two placentals (right) *Rattus norvegicus albinus* no. 210, sec. 324 and *Felis catus* no. 324, sec. 1056. The superior olivary complex is more distinctly elaborated in *Pseudocheirus* and *Felis* than in *Didelphis* and *Rattus*. The basic sensory cell column is represented here by the sensory trigeminal nucleus, and the motor column by the motor trigeminal nucleus. LVN, lateral vestibular nucleus (Deiter's); N Tr B, nucleus of the trapezoid body; Sup Ol, superior olivary complex; L, lateral superior olivary nucleus; M, medial superior olivary nucleus; Tr B, trapezoid body; Trig, trigeminal complex (T, MN, SN); T, descending tract of the trigeminal (5th cranial) nerve; MN, motor trigeminal nucleus; SN, sensory trigeminal nucleus; VCN, ventral cochlear nucleus.

Held, are among the largest in the mammalian brain, and in both their development and final form, they are similar in opossums, rats, rabbits, and cats.

### 4. Nuclei of the Lateral Lemniscus and the Inferior Colliculus

Almost nothing is known of the connections and functions of the nuclei of the lateral lemniscus, which are prominent in marsupials [Fig. 9; see also sections from *Didelphis* in Stokes, 1912, and from *Macropus* in Cowley, 1973].

The lateral lemniscus also connects the superior olive and nucleus of the lateral lemniscus with the inferior colliculus on the posterior roof of the midbrain, in all mammals (Fig. 9). This was described in *Didelphis* by Stokes (1912).

### 5. Medial Geniculate Nucleus of the Thalamus and Auditory Cerebral Cortex

Cells of the inferior colliculus project to the medial geniculate nucleus of the thalamus by way of the brachium of the inferior colliculus (Fig. 9, and see sections from *Didelphis* in Stokes, 1912); this was experimentally verified in *Trichosurus* (Rockel *et al.*, 1972). Processes from the medial geniculate terminate in several regions of cerebral cortex and basal ganglia. Ebner (1967) has analyzed the projections in opossums; they terminate primarily in the auditory cortex in the temporal area (see Section VII, E), but some terminate in other areas of the cortex, and still others project to the putamen and globus pallidus, and the lateral amygdalar nucleus.

The functional role of thalamus and cortex in auditory sensation is far from determined, and the demonstration of centrifugal connections from almost every level of the auditory pathway (Whitfield, 1967; Martin and Megirian, 1972) promises that analysis will be complicated.

## C. Visual System

### 1. Retina

The vertebrate receptors for vision are located in the retina, a layer of cells covering the inside of the back of the eyeball, which develops from a lateral outgrowth of the neural tube. As light enters the mammalian eye, it comes through the transparent cornea at the front of the eyeball and traverses the anterior chamber to the lens, which focuses it onto the retina. The first neural element encountered, but not directly affected, in the light path is the layer of retinal ganglion cells. Next is the layer of amacrine, bipolar and horizontal cells, and finally the layer of receptor cells which are of two types, rods and cones. The rods and cones contain pigmented membranes which react to illumination and set in motion the sensory neural processes.

Light may pass beyond the layer of receptors to a reflective layer, known as the tapetum lucidum, some type of which occurs in many nocturnal animals. It has been speculated that the tapetum increases sensitivity by reflecting light back to the rods, which are specialized for detecting low levels of illumination (Walls, 1942, p. 24). Among the several marsupials possessing one, a most remarkable tapetum is found in *Didelphis virginiana* (Walls, 1939). Enlarged epithelial cells, uniquely free of pigment, are packed with reflective material, and also uniquely, appear to block peripheral circulation entirely from vessels on its internal surface. Walls terms the *Didelphis* tapetum as relatively enormous, as standing alone in having produced a highly effective retinal tapetum which is completely nonocclusible, and which may be decidedly superior to that of other mammals. This peculiarly advanced specialization should serve as a caution against ever considering any extant species as "primitive" or "generalized"; *Didelphis virginiana* is the species most abused by this type of thinking.

The retinal cones of marsupials are remarkable in that many of them are double cones, and many possess oil droplets (Hoffmann, 1876, 1877; O'Day, 1936, 1938; Walls, 1939). These features occur in reptiles, birds, and monotremes, but are not found in any placental mammals (Walls, 1939). In birds the oil droplets are pigmented, and they have been thought to function as filters to heighten perceived contrast of colored objects (Walls, 1942, p. 146). The droplets in marsupials are not pigmented, according to O'Day (1938), although Hoffmann (1876, 1877) reported blue, green, and red droplets in *Macropus rufogriseus frutica* and *M. giganteus*. The function of the oil droplets has not been determined.

The function of other specific retinal elements has not been reported for any marsupials. The retinal ganglion cells carry the output of retinal processing along their axons, which make up the optic nerves and tracts which travel from the eye to the brain. Half or more of these axons, depending upon the species, cross to the contralateral side in the optic chiasma which lies on the midline of the ventral surface of the forebrain (Figs. 9, 14, 19). Both crossed and uncrossed axons terminate in the ventral and dorsal lateral geniculate nuclei (Figs. 10, 14, 20), in the pretectal region at the

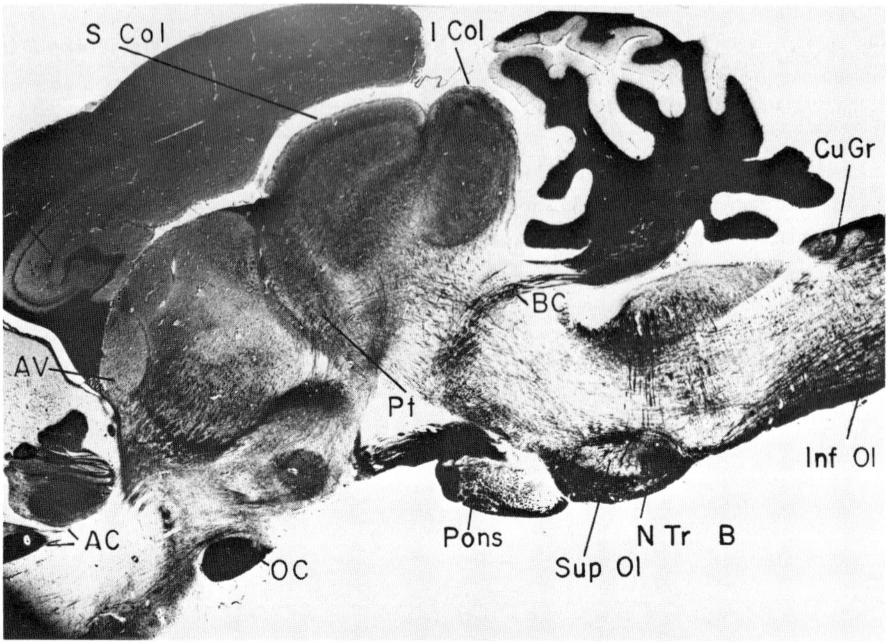
S Col
I Col
CuGr
BC
AV
Pt
Inf Ol
AC
OC
Pons
N Tr B
Sup Ol

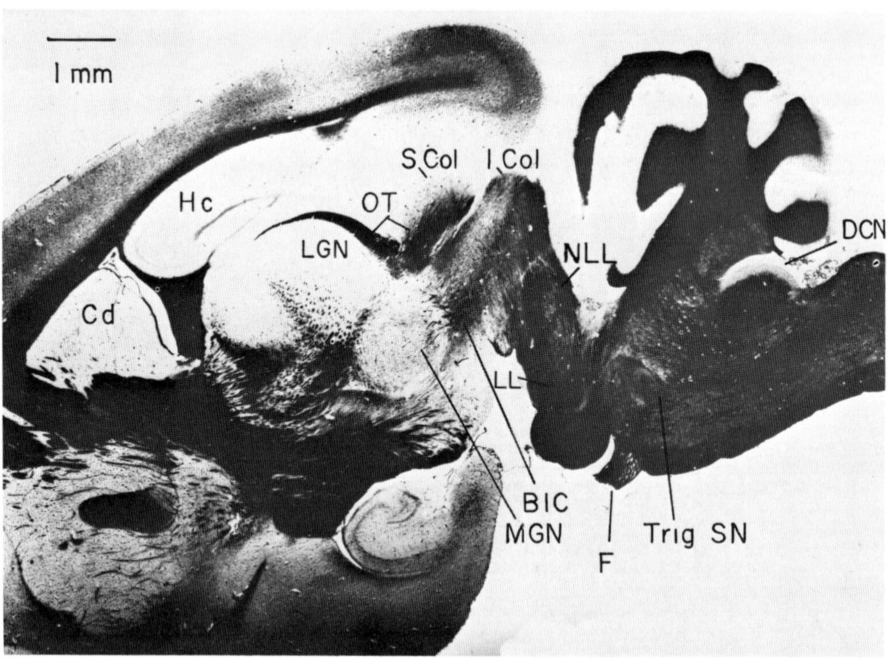
1 mm
S Col
I Col
Hc
OT
LGN
DCN
NLL
Cd
LL
BIC
MGN
F
Trig SN

junction of thalamus and midbrain (Fig. 9), in the superior colliculus on the anterior roof of the midbrain (Figs. 9, 19), and in the suprachiasmatic nucleus of the hypothalamus (Fig. 14); crossed fibers also travel to the nuclei of the accessory optic tract in the midbrain tegmentum. These projections have all been observed in *Didelphis* (Tsai, 1925a; Benevento and Ebner, 1970; Alves Cavalcante *et al.*, 1975; Lent and Rocha-Miranda, 1974).

The marsupial mole *Notorcytes typhlops* possesses at most a very incomplete eye. Receptors, optic nerves, and cranial nerves innervating the eye muscles were all reported to be absent, in contrast to placental moles where the retina and optic nerve are present although poorly developed (Sweet, 1906).

## 2. Lateral Geniculate Nucleus

In several species of the marsupial families Phalangeridae, Petauridae, and Macropodidae, as in several placental families including most Carnivora and Primates, the dorsal lateral geniculate nucleus displays a pronounced lamination of cells and fibers (Fig. 10; Goldby, 1941; Johnson and Marsh, 1969). In *Trichosurus vulpecula* of the Phalangeridae it has been demonstrated that the lamination is related to the distribution of connections from the two eyes (Packer, 1941; Hayhow, 1967) as is the case in

←

Fig. 9. Two parasagittal sections through the brainstem of *Petaurus norfolcensis* no. 28, sec. 191 above and a more lateral sec. 155 below (sections at 25 μm intervals), stained with iron hematoxylin to show myelinated fibers in black. Major features of the auditory pathway can be seen: the dorsal cochlear nucleus with its prominent unmyelinated outermost lamina, the superior olivary complex, the trapezoid body and its nucleus, the lateral lemniscus (enclosing its nucleus) connecting the superior olive with the inferior colliculus, and the brachium of the inferior colliculus leading to the medial geniculate nucleus of the thalamus. AC, components assembling into the anterior commissure as they approach the midline; AV, the ventral nucleus of the anterior thalamic group; BC, brachium conjunctivum, the superior cerebellar peduncle, connecting cerebellar nuclei with the red nucleus in the midbrain; BIC, brachium of the inferior colliculus; Cd, caudate nucleus, anterior to the lateral ventricle; CuGr, cuneate-gracile complex receiving fibers from the overlying dorsal columns of the spinal cord; DCN, dorsal cochlear nucleus; F, fibers of the trigeminal (5th cranial) nerve entering the brainstem; Hc, hippocampus, posterior to the lateral ventricle; I Col, inferior colliculus; LGN, lateral geniculate nucleus; LL, lateral lemniscus; MGN, medial geniculate nucleus; NLL, nucleus of the lateral lemniscus; N Tr B, nucleus of the trapezoid body; OC, optic chiasm; OT, optic tract, branches of which are entering the lateral geniculate nucleus of the thalamus to the left, and the superior colliculus to the right; Pons, pons; Pt, pretectal nucleus between thalamus and midbrain; S Col, superior colliculus; Sup Ol, superior olivary complex; Inf Ol, inferior olivary complex; Trig SN, sensory nucleus of the trigeminal (5th cranial) nerve.

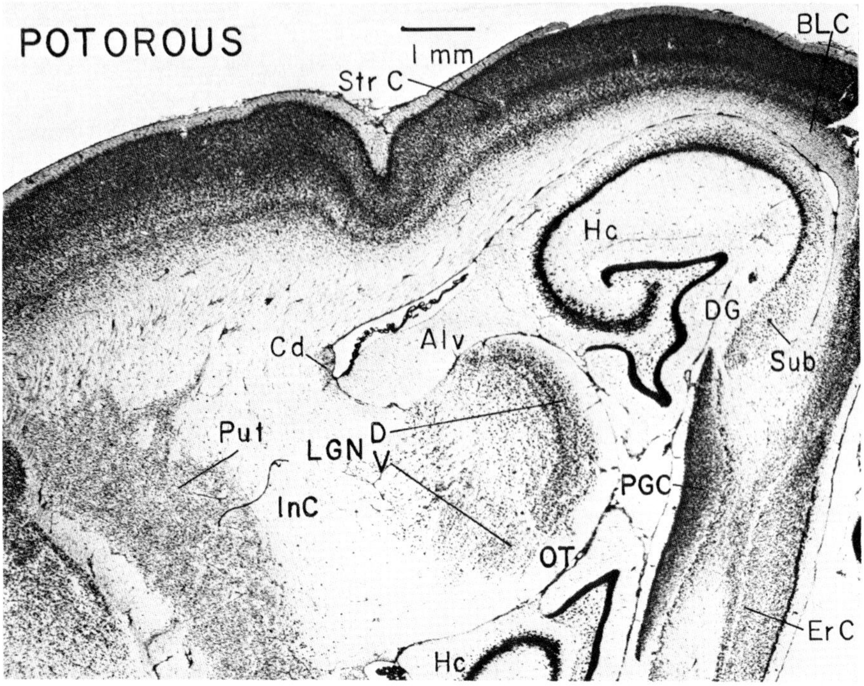
POTOROUS
1 mm
Str C
BLC
Hc
DG
Sub
Cd
Alv
D
V
Put
LGN
InC
PGC
OT
ErC
Hc

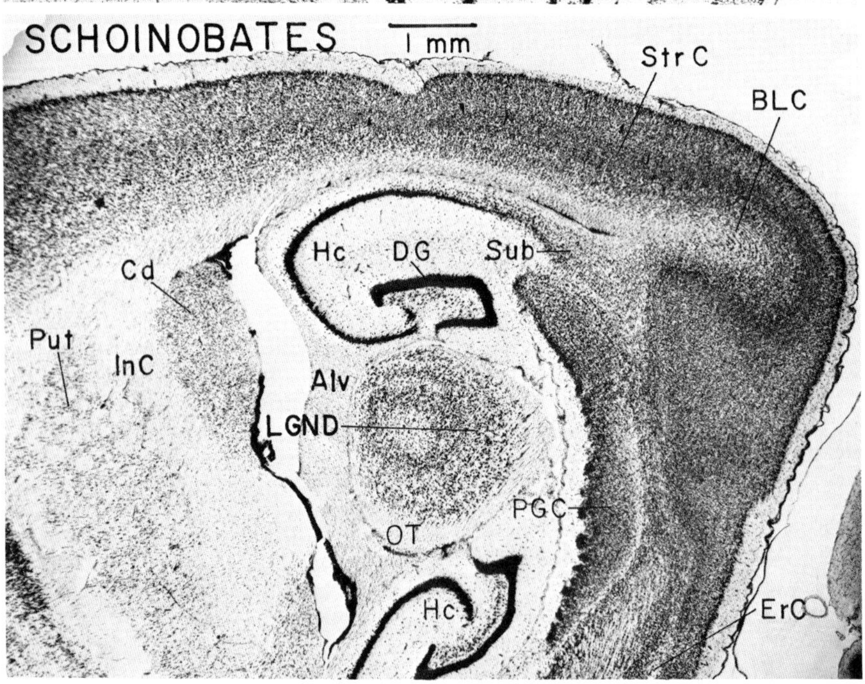
SCHOINOBATES
1 mm
Str C
BLC
Hc
DG
Sub
Cd
Put
InC
Alv
LGND
PGC
OT
Hc
ErC

Carnivora and Primates. An alternating lamination of fibers from the two eyes occurs in animals which show no evident laminar distribution of cells; such "concealed lamination" may mean that layered input from the two eyes is a general characteristic of the dorsal lateral geniculate of all mammals. However, a recent study of two didelphid species challenges this idea (Royce *et al.*, 1975). They found a "concealed" lamination in the dorsal lateral geniculate of *Marmosa*: anterograde degeneration studies showed a clear alternation of laminae delineated by projections from the ipsilateral and contralateral eyes. In *Didelphis*, however, there was extensive overlap of projections from the two eyes, with a slight preponderance of projections from the ipsilateral eye near the center of the nucleus. The functional significance of the lamination, and even of the dorsal and ventral lateral geniculate nuclei themselves, remains mysterious. These nuclei represent the only thalamic projections from the eyes other than those in pretectal regions (Bodian, 1937; Hayhow, 1967).

### 3. Visual Neocortex

Cells of the dorsal lateral geniculate project to a distinctive region of occipital cerebral cortex, where the projections terminate in a layer of such prominence that this region is known as "striate" cortex (Figs. 10 and 18). In opossums, discrete regions of the lateral geniculate, receiving projections from discrete and topologically corresponding parts of the retina, were shown to project topologically to corresponding regions of the striate cortex, but not with the precision and mutual exclusiveness that such geniculocortical projections show in monkeys (Bodian, 1935, 1937; Diamond and Utley, 1963). A similar situation in *Trichosurus* can be inferred from the experiments of Goldby (1943). The striate cortex in macropods shows a degree of elaboration seen elsewhere only in the larger primates

←

Fig. 10. Parasagittal sections, stained with thionine to show cell bodies, through the lateral geniculate nucleus and striate cortex of the visual system, in two phalangeroid marsupials of different families. Above, sec. 188, *Potorous tridactylus* no. 56 (Macropodidae); below, sec. 124, *Schoinobates volans* no. 37 (Phalangeridae). Both show lamination of the dorsal lateral geniculate nucleus, and a pronounced densely packed fourth (granular) layer in the striate cortex which is bounded anteriorly by a sulcus. The presubicular granulous cortex is prominent in these sections. The dentate gyrus of the hippocampus in *Potorous* shows the typical convolution seen in macropods. Alv, alveus of the hippocampus; BLC, basal laminar rows of cells in the lowest level of cerebral cortex; Cd, caudate nucleus anterior to the lateral ventricle; DG, dentate gyrus of the hippocampus; ErC, entorhinal cortex; Hc, hippocampus; InC, fibers of the internal capsule separating the caudate nucleus from the putamen; LGN, lateral geniculate nucleus, (D, dorsal, V, ventral); OT, fibers of the optic tract leading to the lateral geniculate; PGC, presubicular granulous cortex; Put, putamen; Str C, striate cortex; Sub, subiculum.

(Brodmann, 1906; see Section VII, D,2); this is in accord with the highly developed lateral geniculate nuclei of these two groups.

Responses of individual cells in the visual cortex of *Didelphis* have been studied in two species, the North American *D. virginiana* (Christensen and Hill, 1970) and the South American *D. marsupialis* (Rocha-Miranda *et al.*, 1973; Bombardieri, 1974). Both subspecies yielded one class of neural units in opossums that has no counterpart in placental cats and monkeys, which have been the subjects of most studies of visual cortex. These special cells were characterized by large receptive fields, very responsive to stimulation by diffuse light, and without the visual organization into concentric excitatory and inhibitory zones ("center-surround organization"). In addition to these, the South American investigators found, although Christensen and Hill did not, units corresponding to the classic types seen in placentals: simple, complex, and hypercomplex. Simple cells demand generally linear stimuli at a particular location with orientation in the receptive field before they will respond. Complex cells respond best to moving linear stimuli of proper orientation at any location in the receptive field. Hypercomplex cells have a nonexcitable region around the receptive field, but when stimuli encroach upon this region, their normal response is inhibited (thus the stimulus must be "stopped," or have an end, so as not to encroach upon this inhibitory region, or else little or no response will be obtained). All hypercomplex opossum cells were not orientation selective, and thus differ in this respect from their counterparts in cats and monkeys.

### 4. Superior Colliculus of the Midbrain Tectum

As in the dorsal lateral geniculate nucleus of many mammals, the retinal projections to the superior colliculus also terminate in a laminar pattern. Six layers of cells and myelinated fibers are seen in marsupials and placentals. Fibers from the contralateral retina form the third layer from the top, and their terminals extend into the two more dorsal layers (Figs. 9, 19B); fewer fibers project to the same layers from the ipsilateral retina; these findings were similar in *Trichosurus* (Hayhow, 1967) and *Didelphis* (Bodian, 1937; Giolli, 1965; Alves Cavalcante *et al.*, 1965). Using radioautographic techniques in *Didelphis marsupialis*, Alves Cavalcante *et al.* (1975) showed that the ipsilateral projection is to the outermost portion of the top layer only, in a region just superficial to that of the contralateral projections.

In what seems to be the first report of the activity of single neurons in a marsupial visual system, Hill and Goodwin (1968) analyzed response properties of neural units in the superior colliculus of opossums. They reported opossum units to be similar in activity to corresponding units in rabbits.

Projections from the superior colliculus of opossums were traced to the deeper layers of the contralateral superior colliculus; the magnocellular part of the medial geniculate nucleus; the lateral terminal nucleus of the accessory optic system; the ventral lateral geniculate nucleus; the nuclei of the lateral group and the parafascicular and paracentral nuclei of the dorsal thalamus; the pretectal region; the subthalamus and zona incerta; the oculomotor nuclei; the tegmentum and reticular formations of the midbrain and lower brainstem; and the pontine nuclei (Martin, 1969; Benevento and Ebner, 1970; Rafols and Matzke, 1970).

These findings are in general agreement with data from placental mammals, showing an extremely diffuse influence from the superior colliculi upon the rest of the central nervous system.

## 5. Pretectal Nucleus

A third terminal region for fibers from retinal ganglion cells is the pretectal nucleus (Fig. 9). Warner (1970) traced the origins of this nucleus in young *Trichosurus vulpecula*, and found it to be derived from the caudal extremity of the dorsal thalamus, as is the case in the placental mouse.

Cells of the pretectal nucleus in *Didelphis* project to the lateral intermediate nucleus of the thalamus, and receive projections from striate cortex (Benevento and Ebner, 1970). The function of this nucleus is relatively unknown; it plays some role in reflex eye movements.

## 6. Accessory Optic System

Still another group of fibers from retinal ganglion cells travel in the accessory optic system, which has been thoroughly analyzed in marsupials *Trichosurus* (Hayhow, 1966) and *Didelphis* (Giolli, 1965; Lent and Rocha-Miranda, 1974; Alves Cavalcante *et al*., 1975). After crossing in the optic chiasm, these fibers leave the main optic tract and travel across the cerebral peduncle, many on its external surface, and end in the medial and lateral terminal nuclei of the accessory optic tract in the tegmentum of the midbrain. This arrangement is similar to that reported in placentals, although the marsupials are more similar to rodents and lagomorphs in this respect than any of these are to carnivores, artiodactyls, and primates. This pathway is more prominent in reptiles, but little is known of its function in any animal.

### 7. Visual Input to Hypothalamus

Radioautographic techniques showed retinal projections terminating in the suprachiasmatic nucleus (see Fig. 14) of South American opossums *Didelphis marsupialis aurita* (Alves Cavalcante *et al.*, 1975). These projections were reported to be similar to those seen in placental mammals, where the suprachiasmatic nucleus of the hypothalamus has been implicated in the regulation of light-sensitive diurnal cycles.

## D. Olfactory System

### 1. Olfactory Bulbs

These bulbs are a pair of outgrowths which extend from the anterior end of the neural tube; each includes an extension of the ventricle. The olfactory bulbs are exceptionally prominent in all marsupials (Figs. 11, 13, 15), in some cases making up almost half of the forebrain. The structure of the bulbs in marsupials is essentially similar to that in other mammals, but often displays many architectural characteristics more obviously, perhaps reflecting more specialized development (Fig. 12). C. L. Herrick (1892) and C. J. Herrick (1924) have described the detailed anatomy of the bulb in *Didelphis*.

The fibers from the receptive epithelium terminate within glomeruli around the periphery of the bulb. Also terminating in branched endings in the glomeruli are efferent endings of the large mitral and tufted cells of the bulb, which are in synaptic contact with the endings of the axons from the epithelium. The distribution and spatial organization of cell types in the olfactory bulbs of members of the marsupial family Dasyuridae has been analyzed by Switzer (1972). He observed that tufted cells are unevenly distributed in the outer plexiform layer, at times almost forming distinct laminae (Fig. 12). The large mitral cells, as in placentals, form a continuous and distinct layer which is one cell thick, reminiscent of the Purkinje cells in the cerebellum. The axons of the mitral cells travel to the center of the bulb and then into the brain, largely and perhaps entirely through the lateral olfactory tract. Axons of the tufted cells also project through the lateral olfactory tract; this was demonstrated in *Didelphis* (Switzer, 1973). These axons of mitral and tufted cells are thus a major output from the bulb to the rest of the central nervous system. Between the mitral layer and the ventricle are found closely packed granule cells, again recalling cerebellar architecture. But these olfactory granule cells are organized into discrete discoidal bodies made of columns of cells (Switzer,

1972); these discoids are especially prominent in many marsupials (Fig. 12). Other granule cells are found packed around the glomeruli. In several vertebrates, including placental mammals, the granule cells have been shown to contact dendrites of mitral cells with reciprocal synapses; the mitral cell exerts a presumed excitation through a synapse with the granule cell ending, whereupon the granule cell through a neighboring synapse exerts an inhibitory effect on the mitral cell (e.g., Andres, 1970). Granule cells receive some input from cells outside the olfactory bulb (Andres, 1970).

Functional explanations of activity of elements of the olfactory bulb are rare. Phillips and Michels (1964) in opossums found that firing rates in the mitral cell layer in the anterior part of the bulb were increased by certain classes of olfactory stimuli acting on the receptor epithelium; in the posterior part of the bulb firing rates were inhibited by these same stimuli, while rates were increased in the posterior bulb by other stimuli judged by humans to be at the opposite end of an olfactory spectrum from the first group of stimuli. This may indicate regional specialization within the bulb.

## 2. Central Connections of the Olfactory Bulbs

Fibers leave the bulbs in two main groups, which are distributed similarly in placentals and marsupials, as shown by a number of anatomical and physiological studies (Marsupials: *Dasyurus*, Flashman 1910a; *Didelphis*, Putnam and Cone, 1966; *Trichosurus*, Adey, 1953; Placentals reviewed, Adey, 1970).

In the first of these main groups, the lateral olfactory tract, there is a topographical correspondence between the position of cells on the surface of the bulb and the position of their projections in the tract, as determined electrophysiologically in *Didelphis* (Shepherd and Haberly, 1970). Cells on the dorsal part of the medial side of the bulb send fibers through the dorsal part of the tract; more ventral in the tract are found, in regular progression, fibers from cells in the dorsal, then the lateral, the ventral, and finally the medioventral aspect of the bulb.

The lateral olfactory tract travels along the ventrolateral aspect of the forebrain, and terminates in the anterior olfactory "nucleus" or retrobulbar areas of cerebral cortex, the piriform cortex, the cortical amygdaloid nucleus, and the olfactory tubercle. A second group of fibers projects to the anterior continuation of the hippocampus. A large number of fibers, here called the medial olfactopetal tract, cross in the prominent anteroventral portion of the anterior commissure (Figs. 13, 19) to the contralateral retrobulbar area or to the contralateral olfactory bulb.

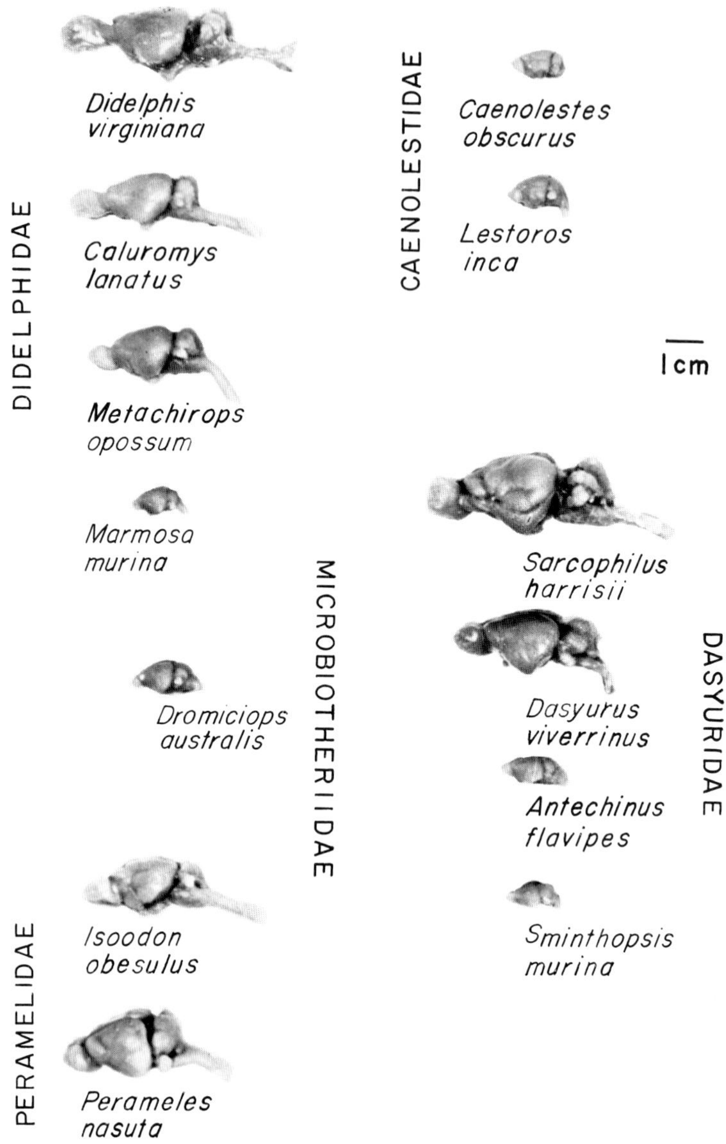


Fig. 11. Range of external morphologies in marsupial brains, illustrated by lateral views of brains of representatives of 9 of the 11 extant polyspecific families, shown in photographs of acrylic casts. The large size of the olfactory bulbs, and the basic pattern of sulci when they occur in the cerebral cortex, are consistent in all the marsupial families.

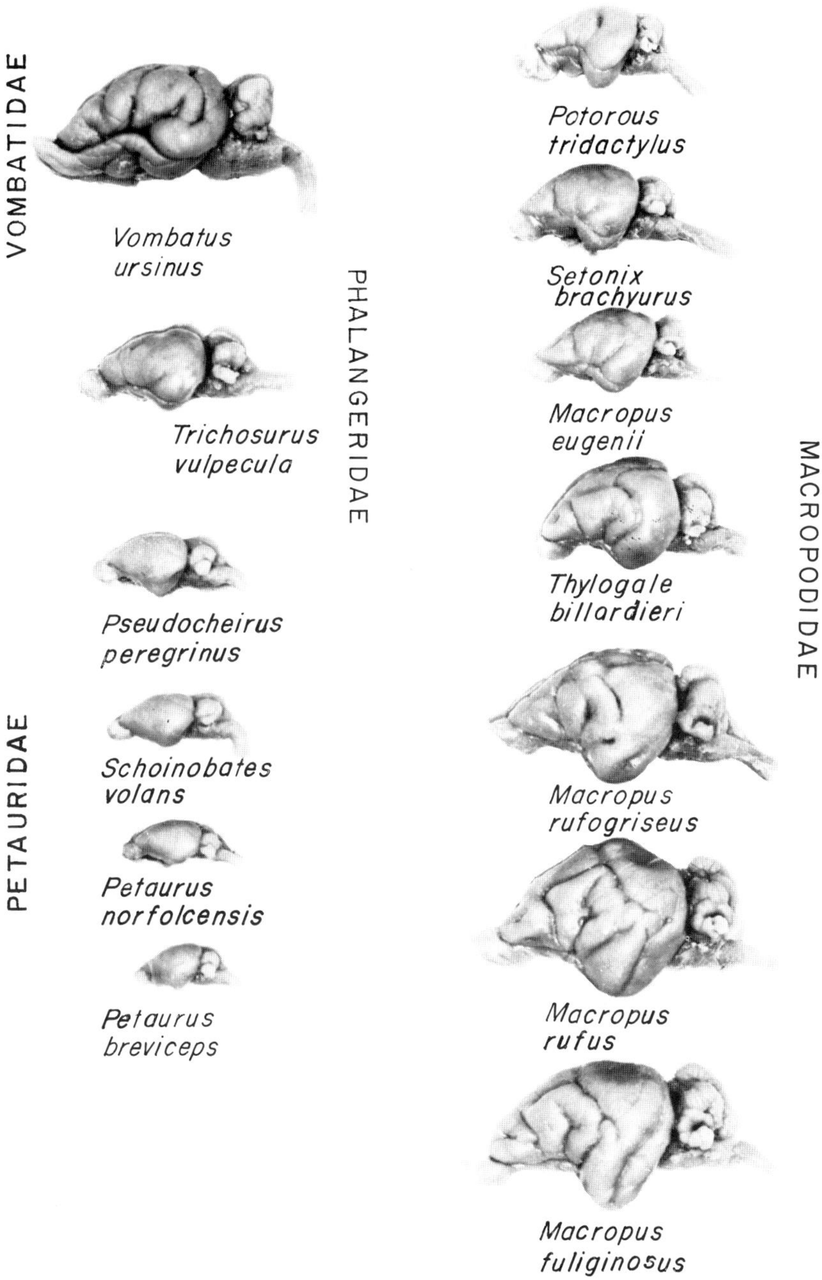
VOMBATIDAE
Vombatus ursinus
PHALANGERIDAE
Trichosurus vulpecula
Pseudocheirus peregrinus
PETAURIDAE
Schoinobates volans
Petaurus norfolcensis
Petaurus breviceps
MACROPODIDAE
Potorous tridactylus
Setonix brachyurus
Macropus eugenii
Thylogale billardieri
Macropus rufogriseus
Macropus rufus
Macropus fuliginosus

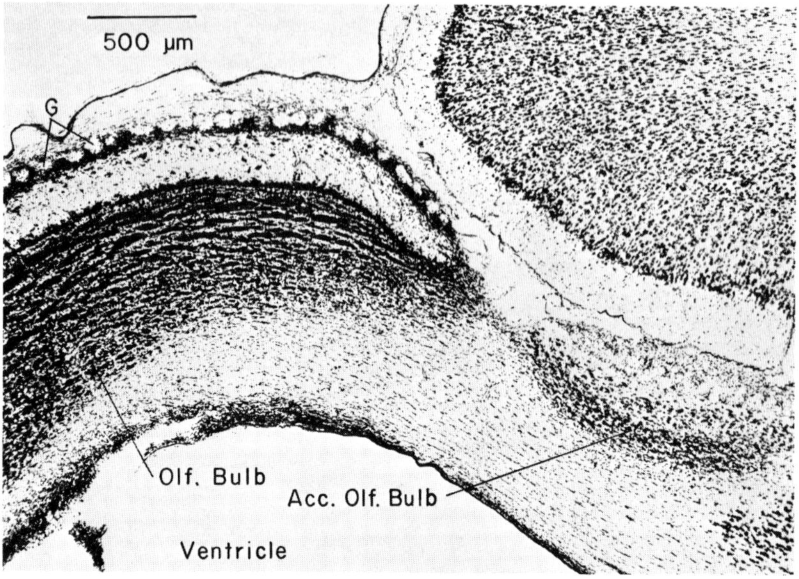
500 µm
G
Olf. Bulb
Acc. Olf. Bulb
Ventricle

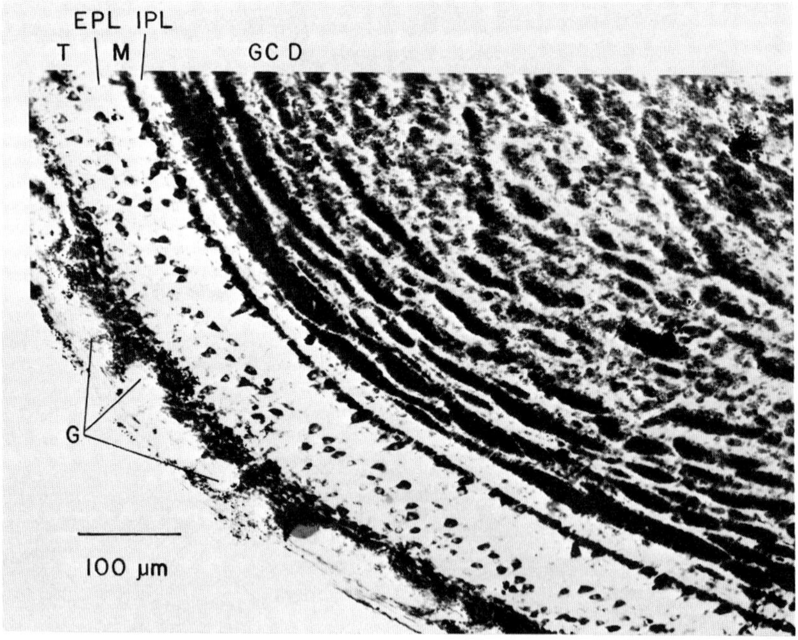
EPL IPL
T
M
GC D
G
100 µm

The terminal regions of these fibers from the olfactory bulbs have wide secondary connections with the thalamus, hypothalamus, hippocampus, and other regions of the brain, but their olfactory function has never been clarified. As in other sensory systems, centrifugal projections have been found (in rodents) from the brain to the bulb; these include processes from the olfactory tubercle and from the preoptic area by the hypothalamus (Adey, 1970).

### 3. Vomeronasal Organ and Accessory Olfactory Bulb

A second sensory epithelium occurs in the vomeronasal organ in the more anterior respiratory passage in many vertebrates, including marsupials (e.g., *Notoryctes*, as described in detail by Sweet, 1904) and placentals, although it is absent in many adult primates, including humans. Tucker (1963) has reported it to be responsive to odors, and it is very like olfactory epithelium in structure, except that its axon-sending cells do not have cilia (Andres, 1970). The axons terminate in the accessory olfactory bulb, and form the vomeronasal nerve (Fig. 15). This was identified in *Didelphis* by McCotter (1912). The accessory olfactory bulb is located in the proximal dorsal surface of the olfactory bulb (Fig. 12). Its internal architecture replicates most of the features of the main olfactory bulb while remaining separate from it (Figs. 12, 15). The mitral cells are not distinct from the tufted cells, nor do they form a single layer.

Projections from the accessory bulb, in comparison with those from the main olfactory bulb, were studied in rabbits, rats, and marsupial opossums

Fig. 12. Parasagittal sections through olfactory bulbs of two marsupials, thionine stain. Below: a detailed enlargement showing the distribution of nerve cell bodies in the periphery of the bulb in the dasyurid *Sminthopsis murina*, no. 17 sec. 174. Above: the olfactory bulb, the accessory olfactory bulb, the overlying neocortex (upper right), and the edge of the retrobulbar area (lower right corner) are seen in sec. 446 of the macropod quokka *Setonix brachyurus* no. 80. Acc. Olf. Bulb, accessory olfactory bulb which shows layers of glomeruli, a band of scattered large cell bodies, less rigidly organized than in the main olfactory bulb; EPL, external plexiform layer between the mitral cell and glomerular layers; G, glomeruli, circular patches containing synaptic contacts between incoming processes from receptors and dendrites of cells of the bulb, here seen outlined by the small granule cell bodies packed around them; GCD, granule cell discoids, made up of tightly juxtaposed columns of internal granule cells, seen here in sections perpendicular to both long axes of each discoid; IPL, internal plexiform layer between the layer of mitral cell bodies and the internal granule cell discoids. IPL is not evident in *Setonix* and many other marsupials and placentals; M, layer of mitral cell bodies, a monolayer (one cell body in thickness) in the main bulb; processes of these cells constitute a major output of the olfactory bulb; Olf. Bulb, main olfactory bulb; T, tufted cell bodies, which approach a laminar arrangement in *Sminthopsis* and other dasyurids. Their processes also pass out of the bulb as an output pathway.

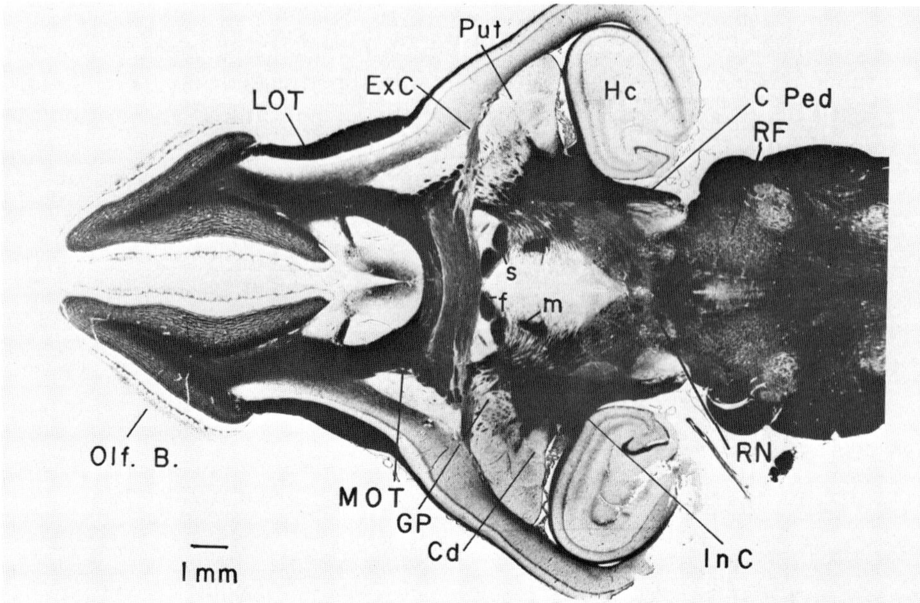


Fig. 13. Horizontal sec. 263 through the ventral part of the forebrain of bandicoot *Isoodon obesulus* no. 32; iron hematoxylin stain for myelin. Olfactory bulbs and their major connections are prominent. Cd, tail of the caudate nucleus; C Ped, cerebral peduncle, containing fibers from the internal capsule extending into the midbrain; ExC, external capsule, from which fibers cross to the other hemisphere in the anterior commissure; f, descending column of the fornix; GP, globus pallidus; Hc, hippocampus; InC, internal capsule; LOT, lateral olfactory tract, continues caudally along the surface of the piriform cortex; m, mammillothalamic tract; MOT, medial olfactopetal tract, continues to the other hemisphere as the anteriormost component of the anterior commissure; Olf. B., olfactory bulb; Put, putamen, which merges posteriorly with the tail of the caudate nucleus Cd; RF, reticular formation of the midbrain; RN, red nucleus; s, stria medullaris.

by Scalia and Winans (1975). In all three species, they found projections from the accessory bulb confined to the bed nucleus of the stria terminalis, the medial amygdaloid nucleus, the posteromedial subdivision of the cortical amygdaloid nucleus, and (in cats and rats) a group of cells between the medial and cortical amygdalar regions which they termed the "bed nucleus of the accessory olfactory tract"—this nucleus was not present in opossums. Projections from the main olfactory bulb in all three species were found in the anterolateral two-thirds of the cortical amygdaloid nucleus, the piriform cortex, the anterior commissure leading to the contralateral olfactory bulb, but not in the nucleus of the lateral olfactory tract as reported by other investigators.

## VI. BASAL TELENCEPHALON

In association with the bilateral extensions of the neural tube into the olfactory bulbs, there have developed in vertebrate brains paired outgrowths on the dorsal and ventral surfaces of the forked end of the tube, becoming collectively the telencephalon. The dorsal structure has been termed the pallium, and consists in a layered mantle of cells, the cerebral cortex (Ariëns Kappers *et al.*, 1936, Vol. 3, p. 1254). The ventral structures form several nuclear aggregations, known collectively as the basal ganglia. Another group of nuclei, located around the junction of the two branches of the anterior neural tube (the junction of the cerebral hemispheres), make up the diencephalon which includes the thalamus, epithalamus, and hypothalamus.

### A. Basal Ganglia

#### 1. Corpus Striatum and Globus Pallidus

As the neural tube bifurcates at its anterior end, so does the central opening; the branches of the central canal form the two lateral ventricles. On the ventrolateral wall of each lateral ventricle, in all vertebrates, is a cellular mass known as the corpus striatum (Ariëns Kappers *et al.*, 1936, Vol. 3, p. 1246, Fig. 533). In mammals this is divided into two segments, the caudate nucleus and the putamen. The caudate nucleus adjoins the ventricle wall throughout its length, and fuses with the putamen at its large front end and at the posterior end of its long tail (cauda). Between these extremes, the fibers of the internal capsule separate the caudate from the putamen. In marsupials (Pilleri, 1962) as in placentals, the caudate and putamen consist of similar masses of small, closely packed cells, with a few large multipolar cells. At its wide rostral end, the corpus striatum merges in the septal region with an ill-defined accumbens nucleus of similar internal architecture (Figs. 15, 19B).

Closely adjoining the internal surface of the more caudal part of the putamen is a distinctive group of large cells in low density. These make up the globus pallidus, which is small in marsupials as compared with many placentals, particularly primates. Similar large cells extend along and into the continuation of the internal capsule into the cerebral peduncles; these make up the entopeduncular nuclei and the substantia nigra.

Figures 14 and 15 illustrate the topographical arrangement of the corpus striatum and the globus pallidus as seen in coronal and horizontal planes in

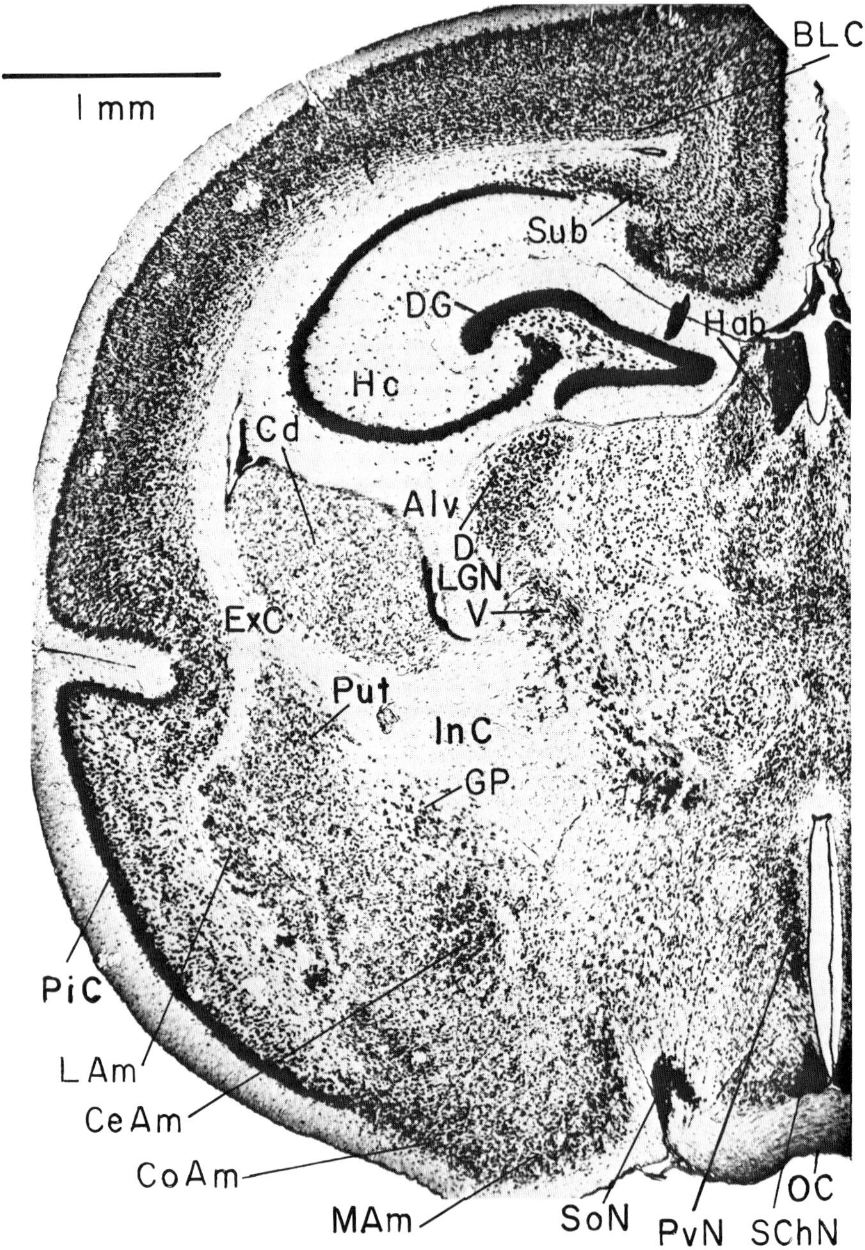
1 mm
BLC
Sub
DG
Hab
Hc
Cd
Alv
D
LGN
V
ExC
Put
InC
GP
PiC
LAm
CeAm
CoAm
MAm
SoN
PvN
SChN
OC

two marsupials; these show the typical mammalian organization. These are large nuclei, with extensive connections throughout the brain, and determining their function remains a major task of neural science. Their injury or destruction produces tremors and disorders of posture, gait, and voluntary control of muscular contraction (observed in *Didelphis* by Rogers, 1924); these disorders are often alleviated by further destruction of these regions.

## 2. Claustrum

Between the putamen and the cerebral cortex lies a group of cells known as the claustrum. It varies in size among mammals and marsupials. For example, it is difficult to distinguish in *Marmosa* (Fig. 14) and *Didelphis* (Gray, 1924), but is a structure of imposing dimensions in *Macropus rufus* (Fig. 15). Its origins, connections, and functions are equally obscure. Pilleri (1961) identified two cytoarchitectural areas in the claustrum of *Didelphis virginiana* and rodents: a dorsal, cell-rich region with three cell types—astroid, spherical, and pyramidal; and a ventral, cell-poor region with the same three cell types plus a fourth type whose cell bodies were larger than the others.

Fig. 14. Coronal sec. 224 through the left cerebral hemisphere of *Marmosa murina* no. 67 showing principal features of forebrain organization; thionine stain. Alv, alveus of the hippocampus, consisting of efferent fibers from the cornu ammonis; BLC, basal laminae of the cerebral cortex, a pronounced feature of marsupial brains: the lowest cells are lined up in rows separated by myelinated fibers; Cd, caudate nucleus; CeAm, central amygdaloid nucleus; CoAm, cortical amygdaloid nucleus; DG, dentate gyrus of the hippocampus, with a striking band of densely packed granule cells overlying scattered pyramidal cells; ExC, external capsule; GP, globus pallidus; Hab, habenular nuclei of the epithalamus; Hc, hippocampus, whose main structure, the cornu ammonis, is a thin dense band of closely packed pyramidal cells; InC, internal capsule; LAm, lateral amygdaloid nucleus; LGN, lateral geniculate nucleus, (D, dorsal LGN of the dorsal thalamus; V, ventral LGN of the ventral thalamus); MAm, medial amygdaloid nucleus; OC, optic chiasm, fibers of the optic nerves crossing to the contralateral hemispheres; PiC, piriform cortex, marked by a dense cell layer, the second layer, immediately under the outermost molecular layer which is relatively free of cells. The rhinal fissure, the only sulcus on the lateral surface, separates piriform cortex from the neocortex. PvN, paraventricular, SChN, suprachiasmatic, SoN, supraoptic, nuclei of the hypothalamus; Sub, subiculum connecting neocortex and hippocampus. Along the midline at this level are the cell masses of the diencephalon, the epithalamic Hab, the thalamus including its LGN, and the hypothalamus including PvN and SoN. Lateral to these are the basal ganglia Cd, Put, and GP and the amygdaloid nuclei LAm, CeAm, CoAm, and MAm. Covering the dorsal and lateral surfaces are the layered cell bodies of the pallium: the paleocortical PiC, and across the rhinal fissue from PiC the long arc of the neocortex with its BLC, then near the midline the archicortical Sub and the folded Hc with its DG.

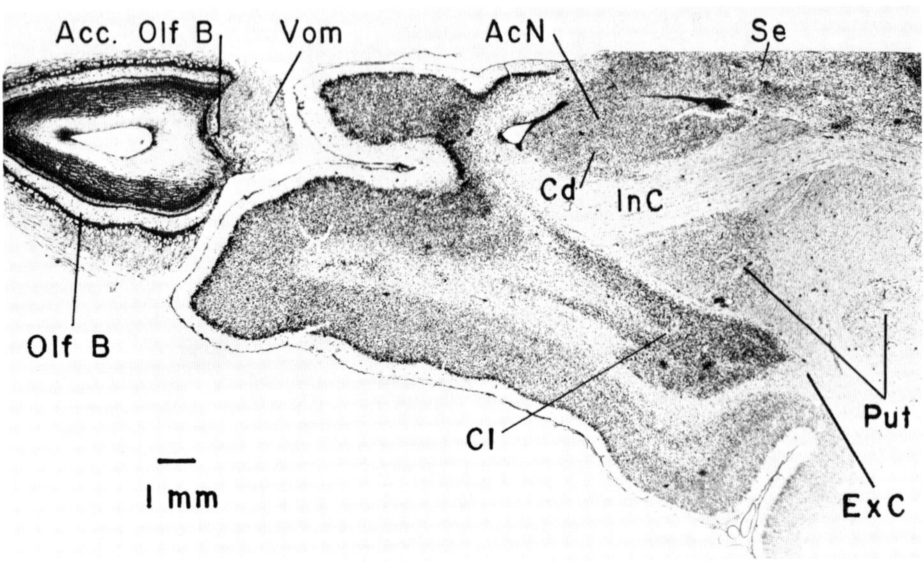


Fig. 15. Horizontal sec. 708 at a ventral level through the left anterior forebrain of red kangaroo *Macropus rufus* no. 21; thionine stain. The midline is at the top and anterior is to the left. The vomeronasal nerve Vom, accessory olfactory bulb Acc Olf B, and the remarkable claustrum Cl are prominent. Acc. Olf B, accessory olfactory bulb; AcN, accumbens nucleus, which merges with the caudate Cd and septal Se nuclei; Cd, head of the caudate nucleus; Cl, claustrum; ExC, external capsule, fibers of which invade the putamen Put on their way to joining the anterior commissure; InC, internal capsule; Olf B, olfactory bulb; Put, putamen, pierced at this level by fibers from the external capsule; Se, septal region; Vom, vomeronasal nerve.

### 3. Amygdaloid Complex

The classic fundamental description of the nuclei of the amygdaloid complex and their connections is that of Johnston (1923) in the opossum. Johnston considered the amygdaloid nuclei in two classes, the older corticomedial group and the more recent basolateral group.

**a. Corticomedial Group.** Johnston believed the corticomedial group to be derived from the lateral olfactory areas of earlier aquatic vertebrates, and this group retains olfactory input and connections with the hypothalamus through the stria terminalis. It includes the nucleus of the lateral olfactory tract and the central amygdaloid nucleus, both of which are prominent in marsupials but reduced in placentals (Fig. 16; Hamel, 1966a), as well as the cortical and medial nuclei at the outer margin of the complex. In both placentals and marsupial opossums, the cortical nucleus has been subdivided into three subregions on cytoarchitectural grounds. Axons from the accessory olfactory bulb were found to project to the medial nucleus and the most medial subdivision of the cortical nucleus; projections from the main olfactory bulb terminate in the two more lateral regions of the cortical nucleus (Scalia and Winans, 1975).

**b. Basolateral Group.** This other group, found in terrestrial vertebrates and their descendants, is made up of the basal and the large lateral amygdaloid nuclei. These nuclei have close connections with the neighboring piriform cortex, with which they share a common embryonic origin. They also connect with the hypothalamus and the external capsule; and rather than through the stria terminalis, they project directly through the temporal component of the anterior commissure (Ariëns Kappers *et al.*, 1936, Vol. 3, p. 1437).

A peculiar branched lamination of very small cells, the massa intercalatus, intervenes between portions of the amygdaloid groups and spreads into the diffuse anterior amygdaloid area. As with the rest of the amygdaloid structures, this formation is much better defined in marsupials than in placentals. In many marsupials it almost entirely encloses the central nucleus and forms a nearly continuous wall between the lateral nucleus and the rest of the amygdala (Fig. 16). In cats, it occurs only as small scattered clumps.

Studies of the function of these separate components of the amygdala would be facilitated by the use of marsupial subjects, where subregions are more readily identified. In placentals, profound effects upon behavior and physiology, concerning aversions, attractions, curiosity and exploration, and rage and attack, have been produced by lesions and stimulations of the amygdala (some of these are reviewed by Johnson *et al.*, 1969), and Hara and Myers (1973) concluded that in *Didelphis* the amygdala was the "forebrain structure most prominently involved in controlling emotional behavior, especially in relation to facial threat and attack," since these behaviors disappeared following destruction of the amygdala.

### B. Paleocortex and Hippocampus

The dorsal and lateral surfaces of the paired hemispheres of the forebrain are covered by the pallium or cerebral cortex. In mammals, this falls into three main divisions. The central division, neocortex, is peculiar to mammals, and it is bordered laterally by paleocortex and medially by archi-

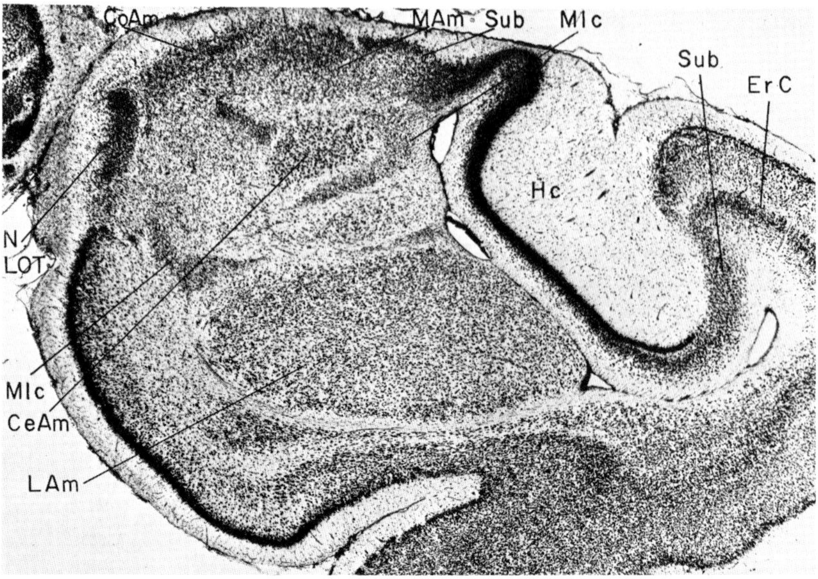


TRICHOSURUS

1 mm

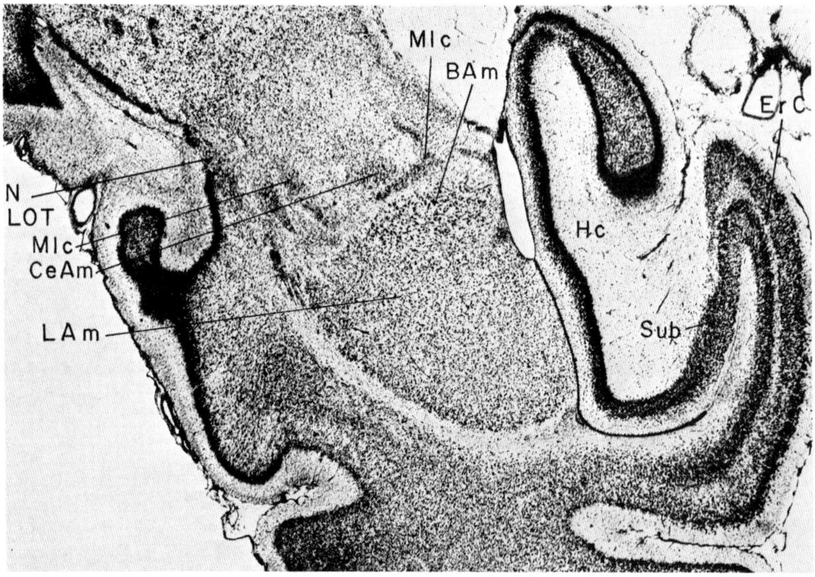


FELIS

1 mm

cortex. Paleocortex and archicortex meet at their anterior and posterior poles, thus completely surrounding the neocortex. The "ring" formed by paleocortex and archicortex around neocortex has been termed the limbic region; this term often includes some of the basal ganglia and peripheral neocortex (Crosby *et al.*, 1962, p. 349).

### 1. Paleocortex

The paleocortex has been described in histological detail in the marsupials *Didelphis virginiana* (Gray, 1924), *Didelphis albiventris* (Rose, 1926–1927), *Caenolestes obscurus* (Obenchain, 1925), *Perameles nasuta* (Abbie, 1942), and a kangaroo whose species was not indicated (Filimonov, 1949). Each used a different terminology and subdivided regions differently, but their descriptions reveal constant features which are also seen in placentals (Crosby *et al.*, 1962, pp. 416–425). All recognized six major regions:

1. Between the olfactory bulb and the rest of the brain is a relatively homogeneous layer of cells, partially occupying the posterior part of the bulb and extending medially and laterally within the cerebral hemisphere proper. This is termed the anterior olfactory nucleus, or the *retrobulbar area*, and its structure was analyzed in detail in *Didelphis* (C. J. Herrick, 1924).
2. On the medial surface of the hemisphere, the retrobulbar area is continuous with the cellular areas of the precommissural or *septal region*, which has not been functionally studied in marsupials.

←

Fig. 16. Horizontal sections through corresponding levels of the amygdaloid nuclear complex in the farthest ventral reaches of the cerebral hemispheres, in placental cat and marsupial brush-tailed possum. Above, sec. 74 from *Trichosurus vulpecula* no. 27; below, sec. 174 from *Felis catus* no. 378; both are stained with thionine. The clearer definition of subnuclei is evident in the marsupial *Trichosurus*. BAm, basal amygdaloid nucleus; CeAm, central amygdaloid nucleus; CoAm, cortical amygdaloid nucleus; ErC, entorhinal cortex, at the medial posterior extremity of the hemisphere at this level; Hc, hippocampus, separated from the amygdaloid nuclei by the lateral ventricle, whose walls adjoin one another at most places in these sections; LAm, lateral amygdaloid nucleus; MAm, medial amygdaloid nucleus; MIc, massa intercalatus, scattered cell clumps in *Felis*, but continuous bands in *Trichosurus*; N LOT, nucleus of the lateral olfactory tract, the anterior limit of the amygdaloid complex; Sub, subiculum connecting the hippocampus with the entorhinal cortex. Not labeled, but prominently bordering the amygdaloid nuclear complex anteriorly in both animals, is the piriform cortex with its dense second layer; it extends from the N LOT to the rhinal fissure, which separates piriform cortex from neocortex. Anterior to N LOT can be seen the edge of the dense and irregular cortex of the olfactory tubercle (Photomicrographs courtesy of G. B. Campos.)

3. Lateral to the septal and posterior to the retrobulbar regions lies the *olfactory tubercle*, which is exceptionally large in *Caenolestes, Perameles*, and the marsupial mole *Notorcytes* (Elliott Smith, 1895; Burkitt, 1938). The cytoarchitecture of this region is quite distinctive, a dense outer layer of cells is convoluted and corrugated, and a less dense inner layer encloses several groupings of very small cells, the "islands" of Calleja.

4. At the posterior end of the olfactory tubercle, just rostral and medial to the amygdaloid complex, a distinct group of cells form the *nucleus of the diagonal band of Broca.*

5. Lateral to the olfactory tubercle and posterior to the retrobulbar area, the *piriform cortex* begins and extends laterally and dorsally around the hemisphere to a rhinal fissure, which is found in all mammals and forms the lateral boundary of neocortex. The piriform cortex is distinguished by a dense outermost layer of cells (Fig. 14), and its anterior surface is covered by the fibers of the lateral olfactory tract (Fig. 13).

6. Posterior and medial to the piriform cortex is found the peculiarly constructed *entorhinal cortex* (Figs. 10, 16). This is schizocortex, so-called because it has a double set of lamina. According to Rose (1926–1927, p. 271) these result from two separate waves of neurogenesis, although this is disputed by Filimonov (1949–1965, p. 13). Entorhinal cortex is bounded by piriform cortex, the subiculum of archicortex, and occipital neocortex. Its position on the hemisphere wall varies greatly from species to species depending upon the relative development of olfactory and neocortical areas (Brodmann, 1909, p. 226, with respect to marsupials). Filimonov (1949, p. 19) considered this region to belong to archicortex rather than paleocortex.

Many of these paleocortical regions, along with the intervening corticomedial group of amygdaloid nuclei, receive fibers from the olfactory bulbs (Adey, 1953, 1970; Putnam and Cone, 1966). Otherwise their function remains obscure, and their connections are complex and diffuse (Crosby *et al.*, 1962, pp. 426–431). The entorhinal area is well connected with the archicortical subiculum and hippocampus and through the latter structures to the habenular and hypothalamic regions. In *Trichosurus*, Adey *et al.* (1956, 1957) found evidence of entorhinal projections to midbrain tegmentum. They also found that ablation of all entorhinal cortex bilaterally resulted in profound behavioral alterations, similar to some components of the "temporal lobe" syndrome seen in monkeys after resection of the temporal lobe, and in other placentals following amygdaloid lesions. Defensive and aggressive reactions disappeared, the animals became very tractable and were extraordinarily interested in or curious about objects, especially moving objects, in their immediate environment.

## 2. Archicortex

The archicortex is coextensive with the hippocampal formation in its broader sense, including the components described in opossum by Loo (1930) and by Hamel (1966b): the dentate gyrus, the cornu ammonis, the subiculum, and the "anterior continuation." The clear and thorough description of these features in the marsupial bandicoot *Perameles nasuta* by Elliot Smith (1897a) is still well worth studying. Tilney (1938) used pouch young opossums to describe the ontogeny of mammalian hippocampus in its least complicated form (not interrupted by a corpus callosum).

The cornu ammonis is characterized by a thin layer of large pyramidal cells, bounded above and below by thick plexiform layers; the whole formation bulges out into the dorsomedial aspect of the lateral ventricle of each cerebral hemisphere (Figs. 10, 14, 16).

At its dorsolateral end, the pyramidal layer widens into a broader band of cells known as the subiculum (Figs. 10, 14, 16), which in turn is continuous with the neighboring neocortex or entorhinal cortex.

The medial end of the pyramidal layer of the cornu ammonis is capped by the dentate gyrus, which contains a band of closely packed granule cells. The scattered pyramidal cells of the dentate gyrus continue anterior to the frontal poles of both the granule cells and the cornu ammonis, making up the "anterior continuation," finally ending on the dorsomedial surface of the retrobulbar area within the olfactory bulb (Hamel, 1966b).

Axons of the granule cells of the dentate gyrus travel for great lengths along the layer of pyramidal cells in the cornu ammonis, making synaptic contact with processes of successive pyramidal cells, again recalling arrangements of granule cells with large cells in the cerebellum and olfactory bulb. Axons of cells in the pyramidal layer send branches to neighboring regions of the cornu ammonis, and they send long processes across the ventricular surface in the alveus, many of these going to the contralateral hippocampus through the hippocampal commissure (Fig. 19). Other fibers from the alveus join the fornix (Figs. 13, 19), a strong band connecting the hippocampus with the mammillary nuclei of the hypothalamus and with the septal region. Other connections exist between the hippocampus, subiculum, and the neighboring neocortex and entorhinal cortex (Elliot Smith, 1897a; Hamel, 1966b).

The hippocampus is prominent in all marsupials and in many placentals, notably humans, it has become more and more restricted to the distal posterior extremity of the lateral ventricle, and is very small in comparison to the total size of the brain and especially of the neocortex. In rodents, bats, insectivores, and marsupials it is almost coextensive with the lateral ventricle. As in placentals, the dentate gyrus varies in size and elaboration among marsupials, becoming convoluted in the Macropodidae (Figs. 10, 20).

Electrical recording from the hippocampus yields a regular oscillation of potential in the frequency range known as the theta rhythm, 4 to 7 cycles per second; hippocampal theta has been reported from *Didelphis albiventris* (Affanni *et al.*, 1967). What these rhythms are, or do, is unknown.

Adey *et al.* (1957) reported evidence of marked influence of electrical stimulation of the hippocampus upon electrophysiological activity in the entorhinal cortex and vice versa, and of stimulation of either hippocampus or entorhinal cortex upon activity in the midbrain in *Trichosurus vulpecula*. Waring *et al.* (1966) interpreted this and other data from *Trichosurus* to constitute evidence that entorhinal and hippocampal regions are involved in processing afferent activity from the midbrain to the forebrain, including that involved in general activation.

### 3. Presubicular Granulous Cortex

Adjacent to the subiculum, but not continuous with it, in Australian marsupials, there is a distinctive cortical region particularly evident in parasagittal sections (Fig. 10, PGC). It consists in a single homogeneous layer of small, densely packed cell bodies beneath a typical outer plexiform layer. The density of the cellular layer is greatest at the junction with the plexiform layer, and in some species such as the potoroo in Fig. 10, this dense edge forms knobs projecting into the plexiform layer. Abbie (1942) described this cortex as granulous in nature, and designated this region parapyriformis 3A in *Perameles nasuta*. The corresponding region seems to be that designated as area praesubicularis by Gray (1924) in *Didelphis virginiana*, by M. Rose (1926, 1927) in *Didelphis albiventris*, and by Brodmann (1909, p. 124, Fig. 82) in *Onychogalea fraenata* ( = Area 27). It is designated here "presubicular granulous cortex" preserving elements of both descriptions. This formation may represent a marsupial specialization; without further study it cannot be properly classified among cerebral cortical regions.

## VII. NEOCORTEX AND DORSAL THALAMUS

The most distinctive feature of mammalian brains, compared with those of other vertebrates, is the development and proliferation of neocortex and the closely associated dorsal thalamus. In extent, the neocortex ranges from a small cap over the rest of the forebrain in placental insectivores such as *Erinaceus* (Ebner, 1969) and marsupial insectivores *Caenolestes* (Obenchain, 1925) and *Notoryctes* (Burkitt, 1938) (see Figs. 11, 17) through an extensively developed structure in placentals such as cats and marsupial wombats and kangaroos (Fig. 11), to the dominant feature of brain morphology in many placentals such as anthropoid apes, cetaceans, and elephants.

### A. Extent and Convolution of Neocortex

The relative extent of neocortex, paleocortex, and archicortex varies greatly from species to species, among placentals as well as marsupials. Probably the only general statement that can be made is that the extreme of neocortical predominance seen in placental primates and cetaceans is not found in any marsupials.

Moeller (1973) assembled ratios of brain weights to body weights in 21 species from seven marsupial families, and described higher ratios as representing a higher evolutionary "grade" owing to expansion of neocortex. He concluded that the families Didelphidae, Dasyuridae, Phalangeridae (including *Phascolarctos*), and Macropodidae, in that sequence, represent successively higher grades; and further that the macropodid stage is as "high" as the "middle" grade of a large group of placentals.

Filimonov (1949, pp. 164, 171, and 191) reports the relative proportions of the three types of cortex, neo-, archi-, and paleo-, in kangaroo to be near the mean value of nine placental brains similarly analyzed (dolphin, seal, man, chimpanzee, macaque, hapale marmoset, dog, anteater, rabbit, hedgehog). The proportion of neocortex for kangaroo was 0.692, the mean for kangaroo plus the nine placentals was 0.729; of archicortex for kangaroo was 0.144, and the mean was 0.113; of paleocortex was 0.093 for kangaroo, and the mean was 0.104. Since kangaroos, among marsupials, have an extremely extensive neocortex (Fig. 11), this would indicate that marsupials are all below this mean in relative extent of neocortex in the pallium. Among the placentals, Filimonov's extreme values of proportion of neocortex were 0.978 for dolphin and 0.959 for man, and 0.324 for hedgehog.

Harman (1947) measured the volume of the brain and of the neocortex in *Didelphis virginiana* and in 21 placental species including 10 primates. His results for *Didelphis*, in $mm^3$, were 1310 for the brain and 162 (12%) for neocortex; corresponding figures for some other species were *Felis* 6690 and 2690 (40%), *Cavia* 867 and 243 (28%), *Macaca* 21,000 and 10,600 (50%), and *Pan* 81,400 and 38,100 (47%). However, no truly macrosmatic species other than *Didelphis* was represented. Thus no conclusions about marsupials can be drawn.

Elias and Schwartz (1969) measured brain volume and surface area of cerebral cortex ("excluding archicortex," without further explanation). For marsupials they gave only common names, and the following data for brain volume in $mm^3$ and cortical area in $mm^2$: mouse opossum 1240 $mm^3$ and 438 $mm^2$; opossum 4250 $mm^3$ (in considerable contrast to Harman's finding) and 1320 $mm^2$; wallaby 25,000 $mm^3$ and 4330 $mm^2$; and kangaroo (body length of 47 cm) 39,200 $mm^3$ and 7490 $mm^2$.

As with cerebellar cortex, cerebral neocortex has a relatively constant thickness. The chief difference in extent of such cortex between one species and another is thus in the other two dimensions: what has been termed the "surface area." Animals with a large extent of cortex accommodate the expanded surface area within the restricted volume inside the skull by infolding, or convolution, of the cortical sheet. The resulting outfoldings and indentations seen on the surface of the forebrain are known, respectively, as gyri and sulci.

As in placentals, the degree of convolution or "gyration" in marsupials is related to relative body size along with other factors. It reaches a maximum in the wombats, large wallabies, and kangaroos (Fig. 11). Neocortical convolution is not restricted to these phalangeroids; the largest of the Marsupicarnivora, *Thylacinus* (Fig. 17) and *Sarcophilus* (Fig. 11), show a degree of convolution appropriate to their body size.

There is a remarkable consistency in the infoldings that develop in the brains of all marsupials. The most universal of these, found in all mammals, is the rhinal fissure at the lateral boundary of neocortex with paleocortex. Even in the very simple brain of the marsupial mole *Notoryctes typhlops*, where there is no clear surface feature, the cellular laminae show an infolding here (Burkitt, 1938; Schneider, 1968). It is the only noticeable folding in the cerebral cortex of *Caluromys, Marmosa, Antechinus*, and *Sminthopsis* (Fig. 11).

The next most constant sulcus is that called $\alpha$ by Ziehen (1897a), whose terminology was also used by Brodmann (1906, p. 360; 1909, p. 213) and Livini (1908). It was termed the orbital sulcus by Elliot Smith (1902a, pp. 158–185), and (unfortunately) most authors since. It is found in all the marsupials studied except the smaller Didelphidae, Dasyuridae, Petauridae, and Burramyidae.

A sulcus enclosing the occipital region is seen in the Phalangeridae and Macropodidae that have been studied (Fig. 10), and indications of a similar configuration were reported in *Didelphis* by Gray (1924, p. 240) and by Benevento and Ebner (1971a, p. 164). In macropods, its dorsal limb was designated $\lambda$ by Ziehen and paramarginal by Elliot Smith; an often separated anterior portion was called $\epsilon$ by Ziehen and suprasylvian by Elliot Smith. A sulcus between the latter and the rhinal fissure was designated $\beta$ by Ziehen and B by Elliot Smith. Between this and the sulcus $\alpha$ (orbital) is a shorter "stem sulcus" around which the others seem to curve; it was called (alas) Sylvian by Ziehen and A by Elliot Smith.

The appellation "orbital sulcus" was used as an example of the needless confusion brought about by using similar names for what are actually vastly different convolutions in different mammalian orders (Edinger, 1948, pp. 2 and 128). A strong case was made for using unique labels for the cerebral features of each separate order, since the convolutions evolved separately in each order. The labels introduced by Ziehen provide a valuable opportunity for doing this for the marsupials (excepting of course his unfortunate use of "Sylvian fissure," which is properly applied only to primates). Although it reverses the historical trend, adoption of this terminology follows venerable precedents and will avoid the chaos which plagues cerebral terminology for other mammals.

The progressive expansion of neocortex has produced other morphological features besides increased folding. Expansion of neocortex through evolutionary time has been best documented for horses (Edinger, 1948). In addition to complete absence of neocortical convolutions, the earliest equines had brains characterized by an exposed midbrain tectum on the dorsal cranial surface between the neocortex and the cerebellum, and a relatively high rhinal fissure on the lateral cerebral surface. Edinger pointed out the similarity between the Eocene equine brain and those of some contemporary marsupials. The exposed midbrain is characteristic of modern *Didelphis, Sarcophilus, Dasyurus, Sminthopsis, Vombatus* (Figs. 2, 11), and probably some others. Although the horse midbrain was progressively covered by continuing expansion of neocortex and cerebellum, Edinger (1964) has discounted midbrain exposure as an index of the stage of neocortical evolution. A preferred indicator (Edinger, 1966) is the downward "forcing" of the rhinal fissure by the expanding neocortex.

Dillon (1963) used the more and more ventral position of the rhinal fissure and the number of neocortical convolutions, together with correlated criteria including the complexity and mass of the internal capsule, the number of cerebellar lobules, and the size of the lateral ventricles, to construct a "phylogenetic series" of increased development of neocortex in six species of Macropodidae. He concluded that potoroos (*Potorous tridactylus*) were the least developed and "left" the stem ancestral group earliest, followed by quokkas (*Setonix brachyurus*), while the wallabies (*Macropus parryi* and *M. rufogriseus*) and kangaroos (*Macropus rufus* and *M. giganteus*) diverged or radiated from a common source in more recent evolutionary history.

The question of the functional significance of neocortical convolutions has aroused recent interest as new information has developed concerning distinctive properties of different neocortical regions. The evidence for relating specific convolutions to functional subdivisions in marsupial neocortex will be considered in later sections of this text, dealing with specialized regions of cortex (Sections VII,C,4; D,E).

### B. Neocortical Lamination

Neocortex has two identifying characteristics: strong reciprocal connections with cells of the dorsal thalamus, and a consistent basic architecture consisting of at least six distinctive laminae (Figs. 10, 14, 18; for other marsupial examples see Brodmann, 1909, pp. 34–42).

Several various systems of designating laminae agree on the characteristics and numerical label of the first five laminae (Crosby *et al.*, 1962, p. 436). The outermost, designated the first, is almost free of cell bodies, except for the distinctive small horizontal cells of Cajal (identified in marsupial opossums by Benevento and Ebner, 1971a). The second layer contains cell bodies near the mean size for neocortex, and in marsupials these form a distinct, dense, darkly staining lamina (Figs. 10, 14, 18). The third layer contains larger cells, more loosely scattered. The fourth, granular layers contains the smallest neocortical cell bodies, and the fifth layer contains large cells.

The remaining cell bodies, between the fifth layer and the underlying white matter, are variously subdivided in different systems of classification. In marsupials there is a distinctive peculiarity of the cell bodies bordering the white matter: they are often arranged in rows, each one cell thick, parallel to the cortical surface (BLC, Figs. 10, 14). This is not commonly seen in placentals; its significance, if any, remains to be determined. It may be related to the path of commissural fibers traveling parallel to the cortical surface through the external capsule, on their way to and from the anterior commissure rather than moving directly to the corpus callosum as they do in placentals.

In addition to cell size and density, there are other features of architecture and connections that distinguish the neocortical laminae. Most of these features have been shown to be common to marsupials and placentals.

The first layer contains dendrites from cells of all lower layers (thus every layer contains dendrites from all layers beneath), as well as terminals from cells in the thalamus and in other regions of the cerebral cortex (Walsh and Ebner, 1970; Benevento and Ebner, 1971a; Ebner, 1969).

Cells of the second layer send axonal projections to lower layers and to neighboring cortex, as do those of the third layer, but the third layer receives some terminals from thalamic cells, which is not true of the second layer (Ebner, 1969; Benevento and Ebner, 1971a). Dendrites from layers two and three are found in layers one, two, and three, but none reach down into layer four; there seems to be a dendrite barrier below layer three for descending dendrites (Walsh and Ebner, 1970).

The fourth and fifth layers contain the preponderance of terminals from cells of the dorsal thalamus (Ebner, 1967; Walsh and Ebner, 1968; Benevento and Ebner, 1969).

Cells of the fifth layer send axons to other cortical regions and to portions of the nervous system far removed from cortex: spinal cord, thalamus, and midbrain.

Morest (1970), analyzing development of neocortex in pouch young opossums, observed an ontogenetic sequence which may be relevant to the origin of neocortical lamination. The earliest cells to become fixed in position by the sending out of processes and the establishment of connections with other cells were those of the lowest layers. Cells developing later took up positions above those already making connections.

## C. Somatic Sensory-Motor Neocortex

Despite the degree to which neocortex is an anatomically homogeneous structure, it has been possible and profitable to subdivide it into specialized regions based on functional activity, variations on the basic architecture, and differences in afferent and efferent connections.

### 1. Excitable Motor Cortex

The first electrical stimulation of motor cortex of marsupials was reported by Herrick and Tight (1890, quoted in C. L. Herrick, 1898), using *Didelphis virginiana*. With *D. marsupialis*, these species have been the subjects of numerous further studies of electromotor cortical excitation (Ziehen, 1897b; Cunningham, 1898; Vogt and Vogt, 1906; Gray and Turner, 1924; Rogers, 1924; Weed and Langworthy, 1925a; Langworthy, 1927; Bromiley and Brooks, 1940; Ward, 1954; Lende, 1963b,c; Bautista and Matzke, 1965; Martin, 1968b), and there have been considerable variations in the findings. Investigations between 1910 and 1935 uniformly reported absence of motor response of hind limb and tail to cortical stimulation, despite their use of the same anesthetic and stimulation techniques, and even the same variations in techniques, as did the earlier and later studies through 1963, which all found hind limb and tail "representations."

This literature is summarized by Lende (1963a), whose map of the motor cortex gives the greatest degree of detail, and is the one we show in Fig. 17. Just posterior to sulcus $\alpha$, a medial forelimb focus is continuous with a very small trunk, tail, and hind limb region caudomedially, and with a large face representation laterally and anteriorly around the ventrolateral end of sulcus $\alpha$.

A new era of hind limb "absence" may be underway. No hind limb or tail movements were obtained by Bautista and Matzke (1965) or Martin (1968b).

A further discrepancy in the findings of these various opossum studies is in the movements obtained from stimulating directly in front of sulcus $\alpha$. Eye movements were obtained by all investigators who stimulated more anteriorly near the frontal pole of neocortex, but nearer the sulcus, vibrissae, nose, or chewing movements were reported by Rogers (1924), Weed and Langworthy (1925a), Langworthy (1927), Bromiley and Brooks (1940), and Martin (1968b). Vogt and Vogt (1906) and Lende (1963b,c) reported head and face muscular movements only from the lateral region behind and at the lateral end of sulcus $\alpha$.

Similar variations in the location of head and face motor regions occurred in studies of other marsupial species. Otherwise, there was a remarkable consistency in results, considering such factors as spreading depression and inhibitory interactions among electrical stimulations which were unrecognized until after 1940. Species investigated included *Trichosurus vulpecula* (Vogt and Vogt, 1906; Goldby, 1939b; Abbie, 1940; Rees and Hore, 1970), *Macropus dorsalis, Petrogale penicillata, Onychogalea fraenata* (Vogt and Vogt, 1906), *Macropus rufogriseus, M. agilis* (Abbie, 1940), *M. eugenii* (Lende, 1963c), *Dasyurus viverrinus* (Flashman, 1906; Abbie, 1940), *Sarcophilus harrisii* (Abbie, 1940), *Perameles nasuta* (Chapman, 1906; Abbie, 1940), and *P.* (= *Isoodon*) *obesulus* (Chapman, 1906).

All of these studies report the motor cortex to lie in close association with sulcus $\alpha$; A forelimb focus was always found just posterior to this sulcus. Hind limb regions, when present, were always just medial to the forelimb region. Face and head responses were obtained from one or the other of two constant locations: either pre-$\alpha$ anterior to the limb regions, or lateral to the limb regions around the ventrolateral limit of sulcus $\alpha$. Eye movements were regularly produced by stimulation near the frontal pole of neocortex.

## 2. Somatic Sensory Projections to Neocortex

Somatic sensory projections to neocortex have been investigated in five marsupial species: *Trichosurus vulpecula* (Adey and Kerr, 1954; Weller and

Haight, 1973; Haight and Weller, 1973), *Didelphis virginiana* (Bodemer and Towe, 1963; Lende, 1963a,c), *Didelphis albiventris* (Magalhães-Castro and Saraiva, 1971), *Macropus eugenii* (Lende, 1963c), and *Vombatus ursinus* (Johnson *et al.*, 1973). In *Didelphis* of both species and *Macropus* the sensory projections were found in the same cortical areas from which muscular movements were elicited by electrical stimulation. Furthermore, for a given cortical locus the sensory projections were from the same part of the body which moved in response to electrical stimulation of that cortical region (Lende, 1963a,c). Thus, these maps of sensory responses are equivalent to the maps of motor representations (Fig. 17).

A second sensory region, corresponding to the SmII region of placentals, was identified in *Trichosurus, Didelphis virginiana,* and *Vombatus*, but it has not been mapped in detail in any marsupial.

### 3. Cytoarchitecture of Sensory–Motor Cortex

From early times, in humans and monkeys, the motor cortex has been considered coextensive with a region of distinctive cytoarchitecture, the "regio praecentralis," which includes "area gigantopyramidalis" (= area 4) and "area frontalis agranularis" (= area 6) according to Brodmann (1909, pp. 128–165 and 308–312). Both areas 4 and 6 are characterized by the absence of cells of the fourth layer (the internal granular layer); area 4 is further distinguished by the presence of many extraordinarily large (giant) pyramidal cells (also called Betz cells) in the fifth layer. Brodmann (1909, p. 213) pointed out the remarkable constancy, in architecture and position, of the giant pyramid area in all mammals studied, including the marsupial *Macropus* (= *Petrogale*?) *penicillata*, although the differentiation of this cortical area was less complete in many animals than in the primates and carnivores. The hedgehog *Erinaceus europaeus* (Insectivora) was an extreme example of a relatively undifferentiated regio praecentralis (Brodmann, 1909, pp. 193–194).

Brodmann (1906, pp. 388–390) located the giant pyramid area in *M.*(*P.*) *penicillata* on the banks of sulcus $\alpha$, extending further out from the caudal bank than from the rostral, and diminishing in anteroposterior extent going laterally from its maximum at the dorsal midline (Fig. 17). Thus it lies within the macropod motor cortex according to Abbie (1940), or overlaps the anterior part of the motor area for the limbs according to Lende (1963c). It includes a region of well-developed granular fourth layer.

The only "agranular" motor neocortex (fourth layer absent) reported in marsupials has been on the banks of sulcus $\alpha$ in *Didelphis virginiana*: a small area on the anterior bank according to Gray (1924) and Langworthy (1927); a small area on the posterior bank according to Walsh and Ebner

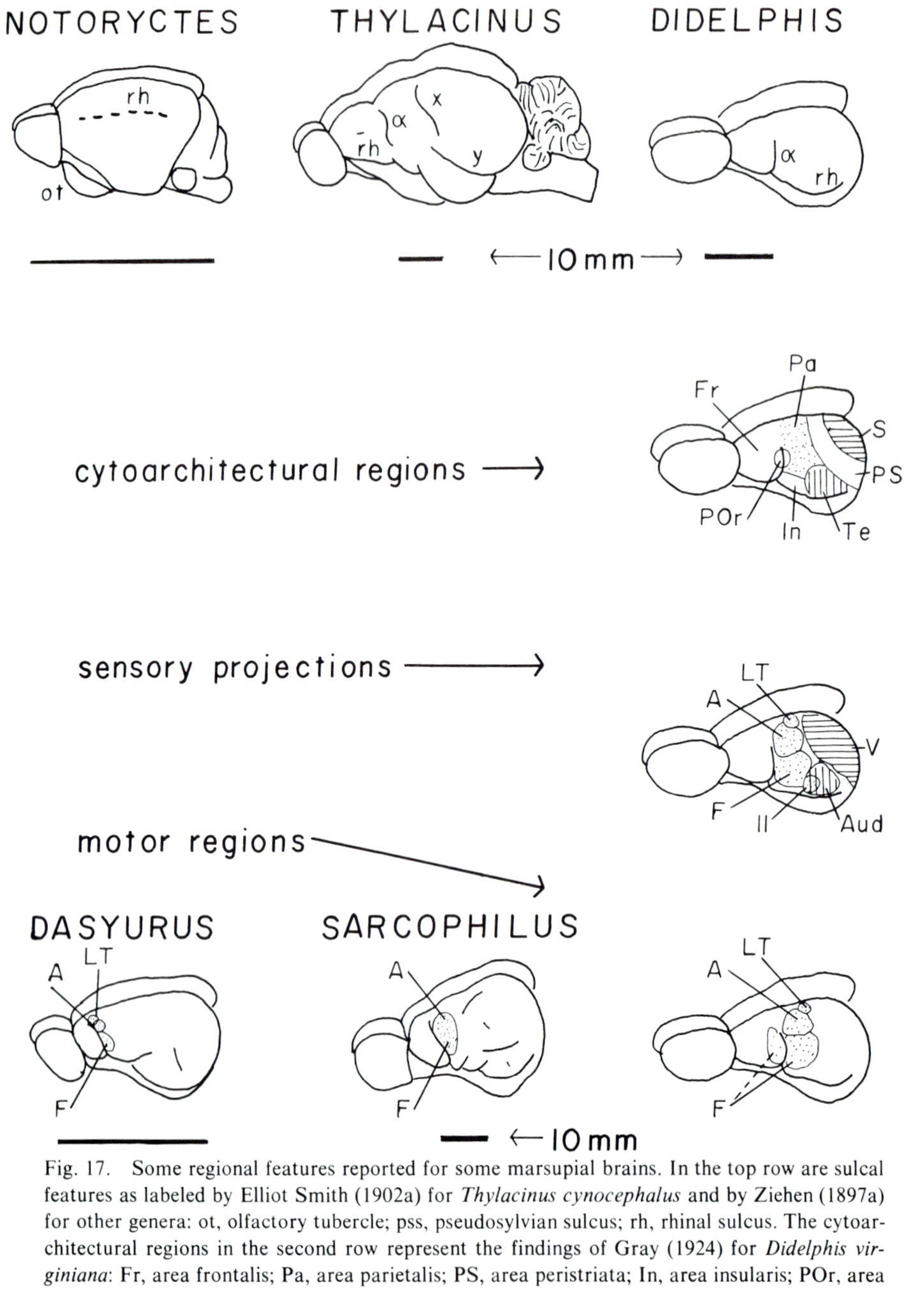


Fig. 17. Some regional features reported for some marsupial brains. In the top row are sulcal features as labeled by Elliot Smith (1902a) for *Thylacinus cynocephalus* and by Ziehen (1897a) for other genera: ot, olfactory tubercle; pss, pseudosylvian sulcus; rh, rhinal sulcus. The cytoarchitectural regions in the second row represent the findings of Gray (1924) for *Didelphis virginiana*: Fr, area frontalis; Pa, area parietalis; PS, area peristriata; In, area insularis; POr, area praeorbitalis; S, area striata; Te, area temporalis; of Abbie (1942) for *Perameles nasuta*: PH, parahippocampal areas; PPy, parapyriform areas; of Goldby (1943) for *Trichosurus vulpecula*, where projection regions are indicated of the following thalamic nuclei: DM, dorsomedial nucleus; PA, pars arcuata of the ventral nucleus; L, lateral nucleus; PE, pars externa of the ventral nucleus; LGN, lateral geniculate nucleus; VA, ventral anterior nucleus; MGN, medial geniculate nucleus (b, field of barrels in layer 4); and of Brodmann (1909) for *Macropus*

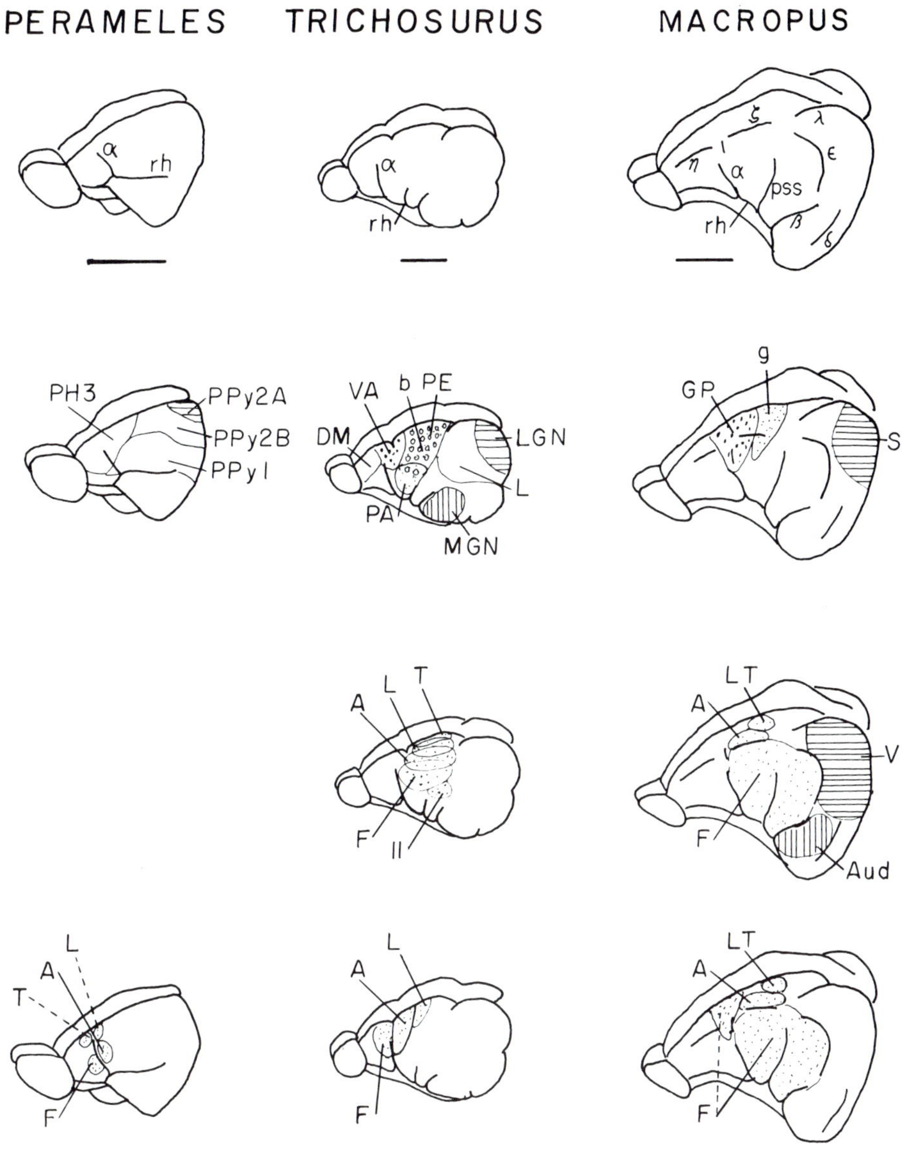


(= *Petrogale*?) *penicillata*: g, area granularis; GP, area gigantopyramidalis; S, area striata. Sensory projections in the third row are the results of Lende (1963a) for *Didelphis virginiana*, Adey and Kerr (1954) for *Trichosurus vulpecula*, and Lende (1936c) for *Macropus eugenii*. Motor regions are those determined by Flashman (1906) for *Dasyurus viverrinus*, Abbie (1940) for *Trichosurus*, and Lende (1963c) for *Macropus eugenii*, with an additional face region found in *M. rufogriseus* and *M. agilis* by Abbie (1940) (dotted line), by Abbie (1940) for *Sarcophilus harrisii*, and by Lende (1963b) for *Didelphis virginiana*, plus an additional face region (dotted line) found by several other investigators, and by Abbie (1940) for *Perameles nasuta*, with additional regions (dotted line) reported by Chapman (1906). Sensory and motor representations are labeled: A, arm (forelimb); L, leg (hind limb); II, second somatic sensory representation; Aud, auditory; T, tail; F, face; V, visual.

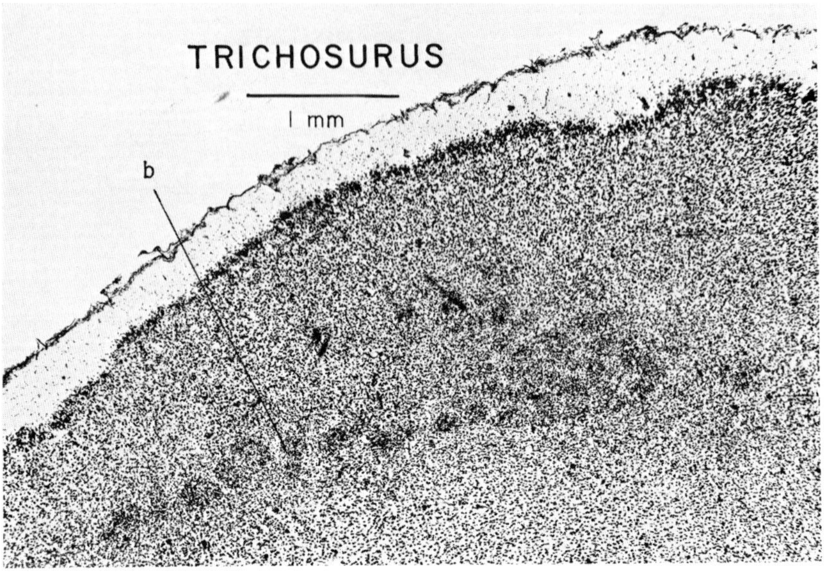
TRICHOSURUS
1 mm
b
1
2
3
4

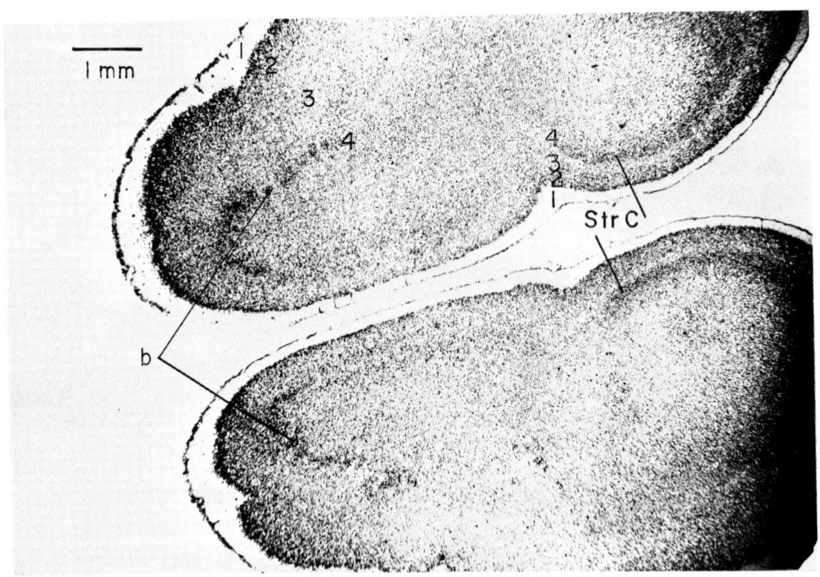
1 mm
1
2
3
4
4
3
2
1
Str C
b

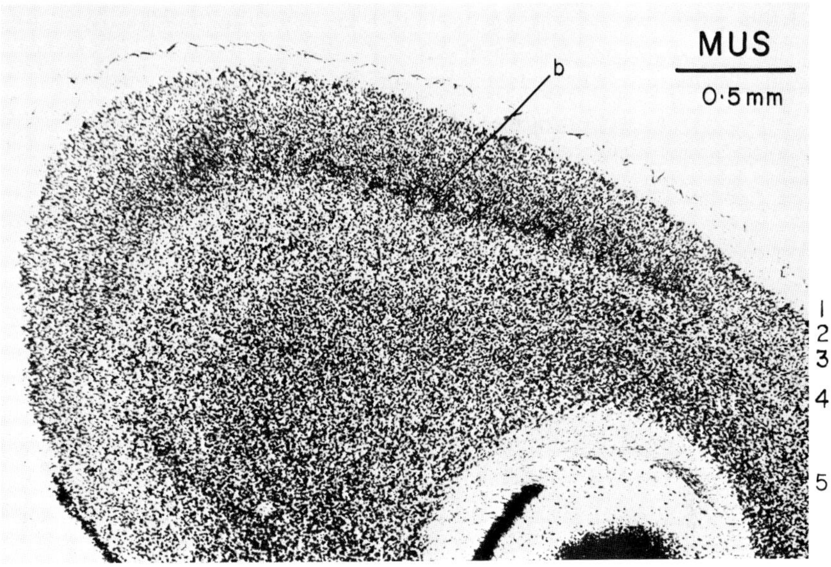


Fig. 18. Barrels in sensory neocortex of marsupial *Trichosurus* and placental *Mus*, as seen in parasagittal sections (above) and horizontal sections from *Trichosurus* (at left). Numerals indicate neocortical layers and b indicates barrel formations in layer 4. The striate cortex Str C and its boundary indentation can be seen in the horizontal sections from *Trichosurus*.

(1970). Excitable motor areas containing unusually large or abundant pyramidal cells in the fifth layer (along with a well-developed fourth layer) have been reported in all marsupials studied cytoarchitectonically: *Didelphis* (Gray, 1924; and Turner, 1924; Langworthy, 1927; Walsh and Ebner, 1970), *Trichosurus* (Goldby, 1939; Abbie, 1940), *Dasyurus, Sarcophilus, Perameles,* and *Macropus* (Abbie, 1940). A safe statement seems to be that within the motor cortex of all marsupials studied, and perhaps all mammals, a well-developed fifth layer with large pyramidal cells will be found. But the cytoarchitectural area and the excitable motor cortex are rarely coterminous.

The somatic sensory projection area has always been found to coincide with a region where the fourth granular layer is exceptionally well developed. In marsupials this has been reported for *Didelphis* (Gray, 1924; Lende, 1963a; Ebner, 1969; Walsh and Ebner, 1970) and *Trichosurus* (Adey and Kerr, 1954).

Within this well-developed, granular fourth layer of sensory–motor cortex in *Trichosurus*, Weller (1971, 1972; Weller and Haight, 1973) has identified an organization of small cells into distinct subgroups (Fig. 18). These clumps of cortical cell bodies in the fourth layer resemble in many ways the "barrels" first described in sensory cortex of mice and related to projections of vibrissae and other regions of high receptor density in rats. Evidence of similar formations was reported in members of the Petauridae: *Petaurus breviceps* and *P. norfolcensis*, *Pseudocheirus peregrinus*, and *Schoinobates volans* (Weller, 1972).

### 4. Convolutions

In *Trichosurus*, convolutions appear in the somatic sensory area in occasional individual animals, and the sulci occur at locations between concentrations of barrels, the barrels thus occupying gyral crowns. This means that in convoluted cortex, the gyral crowns receive projections from peripheral regions where receptors are highly concentrated, with sulci separating these projection regions (Haight and Weller, 1973). Since bodily regions of high receptor density are found frequently on the face and the limb extremities and cortical projections are somatotopically organized, the sulcal pattern of the cortex often forms a facsimile of the outline of the animal's body. This has been documented for marsupial wombats and placental carnivores (Megirian *et al.*, 1972; Johnson *et al.*, 1973).

### 5. Thalamic Connections with Sensory–Motor Cortex

Projections from the dorsal thalamus to the sensory–motor cortex have been identified using retrograde degeneration studies in *Trichosurus* (Goldby, 1943) and *Didelphis* (Bodian, 1942; Diamond and Utley, 1963; Pubols, 1968). All found that cells in the ventral group of thalamic nuclei degenerated following lesions of the sensorimotor cortex. Pubols (1968) reported evidence that specific subregions of cortex are connected to specific subregions of ventrobasal thalamus, in accord with the correspondence of the somatotopically organized cortical and thalamic projection fields discovered by electrophysiological recording.

The afferents from the dorsal thalamus, more specifically from the ventral or ventrobasal nuclei, terminate as usual in neocortex, mostly in the pronounced granular fourth layer and the subjacent pyramidal fifth layer in the sensory–motor region, with a few endings found in the outermost molecular first layer (Ebner, 1969; Walsh and Ebner, 1970). The question of differential projections to separate sensory and motor cortical regions will be considered in Section VII,C,9.

### 6. Corticospinal Fibers

Much remains to be explained about the processes responsible for movements produced by electrical stimulation of the cerebral cortex. One thing which is related to this phenomenon is the projection of fibers from cells in the neocortex directly to cells of the spinal cord. The excitable motor cortex is the source of most, but by no means all, of these corticospinal fibers in mammals so far studied (cats, monkeys, and humans, Wiesendanger, 1969, p. 75), including marsupials *Didelphis* (Bromiley and Brooks, 1940; Martin and Fisher, 1968), *Trichosurus* (Goldby, 1939b; Martin *et al.*, 1970), *Setonix* (Watson, 1971), and *Macropus fuliginosus* and *M. rufus* (Watson, 1972). The giant pyramidal cells give rise to some, but again not all, corticospinal fibers in placentals (Wiesendanger, 1969, p. 76). Electrically stimulating the corticospinal fibers in the medullary pyramids, and recording the resulting antidromically evoked electrical activity in the cortex, in *Trichosurus* (Porter, 1955) showed the cells of origin to lie both anterior and posterior to sulcus $\alpha$, in an area larger than that of the excitable motor cortex as determined by Abbie (1940). Similar studies in *Didelphis virginiana*, some using microelectrodes to record antidromic activity from individual cells, showed the cells of origin to lie in or near the fifth layer in a cortical region posterior to the lateral portion of sulcus $\alpha$ (Towe and Biedenbach, 1969; Biedenbach and Towe, 1970).

In all mammals, the corticospinal fibers travel through the internal capsule of the forebrain, the cerebral peduncles in the midbrain, and from the pyramids on the ventral surface of the lower pons and medulla. In the caudal medulla they ascend into the spinal cord and most cross to the other side. Their course within the spinal cord, and their locus of termination upon cells of the spinal gray matter, vary to some degree from one mammalian order to another.

In marsupials, both crossed and a few uncrossed fibers travel mostly in the base of the dorsal column, with a few fibers at the base of the dorsal horn in the lateral columns. This pattern, seen also in rodents, artiodactyls, and lagomorphs (Ariëns Kappers *et al.*, 1936, Vol. 1, p. 272), has been described in *Pseudocheirus* (Ziehen, 1899, 1901). *Phascolarctos* and *Perameles* (Ziehen, 1897c, 1899, 1901), *Dasyurus* (Flashman, 1901b), *Didelphis* (Turner, 1924; Bautista and Matzke, 1965; Martin and Fisher, 1968), *Trichosurus* (Goldby, 1939b; Martin *et al.*, 1970; Rees and Hore, 1970), *Setonix* (Watson, 1971), *Potorous* (Martin *et al.*, 1972), and *Macropus fuliginosus* and *M. rufus* (Watson, 1972). In primates and carnivores, the bulk of the fibers run in the lateral columns (Petras, 1969).

In all of the studies of marsupials mentioned in the previous paragraph, the corticospinal fibers are present in only the cervical and high thoracic segments, terminating mostly in segments C5 through T1. In *Trichosurus*, some terminals were found down to T10 (Martin *et al.*, 1970; Rees and Hore, 1970), and in *Potorous* terminals from the dorsal tract extended into T12 (Martin *et al.*, 1972). In contrast, in primates and carnivores (Petras, 1969) corticospinal terminals are found in all spinal segments; and a similar arrangement was found in the monotreme echidna *Tachyglossus* (Goldby, 1939a). In placental armadillos, *Dasypus novemcinctus*, of the order Edentata, the corticospinal fibers were limited to cervical and thoracic segments as in the marsupials studied (Fisher *et al.*, 1969). Thus phylogenetic generalizations would be premature.

Most corticospinal terminals in marsupials are in the base of the sensory cell column (dorsal horn) in laminae IV, V, and VI, according to all studies. In addition, a few terminals have been reported in lamina III in opossums (Martin and Fisher, 1968) and quokkas (Watson, 1971), and laminae VII and VIII in *Trichosurus* (Martin *et al.*, 1970; Rees and Hore, 1970). The more caudal extent, and greater number, of corticospinal fibers led to speculation that there is a greater degree of neocortical involvement in motor activities in *Trichosurus* than in *Didelphis*, but that in both marsupials there is far less neocortical involvement than is found in carnivores (especially raccoons) and primates (Martin *et al.*, 1970; Rees and Hore, 1970). According to Biedenbach and Towe (1970) the spectrum of fiber sizes in the medullary pyramids in *Didelphis* was very restricted: all fibers were less than 5 $\mu$m and 99% of them less than 3 $\mu$m in diameter. In this respect they are similar to those of rodents, tree shrews, and some prosimian primates, and they are unlike those of carnivores and monkeys where there are fibers of larger diameter. In this same study, the total number of pyramidal tract fibers in *Didelphis* was estimated to be between 30,000 and 40,000, half the number found in cats.

Fibers from sensory cortex to lower sensory centers may be concerned with feedback control of sensory input. Similar cortical projections terminate in the sensory nuclei of the trigeminal nerve and in the cuneate–gracile complex in all mammals studied (reviewed by Wiesendanger, 1969, pp. 46–49), including marsupials *Didelphis* (Zimmerman and Chambers, 1963; Martin and West, 1967), *Trichosurus* (Martin *et al.*, 1970, 1971; Rees and Hore, 1970), and *Macropus rufus* and *M. fuliginosus* (Watson, 1972). Some cells of origin of corticospinal tract fibers were found to be responsive to electrical stimulation of the paw in *Didelphis*; furthermore, most were driven by stimulation of more than one paw, and stimulation of one paw could prevent the otherwise ready response to stimulation of a different paw, much as was seen in cats (Biedenbach and Towe, 1970). [However, both the number of cells behaving thus, and their responsiveness, were far less in opossums than in cats (Towe and Biedenbach, 1969).] Thus cells of this pyramidal pathway, in both marsupials and placentals, can in some way act in response to a synthesis of somatic sensory sensations. Much remains to be learned about the mechanisms and functional role of this activity, and whether it is less developed in *Didelphis* as suggested by Towe and Biedenbach (1969).

A beginning in the process of unraveling these problems has been made using *Trichosurus* in a series of studies. They have shown that the cortical excitatory influence responsible for electrically induced forelimb movement travel by way of the medullary pyramids, but those facilitating hind limb movements travel by extrapyramidal paths (Hore and Porter, 1971). They further studied facilitation of three hind limb muscle groups by cortical stimulation through extrapyramidal paths, and found that one muscle group, flexor digitorum longus, was influenced by cortical excitation through the pyramids, although pyramidal tract fibers do not reach the region of hind limb motoneurons (Hore and Porter, 1972). This influence is an inhibition of contractions maintaining a grasping postural stance, presumably permitting the animal to assume a new position. The inhibition of the flexor digitorum longus can also be induced by afferent input from the ipsilateral forelimb (Aoki and McIntyre, 1973). Similarly, the corresponding forelimb reflex, that of the flexor digitorum profundus, is inhibited by cortical stimulation acting through the pyramidal tract while antagonistic muscular contractions are facilitated (Aoki and McIntyre, 1973). It appears a general effect of cortical excitation has to do with timing release of postural grasping.

Cutting the pyramids resulted in defects in only the finest controlled movements of the limbs, with some limb misplacement, slipping, loss of grace and confidence in arboreal movements. These defects appeared comparable to disabilities in refined and discrete aspects of motor performance caused by similar lesions in cats and monkeys (Hore *et al.*, 1973).

## 7. Effects of Lesions of Sensory–Motor Cortex on Behavior

Limited effects on behavior have been reported following lesions of sensory–motor cortex in marsupials. In *Didelphis*, Gray and Turner (1924; Turner, 1924) reported no behavioral effects from removal of sensory–motor cortex other than a "favoring" of the forelimb contralateral to the ablation when the animal was climbing. Rogers (1924) noticed difficulty in using the forepaw in grasping and climbing, and some misplacement of the hind feet for a few days following lesions of the sensory–motor cortex.

Bromiley and Brooks (1940) tested placing and hopping reactions before and after lesions of sensory–motor cortex. Placing reactions (the adjustments of feet into supporting positions upon a surface in response to some exteroceptive stimulation) were not as precise nor as easily evoked in normal opossums as they were in cats; and they were completely abolished by lesions of sensory–motor cortex in opossums, as was the case in cats. Hopping reactions (the compensating movements of the limbs in response to changes in the animals' centers of gravity) were also less pronounced in normal opossums than in cats; they were impaired in opossums by sensory–motor cortex lesions, but not as severely as they were in cats.

In *Trichosurus*, Goldby (1936b) reported weakness and clumsiness in both limbs contralateral to sensory–motor cortex lesions, an inability to support the body adequately, tendencies to slip into abnormal postures including lateral rotation of the hind foot, and standing on the dorsal surface of the flexed forepaw. Most of these symptoms disappeared after 1 day, leaving only some disability in climbing; after 1 week the only remaining symptom was flexor weakness in the contralateral forepaw digits with a failure to hold food in that paw. After 3 weeks there were no detectable motor deficits. Similarly, Rees and Hore (1970) observed no pronounced defects in posture and gait due to sensory–motor cortex lesions; there was a failure to reach or grasp with the contralateral forepaw, and there was some abnormal placement of the paw when climbing. These symptoms disappeared after 2 weeks.

These effects in marsupials are quite consistent and in some cases closely resemble symptoms seen in placentals following lesions of the corticospinal fibers (pyramidal tracts) alone: flexor weakness, loss of placing and hopping reactions, delegation of the initiation of grasping to the paw on the unaffected side, loss of discrete movement of separate digits, with eventual recovery of most functions (but not placing and hopping) (Ariëns Kappers *et al.*, 1936, Vol. 3, p. 1597; Wiesendanger, 1969, pp. 81–85). The more severe symptoms seen after motor cortex lesions in primates, including paresis and general hypotonia, due to extrapyramidal processes (cortical projections other than the direct corticospinal fibers) have not been observed in marsupials. The mechanisms affected by any of these lesions have not been elucidated for any animals.

## 8. Development of Sensory–Motor Cortex

The development of the sensory–motor cortex has been observed in relation to the development of normal motor activity in *Didelphis*. Weed and Langworthy (1925a) and Ward (1954) studied responses to electrical stimulation of the cortex in pouch young; Langworthy (1927) and Ward (1954) correlated these observations with neocortical cellular development. Weed and Langworthy observed animals from 23 to 84 days after birth; Ward observed animals for 9 to 50 days postnatally. Langworthy (1927) reported the relative thickness in cortical layers and the increase in spacing (= decrease in density) of cortical cells as criteria of progressive maturation; Ward (1954) observed the growth of fibers from cortical cells to other parts of the brain, particularly through the pyramids to the spinal cord, along with the differentiation of cortical cell types.

At 9 days after birth the animals showed spontaneous movements in a "walking pattern," and a few fibers had entered the internal capsule from the cortex; the cortex showed the usual embryonic arrangement of ependymal, mantle, and marginal layers. By 19 days postnatally, righting and walking reflexes had appeared; spontaneous head and neck movements and finger and tail flexion were observed. There was no movement in response to electrical stimulation of the cortex, which now showed molecular, pyramidal, and polymorph layers and sent fibers into the anterior commissure. At 23 days, cortical stimulation produced an extensor thrust of the foreleg with outspreading digits; a deep first (molecular) layer marked the region of the excitable cortex, but the cells were in such an embryonic state that Langworthy found it difficult to relate possible cortical processes to the movements produced. At 27 to 33 days, animals could support their weight on the forefeet for the first time, hang suspended by the tail, and move the head from side to side. Electrical stimulation produced only foreleg movements. The cortical cells had sorted into recognizable supra (layers 1 to 3) and infra (layers 5 and 6) granular subdivisions. The corticospinal fibers formed the beginnings of the pyramids of the medulla by the thirty-third day. By 41 days, the animals could step, but not walk, supporting their weight. Electrical stimulation yielded flexion and extension of the foreleg, and in one case rotation of the neck. All the cortical layers, except the small cells of the fourth layer, were formed. The forelimb excitatory area appeared more mature (had lower cell density) than did the presumptive "face area" anterior to it. Pyramidal fibers first reached into the contralateral spinal cord at this age (thus the movements produced by cortical stimulation at earlier ages were in the absence of direct corticospinal connections). At 47 days, the animals could crawl and stand supporting their weight, but could not walk. Electrical stimulation of cortex produced only flexion of forelimb and digits. The orbital sulcus appeared on the cortical surface at this time, perhaps related to rapid growth in the forelimb excitatory region according to Langworthy. The only new developments from 54 to 62 days were a further decrease in cell density in the cortex, with the forelimb area still showing the widest spacing between cells. By 76 to 84 days, the animals could run and walk; cortical stimulation produced movements of vibrissae, ears, and eyes in addition to those of the foreleg, and the cortex appeared to be essentially in its adult state.

### 9. Comparison with Placental Sensory–Motor Cortex

In placental primates and carnivores, there are paired sensory and motor fields (Mountcastle, 1968, pp. 1391 and 1791). The motor area is anterior and has the lowest threshold for motor stimulation, and the sensory projections exhibit less stable, less synaptically sure (i.e., more variable) properties than do those in the more posterior sensory area. The "representation" of the body found by mapping projections in the sensory region is a mirror image of, and directly adjoins, the motor representation found by mapping foci where electrical stimulation produces movements of body parts. The motor region is agranular (layer 4 is absent) and includes a region of well-developed pyramidal cells; the sensory region has a hyperdeveloped granular fourth layer. The motor region receives a strong input from the anterior part of the ventral nuclear complex of the dorsal thalamus; this thalamic region in turn receives strong input from the cerebullum, corpus striatum, and globus pallidus. This circuitry all appears involved in the management of muscular movement. The sensory region has its connections with the ventrobasal region of thalamus, which receives the strong somatic sensory input (see Section III).

In decided contrast to this situation in primates and carnivores, in *Didelphis* the sensory and motor regions are not separate (Lende, 1963a,b; Magalhães-Castro *et al.*, 1971); the cerebellar and basal ganglia input to anterior ventral thalamus is sent to the same cortical regions as is the somatic sensory information from the ventrobasal thalamus (Killackey and Ebner, 1973).

This distinction now appears to be another set of variations which exist in both the marsupial and placental subclasses. In placentals, there is only a partial motor–sensory differentiation in *Erinaceus* (Lende, 1969; Brodmann, 1909, p. 193) and *Rattus* (Hall and Lindholm, 1974). In marsupial *Trichosurus* appear many of the differentiated features seen in primates and carnivores: cerebellar input to anterior ventral thalamus (Rockel *et al.*, 1972) is projected to anterior regions of sensory–motor cortex, while somatic sensory projections to ventrobasal thalamus are connected to posterior cortical regions (Goldby, 1943; Ward and Watson, 1973; Haight *et al.*, 1975), and these posterior regions are cytoarchitecturally distinctive (Weller, 1972).

## D. Visual Neocortex

### 1. Visual Projections to Neocortex

The striate cortex at the dorsal occipital pole of the cerebrum of all marsupials studied appears to be homologous with the well-studied visual cortex of carnivores and primates. Visual functions have been demonstrated electrophysiologically in this region in *Didelphis* (Lende, 1963a; Christensen and Hill, 1970; Rocha-Miranda *et al.*, 1973; Bombardieri, 1974), *Macropus eugenii* (Lende, 1963c), and *Vombatus* (Johnson *et al.*, 1973).

Lende mapped the slow wave field potentials on the cortical surface, evoked by flashes of diffuse light to the eye. These visual responses extended to the banks of sulcuses $\lambda$ and $\epsilon$ in *Macropus*, and included all of Gray's striate and peristriate areas in *Didelphis* (Fig. 17). The other studies of *Didelphis* examined activity of single neurons (see Section V,C,3), and did not sample cells outside the striate area.

### 2. Cytoarchitecture of Visual Cortex

Brodmann (1906) used the area striata (= "area 17") as another outstanding example of similar development of cortical cytoarchitecture in different mammalian lines. In particular he pointed out the remarkable similarity in the structure of this region in the marsupial macropods *Petrogale penicillata, Macropus dorsalis,* and *Onychogalea fraenata* with that seen in placental monkeys, apes, and humans (Brodmann, 1906, pp. 360 and 392). In these placental and marsupial groups, the fourth (granular) layer is elaborated such that three laminar subdivisions can easily be distinguished: a layer 4a of small cells in moderate density; a layer 4b sparse in cells and rich in myelinated fibers, which make up the stripe of Baillarger or Gennari or Vicq d'Azyr; and a layer 4c of densely packed granule cells (Brodmann, 1909, p. 36, Fig. 15 illustrates this in *Onychogalea* and p. 113, Fig. 74 does so in *M. dorsalis*).

In *Phalangista* (= *Trichosurus*), Brodmann (1906, p. 360) reported the subdivision of the fourth layer to be less prominent, and even less so in *Didelphis*. Gray (1924, p. 239) designated the dense and prominent fourth layer to be characteristic of his area striata in *Didelphis*; this layer was found to contain great numbers of small stellate cells when Golgi stain was used (Benevento, 1968). In *Perameles*, Abbie (1942) indicated that his area parapyriformis 2B (PPy2B, see Fig. 17) was architecturally equivalent to Gray's area striata, and, as in *Didelphis*, this region in *Perameles* capped the dorsal posterior (occipital) pole of neocortex. The broad dense fourth layer of the area striata can be seen covering the occipital pole in *Trichosurus* (Packer, 1941) and in most other marsupials (e.g., Fig. 10).

Bounding the area striata on its anterior and lateral aspects in *Didelphis* is Gray's area peristriata, wherein the cortex is thinner and has a less well-developed fourth layer than do any of the surrounding areas (area striata, parietalis, and temporalis, see Fig. 170, Gray, 1924). Benevento (1968) described another contrast in *Didelphis*: The area striata shows a predominance of large stellate cells in the fifth layer with only radial dendrites, while the fifth layer in area peristriata shows in contrast many pyramidal cells whose apical dendrites reached to the topmost, first layer. Lende (1963a) in *Didelphis* found visual responses in all parts of the area peristriata as well as in the area striata.

Gray (1924, p. 240) reported in some specimens of *Didelphis* a faint sulcus bounding the rostral border of the area striata. A sulcus bounding the striate formation can be seen in *Schoinobates* (Fig. 10). However, in this same Fig. 10 in the macropod *Potorous*, the striate formation does not reach as far as the rostrally enclosing sulcus. Similarly in the macropod *P. penicillata*, Brodmann (1906, pp. 361–362) found striate cortex stopping far short of the nearest sulci, λ and ϵ. Visual responses also stopped short of the previsual sulcus in *Vombatus* (Johnson *et al.*, 1973). Lende (1963c), however, found visual responses over the entire area enclosed by sulci λ and ϵ in *Macropus eugenii*. In *Didelphis*, Benevento and Ebner (1971a) found a laminar folding resembling an incipient sulcus where area peristriata met area temporalis, where Lende (1963a) found the boundary between visual and auditory responsive areas. It will be of interest to learn if the area in the wombats and macropods between the end of the striate formation and the banks of sulci λ and ϵ is homologous with area peristriata, and further, if the sulci enclosing the occipital region in so many marsupial brains (Section VII,A) are in fact boundaries of a functional visual region of cortex.

### 3. Connections of Dorsal Thalamus with Visual Cortex

The area striata, and more specifically its expanded fourth layer, has been demonstrated to be coterminous with the most prominent projections from the dorsal lateral geniculate nucleus in *Didelphis* (Bodian, 1935, 1937; Diamond and Utley, 1963; Benevento and Ebner, 1969, 1971b) and *Trichosurus* (Packer, 1941; Haight *et al.*, 1975).

There is evidence in *Didelphis* that other as yet unidentified thalamic regions project to the uppermost and lowest layers of striate cortex (Benevento and Ebner, 1969).

The peristriate area in *Didelphis* receives projections from the lateral nuclear group of the dorsal thalamus (Bodian, 1935, 1942; Benevento and Ebner, 1971b) as does the region adjoining striate cortex in *Trichosurus* (Goldby, 1943; see Fig. 17). These lateral nuclei are the nearest dorsal thalamic neighbors of the lateral geniculate nucleus on its medial side in *Didelphis* (Bodian, 1939; Oswaldo-Cruz and Rocha-Miranda, 1967a, 1968) and *Trichosurus* (Goldby, 1941). These nuclei of the lateral group receive projections from the superior colliculus in *Didelphis* (Morest, 1965; Benevento and Ebner, 1970) and *Trichosurus* (Rockel *et al.*, 1972) as do their counterparts in placental cats (Morest, 1965).

In *Didelphis*, cells of both the striate and peristriate areas of cortex were found to project to the lateral nuclear group of the thalamus and to the lateral geniculate nucleus (Martin, 1968b; Benevento and Ebner, 1970), and similar findings have been reported for *Trichosurus* (Packer, 1941; Rockel *et al.*, 1972).

## 4. Interneocortical Connections in Visual Cortex

The visual cortex of *Didelphis virginiana* was the subject of a detailed analysis of corticocortical connections (Benevento and Ebner, 1971a). They found four classes of connections, what we will term here tangential, vertical, association, and commissural.

*Tangential* fibers are axons from cortical cells which never leave the cortex, but travel for short distances parallel to the cortical surface and terminate in layer 1 (concentrated in the outer one-fourth of the layer) and layers 5 and 6 in neighboring cortical regions throughout the striate and peristriate areas.

*Vertical* fibers derive from layer 2 and terminate in layers 5 and 6, mostly in 5, directly beneath their origin.

*Association* fibers derive from cortical cells and descend beneath the cortical layers to the underlying white matter and then ascend into another cortical region in the same hemisphere (often called arcuate or U fibers). Such connections were found from cells of striate cortex to all regions of striate and peristriate cortex, but never to areas beyond the limits of peristriate cortex. Small regions of striate cortex send concentrated projections, in an organized manner, to corresponding small regions in peristriate cortex, thus forming a continuation of the topological correspondence persisting through successive synaptic levels: retina, lateral geniculate nucleus, striate cortex, and now peristriate cortex. This may be the basis of an opossum version of two mirror image projections of the visual field upon the visual cortex such as that described physiologically in cats, rabbits, and monkeys (Mountcastle, 1968, p. 1601). At least some of these association fibers originate in cells of layers 2 or 3, and terminals from these lie in layer 1 over wide areas of ipsilateral striate and peristriate cortex, as well as in layers 3 to 5 of peristriate cortex.

*Commissural* connections are axons from cells in the cortex of one cerebral hemisphere which terminate in the cortex of the other hemisphere. In opossums there was a systematic pattern of commissural terminations in visual cortex which was also found in *Trichosurus* (Heath and Jones, 1971): regions of projections from the central retina, where there is binocular overlap of the visual fields of the two eyes, show strong homotypic projections (from the counterpart region of cortex in the opposite cerebral hemisphere).

Commissural fibers arise in layers 4, 5, and 6 and terminate in all layers in the regions of strong projections. Other commissural terminals were found, restricted to the sixth layer and the inner three-quarters of the first layer, in all striate and peristriate regions.

### 5. Connections from Visual Cortex to Other Regions

In addition to projections to nuclei of the dorsal thalamus (lateral group and lateral geniculate), cells of the visual cortex in *Didelphis* send fibers to the ventral thalamus (zona incerta and ventral lateral geniculate nucleus), the pretectal nucleus, and the superior colliculus (Benevento and Ebner, 1970). In *Trichosurus*, Packer (1941) found fibers from visual cortex projecting to the superior colliculus; Martin and Megirian (1972) confirmed this and in addition found evidence of a topographical ordering of these projections similar to those seen in several placentals. Distinctive terminals from cells of the visual cortex were identified in the pontine nuclei of *Didelphis* (Mihailoff and King, 1975).

## E. Auditory Neocortex

Lende (1963a) found auditory responses in anesthetized *Didelphis* in all parts of and confined to the area temporalis of Gray (1924), including the second somatic sensory area SmII. In *Macropus eugenii*, Lende found auditory projections to a corresponding region ventral to the junction between the visual and somatic sensory regions (Fig. 17). A similar location of auditory projections was found in cortex of *Vombatus* (Johnson *et al.*, 1973). Nieder and Randall (1964) found responses to auditory stimulation in all parts of the neocortex of unanesthetized *Didelphis*.

The medial geniculate nucleus of the dorsal thalamus was found to project primarily to the area temporalis in *Didelphis* (Bodian, 1942; Diamond and Utley, 1963; Ebner, 1967) and to a corresponding region in *Trichosurus* (Goldby, 1943; Haight *et al.*, 1975). In *Didelphis*, Diamond and Utley (1963) found some projections to peristriate and parietal regions as well.

The auditory region in *Macropus* and *Vombatus*, and the medial geniculate projections in *Trichosurus*, are all bounded dorsomedially by a constant sulcus. This was named the *supraauditory sulcus* and may serve, like the convolutions in sensory–motor cortex, as a morphological guide to functional subdivisions (Johnson *et al.*, 1973). A related conformation in *Didelphis* may be the indentation of cortical laminae dorsal to area temporalis (Benevento and Ebner, 1971a, p. 164).

Cytoarchitecturally, Gray described the area temporalis in *Didelphis* as markedly thick cortex, with an unusual fifth layer containing many apparent laminae of small, widely dispersed, pyramidal cells. To Abbie (1942) his parapyriform area PPyl in *Perameles* appeared to correspond to the area temporalis, although whether this was due to architecture or to its position relative to other areas was not made clear.

In *Didelphis*, Ebner (1967) found fibers from the medial geniculate nucleus forming a dense network in layers 4, 5, and 6, with terminals evident in layer 4. Other projection fibers traveled parallel to the cortical surface, but only as far as the dense underlying network.

Functionally, Ravizza and Masterton (1972) found in *Didelphis virginiana* that lesions in auditory neocortex interfered with fine (but not gross) localization of sound sources in the horizontal plane (azimuth) but not in the vertical plane (elevation). As they pointed out, cats suffering lesions of auditory cortex were reported to suffer more serious deficits, but in those studies discrimination was inferred from rather complicated behavioral responses which themselves suffered from the lesions, a common fault of studies attempting to study perception using lesions. In the opossum experiments, an ingenious method of determining perceptual acuity was used: the cessation of an ongoing response (in this case licking from a drinking tube) in response to a novel stimulus (in this case a change in the direction of a sound source). Comparison of this neocortical function across species and subclasses still awaits use of comparable methods of measurement.

### F. Orbitofrontal Neocortex

Just in front of sulcus $\alpha$, on the dorsolateral convexity of the anteriormost part of the cerebral hemisphere, in the American opossum *Didelphis virginiana*, Gray (1924) identified an area praeorbitalis (see Fig. 17, POr). This area is distinguished from the anteriorly neighboring area frontalis (Fr) by its laminar appearance. Area frontalis does not show distinct laminae in Nissl-stained preparations; area praeorbitalis has distinctively large cells in the fifth layer, a distinct condensation of cells into a clear second layer, a band of terminals from the dorsomedial thalamic nucleus demarcating the fourth layer, and a distinctive absence of long apical dendrites in cells of the sixth layer (Tobias and Ebner, 1973).

A corresponding region of neocortex, receiving projections from the dorsomedial thalamic nucleus, was identified in *Trichosurus vulpecula* by Goldby (1943). Retrograde degeneration in the dorsomedial thalamic nucleus was produced by lesions of the frontal pole, anterior to sulcus $\alpha$. More precise location of the projection region was provided by Broomhead (1974), who analyzed corticothalamic and thalamocortical relations of this nucleus in *Trichosurus*. He found reciprocal connections between the nucleus and cortical regions on the orbital and medial surfaces of the frontal pole of the hemisphere. The thalamocortical terminals were in the fourth layer, as they were in the area praeorbitalis of *Didelphis*, where they occur on the external convexity of the frontal pole.

Thus the location of the dorsomedial nucleus projections on the external convexity is not a general feature of marsupial brains, but rather a variable character found in *Didelphis* but not in *Trichosurus*.

Tobias and Ebner (1973) discuss intermammalian differences in cortical projections from the dorsomedial thalamic nucleus, particularly in reference to their possible role as an olfactory pathway to cerebral neocortex. In placental rats, there is a projection of fibers from both the olfactory tubercle and the olfactory receiving area in the piriform cortex to a medial "core" region of the dorsomedial thalamic nucleus. This "olfactory core" projects to a small region on the bank of the rhinal sulcus; other parts of the nucleus send fibers to cortex in the mesial wall of the hemisphere. Unlike *Didelphis*, in rats the external convexity between these regions is occupied by projections from the anterior portions of the ventral nuclear complex of the thalamus, and corresponds to motor cortex; the corresponding projections in *Didelphis* are found caudal to sulcus $\alpha$. This contrast was pointed out by Tobias and Ebner (1973), who suggested that the development of an independent "motor" cortex in some placentals may have "shifted" the olfactory projections ventrolaterally, so that they no longer lie on the convexity of the hemisphere. In *Trichosurus*, however, Ward and Watson (1973) located the cerebellothalamocortical projections characteristic of "motor" cortex close to sulcus $\alpha$, leaving open the question of what projects to the frontal external convexity rostral to sulcus $\alpha$ in *Trichosurus*, and whether this in turn "displaces" possible olfactory input on to the orbital and medial surfaces.

## G. Cingulate Cortex

When large lesions of cerebral cortex invaded the medial wall of the hemisphere where neocortex meets the subiculum, retrograde degeneration was observed in the anterior nuclear group of the thalamus (the anterior ventral, anterior dorsal, and anterior medial nuclei) in both *Didelphis* (Bodian, 1942; Ravizza and Masterton, 1971) and *Trichosurus* (Goldby, 1943), reflecting a familiar projection pattern seen in all placentals studied. In none of these studies was this region of medial wall of the hemisphere ablated by itself, but Bodian achieved some precision by lesioning the anterior thalamic nuclei and tracing the degenerating myelin of projecting fibers to the mesial cortex in *Didelphis* (areas cingularis and retrosplenialis of Gray, 1924). Fibers from the mammillary bodies project to, and en route encapsulate, the anterior thalamic nuclei through the prominent mammillothalamic tract in both marsupials and placentals. The prominent encapsulated anterior ventral (AV) nucleus of the thalamus can be seen in marsupial *Petaurus* in Fig. 9 and in placental *Felis* in Fig. 19A; the mammillothalamic tract is seen in cross section in marsupial *Isoodon* in Fig. 13, and the mammillary bodies in marsupial *Macropus* are illustrated in Fig. 20.

In lesion studies using placentals, lesions of cingulate cortex, with consequent degeneration of the thalamic anterior nuclei, have been associated with deficits in temporal integration of behavioral sequences, and the effects were amplified in novel or strange situations (reviewed in Johnson *et al.*, 1969, pp. 169, 170). This involvement with stimulus novelty was shown again (though the relationship is far from clear) in one study of marsupials: in *Didelphis*, Ravizza and Masterton (1971) found auditory habituation was impaired only when large neocortical lesions invaded the cingulate region and the thalamic anterior nuclei showed retrograde degeneration. Only then did auditory stimuli retain their "novelty" (attention getting) properties despite repeated presentations. A clear picture of the functions of these regions remains to be achieved in both placentals and marsupials.

## H. Thalamocortical Relationships in General

A theory of cortical evolution progressing toward the elaboration of specialized sensory regions was pursued in the thalamic level by Erickson *et al.* (1964), who looked to "primitive" *Didelphis* to demonstrate an unspecialized "general sensory" nuclear region. However, the mass of studies just reviewed indicates that the specialized thalamic nuclei, and the corresponding regions of neocortex to which they project, and even the manner of thalamocortical fiber termination, are by and large all present in marsupials much as they are in placentals. Ebner (1969, p. 255) reports from an extensive series of experiments comparing thalamocortical projections in marsupial opossums and placental hedgehogs that "thalamus projects to . . . all of the neocortex in a very similar manner in both the opossum and the hedgehog."

Furthermore, the cytoarchitectural features of special projection regions, particularly visual cortex, and the sulcal formations, particularly in somatic sensory cortex, are remarkably similar in marsupials and placentals. In thalamic topography and connections, and cortical regions, marsupials are much more similar to placentals than either are to monotremes (Lende, 1969).

Thus in therian mammals there is a thalamic visual nucleus (the dorsal lateral geniculate) at the dorsal posterior aspect of the thalamus, projecting to a corresponding visual neocortex in the dorsal posterior aspect of the hemisphere; a more ventral and posterior auditory thalamic nucleus projects to a more ventral auditory neocortex; going anteriorly in the thalamus there is a ventrally situated nuclear region projecting to sensory–motor cortex more anterior on the hemisphere; anterior to these is the mediodorsal thalamic nucleus projecting to orbitofrontal cortex; the cingulate cortex, bounding neocortex on the medial aspect of the hemisphere, receives projections from the anterior nuclear group of the thalamus. In addition to these consistent distinct thalamocortical regions, there are several others, little understood, of variable extent in placentals, and completely unknown (either nonexistent or as yet unstudied) in marsupials.

## VIII. INTERHEMISPHERIC CONNECTIONS IN FOREBRAIN

A major distinction between the brains of marsupials and placentals is the mode whereby the forebrain structures of one cerebral hemisphere communicate with those of the other hemisphere. This distinction was formally recognized early on by Owen (1837); after further analyses (Symington, 1892; Elliot Smith, 1894, 1896, 1897a,b, 1902b, 1910; Johnston, 1913), the whole picture of the variations seen in mammalian commissures and their phylogenetic significance was succinctly summarized by Abbie (1937, 1939) as follows.

### A. Patterns of Commissural Fiber Bundles

"Polyprotodont" marsupials (orders Polyprotodonta and Paucituberculata of Kirsch, Chapter 1, Table III) show the same pattern of commissural connections as do the monotremes, so we may regard this parsimoniously as a "basic mammalian plan." To this basic plan elaborations have been added in the "diprotodont" marsupials (Kirsch's Order Diprotodonta), and in several degrees by all of the various placental lines.

### 1. Polyprotodont Marsupial and Monotreme (Basic) Arrangement

In the "basic" arrangement there are two large fiber bundles interconnecting pallial structures of the two cerebral hemispheres (Fig. 19A). The more dorsal of these connects the hippocampi of the two sides and has been called the "dorsal commissure" but is better designated the "*hippocampal commissure*" (Elliot Smith, 1897a,b). The more ventral bundle is known as the *anterior commissure* and carries fibers to or from olfactory bulbs, amygdalar regions, and basal ganglia as well as cerebral cortex in placentals and marsupials (Loo, 1931; Martin, 1967; Heath and Jones, 1971). Both the hippocampal and the anterior commissures lie in the "lamina terminalis," the sheet of tissue forming the front of the embryonic neural tube, and the front of the adult third ventricle (see Fig. 1C; Fig. 19; Elliot Smith, 1897b; Johnston, 1913; Akert *et al.*, 1961).

### 2. Arrangement in Diprotodont Marsupials with the Fasciculus Aberrans

To this basic plan in the diprotodont marsupials, there is added another bundle of neocortical commissural fibers (Fig. 19A). In the basic monotreme and polyprotodont plan, fibers from neocortex travel from the corona radiata through the external capsule to join the anterior commissure in their course from one hemisphere to the other. In the diprotodonts, many of the processes from dorsal regions of neocortex (Heath and Jones, 1971) travel instead in the *internal* capsule which pierces the corpus striatum to meet the anterior commissure near the midline in what Elliot Smith (1902b) designated the *fasciculus aberrans* (Symington, 1892; Elliot Smith, 1894, 1902b, 1937; Abbie, 1937; Heath and Jones, 1971). The fasciculus aberrans represents a somewhat more direct path from the dorsal neocortex of one side to that of the other. Elliot Smith used the absence of the fasciculus aberrans in the brain of *Thylacinus* to demonstrate that it was a phylogenetic peculiarity of Diprotodonta rather than a property of any larger marsupial brain. Abbie (1937) similarly showed that the absence of the fasciculus aberrans serves to resolve the problems of relationships between the Perameloidea and Diprotodonta which both have syndactylous

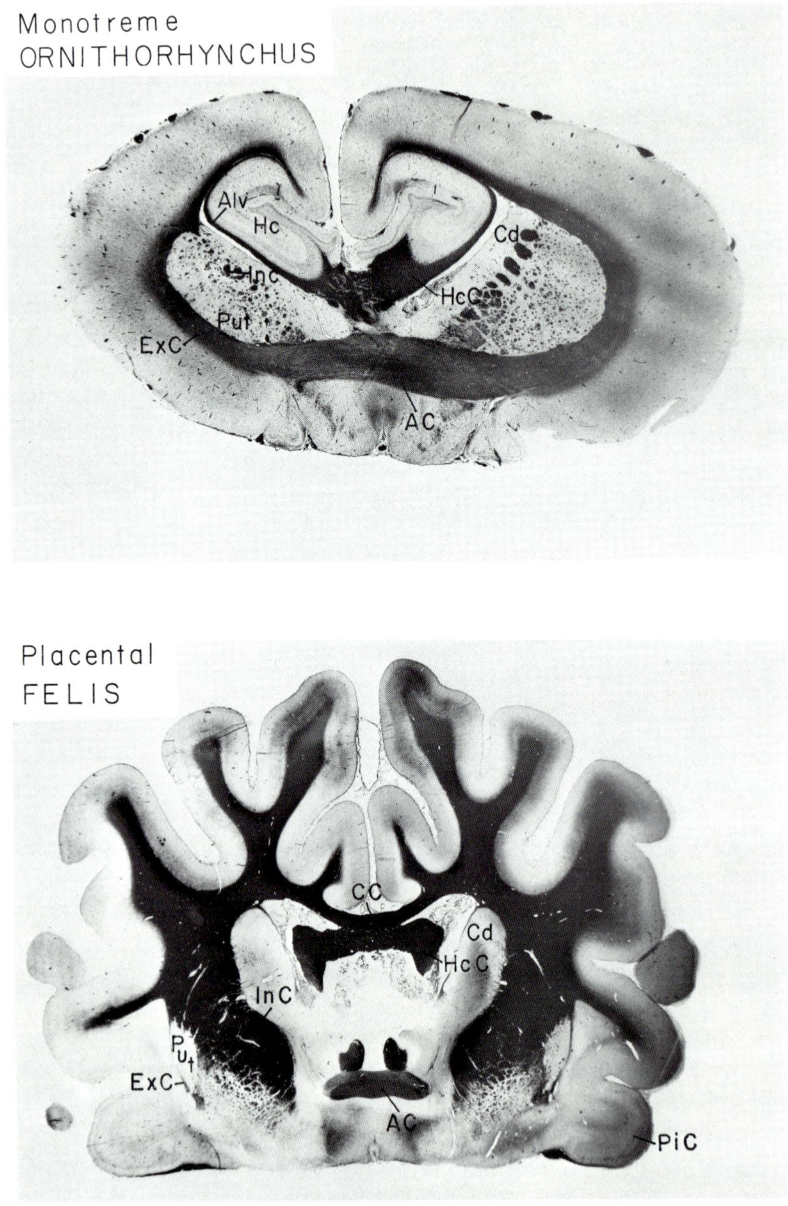


Fig. 19. (A) Four arrangements of mammalian interhemispheric connections. In monotremes and polyprotodont marsupials all cortical connections are via the external capsule and anterior commissure. Diprotodont marsupials have added the fasciculus aberrans going through the internal capsule and the anterior commissure; placentals have added, instead, the corpus

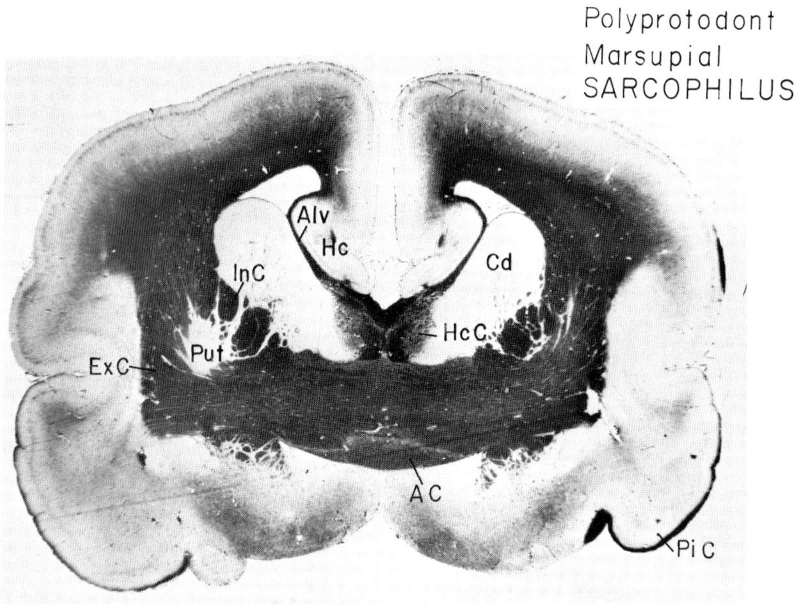


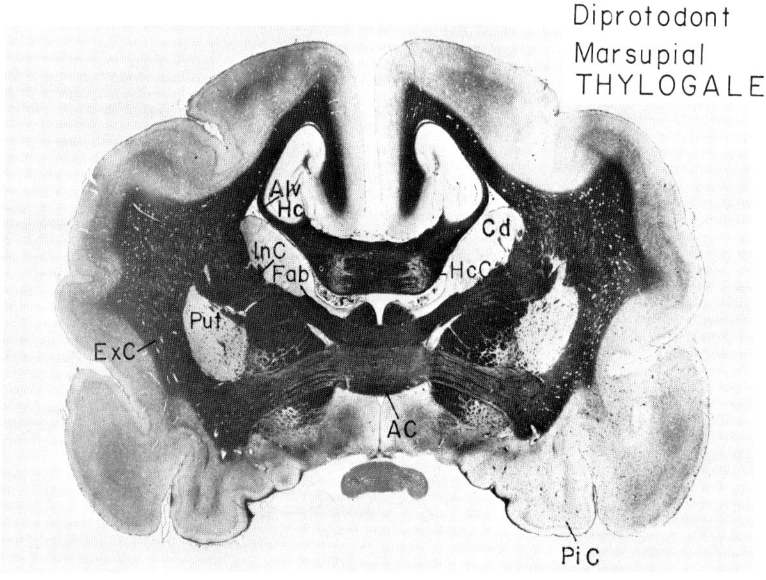


callosum. Sections are coronal and are stained with hematoxylin for myelinated fibers. AC, anterior commissure; Hc, hippocampus; Alv, alveus of the hippocampus; HcC, hippocampal commissure; CC, corpus callosum; InC, internal capsule; Cd, caudate nucleus; PiC, piriform cortex; ExC, external capsule; Put, putamen; Fab, fasciculus aberrans.

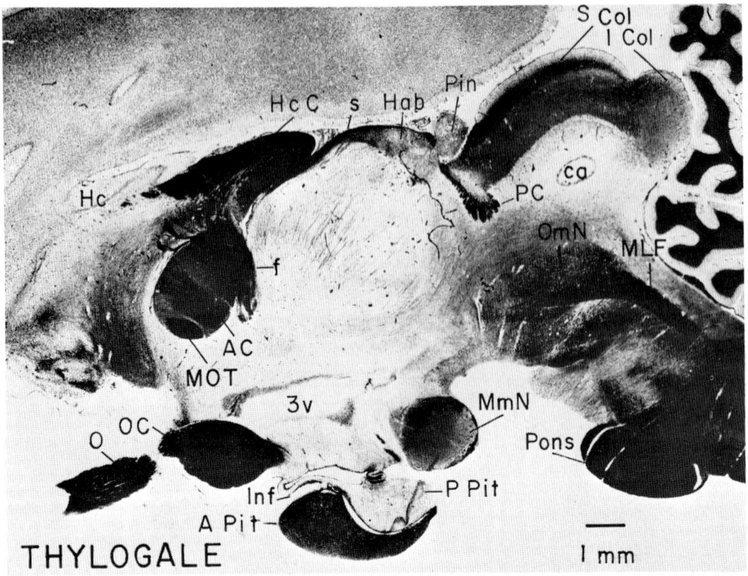
S Col
I Col
Pin
HcC
s
Hab
ca
Hc
PC
OmN
MLF
f
AC
MOT
3v
MmN
OC
O
Pons
Inf
P Pit
A Pit
THYLOGALE
1 mm

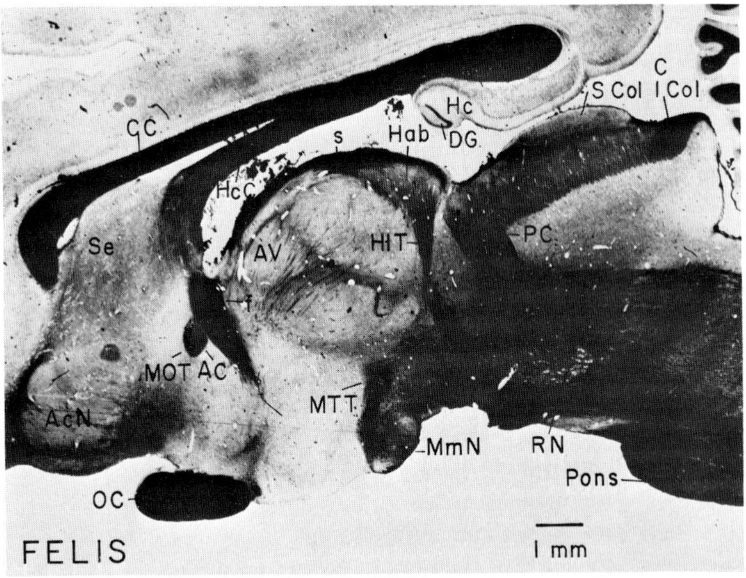
C
S Col
I Col
Hc
CC
DG
s
Hab
HcC
Se
AV
HIT
PC
f
MOT
AC
MTT
AcN
MmN
RN
Pons
OC
FELIS
1 mm

feet, and between the Caenolestoidea and Diprotodonta, which both have diprotodont teeth; neither Perameloidea nor Caenolestoidea have the fasciculus aberrans (Elliot Smith, 1902b; Obenchain, 1925), therefore they can be excluded from Diprotodonta (see also Kirsch, Chapter 1, Section II). This is a striking case of the use of brain characteristics in the determination of phylogenetic affinities.

### 3. Arrangement in Placentals with the Corpus Callosum

In placentals there occurs a shortening of the interneocortical path more drastic than that of the fasciculus aberrans. Fibers from the cerebral cortex of one side plunge directly across the dorsal aspect of the hippocampus and enter immediately the cortex of the other side, forming the corpus callosum (Fig. 19). This is a major brain structure found only in the placental line, wherein its size and shape may vary greatly from one group to another.

## B. Interneocortical Connections

Despite the three diverse pathways characteristic of the different phylogenetic lines, the fibers connecting the neocortices appear to connect the same elements in the same way in terms of function, and in structural considerations other than route of passage. Fibers connecting dorsal regions of the neocortex pass through the dorsalmost commissure available, be it corpus callosum, fasciculus aberrans, or dorsal portion of the anterior commissure (e.g., Ebner, 1967; Heath and Jones, 1971; Martin, 1967). Simi-

Fig. 19. (B) Patterns of commissural connections in placentals and marsupials as seen in sagittal sections from *Felis catus* no. 384 and *Thylogale billardieri* no. 58, sec. 365. Sections are stained with hematoxylin for myelinated fibers. The imposing anteroposterior extent of the corpus callosum and the small size of the anterior commissure in *Felis* contrasts with the large subdivided anterior commissure and the absence of the corpus callosum in *Thylogale*. The stria medullaris is shown in its relation to the habenula in both animals, and the habenulointerpeduncular tract is prominent in the *Felis* section. The connection of the posterior pituitary via the infundibulum appears in the section from *Thylogale*. AC, anterior commissure; AcN, accumbens nucleus; AV, anterior ventral nucleus of the thalamus; ca, cerebral aqueduct; CC, corpus callosum; C I Col, commissure of the inferior colliculus; DG, dentate gyrus of the hippocampus; f, descending column of the fornix; Hab, habenula; Hc, hippocampus; HcC, hippocampal commissure; HIT, habenulointerpeduncular tract; I Col, inferior colliculus; MLF, medial longitudinal fasciculus; MTT, mammillothalamic tract; MmN, medial mammillary nucleus; O, optic nerve; OC, optic chiasm; OmN, oculomotor nucleus; PC, posterior commissure; Pin, pineal body; Pons, pons; RN, red nucleus; s, stria medullaris; Se, septal region; S Col, superior colliculus; 3v, third ventricle.

larly, commissural connections of inferotemporal neocortex pass through the anterior commissure in placentals as they do in marsupials (Heath and Jones, 1971; Ebner, 1967; Martin, 1967).

In *Didelphis virginiana*, Johnston (1913) and Martin (1967) reported connections from dorsal neocortex crossing through the alveus and hippocampal commissure, although this was not supported by later studies (Putnam *et al.*, 1968; Heath and Jones, 1971). The special circulatory problems following cerebral lesions in marsupials (Section XII), and in particular the special vulnerability of the opossum hippocampus to such problems (Scharrer, 1940b), may have produced the degeneration seen in the hippocampal fibers following cortical lesions.

Commissural fibers were found to terminate in all cortical layers in rats and opossums (Heimer *et al.*, 1967), but some cortical areas showed commissural terminals to be restricted to one or another group of laminae (see Section VII, D, 4). The general pattern of commissural terminals in placental hedgehogs was reported by Ebner (1969) to be the same as that found in opossums, despite the very different interhemispheric routes of the projecting fibers. More "specialized" mammals (cats, raccoons, monkeys) showed absence of commissural terminals in primary visual and somatic sensory cortex (Ebner, 1969), while opossums had no cortical areas free of commissural terminals (Ebner, 1967, 1969). Heath and Jones (1971) found *Trichosurus* to be a "specialized" mammal in this regard in that the visual cortex was free of commissural projections. Putnam *et al.* (1968) and Kunze *et al.* (1968), using electrophysiological methods, found a small region of visual cortex in opossums that showed no response to activation of the contralateral hemisphere, so the difference in specialization appears to be one of degree. There is no consistent placental–marsupial difference in this regard.

Nelson and Lende (1965) and Putnam *et al.* (1968) in *Didelphis* found the same pattern of electrical activity in neocortex, evoked by stimulation of the contralateral hemisphere and transmitted through the anterior commissure (from where it could also be recorded). This was a positive–negative wave of change in electrical potential of latency (5–15 msec) and duration (0.1 sec) comparable to that of the "transcallosal" response seen in placental cats and monkeys. The response could be evoked at any point on the neocortex (except the small region of visual cortex previously mentioned) by stimulating the homotopic point on the opposite cortex.

Two other types of interneocortical response could be evoked which were unaffected by sectioning the anterior and hippocampal commissures. The first of these was a later positive–negative wave similar to one found in cats and monkeys whose latency varied from region to region (from 60 to 80 msec), and which was much less stable than the early response through the anterior commissure; the late response was attenuated by high frequency stimulation, asphyxia, or intravenous pentobarbital. The second late response, not mediated by anterior or hippocampal commissures, was a repetitive wave of potential change, a "rhythmic afterdischarge" not found in placentals. These late responses also survived midbrain transections, and thus were believed to be mediated commissurally through the thalamic region. As with the early response, both late responses were activated by stimulating homotopic loci on the contralateral hemisphere. Thus in neocortex there is a point to point functional interhemispheric projection through thalamus as well as through the anterior commissure.

Affanni and Morita (1968) found that cutting the anterior commissure in *Didelphis albiventris* had the same effect as cutting the corpus callosum in cats. Electroencephalographic symmetry was abolished: the two hemispheres no longer showed identical types, amplitudes, durations, and phase shifts of potential fluctuations during waking and the two phases of sleep.

Using microelectrodes to study neuronal action potentials in striate cortex in *Didelphis virginiana,* Kunze *et al.* (1968) found direct commissural connections in all layers. Some of these cells responded to a light flashed in the contralateral eye as well as to homotopic electrical stimulation of the opposite hemisphere, which could form a basis for neocortical binocular integration of information. The latencies of impulses recorded in response to contralateral hemisphere stimulation were evenly distributed across those of both the early and the delayed responses described by Nelson and Lende (1965) and Putnam *et al.* (1968). Some of the long latency cells appeared to be multisynaptically activated, perhaps through the thalamus, but others were direct connections apparently through small, slowly conducting fibers.

### C. Hippocampal Commissure

The hippocampal commissure in marsupials is distinctive in that it crosses the midline, forming a projecting bulge into the anterior wall of the third ventricle (Johnston, 1913). In placental brains it is contained within the wall.

Another feature distinguishing the hippocampal commissures of placentals from those of marsupials is the separation of dorsal and ventral portions (Abbie, 1939). In placentals, many of the crossing fibers traverse the dorsal aspect of the hippocampal formation, forming the dorsal hippocampal commissure. The other fibers collect in the ventral aspect of the hippocampus to cross, forming the ventral hippocampal commissure. All of the crossing fibers in marsupials pass ventral to the hippocampus; thus marsupials do not show a dorsal hippocampal commissure.

### D. Diencephalic Commissures

One suggested pathway for the delayed interhemispheric response (Nelson and Lende, 1965; Putnam *et al.*, 1968) is the set of thalamic commissural fibers described by Bodian (1940) in *Didelphis*. He distinguished six groups of fibers, passing by way of the internal medullary lamina of the dorsal thalamus and connecting, respectively, the (1) parataenial, (2) anterior dorsal, (3) central and paracentral, (4) ventral anterior, (5) ventral (including ventrobasal), and (6) parafascicular thalamic nuclear regions of the two sides.

Another hemispheric interconnection in the diencephalon is the optic chiasm (see Figs. 9, 13, 19, OC), where fibers from the retinas of both eyes cross to the other side of the brain.

A third group of fibers crossing the diencephalic midline are the mysterious "suprachiasmatic commissures." These include those described in *Didelphis* by Loo (1931, p. 136) as the anterior suprachiasmatic commissure and by Bodian (1940) as the commissures of Ganser, Meynert, and Gudden, long recognized but never explained in placental mammals (Crosby *et al.*, 1962, p. 322).

A most prominent connection is the posterior commissure seen in all mammals. A marsupial example is prominent in the parasagittal sections from *Thylogale* in Figs. 19 and 20. Despite its size and universality, little can be said of this striking group of fibers other than that it includes interconnections between the pretectal and other cell groups of the mesodiencephalic junction (Crosby *et al.*, 1962, p. 267).

The habenular commissure interconnects the habenulae of the two sides, which are treated in Section IX,B.

## IX. HYPOTHALAMUS, EPITHALAMUS, AND CIRCUMVENTRICULAR ORGANS

A number of specialized organs have developed around the lining of the front end of the neural tube whose function is only gradually coming to light; they appear to have the common feature of serving as functional interfaces among the three major systems: nervous, blood circulatory, and brain ventricular, which circulates cerebrospinal fluid. Hofer's term *circumventricular* delineates these structures well, although they are not, strictly speaking, organs (Weindl, 1973). They include many demonstrated, and more suspected, sites of action—by hormones in the blood upon cells in the brain, and conversely, by cells in the brain secreting hormones into the blood or the cerebrospinal fluid. These types of activity have been well documented in the hypothalamus and its accessory, the hypophysis or pituitary gland. The epithalamus, including the habenulae and pineal body, and the neighboring subcommissural body are now appearing to conduct similar functions, as is the previously mentioned subfornical body and the area postrema lining the back of the fourth ventricle in the medulla. These structures are illustrated in Fig. 20.

In *Didelphis*, Wislocki (1940) found the posterior pituitary and its infundibular stalk, the subfornical body, and the area postrema all showed a fine capillary network of interconnected blood vessels, in contrast to the arteriovenous end loops characteristic of the rest of the marsupial brain (see Section XII). Furthermore, stains penetrated these regions from the blood, indicating they are not isolated from the bloodstream by the usual blood–brain barrier and are thus favored sites for hemoneural interactions.

### A. Hypothalamus, Hypophysis (Pituitary), and Autonomic Nervous System

In Weindl's (1973) apt terminology, central hemoneural receptor functions combined with neurohemal release of neurohormones constitutes an efficient apparatus for regulation of many features of the body's internal milieu. Several such mechanisms have been localized in the hypothalamus–hypophysis. Osmoreceptors sensitive to water content of the blood are related to hypothalamic cells which secrete antidiuretic hormone through their hypophyseal endings; hypothalamic receptors sensitive to the temperature of the blood are in spatial association with regions governing heating and cooling activities of the body; nutritive contents of the bloodstream appear related to hunger and feeding mechanisms through hypothalamic processes; physiological and behavioral sensitivity to sexual hormones as well as the regulation of the production of these hormones reside in hypothalamic components. These are only a few examples.

The anatomy of the hypothalamus of *Didelphis* has been described in detail periodically (Tsai, 1925b; Warner, 1929; Bodian, 1939, 1940; Thompson, 1942; Oswaldo-Cruz and Rocha-Miranda, 1967b), and none of these studies reported any significant deviation from the general picture seen in placentals.

The typical mammalian arrangement of hypothalamus and hypophysis (pituitary) is seen in a marsupial manifestation in advantageous parasagittal sections from *Thylogale billardieri* in Figs. 19B and 20A. The hypothalamus extends from the preoptic region anterior to the optic chiasm and below the anterior commissure to its posteriormost component, the medial mammillary nucleus. Axons of hypothalamic cells pass through the slender stalk of the infundibulum to terminate in the posterior pituitary. Adjacent to this is the glandular anterior pituitary, which secretes a great variety of hormones under the direct influence of secretions from the hypothalamus transported through local blood vessels.

### 1. Posterior Pituitary Secretions and Water Regulation

Large cells, including those of the supraoptic and paraventricular nuclei which give rise to axons going to the posterior pituitary, are particularly plentiful in *Didelphis* (Bodian, 1939; Zeballos *et al.*, 1967), and their endings in the pituitary are well defined as is the lobulation of the posterior pituitary (Bodian, 1951, 1963; Roth, 1963). Many, if not all, of these cells respond electrophysiologically to osmotic stimulation and histologically to stains for neurosecretory substances, as do their counterparts in placentals (Zeballos *et al.*, 1967). Wang (1969) made use of this combination of characteristics in *Didelphis* to study the relative distributions of osmoreceptors and neurosecretors of antidiuretic hormones to support the hypothesis that the same cells carry out both functions.

In general, glial astrocytes in the hypothalamus of *Didelphis* were fewer and smaller than those in placental cats and armadillos; but in the supraoptic nucleus the astrocytes in opossums were larger than those in armadillos and cats (Royce, 1971).

Wislocki (1940) found the looped end vessels in the supraoptic and paraventricular nuclei in *Didelphis* hypothalamus to be twice as numerous and twice as dense as in other hypothalamic regions, again favoring a most intimate connection between blood composition and neural activity.

Hormones secreted by the neurons of this region, through their terminals in the posterior pituitary, have been investigated in at least five marsupial species. Antidiuretic activity was ascribed to arginine vasopressin, a hormone peculiar to mammals, in *Didelphis virginiana* (Sawyer *et al.*, 1960), *Trichosurus vulpecula, Setonix brachyurus, Macropus rufus* (Ferguson and Heller, 1965), and *Phascolarctos cinereus* (Yesberg *et al.*, 1967); this same hormone is similarly effective in all placentals except the pigs and their relatives, whose counterpart hormone is lysine vasopressin (Ferguson and Heller, 1965). Ferguson and Heller also found oxytocin, the other major hormone of the posterior pituitary, in their marsupials, but in a ratio to vasopressin lower than that of placental mammals studied. They suggested this analysis as one biochemical means to taxonomic arrangement of mammalian species.

## 2. Temperature Regulation

In pouch young *Didelphis virginiana*, Petajan and Morrison (1962) noted the simultaneous appearance of shivering, curling, and a rise in metabolism following cold stimulation and related to the beginning of temperature regulation; hissing, opening the mouth, and jabbing with the head in response to painful stimulation appeared at the same time, between 73 and 80 days after birth. Petajan *et al.* (1962) found evidence that the temperature regulation depended upon a sensing and a correcting system within the central nervous system, as opposed to mechanisms within peripheral skin and muscle. They suggested the anterior hypothalamus and nearby preoptic region as likely loci for these mechanisms, since these regions showed significant developments in their anatomical differentiation at the same age at which behavioral thermoregulation began. In adult opossums, Roberts *et al.* (1969) obtained relaxation, lying down, grooming, spreading saliva over the fur, and panting in response to local warming of these brain regions.

## 3. Male Sexual Behavior

The anterior hypothalamic and preoptic regions are favored sites for the cellular uptake of sexual hormones; lesions in these regions disrupt sexual behavior; implantation of hormones here affects sexual activities, as does electrical stimulation (e.g., Johnson *et al.*, 1969, p. 165).

Roberts *et al.* (1967) observed dramatically complete sequences of male mating behavior following electrical stimulation of discrete loci in these regions in *Didelphis virginiana*. Since these activities, including mounting, rubbing, biting, clicking, and penile erection, occurred only in the presence of a somewhat suitable object animal or surrogate, it was concluded that the stimulation did not activate responses automatically as part of a sensory or motor pathway. Rather, they were interpreted as indicators of "facilitation, or reduction of tonic inhibition in specialized sensory–motor mechanisms located outside the area explored" electrically, by means of activation of cells or fibers of passage within 1.5 mm of electrode loci, which were restricted to near the midline in the anterior hypothalamic and preoptic vicinity. The fact that the same responses to electrical stimulation were obtained in males and females, despite the fact the females normally rarely display such behaviors, constitutes evidence that a relatively complete male response mechanism resides within the brain of both sexes, but is more easily elicited in males. Furthermore, the behavioral sequence following electrical stimulation was found to have intrinsic reward properties for both males and females. Bergquist (1970) found that the medial forebrain bundle, descending from hypothalamus to midbrain, was an essential component in this mechanism in opossums. From earlier studies, Roberts *et al.* (1967) believed that counterpart mechanisms had a similar location in placental rats, cats, and goats.

### 4. Other Behavioral Correlates

Roberts *et al.* (1967) found several other integrated behavioral sequences that could be elicited by electrical stimulation of *Didelphis* hypothalamus, and these finding paralleled results obtained in cats. The medial forebrain bundle path to midbrain was essential to all of these mechanisms (Bergquist, 1970). These included sequences, or portions of sequences, of biting attack, defensive threat, eating, grooming, and various "exploratory" and "escapelike" activities. Most of these were elicited from specific limited regions of the hypothalamus.

### 5. Autonomic Nervous System

The hypothalamus has been regarded (e.g., Brodal, 1969, p. 575) as the main integrator of functions of the extensive autonomic nervous system, one of the major subdivisions of the vertebrate nervous system. Through hypothalamic processes, the autonomic functions which accomplish the nervous control of the visceral organs and their "vegetative" functions are modulated and correlated with events in the rest of the brain.

There is an extraordinary dearth of investigations of the autonomic system, with its sympathetic and parasympathetic components, in marsupials, beyond cursory anatomical descriptions of the autonomic nuclei in general descriptions of regions of the brain.

Marsupials may constitute a fertile field for autonomic investigation. For example, although they studied peripheral nerves, technically outside the scope of this review, Koizumi and Sato (1969) found in opossums a possible unique, or at least especially well developed, sympathetic nerve supply to carotid baroreceptors. Autonomic reactions caused changes in blood pressure to a degree not seen in cats. It was speculated that opossums had a powerful mechanism for regulating fright-induced increases in blood pressure, and it was concluded that here the autonomic system worked quite differently in opossums when compared with placental mammals.

## B. Epithalamus: Habenula and Pineal Body

Just as the hypothalamus is found suspended between the third ventricle and the exterior cerebrospinal fluid below the thalamus, so the epithalamus intervenes between the third ventricle and the extracerebral fluid above the thalamus. The most prominent epithalamic structures are the medial and lateral habenular nuclei, the fibrous habenular commissure crossing between the habenulae of the two hemispheres, and the glandlike pineal body found in association with the habenular commissure.

The habenular nuclei are another of the long standing puzzles in neural science. They are present in some form throughout virtually all the vertebrates, but they have resisted all attempts to determine their function. In mammals, including marsupials, they are positioned between two very prominent bundles of myelinated fibers, the stria medullaris between the septal region and the habenulae, and the so-called habenulointerpeduncular tract which goes from the habenulae to the vicinity of the equally mysterious interpeduncular nuclear complex on the ventral surface of the midbrain.

Way and Kaelber (1969) noted the remarkable size of the habenulae in *Didelphis*, and analyzed their efferent projections. They found that the main projections via the habenulointerpeduncular tract (HIT in Figs. 13 and 19) did not terminate in the interpeduncular region where the tract disappears, rather the fibers separate and continue on through varied courses to terminate in many regions in midbrain tegmental and raphe regions, a few even reaching the superior colliculi. Other fibers followed the stria medullaris rostroventrally to terminate in the preoptic region and the olfactory tubercle. Whatever habenular activity does, its influence is widely spread.

The pineal body in our times is thought to be a glandular entity responsive to illumination, and influencing diurnal activity cycles and photoperiodic gonadal activity (e.g., Schapiro and Salas, 1971). Among the various marsupials it is highly variable in shape and position. In fact the pineal can be quite variable within a single species, as Jordan (1911) early pointed out in *Didelphis*, and Machado (1965) confirmed in several didelphids. Pineal tissue in opossums is found around the habenular commissure and as a greater or lesser accumulation along the ventricular wall leading from the habenular to the posterior commissure. In Fig. 20 the pineal tissue can be seen investing the habenular commissure like a doughnut in *Caenolestes obscurus*, and forming a more typically mammalian protuberance extending between the superior colliculi in the macropods *Thylogale billardieri* and *Macropus fuliginosus*. In red and gray kangaroos, Quay (1966) reported what he considered the most striking features of any mammals studied: a pineal recess especially deep and incompletely lined by ependymal cells, fascicles of both the habenular and posterior commissures embedded between lobules of pineal parenchyma, a rich pineal vascularization with coiled anastomatic chains of capillaries or sinusoids between the lobules, and a prominent zonation of parenchymal cell types.

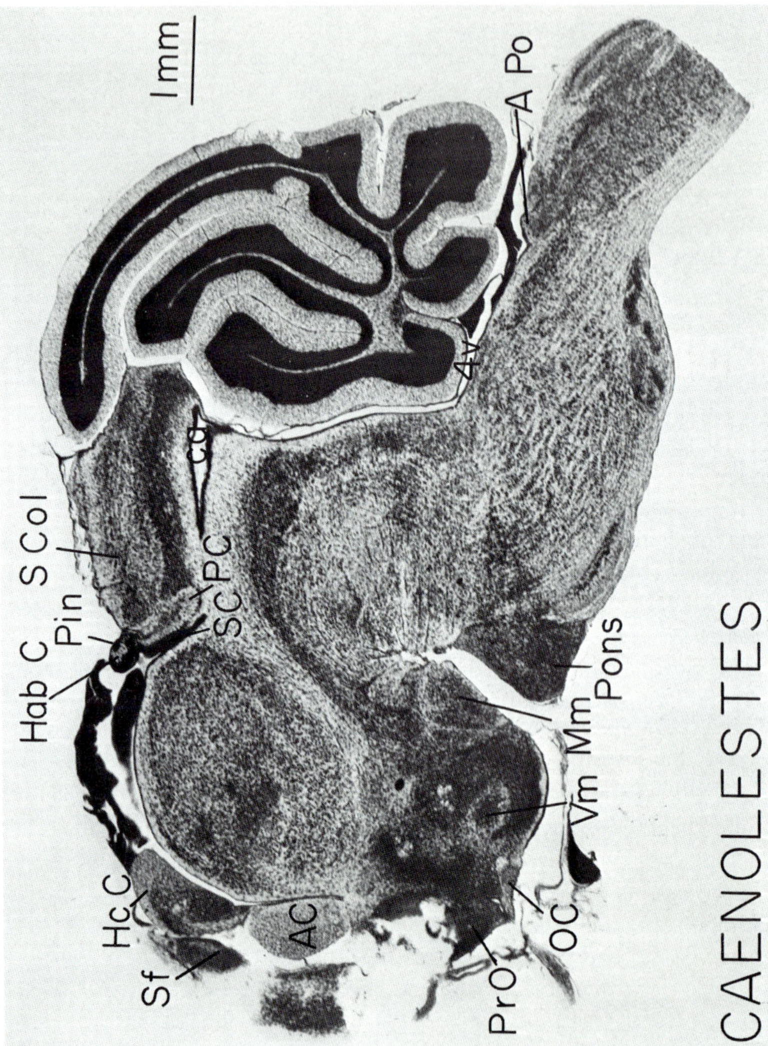
1mm
A Po
4v
Hab C
Pin
S Col
SC
PC
Pons
Mm
Vm
OC
PrO
AC
Hc C
Sf
CAENOLESTES

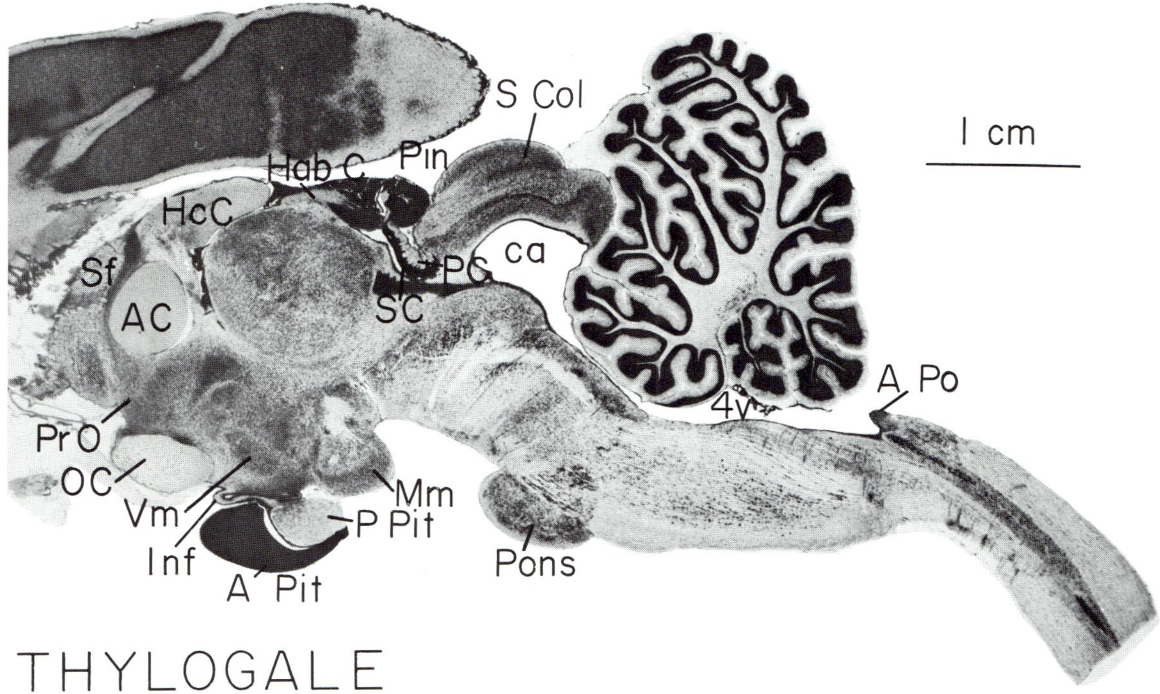


Fig. 20. (A) Some circumventricular organs and other midline structures seen in two sagittal sections from marsupials *Caenolestes obscurus* no. 89, sec. 186 and *Thylogale billardieri* no. 58, sec. 342. Note the contrast in the amount of pineal tissue. Sections are stained with thionine. AC, anterior commissure; A Pit, anterior pituitary; A Po, area postrema; ca, cerebral aqueduct; Hab C, habenular commissure; HcC, hippocampal commissure; Inf, infundibulum; Mm, mammillary nuclei; OC, optic chiasm; PC, posterior commissure; Pin, pineal body; P Pit, posterior pituitary; Pr O, preoptic region; SC, subcommissural body; S Col, superior colliculus; Sf, subfornical body; Vm, ventromedial hypothalamic nucleus; 4v, fourth ventricle.

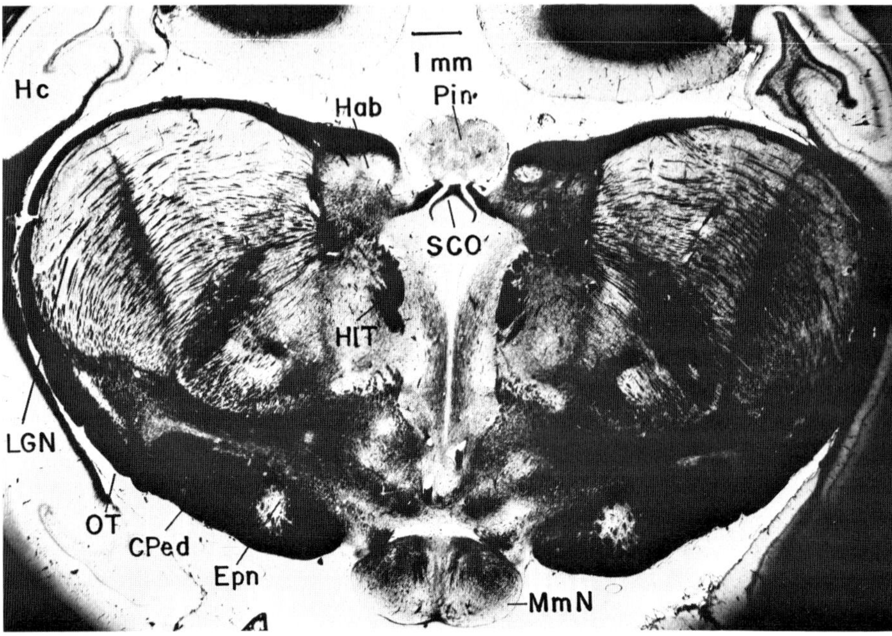


Fig. 20. (B) Coronal section through the posterior diencephalon of *Macropus rufus* no. 22, sec. 1007, stained with hematoxylin for myelinated fibers. CPed, cerebral peduncle; Epn, entopeduncular nucleus; Hab, habenula; Hc, hippocampus; HIT, habenulointerpeduncular tract; LGN, lateral geniculate nucleus; MmN, medial mammillary nucleus; OT, optic tract; Pin, pineal body; SCO, subcommissural body.

Upon superficial observation, our own specimens of *Petaurus breviceps* and *Petaurus norfolcensis* appear to have no pineal tissue at all, recalling the situation in some placentals in the order Edentata (Machado, 1965). Their near petaurid relative *Pseudocheirus*, and the more distantly related *Trichosurus,* show a modest amount of pineal tissue confined on the ventricular surface between the habenular and posterior commissures. This interspecific variability should provide unusual opportunities for the study of variable pineal function in phylogenetically related animals.

### C. Subcommissural and Subfornical Bodies

Apparently continuous with the pineal tissue is the thick ependymal tissue underlying the posterior commissure (see Fig. 20), called the *subcommissural body.* Its function is unknown. This structure is quite prominent in kangaroos (Fig. 20B) and continues along the roof of the cerebral aqueduct back to the level of the inferior colliculi.

Marsupials show a distinctive positioning of the *subfornical body*, a peculiar structure of unknown function on the ventricular lining, found in association with the hippocampal commissure in all mammals (Akert *et al.*, 1961). In placentals the subfornical body is attached to the posteroventral surface of the hippocampal commissure around the midline, lying between that corner of the commissure and the ventricular fluid-filled space. In marsupials, the subfornical body is located dorsal and anterior to the hippocampal commissure, in the ventricular "outpocketing" formed by the projection of the commissure into the ventricle. The subfornical body is separated from the commissure by a narrow cleft, rather than applied directly to it as in placentals. Akert *et al.* (1961) called attention to the fact that the anterior location, and the absence of the corpus callosum, enables direct visualization of, and surgical approach to, the marsupial subfornical body through the membranous roof of the third ventricle when the cerebral hemispheres are gently retracted.

The subfornical body may play some role in the regulation of water and electrolyte balance. Drinking was elicited in water-sated rats by application of angiotensin to the subfornical body (Simpson and Routtenberg, 1973). Angiotensin is a stimulator of the production of aldosterone which increases sodium absorption in the kidneys, which calls forth drinking to increase the plasma volume. Angiotensin is produced normally when renin is secreted from the kidneys in response to low plasma volume or induced sodium deficiency, as was observed by Reid and McDonald (1969) in *Trichosurus*. They concluded that *Trichosurus* showed the same type of renin–angiotensin system as is found in placentals.

## X. DESCENDING SYSTEMS

A major feature of mammalian, including marsupial, brains is the high degree of "prosencephalization of motor control." In mammals, compared with other vertebrates, processes from the forebrain "reach out" and assume a great deal of control over, and monitoring of, functions otherwise carried out at lower levels, including those regulating muscular movements. The most obvious case of this is the corticospinal connection described in Section VII,C,6, unequivocally a function of the forebrain. But other descending processes exist, involving both forebrain processes and components from other levels. The most extensively investigated of these, at least in marsupials, are those involving the red nucleus of the midbrain along with cerebellum, cerebral cortex, and spinal cord. Others deal with the reticular regions of the brainstem and their influence upon spinal processes.

### A. The Red Nucleus: Rubral Connections and Functions

This prominent cell group in the ventral midbrain has been extensively studied in *Didelphis virginiana*, with corroborating findings in *Trichosurus vulpecula*, and in every case the results indicate a structure and function very similar to those reported for placental cats, although there are many intraplacental differences, chiefly between primates and other placentals (Martin *et al.*, 1974b). The location of the nucleus in the midbrain can be seen in the sections shown in Figs. 13 and 19B.

#### 1. Cytoarchitecture

King *et al.* (1971) identified three types of neurons in the opossum red nucleus: (1) giant cells, confined to the caudomedial third of the nucleus; (2) numerous large medium cells, and (3) few small cells, both distributed throughout the nucleus. The small cells were identified as "interneurons" with all their processes confined within the nucleus (King *et al.*, 1974). The giant cells serve to demarcate a basic subdivision of the nucleus into two parts: (1) the caudomedial third of the nucleus, where both giant and large medium cells show intense acetylcholinesterase activity and project mainly to the spinal cord; and (2) the rostrolateral two-thirds of the nucleus where there are no giant cells, less acetylcholinesterase activity, and the large medium cells project to brainstem regions as well as to the spinal cord. The caudomedial division receives massive input from the cerebellum and virtually none from the cerebral cortex, while the anterolateral division receives less input from cerebellum and nearly all the rubral input from the cerebral cortex (Martin *et al.*, 1974b).

#### 2. Cerebellorubral Connections

In opossums, as in cats, the largest input to the red nucleus from any source comes from the interpositus nucleus of the cerebellum (King *et al.*, 1973). A small input from the lateral cerebellar nucleus projects to a small dorsal rostral region in the red nucleus. The cerebellorubral fibers may be collaterals of cerebellothalamic axons. Terminals of cerebellar cells contact the soma and proximal dendrites of large medium and giant cells in the red nucleus, indicative of a powerfully influential connection. In cats these projections showed a tonic excitatory effect on rubrospinal neurons.

### 3. Corticorubral Connections

Input from the cerebral cortex to the opossum red nucleus is more restricted than that from the cerebellum in two ways (King *et al.*, 1972): (1) it is largely confined to the rostral two-thirds of the nucleus (thus few cortical fibers reach giant cells), and (2) it is confined to more distal locations on the dendrites and thus may be thought to modulate the cerebellorubral interactions. King *et al.* (1972) believed that corticorubral fibers may be collaterals of corticotectal and corticospinal fibers, and the corticospinal collaterals might contact the small interneurons. These interneurons may be a means for cortical inhibition of rubral large medium neurons, while the terminals on the large medium cells directly from cerebral cortex could be excitatory. Most of the corticorubral fibers originate in the forelimb region of sensory–motor neocortex, with some coming from the hind limb region.

In *Trichosurus,* Martin and Megirian (1972) found corticorubral projections, as in *Didelphis,* to all parts of the red nucleus except for a ventromedial region containing large neurons. At least a large portion of the corticorubral fibers originated in the fore- and hind limb regions of sensory–motor neocortex.

### 4. Rubrospinal Projections

Giant cells from the caudomedial division, and large medium cells from all regions, of the opossum red nucleus send axons to the gray matter of the spinal cord (Martin *et al.*, 1974b). All levels of the spinal cord receive rubral fibers (Martin and Dom, 1970a), as is true in *Trichosurus* and placental cats (Warner and Watson, 1972). In opossums, medial rubral neurons projected to cervical cord levels, while more lateral neurons projected to more caudal levels, similar to previous descriptions in placental cats and rats (Martin and Dom, 1970a; Martin *et al.*, 1974b).

In all these species the rubrospinal terminals were located in the lateral regions of laminae V, VI, and VII of the contralateral spinal gray matter. The rubral projections were partially overlapping, but not coextensive, with the fields of neocorticospinal projections which are concentrated more dorsomedially in the spinal gray, at those levels where both corticospinal and rubrospinal projections are found together (corticospinal projections are restricted to cervical and thoracic levels in these marsupials, see Section VII,C,6).

Electrical stimulation of the opossum red nucleus yielded contractions of contralateral, paraxial, and limb muscles (Martin and Dom, 1970a). Thus the prominent rubrospinal path in marsupials represents a possible indirect mechanism whereby neocortical as well as cerebellar control might be exercised over spinal mechanisms of muscular movement.

### 5. Other Rubral Projections

In *Didelphis* (Martin and Dom, 1970b; Martin *et al.*, 1974b) and *Trichosurus,* as in placental *Felis* (Watson and Warner, 1973), contralateral projections from the red nucleus were found in various of the sensory trigeminal nuclei, in the lateral reticular nucleus, and in portions of the external cuneate and cuneate–gracile nuclei; there were also ipsilateral projections to the pontine reticular formation and the inferior olive. While all three species showed contralateral projections to the facial nucleus, *Trichosurus* was distinctive in that the rubral terminals were medial in the nucleus, while in cats and opossums they were located laterally. Watson and Warner (1973) speculated that this distinction might be related to the vibrissa specialization reflected in the neocortical barrels (Fig. 18 and Section VII,C,3).

In their more detailed studies of rubral projections in *Didelphis,* Martin *et al.* (1974b) found many additional rubral projections, including fibers *to* the cerebellar interpositus nucleus which is the major source of fibers to the red nucleus itself, constituting yet another example of the reciprocity of connections so frequently seen in the mammalian brain.

A coherent scheme for explaining the possible functions of all these rubral connections is yet to be formulated.

## B. Reticulospinal Projections

Cells in many locations throughout the reticular formation of the opossum brainstem send fibers to the spinal cord, much as they do in cats (Beran and Martin, 1971; Martin and Dom, 1971). There are projections to reticular cells from red nucleus and cerebral cortex as well as a great number of other locations in the nervous system; the relation of these projections to the spinally projecting cells is yet to be determined. The reticulospinal fibers terminate in laminae VII, VIII, and in some cases IX of the spinal gray at all levels, and were reported to be more plentiful than tectospinal, rubrospinal, or corticospinal fibers. They doubtless represent the major supraspinal influence upon the ventral columns of the spinal gray matter.

It can be speculated that these reticulospinal influences, together with vestibulospinal input (see Section V,A), are responsible for the "decerebrate rigidity" seen in opossums (Rogers, 1923; Weed and Langworthy, 1925b) as in placental mammals following removal of the brain anterior to the inferior colliculi (thus removing cortical, tectal, and rubral controls from both reticular and spinal cells).

### C. Other Descending Connections

There are a great number of other descending processes in mammalian and marsupial brains. For example, a great variety of corticomesencephalic connections have been described in *Didelphis* (Martin, 1968a) and *Trichosurus* (Martin and Megirian, 1972), and a similarly diverse set of tectofugal connections were reported for *Didelphis* (Martin, 1969). Organization of these connections into manageable functional conceptual schemes is so far even less advanced than for the sketchy systems considered in Section X,A and B. Within the spinal cord itself there are descending processes, including those responsible for the baffling centrifugal discharges through the sensory roots in response to somatic sensory and vestibular stimulation, which were especially prominent in *Trichosurus* (Megirian, 1968, 1971).

## XI. CEREBRAL CORRELATES OF SLEEP AND AROUSAL

Among marsupials, there have been systematic electroencephalographic (EEG) studies of two species of *Didelphis, D. virginiana* and *D. albiventris*. Barratt (1965) identified five EEG states correlated with behavioral stages: (1) sleep, with EEG consisting of large amplitude slow waves as in man; (2) waking–alert, with low amplitude fast waves; (3) waking–alert–searching, with short bursts of the 8 to 12 per second rhythmic waves known as alpha activity interspersed with the low amplitude fast waves; (4) waking–alert–fixating, with longer periods of alpha rhythm which could be blocked by sudden sensory stimulation; and (5) tonic immobility, the "feigned death" of this species "playing 'possum," with very low amplitude fast activity preceded by alpha bursts. Norton *et al.* (1964) specifically analyzed the EEG during induced states of "feigned" death or sleep, the waxy flexibility of muscles adopted by *Didelphis* when attacked or severely threatened; they also found extremely low voltage fast EEG activity just as occurs in states of high behavioral arousal in opossums and placentals.

Two distinctive phases of sleep have been identified in placental mammals, based largely on EEG information and confirmed by several lines of behavioral, pharmacological, and physiological investigations. The first is that described by Barrat in *Didelphis* and has acquired the name "quiet sleep," also known as "slow wave sleep" from its EEG correlate. The second, known as "active sleep," has also been known as "paradoxical sleep" since the EEG pattern in the sleeping animal in this state is the same as that seen in the alert waking animal; it has also been called "REM sleep" from the rapid eye movements which are frequently observed in this state. In placentals, quiet sleep has been shown to have some dependence upon the activity of serotonergic cells in the raphe nuclei of the midline of the pons, and active sleep has some relation to the far reaching projections of noradrenergic cells of the locus coeruleus near the junction of pons and midbrain lateral to the fourth ventricle.

Affanni and Vaccarezza (1964) identified active sleep in *D. albiventris* and Snyder (1965) and Van Twyver and Allison (1970) analyzed it in *D. virginiana.* As in placentals, cerebral cortical EEG showed high frequency, low amplitude fluctuations in states of waking and active sleep; while in quiet sleep there were high amplitude, slow EEG waves interspersed with what Van Twyver and Allison called the distinctive feature of quiet sleep, spindling. Spindling denotes bursts of high amplitude fluctuations, cyclic with a frequency of 8 to 11 per second, with a spindle-shaped wave envelope due to the rapid increase and then decrease of the amplitude of successive peaks of individual waves. Spindling is typical of quiet sleep in placentals as well.

EEG recording from hippocampus shows a prominent rhythmic fluctuation at a frequency of 6 to 7 per second; this is known as the theta rhythm. All of the just-mentioned investigations found prominent hippocampal theta activity in opossum active sleep, and Affanni *et al.* (1967) found that it could be recorded along with the cortical activity with a transverse placement of a single pair of electrodes. This made it possible to distinguish between the waking EEG and the active sleep EEG, since in active sleep the waking pattern was superimposed upon the underlying hippocampal theta activity. Theta activity did not survive section of the anterior and hippocampal commissures in *D. albiventris* (Affanni and Morita, 1968). Van Twyver and Allison (1970) noted bursts of theta activity as a characteristic of aroused states.

These last-named investigators were specifically prompted to study opossums by their failure to find active sleep in the monotreme spiny anteater *Tachyglossus*, and they concluded that active sleep is not a general mammalian characteristic, but seems to be present in all therians (Van Twyver and Allison, 1970). Using behavioral criteria, Cicala *et al.* (1970) identified active sleep in kangaroos *Macropus rufus*. Vaccarezza and

Affanni (1968) found active sleep reduced in *D. albiventris* by removing nonolfactory input from olfactory bulbs.

## XII. CEREBRAL CIRCULATION

Accounts of the architecture of the vessels carrying blood to and from the brain of *Didelphis* include those of Voris (1928a), Scharrer (1940a,b), Dom *et al.* (1970), and Gillilan (1972). The remarkable feature of the circulation in the central nervous system of opossums, and apparently of all marsupials, is that discovered by Wislocki and Campbell (1937): the precisely paired branching of arteries and veins and their terminal communication in arteriovenous end loops.

In placental mammals and in monotremes (Sunderland, 1941), as in most vertebrates, the arteries branch into anastamosing capillary networks from which the blood passes into fine branches of veins. This is true of extracerebral regions of the opossum circulatory system as well, and of those regions in the brain but outside the blood–brain barrier (see Section VIII). But when blood vessels enter the brain or spinal cord in marsupials, they do so in pairs consisting of one artery and a corresponding vein. These two branch together as far as, and including, the finest capillaries. Here both arterial and venous members are enclosed by a single basement membrane, and are separated from each other by an "intercapillary cell" which may act as a buffer to equalize osmotic and pressure differences between the two vessels (Bubis and Luse, 1964). Examples of paired vessels are shown in Fig. 21. After the finest branching, the arterial and venous component of each limb meet and are continuous with one another, forming a terminal end loop. Therefore the central nervous system is supplied by distinct and separate vascular units which do not communicate with each other. Similar arrangements are seen only in a certain few reptiles and amphibians, and appear to have arisen independently in each case (Wislocki and Campbell, 1937).

This special ordering results not from a differentiation out of the usual embryonic anastomosing network, but from paired sprouts from the pial vascular plexus whose tips eventually unite to form the loops (Wislocki, 1939). Following injury, the cut capillaries regenerate to reform their individual loops, and there is no increase in vascularity around the site of injury as happens in placentals (Scharrer, 1939c). The loops form even when they grow into implants of placental brain tissue, indicating an inherent property of the vascular tissue (Scharrer, 1940c). One consequence of this arrangement is that occlusion or severing of even a single capillary deprives at least

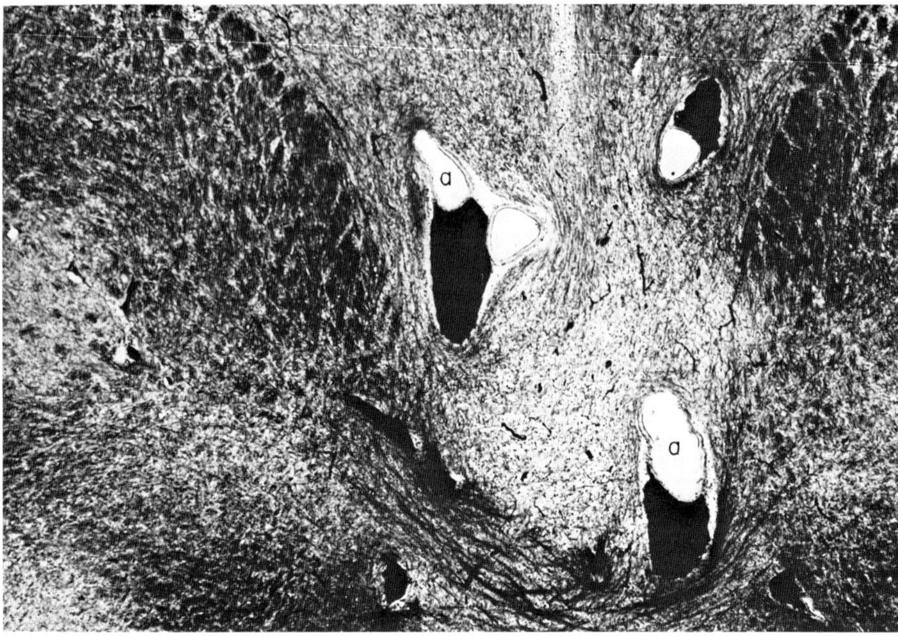


Fig. 21. Examples of paired blood vessels in diencephalon of *Macropus rufus* (enlargement of section from Fig. 20). The veins appear black since they are filled with blood; each is accompanied by a clear artery (a) from which blood was cleared.

some nerve cells of their only blood supply (Scharrer, 1939a,b). This has far reaching consequences for studies utilizing lesions of the central nervous system, since damage occurs not only at the site of the intended lesion, but also through the entire course of blood vessels emanating from that site.

Following its discovery in *Didelphis*, presence of this feature of vascular organization was confirmed in several other marsupials: *Macropus rufogriseus* (Craigie, 1938), *Dasyurus viverrinus*, *Macropus giganteus*, *Pseudocheirus*, *Trichosurus*, and *Phascolarctos* (Sunderland, 1941). Thus it appears to represent a specialization of the entire marsupial subclass.

## XIII. OVERVIEW

This review has covered the years 1890 through 1970 with a reasonable claim to comprehensive coverage. Much of the literature from 1971 through 1975 is also included, although the search did not attempt to be exhaustive for these years. Looking over the sum of what has been learned about the brains of marsupials in the past four score and seven years, four topics of special interest emerge. First, it is impressive how little marsupial brains differ from those of placental mammals in their structure and function. Second, since they develop after animals are born, during pouch life (Ulinski, 1971), they are superb subjects for experimental studies of mammalian neurogenesis. Third, since so many features have evolved separately in marsupial and placental brains to produce convergently similar results, marsupials are also the subjects of choice, in fact our only available subjects in most cases, for the study of converging responses to environmental selective pressures in evolving mammalian brains. Fourth and finally, the relative incidence of investigations of marsupial neurology through the years provides a fascinating look at the growth of a select corner of biological science, which may serve as a paradigm for the recent history of broader fields of science.

### A. Differences between Marsupial and Placental Brains

We have encountered six respects in which marsupial brains are all distinct from all placental brains, according to data now available. Only two of these seem to imply a true functional difference: in the unique blood circulation (see Section XII) which represents a marsupial specialization, and in the retinal cones (see Section V,C,1) where the absence of double cones and oil droplets may represent a specifically placental development. The remaining four differences seem to be more accidents of placement rather than different mechanisms of operation. These include the different paths of neocortical commissural connections (Section VIII), the different shape and placement of the hippocampal commissure and subfornical body (Section VIII,C, IX,B,3), the lateral position of the ventral nucleus of the inferior olive in marsupials (Section V,A,3), and the different branching of spinocerebellar fibers (Section V,A,1).

This last may yet be found to be an interspecies rather than an intersubclass difference. Many other differences, once thought to distinguish marsupials from placentals, have, with further study, been shown to represent intrasubclass variability present in both Eutheria and Metatheria. These include the restriction of corticospinal fibers to cervical and thoracic regions (Section VII,C,5), the "overlap" of sensory and motor neocortex (Section VII,C,9), and the "displacement" to ventral and medial frontal neocortex of input from the dorsomedial thalamic nucleus (Section VII,F). The difference between cats and opossums in cerebelloolivary circuitry (Section IV,A,3) may prove to be a placental distinction from marsupials, but may as well be a specialization of one or the other species. The presubicular granulous cortex (Section VI,B,3) and the striations in the lowest neocortical layer (Section VII,B) could be additional instances of distinctive marsupial placement of processes shared with placentals.

### B. Marsupial Advantages for Developmental Neurology

The excellent opportunities marsupials offer for studies of brain development remain largely unexploited. Development of auditory structure and function was most effectively studied using *Didelphis* pouch young (Section V,B,1). Morest's (1968, 1969a,b, 1970) studies of cellular growth and migration, and formation of connections, also used *Didelphis* young to good advantage. The same species was the subject of studies of cuneate–gracile nuclear development which yielded the information that receptors in some ways govern the growth and differentiation of brain tissue (Johnson *et al.*, 1972). These few examples suggest that much valuable information could be obtained from this largely neglected resource.

### C. Evolutionary Convergence in Placental and Marsupial Brains

Striking convergences have been reported in at least three regional brain specializations, where the independent paths of marsupial and placental evolution have produced remarkably similar outcomes. The multilaminated lateral geniculate nucleus seen in primate placentals and phalangeroid marsupials (Section V,C,2) raises many hypotheses about arboreality and visual evolution. The barrels in neocortex of rodents and again the phalangeroid marsupials (Fig. 18 and Section VII,C,3) generate hypotheses about principles of organization of highly developed sensory neocortex. The formation of gyri and sulci at predictable loci in the sensory representation in neocortex (Section VII,C,4) similarly points to a basic feature of neocortical evolution and growth. Marsupials represent the great alternative case of mammalian adaptive radiation, and when the same result happens in two such separate phylogenetic lines, we can begin to identify the determining factors in brain evolution.

### D. The Historical Course of Marsupial Brain Research

Figure 22 shows the incidence of publications dealing with marsupial central nervous systems from 1890 through 1974. Growing activity seems to

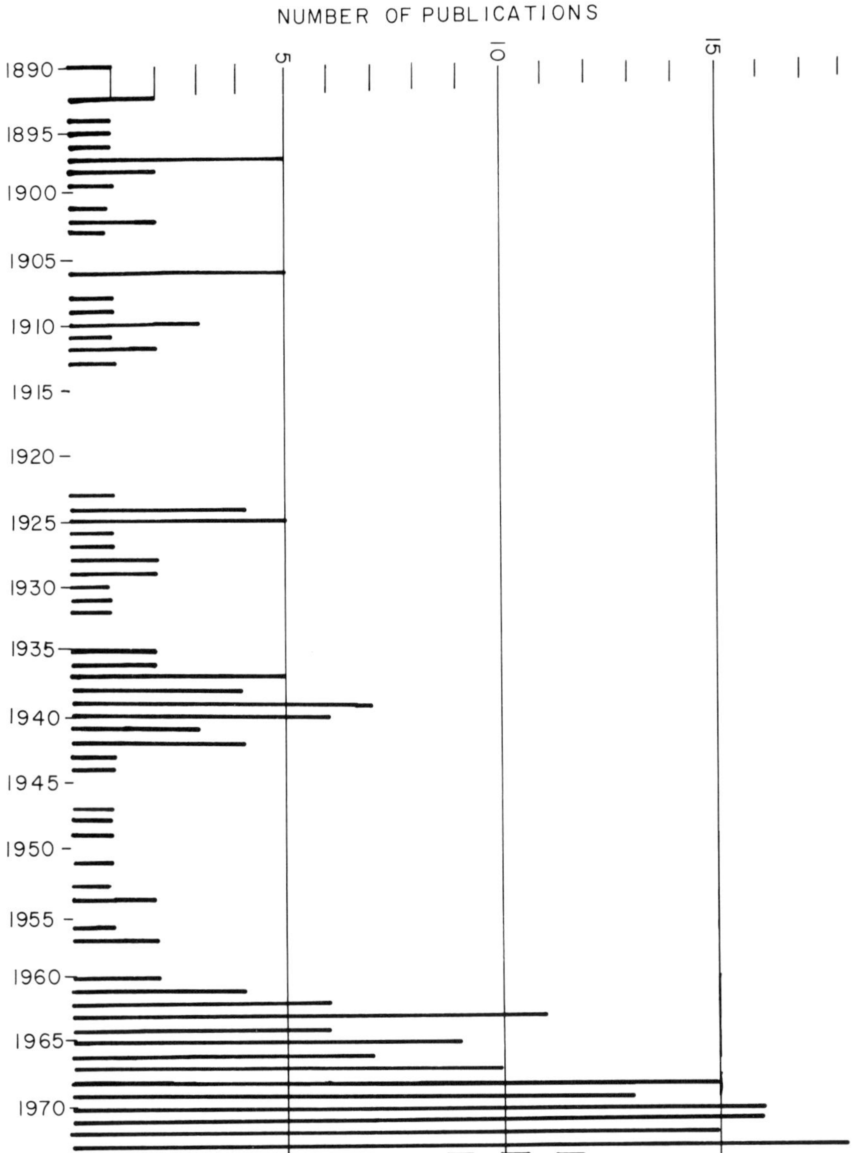


Fig. 22. Incidence of publications on marsupial neurology 1890–1974. The number from 1974 is probably not complete, hence is indicated by an interrupted bar.

have been cut short by the two world wars and took some time to recover each time, and there has been an overwhelming positive acceleration in the past decade. This all may reflect overall scientific activity, rather than indicating anything about interest in marsupial neurology as such. The relative frequencies may represent the amount of resources available for research, including the number of scientists on hand, at a given time. Thus this sample may be representative of events in the larger field of neural, or perhaps of biological, science.

One conclusion can be drawn: if the accelerating trend continues, it will not again be practically possible to review *all* the studies in the field, as has been attempted here.

## ACKNOWLEDGMENTS

The preparation of this chapter was aided in many ways by support from NSF grants GB 43236 and GB 30783, and NIH grants HD K3 11,655 and MH 10116. Assistance and facilities were provided by the Department of Physiology, University of Sydney; the Departments of Anatomy and Physiology, University of Tasmania; and the Department of Psychology, Yale University. Special thanks for advice, suggestions, information, and encouragement are due to the late Richard A. Lende who provided his extensive bibliographic resources, to Robert W. Rodieck, Gilberto B. Campos, Charles R. R. Watson, Peter O. Bishop, Angelo B. M. Machado, John A. W. Kirsch, George F. Martin, C. Geoffrey Wood, Glenn I. Hatton, Pamela Parker, William C. Blumer, and many others. Neal Brophy contributed much to the photography of the illustrations. Most is owed to my students, W. Lee Weller, Philip S. Ulinski, Robert C. Switzer, Edwin W Rubel, Daniel D. Lyons, Paul Herron, John R. Haight, Anthony C. Bonduki, and Rocco A. Bombardieri for their copious enthusiasm, substantial contributions, and vital assistance in so many phases of this work.

## REFERENCES

The following list includes a comprehensive bibliography on marsupial central nervous systems from 1890 to 1970, and many additional sources from more recent years. Some references have been omitted when the same material has been restated in later, more comprehensive sources (such as the many Larsell articles whose contents later appeared in his 1970 book). References with asterisks do not contain work on marsupials as such, and are included because they are cited in the text for other reasons.

Abbie, A. A. (1937). Some observations on the major subdivisions of the Marsupialia, with especial reference to the position of the Peramelidae and Caenolestidae. *J. Anat.* **71,** 429–436.

Abbie, A. A. (1939). The origin of the corpus callosum and the fate of the structures related to it. *J. Comp. Neurol.* **71,** 9–44.

Abbie, A. A. (1940). The excitable cortex in *Perameles, Sarcophilus, Dasyurus, Trichosurus,* and *Wallabia* (*Macropus*). *J. Comp. Neurol.* **72,** 469–487.

Abbie, A. A. (1942). Cortical lamination in a polyprotodont marsupial, *Perameles nasuta. J. Comp. Neurol.* **76,** 509–536.

Adey, W. R. (1953). An experimental study of the central olfactory connexions in a marsupial (*Trichosurus vulpecula*). *Brain* **76,** 311–330.

Adey, W. R. (1970). Higher olfactory centers. *Taste Smell Vertebr. Ciba Found. Symp., 1969* p. 357.

Adey, W. R., and Kerr, D. I. B. (1954). The cerebral representation of deep somatic sensibility in the marsupial phalanger and the rabbit; an evoked potential and histological study. *J. Comp. Neurol.* **100,** 597–626.

Adey, W. R., Merrillees, N. C. R., and Sunderland, S. (1956). The entorhinal area: Behavioural, evoked potential, and histological studies of its relationships with brain-stem regions. *Brain* **79,** 414–439.

Adey, W. R., Sunderland, S., and Dunlop, C. W. (1957). The entorhinal area; electrophysiological studies of its interrelations with rhinencephalic structures. *Electroencephalogr. Clin. Neurophysiol.* 309–324.

Affanni, J. M., and Morita, E. (1968). Asymétrie bioélectrique du cortex des hémisphères cérébraux chez le Marsupial *Didelphis azarae.* Effets de la section des commissures télencéphaliques. *C. R. Seances Soc. Biol. Ses Fil.* **162,** 1608–1609.

Affanni, J. M., and Vaccarezza, O. L. (1964). Observaciones sobre el sueño del marsupial *Didelphis azarae* (comadreja o zarigüeya). *Rev. Soc. Argent. Biol.* **40,** 2–8.

Affanni, J. M., Vaccarezza, O. L., and Aballone, J. C. (1967). Difference in electrical activity during wakefulness and the phase of 'sleep with muscular twitches' as recorded from the cortex of the marsupial *Didelphis azarae* (South American Opossum). *Experientia* **23,** 216–217.

Akert, K., Potter, H. D., and Anderson, J. W. (1961). The subfornical organ in mammals. I. Comparative and topographical anatomy. *J. Comp. Neurol.* **116,** 1–13.

Alves Cavalcante, L., Rocha-Miranda, C. E., and Lent, R. (1975). Hypothalamic, tectal, and accessory optic projections in the opossum. *Brain Res.* **84,** 302–307.

*Andres, K. H. (1970). Anatomy and ultrastructure of olfactory bulb in fish, amphibia, reptiles, birds and mammals. *Taste Smell Vertebr., Ciba Found. Symp., 1969* pp. 177–196.

Aoki, M., and McIntyre, A. K. (1973). Pyramidal effects on some forelimb motoneurone populations of the arboreal brush-tailed possum (*Trichosurus vulpecula*). *Brain Res.* **60,** 485–488.

Ariëns Kappers, C. U., Huber, G. C., and Crosby, E. C. (1936). "The Comparative Anatomy of the Nervous System of Vertebrates, Including Man," 3 vols. Hafner, New York (reprinted in 1960).

Barratt, E. S. (1965). EEG correlates of tonic immobility in the opossum (*Didelphis virginiana*). *Electroencephalogr. Clin. Neurophysiol.* **18,** 709–711.

Bautista, N. S., and Matzke, H. A. (1965). A degeneration study of the course and extent of the pyramidal tract of the opossum. *J. Comp. Neurol.* **124,** 367–376.

Benevento, L. A. (1968). Organization of visual cortex in the opossum. *Anat. Rec.* **160,** 313 (abstr.).

Benevento, L. A., and Ebner, F. F. (1969). Lateral geniculate projections to neocortex in the opossum (abstr.). *Anat. Rec.* **163,** 294

Benevento, L. A., and Ebner, F. F. (1970). Pretectal, tectal, retinal and cortical projections to thalamic nuclei of the Virginia opossum in stereotaxic coordinates. *Brain Res.* **18,** 171–175.

Benevento, L. A., and Ebner, F. F. (1971a). The areas and layers of corticocortical terminations in the visual cortex of the Virginia opossum. *J. Comp. Neurol.* **141,** 157–190.

Benevento, L. A., and Ebner, F. F. (1971b). The contribution of the dorsal lateral geniculate nucleus to the total pattern of thalamic terminations in striate cortex of the Virginia opossum. *J. Comp. Neurol.* **143,** 243–260.

Beran, R. L., and Martin, G. F. (1971). Reticulospinal fibers of the opossum, *Didelphis virginiana* I. Origin. *J. Comp. Neurol.* **141,** 453–466.

Bergquist, E. H. (1970). Output pathways of hypothalamic mechanisms for sexual, aggressive, and other motivational behaviors in opossum. *J. Comp. Physiol. Psychol.* **70,** 389–398.

Biedenbach, M. A., and Towe, A. L. (1970). Fiber spectrum and functional properties of pyramidal tract neurons in the American opossum. *J. Comp. Neurol.* **140,** 421–430.

Bischoff, E. (1899). Zur Anatomie der Hinterstrangskerne bei Säugethieren. *Jahrb. Psychiatr. Neurol.* **18,** 371–384.

Blumer, W. F. C. (1963). Ascending and descending spinal tracts of the quokka (*Setonix brachyurus*). *J. Anat.* **97,** 490.

Bodemer, C. W. and Towe, A. L. (1963). Cortical localization patterns in the somatic sensory cortex of the opossum. *Exp. Neurol.* **8,** 380–394.

Bodian, D. (1935). The projection of the lateral geniculate body on the cerebral cortex of the opossum, *Didelphis virginiana. J. Comp. Neurol.* **62,** 169–494.

Bodian, D. (1937). An experimental study of the optic tracts and retinal projection of the Virginia opossum. *J. Comp. Neurol.* **66,** 113–144.

Bodian, D. (1939). Studies on the diencephalon of the Virginia opossum. Part I. The nuclear pattern in the adult. *J. Comp. Neurol.* **71,** 259–323.

Bodian, D. (1940). Studies on the diencephalon of the Virginia opossum. Part II. The fiber connections in normal and experimental material. *J. Comp. Neurol.* **72,** 207–297.

Bodian, D. (1942). Studies on the diencephalon on the Virginia opossum. Part III. The thalamo-cortical projection. *J. Comp. Neurol.* **77,** 525–575.

Bodian, D. (1951). Nerve endings, neurosecretory substance, and lobular organization of the neurohypophysis. *Bull. Johns Hopkins Hosp.* **89,** 354–376.

Bodian, D. (1963). Cytological aspects of neurosecretion in opossum neurohypophysis. *Bull. Johns Hopkins Hosp.* **113,** 57–93.

Bombardieri, R. A. (1974). Unit analysis of visual cortex in the opossum *Didelphis marsupialis aurita.* Ph.D. Thesis, Michigan State University (Zoology Dept.), East Lansing.

Bombardieri, R. A., Johnson, J. I., and Campos, G. B. (1975). Species differences in mechanosensory projections from the mouth to the ventrobasal thalamus. *J. Comp. Neurol.* **163,** 41–64.

Bowman, M. H., and King, J. S. (1974). The conformation, cytology and synaptology of the opossum inferior olivary nucleus. *J. Comp. Neurol.* **148,** 491–524.

*Bowsher, D. (1965). The anatomophysiological basis of somatosensory discrimination. *Int. Rev. Neurobiol.* **8,** 37–76.

*Brodal, A. (1969). "Neurological Anatomy in Relation to Clinical Medicine." Oxford Univ. Press, London and New York.

Brodmann, K. (1906). Beitrage zur histologische Lokalisation der Groszhirnrinde. Funfte Mitteilung: Uber der allgemeinen Bauplan des Cortex pallii bei den Mammaliern und Zwei homologe Rindenfelder im besonderen. Zugleich ein Beitrag zur Furchenlehre. *J. Psychol. Neurol.* **6,** Suppl., 275–400.

Brodmann, K. (1909). "Vergleichende Lokalisationslehre der Groszhirnrinde." Barth, Leipzig.

Bromiley, R. B., and Brooks, C. McC. (1940). Role of neocortex in regulating postural reactions of the opossum (*Didelphis virginiana*). *J. Neurophysiol.* **3,** 339–346.

Broomhead, A. (1974). The mediodorsal thalamic nucleus of the brush-tailed possum, *Trichosurus vulpecula*. *J. Anat.* **118,** 392 (abstr.).

Bubis, J. J., and Luse, S. A. (1964). An electron microscope study of the cerebral blood vessels of the opossum. *Z. Zellforsch. Mikrosk. Anat.* **62,** 16–25.

Burkitt, A. N. (1938). The external morphology of the brain of *Notoryctes typhlops*. *Proc. K. Ned. Akad. Wet.* **41,** 3–15.

Chapman, H. G. (1906). Note on cerebral localization in the bandicoot (*Perameles*) *Proc. Linn. Soc. N. S. W.* **31,** 493–494.

Christensen, J. L., and Hill, R. M. (1970). Response properties of single cells of a marsupial visual cortex. *J. Optometry Arch. Am. Acad. Optometry* **47,** 547–556.

Cicala, G. A., Albert, I. B., and Ulmer, F. A. (1970). Sleep and other behaviors of the red kangaroo (*Megaleia rufa*). *Anim. Behav.* **18,** 787–790.

Clezy, J. K. A., Dennis, B. J., and Kerr, D. I. B. (1961). A degeneration study of the somaesthetic afferent systems in the marsupial phalanger. *Aust. J. Exp. Biol. Med. Sci.* **39,** 19–28.

Cowley, A. R. (1973). The nuclei of the cochlear nerve of the red kangaroo, *Megaleia rufus*. *J. Hirnforsch.* **14,** 287–301.

Craigie, E. H. (1938). The blood vessels in the central nervous system of the kangaroo. *Science* **88,** 359–360.

Crosby, E. C., Humphrey, T., and Lauer, E. W. (1962). "Correlative Anatomy of the Nervous System." Macmillan, New York.

Culberson, J. L., and Kimmel, D. L. (1972). Central distribution of primary afferent fibers of the glossopharyngeal and vagal nerves in the opossum, *Didelphis virginiana*. *Brain Res.* **44,** 325–335.

Cunningham, R. H. (1898). The cortical motor centres of the opossum, *Didelphis virginiana*. *J. Physiol.* (*London*) **22,** 264–269.

Dennis, B. J., and Kerr, D. I. B. (1961a). Somaesthetic pathways in the marsupial phalanger *Trichosurus vulpecula*. *Aust. J. Exp. Biol. Med. Sci.* **39,** 29–42.

Dennis, B. J., and Kerr, D. I. B. (1961b). An afferent system in the central tegmental fasciculus. *Aust. J. Exp. Biol. Med. Sci.* **39,** 43–56.

Diamond, I. T., and Utley, J. D. (1963). Thalamic retrograde degeneration study of sensory cortex in opossum. *J. Comp. Neurol.* **120,** 129–160.

Dillon, L. S. (1963). Comparative studies of the brain in the Macropodidae. *J. Comp. Neurol.* **120,** 43–52.

Dom, R., Fisher, B. L., and Martin, G. F. (1970). The venous system of the head and neck of the opossum (*Didelphis virginiana*). *J. Morphol.* **132,** 487–496.

Dom, R., King, J. S., and Martin, G. F. (1973). Evidence for two direct cerebello-olivary connections. *Brain Res.* **57,** 498–501.

DuBois, F. (1929). The tractus solitarius and attendant nuclei in the virginian opossum (*Didelphis virginiana*). *J. Comp. Neurol.* **47,** 189–224.

Ebner, F. F. (1967). Afferent connections to neocortex in the opossum (*Didelphis virginiana*). *J. Comp. Neurol.* **129,** 241–268.

Ebner, F. F. (1969). A comparison of primitive forebrain organization in metatherian and eutherian mammals. *Ann. N.Y. Acad. Sci.* **167,** 241–257.

*Eccles, J. C., Ito, M., and Szentágothai, J. (1967). "The Cerebellum as a Neuronal Machine." Springer-Verlag, Berlin and New York.

*Edinger, T. (1948). Evolution of the horse brain. *Mem., Geol. Soc. Am.* **25,** 1–177.

*Edinger, T. (1964). Midbrain exposure and overlap in mammals. *Am. Zool.* **4,** 5–19.

*Edinger, T. (1966). Brains from 40 million years of Camelid history. *In* "Evolution of the

Forebrain" (R. Hassler and H. Stephan, eds.), pp. 153–161. Thieme, Stuttgart (also published by Plenum, New York, 1967).

Elias, H., and Schwartz, D. (1969). Surface areas of the cerebral cortex of mammals determined by stereological methods. *Science* **166,** 111–113.

Elliot Smith, G. (1894). Preliminary observations on the cerebral commissures of the Mammalia with special reference to the Monotremata and Marsupialia. *Proc. Linn. Soc. N. S. W.* **9,** 635–657, reprinted in *J. Anat.* **71,** 528–593 (1937).

Elliot Smith, G. (1895). The comparative anatomy of the cerebrum of *Notorcytes typhlops*. *Trans. Proc. R. Soc. South Aust.* **19,** 167–193.

Elliot Smith, G. (1896). The 'fornix superior.' *J. Anat. Physiol.* **31,** 80–94.

Elliot Smith, G. (1897a). The origin of the corpus callosum: A comparative study of the hippocampal region of the cerebrum of Marsupialia and certain Cheiroptera. *Trans. Linn. Soc. London, Zool.* **7,** 47–69.

Elliot Smith, G. (1897b). The relation of the fornix to the margin of the cerebral cortex. *J. Anat. Physiol.* **32,** 23–58.

Elliot Smith, G. (1902a). "Descriptive and Illustrated Catalogue of the Physiological Series of Comparative Anatomy Contained in the Museum of the Royal College of Surgeons of England," Physiol. Ser. Vol. 2, 2nd ed. Taylor & Francis, London.

Elliot Smith, G. (1902b). On a peculiarity of the cerebral commissures in certain Marsupialia, not hitherto recognized as a distinctive feature of the Diprotodontia. *Proc. R. Soc. London* **70,** 226–231.

Elliot Smith, G. (1910). The Arris and Gale lectures on some problems relating to the evolution of the brain. *Lancet* **1,** 1–6, 147–153, and 221–227.

Erickson, R. P., Jane, J. A., Waite, R., and Diamond, I. T. (1964). Single neuron investigation of sensory thalamus of the opossum. *J. Neurophysiol.* **27,** 1026–1047.

Ferguson, D. R., and Heller, H. (1965). Distribution of neurohypophyseal hormones in mammals. *J. Physiol.* (*London*) **180,** 846–863.

Fernandez, C., and Schmidt, R. S. (1963). The opossum ear and evolution of the coiled cochlea. *J. Comp. Neurol.* **121,** 151–159.

Filimonov, I. N. (1949). "Comparative Anatomy of the Cerebral Cortex of Mammalians. Paleocortex, Archicortex, and Interstitial Cortex." Clearinghouse for Federal Scientific and Technical Information, U.S. Dept. of Commerce, Washington, D.C. (1965 translation by V. E. Dukoff from Russian edition, 1949. Publ. House Acad. Med. Sci., Moscow, USSR).

*Fisher, A. M., Harting, J. K., Martin, G. F., and Stuber, M. I. (1969). The origin, course and termination of corticospinal fibers in the armadillo (*Dasypus novemcinctus mexicanus*). *J. Neurol. Sci.* **8,** 347–361.

Flashman, J. F. (1906). A preliminary note on the motor areas in the cerebral cortex of marsupials; motor areas in the cerebral cortex of *Dasyurus viverrinus*. *Rep. Pathol. Lab. Lunacy Dep. N. S. W.* **1,** Part 2, 23–26.

Flashman, J. F. (1910a). Degenerations following resection of olfactory lobe in *Dasyurus viverrinus*. *Rep. Pathol. Lab. Lunacy Dep. N. S. W.* **2,** Part 1, 103–106.

Flashman, J. F. (1910b). The cortico-spinal tracts in *Dasyurus viverrinus*. *Rep. Pathol. Lab. Lunacy Dep. N. S. W.* **2,** Part 1, 107–111.

Foltz, F. M., and Matzke, H. A. (1960). An experimental study in the origin course, and termination of the cerebellifugal fibers in the opossum. *J. Comp. Neurol.* **114,** 107–125.

Gillilan, L. A. (1972). Blood supply to primitive mammalian brains. *J. Comp. Neurol.* **145,** 209–222.

Giolli, R. A. (1965). An experimental study of the accessory optic system and of other optic fibers in the opossum, *Didelphis virginiana*. *J. Comp. Neurol.* **124,** 229–242.

*Goldby, F. (1939a). An experimental investigation of the motor cortex and pyramidal tract of *Echidna aculeata*. *J. Anat.* **73,** 509–524.

Goldby, F. (1939b). An experimental investigation of the motor cortex and its connexions in the phalanger, *Trichosurus vulpecula*. *J. Anat.* **74,** 12–33.

Goldby, F. (1941). The normal histology of the thalamus in the phalanger (*Trichosurus vulpecula*). *J. Anat.* **75,** 197–225.

Goldby, F. (1943). An experimental study of the thalamus in the phalanger, *Trichosurus volpecula*. *J. Anat.* **77,** 195–224.

Gray, P. A. (1924). The cortical lamination pattern of the opossum *Didelphys virginiana*. *J. Comp. Neurol.* **37,** 221–263.

Gray, P. A., and Turner, E. L. (1924). The motor cortex of the opossum. *J. Comp. Neurol.* **36,** 375–385.

Haight, J. R., and Weller, W. L. (1973). Neocortical topography in the brush-tailed possum: Variability and functional significance of sulci. *J. Anat.* **116,** 473–474 (abstr.).

Haight, J. R., Neylon, L. and Weller, W. L. (1976). Horseradish peroxidase determination of thalamocortical projections in the brush-tailed possum, *Trichosurus vulpecula*. *J. Anat.* **121,** 416–417 (abstr.).

*Hall, R. D., and Lindholm, E. P. (1974). Organization of motor and somatosensory neocortex in the albino rat. *Brain Res.* **66,** 23–38.

Hamel, E. G. (1966a). The amygdaloid complex in the kangaroo and the North and South American opossum. *Anat. Rec.* **154,** 353.

Hamel, E. G. (1966b). A study of the hippocampal formation in the opossum, *Didelphis virginiana*. *In* "Evolution of the Forebrain" (R. Hassler and H. Stephan, eds.), pp. 81–91. Thieme, Stuttgart.

Hamilton, T. C., and Johnson, J. I. (1973). Somatotopic organization related to nuclear morphology in the cuneate–gracile complex of the opossum *Didelphis marsupialis virginiana*. *Brain Res.* **51,** 125–140.

Hara, K., and Myers, R. E. (1973). Role of the forebrain structures in emotional expression in opossum. *Brain Res.* **52,** 131–144.

Harman, P. J. (1947). Quantitative analysis at the brain–isocortex relationship in Mammalia. *Anat. Rec.* **97,** 343.

Hayhow, W. R. (1966). The accessory optic system in the marsupial phalanger, *Trichosurus vulpecula*. An experimental degeneration study. *J. Comp. Neurol.* **126,** 653–672.

Hayhow, W. R. (1967). The lateral geniculate nucleus of the marsupial phalanger, *Trichosurus vulpecula*. An experimental study of cytoarchitecture in relation to the intranuclear optic nerve projection fields. *J. Comp. Neurol.* **131,** 571–604.

Hazlett, J. C., Martin, G. F., and Dom, R. (1971). Spino-cerebellar fibres of the opossum *Didelphis marsupialis virginiana*. *Brain Res.* **33,** 257–271.

Heath, C. J., and Jones, E. G. (1971). Interhemispheric pathways in the absence of a corpus callosum. *Anat.* **109,** 253–270.

Heimer, L., Ebner, F. F., and Nauta, W. J. H. (1967). A note on the termination of commissural fibers in the neocortex. *Brain Res.* **5,** 171–177.

Herrick, C. J. (1924). The nucleus olfactorius anterior in the opossum. *J. Comp. Neurol.* **37,** 317–359.

Herrick, C. L. (1892). The cerebrum and olfactories of the opossum, *Didelphys virginiana*. *J. Comp. Neurol.* **2,** 1–20.

Herrick, C. L. (1898). The cortical motor centres in lower mammals. *J. Comp. Neurol.* **8,** 92–98.

Hill, R. M., and Goodwin, H. (1968). Visual receptive fields from cells of a marsupial (*Didelphis virginiana*). *Experientia* **24,** 559–560.

Hoffmann, C. K. (1876–1877). Zur Anatomie der Retina. II. Ueber den Bau der Retina bei den Beutelthieren. *Nied. Arch. Zool.* **3,** 195–198.

Holmes, G. M. (1903). On the comparative anatomy of the nervus acusticus. *Trans. R. Ir. Acad. Ser. B* **32,** 101–144.

Hore, J., and Porter, R. (1971). The role of the pyramidal tract in the production of cortically evoked movements in the brush-tailed possum (*Trichosurus vulpecula*). *Brain Res.* **30,** 232–234.

Hore, J., and Porter, R. (1972). Pyramidal and extrapyramidal influences on some hindlimb motoneuron populations of the arboreal brush-tailed possum *Trichosurus vulpecula. J. Neurophysiol.* **35,** 112–121.

Hore, J., Phillips, C. G., and Porter, R. (1973). The effects of pyramidotomy on the motor performance in the brush-tailed possum (*Trichosurus vulpecula*). *Brain Res.* **49,** 181–184.

Johnson, J. I., and Marsh, M. P. (1969). Laminated lateral geniculate in the nocturnal marsupial *Petaurus breviceps* (Sugar glider). *Brain Res.* **15,** 250–254.

*Johnson, J. I., Hatton, G. I., and Goy, R. W. (1969). The physiological analysis of animal behavior. *In* "Behavior of Domestic Animals" (E. S. E. Hafez, ed.), Chapter 7. pp. 131–191. Baillière, London.

Johnson, J. I., Hamilton, T. C., Hsung, J.-C., and Ulinski, P. S. (1972). Gracile nucleus absent in adult opossums after leg removal in infancy. *Brain Res.* **38,** 421–424.

Johnson, J. I., Haight, J. R., and Megirian, D. (1973). Convolutions related to sensory projections in cerebral neocortex of marsupial wombats. *J. Anat.* **114,** 153 (abstr.).

Johnston, J. B. (1913). The morphology of the septum, hippocampus, and pallial commissures in reptiles and mammals. *J. Comp. Neurol.* **23,** 371–478.

Johnston, J. B. (1923). Further contributions to the study of the evolution of the forebrain. *J. Comp. Neurol.* **35,** 337–482.

Jordan, H. E. (1911). The microscopical anatomy of the epiphysis in the opossum. *Anat. Rec.* **5,** 325–338.

Killackey, H., and Ebner, F. (1973). Convergent projection of the three separate thalamic nuclei on to a single cortical area. *Science* **179,** 283–285.

King, J. S., Martin, G. F., and Biggert, T. P. (1968). The basilar pontine gray of the opossum (*Didelphis virginiana*) I. Morphology. *J. Comp. Neurol.* **133,** 439–446.

King, J. S., Bowman, M. H., and Martin, G. F. (1971). The red nucleus of the opossum (*Didelphis marsupialis virginiana*): A light and electron microscopic study. *J. Comp. Neurol.* **143,** 157–184.

King, J. S., Martin, G. F., and Conner, J. B. (1972). A light and electron microscopic study of the corticorubral projections in the opossum, *Didelphis marsupialis virginiana. Brain Res.* **38,** 251–265.

King, J. S., Dom, R. M., Conner, J. B., and Martin, G. F. (1973). An experimental light and electron microscopic study of the cerebellorubral projections in the opossum, *Didelphis marsupialis virginiana. Brain Res.* **52,** 61–78.

King, J. S., Dom, R. M., and Martin, G. F. (1974). Anatomical evidence for an intrinsic neuron in the red nucleus. *Brain Res.* **67,** 317–323.

King, J. S., Martin, G. F., and Bowman, M. H. (1975). The direct spinal receiving area of the inferior olivary nucleus. *Exp. Brain Res.* **22,** 13–24.

Koizumi, K., and Sato, A. (1969). Influence of sympathetic innervation on carotid sinus baroreceptor activity. *Am. J. Physiol.* **216,** 321–329.

*Kruger, L., Siminoff, R., and Witkovsky, P. (1961). Single neuron analysis of dorsal column nuclei and spinal nucleus of trigeminal in cat. *J. Neurophysiol.* **24,** 333–349.

Kunze, D., Putnam, S. J., and Manning, J. W. (1968). Transcortical striate connections in the opossum. *J. Comp. Neurol.* **132,** 463–468.

Langworthy, O. R. (1927). Correlated physiological and morphological studies on the development of electrically responsive areas in the cerebral cortex of the opossum. *Carnegie Inst. Washington Publ.* **380,** Contrib. Embryol. No. 103 177–207.

Larsell, O. (1970). "The Comparative Anatomy and Histology of the Cerebellum from Monotremes through Apes." Univ. of Minnesota Press, Minneapolis.

Larsell, O., McCrady, E., and Zimmermann, A. A. (1935). Morphological and functional development of the membranous labyrinth in the opossum. *J. Comp. Neurol.* **63,** 95–118.

Larsell, O., McCrady, E. M., and Larsell, J. F. (1944). The development of the organ of Corti in relation to the inception of hearing. *Trans. Am. Acad. Ophthalmol. Otolaryngol.* **48,** 333–357.

Lende, R. A. (1963a). Sensory representation in the cerebral cortex of the opossum (*Didelphis virginiana*). *J. Comp. Neurol.* **121,** 395–403.

Lende, R. A. (1963b). Motor representation in the cerebral cortex of the opossum (*Didelphis virginiana*). *J. Comp. Neurol.* **121,** 405–415.

Lende, R. A. (1963c). Cerebral cortex: A sensorimotor amalgam in the Marsupialia. *Science* **141,** 730–732.

Lende, R. A. (1969). A comparative approach to the neocortex: Localization in monotremes, marsupials and insectivores. *Ann. N.Y. Acad. Sci.* **167,** 262–276.

Lent, R., and Rocha-Miranda, C. E. (1974). Survival times and patterns of degeneration in the visual system of the opossum. *Brain Res.* **72,** 294–299.

Livini, R. (1908). Il proencefalo di un marsupiale (Hypsiprymnus rufescens). *Arch. Ital. Anat. Embriol.* **6,** 549–584.

Loo, Y. T. (1930). The forebrain of the opossum, Didelphis virginiana. Part I. Gross anatomy. *J. Comp. Neurol.* **51,** 13–64.

Loo, Y. T. (1931). The forebrain of the opossum, Didelphis virginiana. Part II. Histology. *J. Comp. Neurol.* **52,** 1–148.

Lyons, D. D. (1972). Subdivisions of the superior olivary nuclear complex: Marsupial versions with placental comparisons. M.S. Thesis, Michigan State University (Zoology Dept.), East Lansing.

Machado, A. B. M. (1965). Dados histológicos e embriológicos sobre a região pineal em alguns marsupiais brasileiros. *Cienc. Cult.* (*Sao Paulo*) **17,** 249.

McCotter, R. E. (1912). The connections of the vomeronasal nerves with the accessory olfactory bulbs in opossum and other mammals. *Anat. Rec.* **6,** 299–318.

McCrady, E., Wever, E. G., and Bray, C. W. (1937). The development of hearing in the opossum. *J. Exp. Zool.* **75,** 503–517.

Magalhães-Castro, B., and Saraiva, P. E. S. (1971). Sensory and motor representation in the cerebral cortex of the marsupial *Didelphis azarae azarae. Brain Res.* **34,** 291–299.

Magalhães-Castro, B., Saraiva, P. E. S., and Magalhães-Castro, H. H. (1971). Organizacao somatotópica do complexo nuclear gracilis-cuneatus no *Didelphis azarae azarae. Cienc. Cult.* (*Sao Paulo*) **23,** Suppl., R-82.

Magni, F., and Oscarsson, O. (1962a). Comparison of ascending spinal tracts activated by group I muscle afferents in the phalanger, rabbit, and cat. *Acta Physiol. Scand.* **54,** 37–52.

Magni, F., and Oscarsson, O. (1962b). Principal organization of coarse-fibred ascending spinal tracts in phalanger, rabbit, and cat. *Acta Physiol. Scand.* **54,** 53–64.

Martin, G. F. (1967). Interneocortical connections in the opossum, *Didelphis virginiana. Anat. Rec.* **157,** 607–616.

Martin, G. F. (1968a). The pattern of neocortical projections to the mesencephalon of the opossum, *Didelphis virginiana. Brain Res.* **11,** 593–610.

Martin, G. F. (1968b). Some efferent cortical pathways of the opossum. *J. Hirnforsch.* **10,** 55–78.

Martin, G. F. (1969). Efferent tectal pathways of the opossum (*Didelphis virginiana*). *J. Comp. Neurol.* **135,** 209–224.

Martin, G. F., and Dom, R. (1970a). The rubro-spinal tract of the opossum (*Didelphis virginiana*). *J. Comp. Neurol.* **138,** 19–30.

Martin, G. F., and Dom, R. (1970b). Rubro-bulbar projections of the opossum (*Didelphis virginiana*). *J. Comp. Neurol.* **139,** 199–214.

Martin, G. F., and Dom, R. (1971). Reticulospinal fibers of the opossum, *Didelphis virginiana*. II. Course, caudal extent and distribution. *J. Comp. Neurol.* **141,** 467–484.

Martin, G. F., and Fisher, A. M. (1968). A further evaluation of the origin, the course and the termination of the opossum corticospinal tract. *J. Neurol. Sci.* **7,** 177–188.

Martin, G. F., and King, J. S. (1968). The basilar pontine gray of the opossum (*Didelphis virginiana*). II. Experimental determination of neocortical input. *J. Comp. Neurol.* **133,** 447–462.

Martin, G. F., and Megirian, D. (1972). Corticobulbar projections of the marsupial phalanger (*Trichosurus vulpecula*) II. Projections to the mesencephalon. *J. Comp. Neurol.* **144,** 165–192.

Martin, G. F., and West, H. J. (1967). Efferent neocortical projections to sensory nuclei in the brain stem of the opossum (*Didelphys virginiana*). *J. Neurol. Sci.* **5,** 287–302.

Martin, G. F., Megirian, D., and Roebuck, A. (1970). The corticospinal tract of the marsupial phalanger (*Trichosurus vulpecula*). *J. Comp. Neurol.* **139,** 245–257.

Martin, G. F., Megirian, D., and Roebuck, A. (1971). Corticobulbar projections of the marsupial phalanger (*Trichosurus vulpecula*). I. Projections to the pons and medulla oblongata. *J. Comp. Neurol.* **142,** 275–296.

Martin, G. F., Megirian, D., and Conner, J. B. (1972). The origin, course and termination of the corticospinal tracts of the Tasmanian potoroo (*Potorous apicalis*). *J. Anat.* **111,** 263–281.

Martin, G. F., King, J. S., and Dom, R. (1974a). The projections of the deep cerebellar nuclei of the opossum *Didelphis marsupialis virginiana. J. Hirnforsch.* **15,** 545–573.

Martin, G. F., Dom, R., Katz, S., and King, J. S. (1974b). The organization of projection neurons in the opossum red nucleus. *Brain Res.* **78,** 17–34.

Martin, G. F., Dom, R., King, J. S., RoBards, M. J., and Watson, C. R. R. (1975). The inferior olivary nucleus of the opossum (*Didelphis marsupialis virginiana*), its organization and connections. *J. Comp. Neurol.* **160,** 507–534.

Megirian, D. (1968). Centrifugal discharges in cutaneous nerve fibers evoked by cutaneous afferent volleys in the acutely spinal phalanger, *Trichosurus vulpecula. Arch. Ital. Biol.* **106,** 343–352.

Megirian, D. (1971). Vestibular and somatosensory evoked centrifugal cutaneous nerve discharges in the decerebrate–decerebellate phalanger, *Trichosurus vulpecula. Arch. Ital. Biol.* **109,** 152–165.

Megirian, D., Johnson, J. I., and Haight, J. R. (1972). Le rapport entre les circonvolutions du cortex cérébral et les projections sensorielles chez le Wombat *Vombatus ursinus. J. Physiol.* (*Paris*) **65,** 448A.

Mehler, W. R. (1957). The mammalian "pain tract" in phylogeny. *Anat. Rec.* **127,** 332.

Mehler, W. R. (1966). Some observations of secondary ascending afferent systems in the central nervous system. *In* "Pain" (R. S. Knighton and P. R. Dumke, eds.), pp. 11–32. Little, Brown, Boston, Massachusetts.

Mihailoff, G. A., and King, J. S. (1975). The basilar pontine gray of the opossum: A correlated light and electron microscopic analysis. *J. Comp. Neurol.* **159,** 521–552.

Moeller, H. (1973). Zur Evolutions höhe des Marsupialia gehirns. *Zool. Jahrb., Abt. Anat. Ontog. Tiere* **91,** 434–448.

Morest, D. K. (1965). Identification of homologous neurons in the posterolateral thalamus of rat and Virginia opossum. *Anat. Rec.* **151,** 390.

Morest, D. K. (1968). The growth of synaptic endings in the mammalian brain: A study of the calyces of the trapezoid body. *Z. Anat. Entwicklungsgesch.* **127,** 201–220.

Morest, D. K. (1969a). The differentiation of cerebral dendrites: A study of the post-migratory neuroblast in the medial nucleus of the trapezoid body. *Z. Anat. Entwicklungsgesch.* **128,** 271–289.

Morest, D. K. (1969b). The growth of dendrites in the mammalian brain. *Z. Anat. Entwicklungsgesch.* **128,** 290–317.

Morest, D. K. (1970). A study of neurogenesis in the forebrain of opossum pouch young. *Z. Anat. Entwicklungsgesch.* **130,** 265–305.

*Mountcastle, V. B., ed. (1968). "Medical Physiology," 12th ed. Mosby, St. Louis, Missouri.

Nelson, L. R., and Lende, R. A. (1965). Interhemispheric responses in the opossum. *J. Neurophysiol.* **28,** 189–199.

Nieder, P. C., and Randall, W. (1964). Sound evoked potentials in neocortex of unanesthetized opossum. *Science* **144,** 429–430.

Norton, A. C., Bearn, A. V., and Misrahy, G. A. (1964). Electroencephalograph during 'feigned' sleep in the opossum. *Nature (London)* **204,** 162–163.

Obenchain, J. B. (1925). The brain of the South American marsupials Caenolestes and Orolestes. *Field Mus. Nat. Hist. Publ., Zool. Ser.* **14,** 1–148.

O'Day, K. J. (1936). A preliminary note on the presence of double cones and oil droplets in the retina of marsupials. *J. Anat.* **70,** 465–467.

O'Day, K. (1938). The retina of the Australian mammal. *Med. J. Aust.* **1,** 326–328.

Oswaldo-Cruz, E., and Rocha-Miranda, C. E. (1967a). The diencephalon of the opossum in stereotaxic coordinates. I. The epithalamus and dorsal thalamus. *J. Comp. Neurol.* **129,** 1–38.

Oswaldo-Cruz, E., and Rocha-Miranda, C. E. (1967b). The diencephalon of the opossum in stereotaxic coordinates. II. The ventral thalamus and hypothalamus. *J. Comp. Neurol.* **129,** 39–48.

Oswaldo-Cruz, E., and Rocha-Miranda, C. E. (1968). "The Brain of the Opossum (Didelphis marsupialis)." Instituto de Biofísica, Universidade Federal do Rio de Janeiro, Rio de Janeiro.

Oswaldo-Cruz, E., Pagani, R., and Rocha-Miranda, C. E. (1965). Lumbar and sacral dermatomes in the opossum (*D. aurita,* Wied). *An. Acad. Bras. Cienc.* **37,** 337–344.

Owen, R. (1837). On the structure of the brain in marsupial animals. *Phil. Trans. R. Soc. London* **127,** 87–96.

Packer, A. D. (1941). An experimental investigation of the visual system in the phalanger, *Trichosurus vulpecula. J. Anat.* **75,** 309–329.

Papez, J. (1929). "Comparative Neurology." Crowell-Collier, New York.

Petajan, J. H., and Morrison, P. (1962). Physical and physiological factors modifying the development of temperature regulation in the opossum. *J. Exp. Zool.* **149,** 45–58.

Petajan, J. H., Morrison, P., and Akert, K. (1962). Localization of central nervous control of temperature regulation in the opossum. *J. Exp. Zool.* **150,** 225–231.

*Petras, J. M. (1969). Some efferent connections of the motor and somatosensory cortex of simian primates and felid, canid and procyonid carnivores. *Ann. N.Y. Acad. Sci.* **167,** 469–505.

Phillips, D. S., and Michels, K. M. (1964). Selective stimulation and electrophysiological responses of the olfactory bulb of the opossum. *Percept. Mot. Skills* **8,** 63–69.

Pilleri, G. (1961). Zur Struktur des Claustrum bei Didelphis marsupialis Linnaeus (Marsupialia). *Acta Anat.* **45,** 310–314.

Pilleri, G. (1962). Zur feinerem Struktur und vergleichenden Anatomie des Corpus striatum primitiver Beutel- und Nagetiere. *Acta Anat.* **48,** 347–367.

Porter, R. (1955). Antidromic conduction of volleys in the pyramidal tract. *J. Neurophysiol.* **18,** 138–150.

Pubols, B. H. (1968). Retrograde degeneration study of somatic sensory thalamocortical connections in brain of Virginia opossum. *Brain Res.* **7,** 232–251.

Pubols, B. H., and Pubols, L. M. (1966). Somatic sensory representation in the thalamic ventrobasal complex of the Virginia opossum. *J. Comp. Neurol.* **127,** 19–34.

Putnam, S. J., and Cone, D. M. (1966). Terminal connections of the olfactory tract fibers in the opossum. *Anat. Rec.* **154,** 405.

Putnam, S. J., Megirian, D., and Manning, J. W. (1968). Marsupial interhemispheric relation. *J. Comp. Neurol.* **132,** 227–234.

Quay, W. B. (1966). Pineal structure and composition in red and gray kangaroos. *Anat. Rec.* **154,** 405.

Rafols, J. A., and Matzke, H. A. (1970). Efferent projections of the superior colliculus in the opossum. *J. Comp. Neurol.* **138,** 147–160.

Ravizza, R. J., and Masterton, B. (1971). The habituation of auditory reflexes in the decorticate opossum (*Didelphis virginiana*). *Physiol. Behav.* **6,** 717–722.

Ravizza, R. J., and Masterton, B. (1972). Contribution of neocortex to sound localization in opossum (*Didelphis virginiana*). *J. Neurophysiol.* **35,** 344–356.

Rees, S., and Hore, J. (1970). The motor cortex of the brush-tailed possum (*Trichosurus vulpecula*): motor representation, motor function and the pyramidal tract. *Brain Res.* **20,** 439–452.

Reid, I. A., and McDonald, I. R. (1969). The renin–angiotensin system in a marsupial. *J. Endocrinol.* **44,** 231–240.

*Rexed, B. (1952). The cytoarchitectonic organization of the spinal cord in the cat. *J. Comp. Neurol.* **96,** 415–496.

Roberts, W. W., Steinberg, M. L., and Means, L. W. (1967). Hypothalamic mechanisms for sexual, aggressive, and other motivational behaviors in the opossum, *Didelphis virginiana. J. Comp. Physiol. Psychol.* **64,** 1–15.

Roberts, W. W., Bergquist, E. H., and Robinson, T. C. L. (1969). Thermoregulatory grooming and sleep-like relaxation induced by local warming of preoptic area and anterior hypothalamus in opossum. *J. Comp. Physiol. Psychol.* **67,** 182–188.

Rocha-Miranda, C. E., and Oswaldo-Cruz, E. (1965). Frequency distribution of fibre diameters in ventral root and muscle nerve of *Didelphys aurita* (Wied). *An. Acad. Bras. Cienc.* **37,** 331–335.

Rocha-Miranda, C. E., Bombardieri, R. A., Monasterio, F. M., and Linden, R. (1973). Receptive fields in the visual cortex of the opossum. *Brain Res.* **63,** 362–367.

Rockel, A. J., Heath, C. J., and Jones, E. G. (1972). Afferent connections to the thalamus in the marsupial phalanger and the question of sensory convergence in the thalami of "primitive" mammals. *J. Comp. Neurol.* **145,** 105–129.

Rogers, F. T. (1923). On the relations of cortical and sub-cortical cerebral lesions to spastic phenomena in the marsupial. *Am. J. Physiol.* **63,** 433–434.

Rogers, F. T. (1924). An experimental study of the cerebral physiology of the Virginian opossum. *J. Comp. Neurol.* **37,** 265–315.

Rose, M. (1926–1927). Der allocortex bei Tier und Mensch. *J. Psychol. Neurol.* **34,** 1–111 and 261–401.

Roth, L. M. (1963). The fine structure of the neurohypophysis of the opossum, *Didelphis virginiana. Anat. Rec.* **145,** 278.

Royce, G. J. (1971). Morphology of neuroglia in the hypothalamus of the opossum (*Didelphis virginiana*), armadillo (*Dasypus novemcinctus mexicanus*) and cat (*Felis domestica*). *J. Morphol.* **134,** 141–180.

Royce, G. J., Ward, J. P., Bade, B. B., and Harting, J. K. (1975). Retinogeniculate pathways in two marsupial opossums, *Didelphis virginiana* and *Marmosa mitis*. *Anat. Rec.* **181,** 467–468.

Sawyer, W. H., Munsick, R. A., and van Dyke, H. B. (1960). Pharmacological characteristics of neurohypophyseal hormones from a marsupial (*Didelphis virginiana*) and monotreme (*Tachyglossus* (Echidna) *aculeatus*). *Endocrinology* **67,** 137–138.

Scalia, F., and Winans, S. S. (1975). The differential projections of the olfactory bulb and accessory olfactory bulb in mammals. *J. Comp. Neurol.* **161,** 31–57.

*Schapiro, S., and Salas, M. (1971). Effects of age, light and sympathetic innervation on electrical activity of the rat pineal gland. *Brain Res.* **28,** 47–55.

Scharrer, E. (1939a). Embolic lesions in the opossum brain. *Anat. Rec.* **73,** 45–46.

Scharrer, E. (1939b). The functional significance of the capillary bed in the brain of the opossum. *Anat. Rec.* **75,** 319–339.

Scharrer, E. (1939c). The regeneration of end-arteries in the opossum brain. *J. Comp. Neurol.* **70,** 69–76.

Scharrer, E. (1940a). Arteries and veins in the mammalian brain. *Anat. Rec.* **78,** 173–196.

Scharrer, E. (1940b). Vascularization and vulnerability of the cornu ammonis in the opossum. *Arch. Neurol. Psychiatry* **44,** 483–506.

Scharrer, E. (1940c). Further experiments on the regeneration of end-arteries in the brain of the opossum. *J. Exp. Zool.* **85,** 365–381.

Schneider, C. (1968). Beitrag zur Kenntnis des Gehirnes von *Notoryctes typhlops*. *Anat. Anz.* **123,** 1–24.

Shepherd, G. M., and Haberly, L. B. (1970). Partial activation of olfactory bulb: Analysis of field potentials and topographical relation between bulb and lateral olfactory tract. *J. Neurophysiol.* **33,** 643–653.

*Simpson, J. B., and Routtenberg, A. (1973). Subfornical organ: Site of drinking elicitation by angiotensin II. *Science* **181,** 1172–1175.

Snyder, F. (1965). Toward an evolutionary theory of dreaming. *In* "Psychoanalysis and Current Biological Thought" (N. S. Greenfield and W. C. Lewis, eds.), pp. 275–315. Univ. of Wisconsin Press, Madison.

Sousa, A. P. B., Oswaldo-Cruz, E., and Gattass, R. (1971). Somatotopic organization and response properties of neurons of the ventrobasal complex in the opossum. *J. Comp. Neurol.* **142,** 231–248.

Sreesai, M. (1974). Cerebellar cortical projections of the opossum (*Didelphis marsupialis virginiana*). *J. Hirnforsch.* **15,** 529–544.

Stokes, J. H. (1912). The acoustic complex and its relations in the brain of the opossum (*Didelphys virginiana*). *Am. J. Anat.* **12,** 401–445.

Sunderland, S. (1941). The vascular pattern in the central nervous system of the monotremes and Australian marsupials. *J. Comp. Neurol.* **75,** 123–129.

Sweet, G. (1904). Contribution to our knowledge of the anatomy of *Notoryctes typhlops*, Stirling, Parts I and II. Part I. Nose, with organ of Jacobson and associated parts. Part II. Blood vascular system.

Sweet, G. (1906). Contribution to our knowledge of the anatomy of *Notoryctes typhlops*, Stirling. Part III. The eye. *Q. J. Microsc. Sci.* [N.S.] **50,** 547–572.

Switzer, R. C. (1972). Quantitative relationships of cell population in olfactory bulbs of mammals. M.S. Thesis, Michigan State University (Biophysics Dept.), East Lansing.

Switzer, R. C. (1973). Relationships of the tufted cells of the olfactory bulb to the lateral olfactory tract. Ph.D. Thesis, Michigan State University (Biophysics Dept.), East Lansing.

Symington, J. (1892). The cerebral commissures in the Marsupialia and Monotremata. *J. Anat.* **27**, 69–84.

Thompson, E. L. (1942). The dorsal longitudinal fasciculus in *Didelphis virginiana. J. Comp. Neurol.* **76**, 239–281.

Tilney, F. (1938). The hippocampus and its relations to the corpus callosum. *Bull. Neurol. Inst. N.Y.* **7**, 1–77.

Tobias, T. J., and Ebner, F. F. (1973). Thalamocortical projections from the mediodorsal nucleus in the Virginia opossum. *Brain Res.* **52**, 79–96.

Towe, A. L., and Biedenbach, M. A. (1969). Observations on the primitive pyramidal system of the American opossum. *Brain, Behav. Evol.* **2**, 498–529.

Tsai, C. (1925a). The optic tracts and centers of the opossum, *Didelphis virginiana. J. Comp. Neurol.* **39**, 173–216.

Tsai, C. (1925b). The descending tracts of the thalamus and midbrain of the opossum, *Didelphis virginiana. J. Comp. Neurol.* **39**, 217–248.

*Tucker, D. (1963). Physical variables in the olfactory stimulation process. *J. Gen. Physiol.* **46**, 453–489.

Turner, E. L. (1924). The pyramidal tract of the Virginian opossum (*Didelphis virginiana*). *J. Comp. Neurol.* **36**, 387–397.

Ulinski, P. S. (1971). External morphology of pouch young opossum brains: A profile of opossum neurogenesis. *J. Comp. Neurol.* **142**, 33–58.

Vaccarezza, O. L., and Affanni, J. M. (1968). Influence des bulbes olfactifs sur le sommeil du Marsupial (*Didelphis azarae*). *C. R. Seances Soc. Biol. Ses Fil.* **162**, 1609–1610.

Van Twyver, H., and Allison, T. (1970). Sleep in the opossum *Didelphis marsupialis. Electroencephalogr. Clin. Neurophysiol.* **29**, 181–189.

Vogt, C., and Vogt, O. (1906). Zur Kentniss der elektrisch erregbaren Hirnrindgebiet bei der Säugetieren. *J. Psychol. Neurol.* **8**, 277–456.

Voris, H. C. (1928a). The arterial supply of the brain and spinal cord of the Virginia opossum (*Didelphis virginiana*). *J. Comp. Neurol.* **44**, 403–423.

Voris, H. C. (1928b). The morphology of the spinal cord of the Virginian opossum (*Didelphis virginiana*). *J. Comp. Neurol.* **46**, 407–459.

Voris, H. C., and Hoerr, N. L. (1932). The hindbrain of the opossum, *Didelphis virginiana. J. Comp. Neurol.* **54**, 277–356.

Walls, G. L. (1939). Notes on the retinae of two opossum genera. *J. Morphol.* **64**, 67–87.

Walls, G. L. (1942). "The Vertebrate Eye and its Adaptive Radiations" (reprinted by Hafner, New York, 1963).

Walsh, T. M., and Ebner, F. F. (1968). Organization of somatic-motor cortex in the opossum. *Anat. Rec.* **160**, 446.

Walsh, T. M., and Ebner, F. F. (1970). The cytoarchitecture of somatic sensory–motor cortex in the opossum (*Didelphis marsupialis virginiana*), a Golgi study. *J. Anat.* **107**, 1–18.

Walsh, T. M., and Ebner, F. F. (1973). Distribution of cerebellar and somatic lemniscal projections in the ventral nuclear complex of the Virginia opossum. *J. Comp. Neurol.* **147**, 427–446.

Wang, M. B. (1969). The distribution and control of osmosensitive cells within the hypothalamus of the opossum (*Didelphis virginiana*). *Neuroendocrinology* **4**, 51–63.

Ward, J. N. (1954). The development of the cortico-spinal tract in the pouch-young of the Virginia opossum, *Didelphys virginiana. J. Comp. Neurol.* **101**, 483–494.

Ward, L., and Watson, C. R. R. (1973). An experimental study of the ventro-lateral nucleus of the brush-tailed possum. *J. Anat.* **116**, 472.

Waring, H., Moir, R. J., and Tyndale-Biscoe, C. H. (1966). Comparative physiology of marsupials. *Adv. Comp. Physiol. Biochem.* **2,** 237–376.

Warner, F. J. (1929). The hypothalamus of the opossum (*Didelphis virginiana*). *J. Nerv. Ment. Dis.* **70,** 485–494.

Warner, F. J. (1970). The development of the pretectal nuclei in *Trichosurus vulpecula. Okajimas Folia Anat. Jpn.* **47,** 73–100.

Warner, G., and Watson, C. R. R. (1972). The rubro-spinal tract in a diprotodont marsupial (*Trichosurus vulpecula*). *Brain Res.* **41,** 180–183.

Watson, C. R. R. (1971). The corticospinal tract of the quokka wallaby (*Setonix brachyurus*). *J. Anat.* **109,** 127–133.

Watson, C. R. R. (1972). An experimental study of the corticospinal tract of the kangaroo. *J. Anat.* **110,** 491.

Watson, C. R. R. (1973). Studies on long spinal tracts in diprotodont marsupials. M.D. Thesis, University of New South Wales, Kensington.

Watson, C. R. R., and Herron, P. (1974). The inferior olivary complex—a comparison between marsupial and placental mammals. *Program Abstr. Soc. Neurosci. 4th Annu. Meet.* Abstract 720, p. 468.

Watson, C. R. R., and Symons, M.-C. (1972). Ascending pathways to the brain stem of the phalanger (*Trichosurus vulpecula*). *J. Anat.* **110,** 501.

Watson, C. R. R., and Warner, G. (1973). Rubrobulbar projections in the possum (*Trichosurus vulpecula*). *Proc. Aust. Physiol. Pharmacol. Soc.* **4,** 188.

Way, J. S., and Kaelber, W. W. (1969). A degeneration study of efferent connections of the habenular complex in the opossum. *Am. J. Anat.* **124,** 31–46.

Weed, L. H., and Langworthy, O. R. (1925a). Developmental study of excitatory areas in the cerebral cortex of the opossum. *Am. J. Physiol.* **72,** 8–24.

Weed, L. H., and Langworthy, O. R. (1925b). Decerebrate rigidity in the opossum. *Am. J. Physiol.* **72,** 25–38.

Weindl, A. (1973). Neuroendocrine aspects of circumventricular organs. *In* "Frontiers in Neuroendocrinology" (W. F. Ganong and L. Martini, eds.), pp. 322. Oxford Univ. Press, London and New York.

Weller, W. L. (1971). Patterns of mechanoreceptor projections to the trigeminal nuclei of the opossum. M.S. Thesis, Michigan State University (Biophysics Dept.), East Lansing.

Weller, W. L. (1972). Barrels in somatic sensory neocortex of the marsupial *Trichosurus vulpecula* (brush-tailed possum). *Brain Res.* **43,** 11–24.

Weller, W. L., and Haight, J. R. (1973). Barrels and somatotopy in SI neocortex of the brush-tailed possum. *J. Anat.* **118,** 474.

*Whitfield, I. C. (1967). "The Auditory Pathway." Arnold, London.

*Wiesendanger, M. (1969). The pyramidal tract. Recent investigations on its morphology and function. *Ergeb. Physiol., Biol. Chem. Exp. Pharmakol.* **61,** 72–136.

Wislocki, G. B. (1939). The unusual mode of development of the blood vessels of the opossum's brain. *Anat. Rec.* **74,** 409–427.

Wislocki, G. B. (1940). Peculiarities of the cerebral blood vessels of the opossum: Diencephalon, area postrema and retina. *Anat. Rec.* **78,** 119–131.

Wislocki, G. B., and Campbell, A. C. P. (1937). The unusual manner of vascularization of the brain of the opossum (*Didelphys virginiana*). *Anat. Rec.* **67,** 177–191.

Yesberg, N. E., Budtz-Olsen, O. E., and Sharples, K. (1967). Antidiuretic activity in the pituitary gland of the koala bear. *Nature* (*London*) **216,** 916.

Yuen, H., Dom, R. M., and Martin, G. F. (1974). Cerebellopontine projections in the American opossum. A study of their origin, distribution and overlap with fibers from the cerebral cortex. *J. Comp. Neurol.* **154,** 257–286.

Zeballos, G. A., Wang, M. B., Koizumi, K., and Brooks, C. McC. (1967). Studies of the anterior hypothalamus of the opossum (*Didelphis virginiana*). *Neuroendocrinology* **2,** 88–98.

Ziehen, T. (1897a). Das Centralnervensystem der Monotremen und Marsupialier. I. Theil: Makroskopische Anatomie. *In* "Zoologische Forschungsreisen in Australien und dem Malayischen Archipel" (R. Semon, ed.), Vol. 3, pp. 1–187. *Denkschriften der Medicinisch-Naturwissenschaftlichen Gesellschaft zu Jena* **6,** 1–187. Fischer, Jena.

Ziehen, T. (1897b). Ueber die motorische Rindenregion von *Didelphys virginiana. Zentralbl. Physiol.* **11,** 457–461.

Ziehen, T. (1897c). Der aufbau des Cervicalmarks und der Oblongata bei Marsupialiern und Monotremen. *Anat. Anz.* **13,** 171–174.

Ziehen, T. (1899). Zur vergleichenden Anatomie der Pyramidenbahn. *Anat. Anz.* **16,** 446–452.

Ziehen, T. (1901). Das Centralnervensystem der Monotremen und Marsupialer. II. Theil. Mikroskopische Anatomie. *In* "Zoologische Forschungsreisen in Australien und dem Malayischen Archipel" (R. Semon, ed.), Vol. 3, pp. 677–728. *Denkschriften der Medicinisch-Naturwissenschaftlichen Gesellschaft zu Jena* **6,** 677–728. Fischer, Jena.

Zimmerman, E. A., and Chambers, W. W. (1963). Cortical projections to sensory relay nuclei in the brain stem of the opossum and rat. *Anat. Rec.* **145,** 304.

# 5

# Behavior of New World Marsupials

DON HUNSAKER II and DONALD SHUPE



## I. INTRODUCTION

Behavioral studies of any vertebrate are difficult, especially those conducted in the natural habitat of the species. The task becomes even more formidable when the animal is nocturnal, lives in a variety of habitats, and is partially nomadic. Such is the case with the New World marsupials, even the common American opossum *Didelphis virginiana,* but the degree of difficulty is intensified when one begins to study the small murine opossums like *Marmosa* in a tropical rainforest. It is difficult to convince oneself that the natural behavior of murine opossums may be understood by direct observation when the average length of observation time in nature is less than a minute. Usually this minute is spent looking 20 feet up in a tree at a pair of eyes reflecting back the light of a head lamp. Because of these problems, few behavioral studies have been made on opossums in their natural habitat. Most of our understanding of didelphid behavior in nature comes from years of trapping data by investigators like Reynolds, and research on captive animals by Hartman and McManus. We are fortunate to have excellent field studies of *Didelphis virginiana* by Reynolds, Lay, Fitch and Sandidge, and others, who trapped, marked and recaptured, measured home ranges and dispersal distances, and conducted food and habitat selection studies.

Early investigators (i.e., Goldman, Tate, and Enders) referred to behavior of these species while working on distribution and taxonomic or other research. As a result, similar studies of the Central and South American species are rare. More recently, research by Fleming on Central American forms provides us with valuable insight into some of the lesser known species. Experimental behavioral work has been conducted in the laboratories of James, Langworthy, Friedman and Marshall, Tilley, Doolittle, Mason, and other investigators. Their laboratory studies in psychological research of conditioned responses, behavior in mazes and puzzle boxes, and reversal training provide us with good comparative data with which we can evaluate the performance of marsupials with a variety of other vertebrates that have been observed under similar conditions.

The sum of the results of these investigations indicate that behavior patterns of the opossums are as variable and generalized as the rest of their characteristics. The unpredictable and often aggressive responses of an animal when first approached, and the problems one encounters when trying to tame the animals in a laboratory situation, suggest that this family of vertebrates has been successful in their competition with other animals because of their behavioral ability to adapt to a variety of conditions. Because of the plasticity, variability, and generality of the behavior patterns that have been observed, research projects in almost any area of marsupial behavior are greatly needed.

## II. SOCIAL BEHAVIOR

Most of the information on the social behavior of didelphids in nature has been collected in field studies involving trapping, marking, and observing the movements of a local population of opossums. Works on the North American *Didelphis* by Lay (1942), Fitch and Sandidge (1953), Llewellyn and Dale (1964), Reynolds (1945), Wiseman and Hendrickson (1950), and Shirer and Fitch (1970) provide us with substantial information on population structure and spatial interactions between males, females, and young animals. Field studies on Central and South American forms have been reported by Enders (1935), Davis (1945), and Fleming (1970, 1972, 1973). When combined with information from Goldman (1920), Cabrera and Yepes (1940), Sanderson (1955), Walker (1964), Hall and Kelson (1959), and Wiedorn (1954), we find only a limited amount of information on behavior of the group has been formulated. In general, investigators have limited data on only three tropical species: *Didelphis marsupialis, Philander opossum,* and *Marmosa robinsoni.* Some behavioral information can be gleaned from reports by Davis (1945a,b) and others who were primarily concerned with field studies of yellow fever, malaria, and other health problems. Enders' (1935) work on Barro Colorado Island was oriented toward collecting reproductive biology data of *Marmosa* in conjunction with his and Hartman's research work, but he provided a wealth of information on the mammalian fauna of the area in addition to his work on reproductive physiology. All in all, we have more than basic information on the behavior of only the single species *Didelphis virginiana* in this very diverse family.

Since most of our experience with marsupials in North America stems from *Didelphis virginiana,* it is very easy for one to picture the entire family as a slow moving, primitive group of beasts with remarkable dentition and some objectionable personal habits that include biting the investigator. If unable to make tooth contact, they are just as liable to defecate and spray him with evil smelling fluids. In contrast to this species, however, the active, curious, and intelligent acting woolly opossums (*Caluromys*) are highly reminiscent of lemurs bounding from limb to limb investigating and manipulating every unusual thing they encounter. These contrasts in behavior leave little doubt that the didelphids represent a virtually untouched group which have a great potential for studies in convergent evolution and comparative behavior.

## A. Territoriality

The social behavior of *Didelphis virginiana* indicates that it is a solitary species rather than one which forms permanent social bonds or is territorial in the strictest sense of the word. Numerous field and laboratory studies of *Didelphis* indicate that the animals will stay in a particular area only for limited periods of time (6 months to a year), with the females being more sedentary than the males. During the reproductive season, females can be found nesting in a particular den site or moving from den to den within a reasonably well-defined home range.

*Didelphis* is basically a solitary or antisocial animal and will either actively avoid or agress toward another individual in close proximity. It is not territorial in the sense that it defends a particular area and advertises its presence, but rather the challenge to any interloper is more a case of defending the interpersonal distance surrounding the animal wherever it may be.

One of the most characteristic behavior patterns to be observed between individuals is a high level of agonistic behavior. Almost invariably when two male opossums meet there is threat behavior, including aggressive vocalization with the mouth widely opened displaying fearsome looking dentition. Rapid slashing bites and snapping occur in aggressive encounters. Evans (1959) reported up to four bites per second in two male opossums that were fighting. The aggressive nature of the interaction changes if it is a male–female encounter; it begins with an aggressive display and almost immediately shifts into a courtship pattern by the male. Generally, if the female is receptive, she reduces her aggressive interaction and permits the male to continue into precopulatory and copulatory behavior. On the other hand, if the female is not receptive, she maintains an aggressive attitude with a considerable amount of growling, hissing, and biting. Unless the male actively pursues courtship with the nonestrous female, this behavior is not as highly aggressive as the encounter would be between two males. If in the process of the association the male and female stay together, which happens in captive or semifree ranging animals, the male begins to reduce the level of courtship behavior during the ensuing days. After 6 or 7 days have elapsed, the male shows very little reproductive interest in the female. If he does not reduce the level of courtship activity, the female becomes aggressive and may even kill the male; this occurs mostly in smaller cages where the male cannot avoid the contact. There is either a satiation of the courtship–copulatory drive by prolonged association with the female or the male habituates to the olfactory, tactile, or visual stimuli that the female presents to him. In captive situations, the male will become responsive again after a separation period of a week or two.

*Didelphis* females are considerably more sedentary than the males, as is true of *Marmosa* and probably other genera. In the process of protecting the integrity of their individual distance, the females' limited area of activity would tend to exclude any other members of the species. This would be especially true during the period that she is with young. The absence of territoriality and habituation to other individuals in nature is noted by the occurrence of multiple individuals in a particularly attractive denning site. It is not uncommon for *Didelphis* to be found in association with other opossums or other species of vertebrates (i.e., raccoons) in dens. Very little observational work has been done on the South American species, so it is unknown whether there are multiple or single individuals in one denning site. Since the tropical forms tend to build nests in trees and shrubs and are not dependent upon natural nesting and denning sites as much as the more temperate forms, they probably tend to be more solitary. The rocky outcrops in some parts of the country are used as dens by *Didelphis* and tend to concentrate individuals in particular areas. Field investigations of the other South American didelphids are lacking, but there are no indications in the scattered references and observations by the authors that their social behavior differs significantly from that found in *Didelphis*. Behavior studies in the authors' laboratories also confirm that the basic social behavior of *Didelphis, Caluromys,* and *Marmosa* are very similar. Most reports relate to a single individual or pairs seen only for a short time; the majority of references on the behavior of these forms are generally about captive specimens.

Boggs (1969) studied the social behavior of *Marmosa robinsoni* and found that they easily formed social hierarchies in captivity. Social dominance of the males was dependent more on experience than on size or age. In large cages, 3 × 3 × 3 m, the dominant males would mark the habitat with their sternal glands. These glands produce a viscous, oily secretion which apparently is used to mark the animal's area of activity. *Didelphis virginiana* males have also been observed to mark objects with saliva and the side of their heads. Boggs, Hunsaker, and other authors have not observed any obvious reaction in *Marmosa* when another male or female smells the secretions. *Didelphis virginiana* males have a very characteristic reaction to a marked object, so there are definite indications that this species does rely on olfactory cues for identification of a particular area. The sternal glands are well developed in some other species of *Marmosa* (*M. cinerea, M. elegans, M. murina, M. mexicana,* etc.) and appear to be undeveloped in other species (*M. fuscata* and *M. incana*). The glandular secretion is very prominent in *Monodelphis* and has a strong odor. The glands are probably active in other genera and are used by the males to mark an area during reproductive periods. These marking activities suggest advertisement of the locality of males, but since field studies have not found males residing in a particular area for a significant period of time, the evidence suggests more of a nomadic, solitary life rather than a territorial, fixed home range for the didelphids.

### B. Reproductive Behavior

Female *Didelphis virginiana* enter estrus with the associated behavior patterns in the first few weeks of the calendar year, with the peak of the breeding season depending upon the latitude of the population's area. The earliest breeding in the northeastern United States was reported by Grote and Dalby (1973), who found a female in Columbus, Ohio on February 25, 1973 with pouch young a mean length of 57.8 mm. They estimated that she bred on the third or fourth of January. This corresponds with the earliest breeding in the West, which is the last week of December in southern California. Hunsaker and Cadzow have collected estrous females in San Diego and Los Angeles during this time. Hartman (1928), Reynolds (1945, 1952), and other investigators determined that the third week in January is the approximate beginning of estrus in the central United States, with the greatest number of females in estrus during the fifth and sixth weeks of the new year. Usually by the middle of February most of the females are pregnant or in an anestrous state. Burns and Burns (1957) observed females that gave birth the last week of January in northern Florida. Depending upon the location, most *Didelphis virginiana* in the United States produce two litters a year. In southern Texas and southern California there are records of three litters being produced in one calendar year. Opossums in southern Florida could also have three litters a year. There appears to be very little difference in the estrous cycles of wild captured *Didelphis* and animals in captivity, with the majority of research colonies reporting that their females are in estrus during the latter part of January and the first part of February. The early estrus produces young that are weaned in the summer when food is plentiful and environmental conditions are excellent. These factors enhance the survival rate of the young animals. The second and third litters are weaned during the fall and early winter months which, in areas other than the very warm parts of southern Texas, southern California, and Florida, produce a very reduced level of population survival. In general, *Didelphis virginiana* breeds less often and has larger litters in the north and, conversely, has about the same net production by having smaller litter sizes in the south but with more litters. Reynolds (1952) stated that a 9-month breeding season in California does not allow time for the earliest female breeders to produce three litters a year. Audubon and Bachman (1851) stated that South Carolina had three breeding seasons occasionally, and Hartman (1928) stated that a small percentage did have a third litter in Texas. Lay (1942) suggested that, based on his collecting data in eastern Texas, there was not time for three litters a year. From this information, it is apparent that there is breeding behavior in *Didelphis virginiana* throughout most of the year which is largely confined to two seasonal peaks in February and early summer.

*Didelphis marsupialis* begins to breed late in January and February (early dry season) in Central and South America, and might breed throughout the year in some localities. Sufficient data are unavailable to know how many litters a year are produced in the equatorial regions. Fleming (1973) stated that *Didelphis marsupialis* was in a reproductive state from February through October in Panama. Tyndale-Biscoe and Mackenzie (1976) surveyed the literature and examined numerous specimens and determined that the same correlation between latitude and litter size that is found in *Didelphis virginiana* is exhibited by *D. marsupialis*, except it is reversed in southern latitudes.

The more tropical species, *Marmosa robinsoni*, is a definite seasonal breeder with females pregnant or with young present from February to September. Hunsaker (1970) pointed out that the breeding season in Colombia was closely correlated with rainfall patterns; it breeds as early as December in southern Colombia and Equador. Enders (1966) reported *Marmosa* begins a distinct onset of the breeding season in early February in Panama. Fleming (1973) stated that *M. robinsoni* has one or two litters a year in Panama. Tate (1933) stated that *Marmosa* breeds from one to three times a year. *Marmosa quichua* has three broods a year and breeds all year. Other species breed all year and yet have major and/or minor peaks during the year. There are probably all types of annual breeding cycles in *Marmosa* because it is such a widespread genus. As a general rule, the greater the distance from the equator, the more fixed are the seasons for breeding.

*Caluromys derbianus* breeds throughout the year in Central America. Biggers (1967) and Enders (1966) reported woolly opossums breeding from January to June (during the dry season) and for a short period into the wet season. Phillips and Jones (1968) determined the species bred in December in Nicaragua, which provides evidence for breeding all year. *Philander opossum* probably breeds all year in some areas. Hall and Dalquest (1963) stated that the species reproduced throughout the year in Vera Cruz, Mexico, and Biggers (1966) found anestrous females in November and December (late rainy season) in Nicaragua. Fleming (1973) reported the *P. opossum* breeding season to be from February to November in Panama. Hunsaker has collected *Philander* with pouch young in Colombia from May through August, but did not trap for them extensively during the latter part of the year.

The reproductive behavior and seasons of the Caenolestidae is virtually unknown. *Caenolestes obscurus* breeds in June and July (at least) in Colombia, but additional field work is needed to establish the seasons with any accuracy.

Collins (1973) has reported on the breeding seasons of New World marsupials from data largely collected from zoos. All the available data on reproductive periods are summarized by Hunsaker (Chapter 3).

Courtship behavior in *Didelphis* and other New World marsupials involves vocal, tactile, olfactory, and visual cues. The onset of the breeding season involves an increase in the activity levels of both males and females and the appearance of signpost marking by the males. Adult males will select particular objects to mark by licking and rubbing their head and neck region against it. This behavioral pattern is evident in wild as well as captive-born individuals. Olfactory cues associated with the saliva and possibly the skin glands of the head and neck region serve to identify the presence of a male in an area. In all probability, the olfactory perception of a female is sufficiently well developed to identify individual males by the markers. This behavior also suggests that territorial marking could possibly serve to repel other males. There is no evidence that these olfactory signals will repel other males but they do elicit a definite aggressive response. Upon encountering a marked signpost of another male, an opossum will begin to display an aggressive behavior pattern described by Reynolds (1952) as the "fighting dance." This suggests that these pheromones do serve to keep individuals informed of the presence of other males and possibly serve to attract females who are in estrus. Since the female is in receptive estrus for only 36 to 48 hr, it would benefit the population to have both sexes searching for available mates during this period of fertility. As the home ranges are so overlapping, this probably serves to attract females more than to repel a competitive male. In captive colonies, the female will often spend more time smelling and associating with the signpost of a new male than with one that had been well marked by a male she had been associating with previously. This relates to the fact that female opossums generally become unresponsive to cage mates that they have been with for some time. Females have been known to rub their head against an object also, but it is an uncommon occurrence. In no case has the author seen, or any author mentioned observing, a female that has been licking and rubbing an object to the extent that the entire side of her head and shoulders are covered by saliva, as observed in many captive males who will spend hours rubbing on a signpost. In all probability the occurrence of male marked signposts and the presence of urine marking from estrous females would tend to clump a population during the breeding season to insure ready access to a mate.

Although somewhat tolerant of other individuals, the basic social behavior of female *Didelphis* is to aggressively assert her independence from the advances of any male, and to avoid the presence of any female. When a male and female encounter each other, the male is usually the initiator of the social contact. A male will approach any other opossum it observes and identification of the sex of the other animal is made within a meter. Visual cues are used first and olfactory cues are probably used in close proximity. The approach to a male becomes aggressive while toward a female it includes courtship, nuzzling, and following by the male. Unless the female is in estrus and physiologically ready to have her ova fertilized, she aggressively avoids the courtship overtures of the male. In the majority of cases observed and reported for the didelphids in general, the female will aggressively hiss, vocalize, and physically attack the male to discourage the advances when she is not in estrus. As estrus approaches, these attacks become less vicious and the moment eventually arrives when he can make advances to the willing female with copulation ensuing. Aggressive behavior patterns by the female include mouth threats with the mouth wide open, attempted bites, hissing in which air is forcefully exhaled through the mouth, tooth clicking or chattering in which the teeth, tongue, and lips are included in sound production, and actual biting and grabbing with the hands. The usual procedure is for the male to approach the female, sniff the genital region, and initiate body contact. Receptive females are very passive and allow the male to approach closely, sniff the genital region, initiate mounting, and proceed with the copulatory behavior. Submissive gestures include pointing the head down and turning it at an angle so that there is little or no direct eye contact, and having its head turned away from its body which is laterally displayed to the other animal. Table I lists the postures that are involved in dominance behavior. The male usually employs a pursuit of the female by attempted mounts and submissive behavior to stay in close association with her until she permits copulation.

Copulation in *Didelphis virginiana* involves the male mounting the female on the back, grasping her neck in his jaws and her shoulder region with his forefeet, and falling over on the right side for intromission and copulation. In the mounting process, the female's hind legs are grasped by the male's hind feet which effectively renders her immobile during intromission.

The complete act of copulation lasts for approximately 20 min, with a range of 5 to 30 min. Intromission occurs after the paired animals fall to their sides and a series of slow pelvic thrusts are initiated by the male, with each thrust lasting from 1 to 2 sec spaced 15 to 20 sec apart. There is no well-demarcated ejaculatory response that can be determined by behavioral observation. The female lies passive and, after pelvic thrusts of the male and ejaculation cease, the male will release his hold on her. Ejaculation apparently occurs about 15 min after intromission, since the thrusts increase in force and then cease. The female then disentangles herself from the released embrace of the male and moves away.

Postcopulatory behavior of the female involves her returning to her normal feeding and exploratory or resting behavior with little or no genital grooming or similar activities. Postcopulatory behavior of the male involves stretching, arching of the back, and some genital grooming involving licking the penis or inguinal areas. Reynolds (1952) stated that opossums had to copulate on their right side for fertilization to occur. Not enough investigators have observed mating and subsequent parturition of opossums for us to

**TABLE I**

Social Dominance, Submission, and Agonistic Behavioral Patterns of Didelphids

| Behavior | Significance |
|---|---|
| Neutral | |
| Approach | General investigative behavior; involves the approach to any new individual observed |
| Freeze | Low level response to approach or visual sighting of another animal or novel stimulus; component of submissive behavior repertoire |
| Stare | Eye fixation and stare directed to approaching animal |
| Face | Accompanies stare, face pointed forward, directly at intruder |
| Aggressive | |
| Mouth display | Lowest level of threat behavior; mouth is partially opened and the lips are posteriorly contracted to expose the incisors, canines, and premolars; salivation with varying amounts of drooling usually accompanies this threat |
| Tooth display | Mouth fully open, all teeth displayed; intermediate level of threat behavior |
| Ears flattened | Usually accompanies threat behavior or associated with pain in absence of threat |
| Vibrissa flattened | Associated with tooth display; threat behavior |
| Nose feinting | Usually associated with combating males, where animals stand nose to nose and sequentially lunge at each other |
| Bite | Actual clamping of jaws on another animal; high level threat |
| Vocal hiss | Low level vocal threat |
| Vocal growl | Intermediate vocal threat |
| Vocal screech | Highest level of vocal threat |
| Fighting dance | A shuffling walk by male *Didelphis*. The hind quarters are depressed and the hind feet kicked laterally. Tail is straight. Dominance display |
| Submissive | |
| Vocal click | Courtship in males, maternal in females; species identification noise; more in males than in females |
| Death feign | Observed in wild caught animals; maximum submissive pattern (*Didelphis virginiana*) |
| Head turn | Submissive posture; avoids direct eye contact. |
| Lateral position | Submissive posture; turns body partially away from head to head position |
| Retreat | Moving away from an intruding animal; submissive behavior |

be able to answer this question with certainty. McManus (1970) observed one mating on the left side which did not result in fertilization. It is obvious that the right side is the preferred one for the species and left-side copulation could reduce the probability of fertilization, but it is doubtful that it would preclude fertilization in all cases.

Other species of didelphids have very similar reproductive behavior. The murine opossums, *Marmosa*, take some time for individuals to become accustomed to each other before courtship activities are initiated. Barnes (1968a,b) has described the copulatory behavior and husbandry of *Marmosa robinsoni*. Barnes and Barthold (1969) introduced the female into the male's cage for breeding in their colony. If nest boxes, tubes, or other hiding places are available, the female will hide in them and courtship behavior will occur from several hours to 2 days later. If there are no nesting tubes, courtship is usually initiated by the male within minutes. If the female is completely uninterested or the male is distracted, courtship will not occur. If the female is in estrus, the male will begin an active series of precoital activities: mutual investigations, smelling, and genital stimulation, in that order. The pair will circle while licking the genital area of each other. The female slows down and stops when the male begins to grab her shoulders with his forefeet and bites her neck region. Almost simultaneously with the neck and shoulder grab, the male feels around the habitat with his tail and grasps the nearest object that will support his weight.

*Marmosa* almost invariably utilizes its prehensile tail to give the male a firm attachment during copulation. The male continues the grasp of the female by the shoulders and the biting of the neck region. Anchoring himself by his tail, he grabs the female's hind feet with his and copulation ensues. This specialized behavior is an adaptation for arboreal life, since on occasion the males have been observed to complete the copulation sequence with both the male and female completely suspended by the tail of the male. Barnes and Barthold (1969) suggest that the erection of the penis is dependent upon the extension of the tail. They have observed normal copulation in captive animals in cages, but have not observed intromission when the animals were in smooth-sided cages and unable to securely attach themselves by their tails. At the present time it is not known if the tail attachment is an anatomical necessity for penile erection or if it represents only a preferred position for copulation. The copulatory behavior of *Caluromys* is similar to *Didelphis*. The tail is sometimes used to secure a hold on the substrate but does not appear to be as extensively used nor as much a necessity as it is in *Marmosa*.

During copulation, *Marmosa robinsoni* methodically rotates the shoulder skin of the female and begins to massage the lateral and abdominal areas as well. Insertion of the penis takes place as the male brings the female back toward and under his body. Pelvic thrusts penetrate only about 10 mm into the female, a little less than 50% of the 2- to 2.5-cm length of the penis. Pelvic thrusting occurs for 15 to 20 min at which time ejaculation takes place. The ejaculatory sequence is not well defined and probably takes place during the most active thrusting period. In captivity, some *Marmosa* will copulate for more than 6 hr with the female disengaging and then resuming copulation almost immediately. It is not known if this occurs in the natural environment or not. Postcopulatory behavior is not complex; the animals separate and groom themselves.

The courtship of *Marmosa* and especially *Caluromys* involves a pursuit pattern of the male following the female during the courtship activities. In our laboratories, *Marmosa robinsoni* males would pursue the females for several minutes in a snout to tail position. Since they are in small cages, the male would usually catch the female in a relatively short time. Females who were in the peak of estrus would still move away from the male for the equivalent distance of 10 m or so.

*Caluromys* were paired for breeding in our lab in a large cage, 2 × 2 × 3 m, and exhibited a very extensive and active chasing behavior during courtship. They were fond of a complex arboreal habitat and the pursuit by the male was very fast and animated. The female would cover the equivalent of a 100- to 1000-m distance in her retreat. We consider the precopulatory pursuit to be a normal part of their courtship; it did not appear to be a matter of fearful escape by the female.

### C. Birth

There are few phenomena in biology that have had more anecdotal reports associated with it than the reproductive biology of marsupials. It was assumed, and hotly defended, in the early 1800's that a female opossum removed the babies from her vagina, and carefully and tenderly encasing them with her lips, put them inside the marsupium whereby they become attached to the teats. The first person to observe and record the birth of an opossum was Barton, who in 1823 wrote a letter to a friend in Paris and described "the event which I was so fortunate to witness, is, in my opinion, one of the most interesting in the whole science of Zoology; and so far as I know, it has never been noticed by any other naturalist but myself." His description of the birth process in which the young embryos exit from the vagina and crawl unassisted by the female into the marsupium and attach themselves to the teats was vague, but it does represent the first scientific notation of the process.

The normal gestation period in the Virginia opossum is considered to be 13 days ± 6 hr. Reynolds (1952) estimated this period to be about the average for the majority of females. Jurgelski and Porter (1974), the authors, and other researchers have used this figure with considerable success.

Birth is preceded by some degree of pain in the female. As general activity levels increase, the female begins to groom and clean the pouch area and, all in all, appears to be in the birth process for less than 30 min. Overt behavior indicating internal pain, which is probably early labor contractions, consists of laying the ears back and grimacing. This can be observed from 30 to 14 min before birth. Generally, the female will sit in a hunched-over position during these early phases of labor. An interesting behavior pattern that should be watched for at this time is the female directing her tail forward between her legs. This behavior is observed only when the female is in labor. It does not always occur, but when it does it is an indication of imminent parturition. Actual birth is preceded by her assuming an upright sitting position with the genital area extended as far forward as comfortable. This position results in the shortest distance possible between the vulva and the pouch for the young to crawl over. The tail is usually directed forward between her legs and the female rests her weight on the base of the tail. Pouch cleaning ensues, the time spent being quite variable from one female to another. During this period, a series of abdominal contractions begins to take place. The contractions begin about 15 min before parturition occurs, with each contraction spaced about 3 to 4 min apart.

The embryons, sometimes encased in fetal membranes, appear at the vulva entrance. If intact, the membranes are separated by the female in the process of licking and cleaning, and they then crawl unassisted into the pouch. The only real maternal behavior exhibited by the mother during the process is the breaking of the fetal membranes, and this appears to be a by-product of the intensive licking rather than a maternally directed response to aid the embryon. Neonates have been observed to drown if the fetal membrane is not broken by the female. This was reported by Hartman (1952) with an anesthesized female. Most embryons are born free of any membranes, and the cleaning process is not usually necessary. The embryon's journey to the nipple requires about 2 minutes and the entire parturition process is over in 15 min. The female exhibits very little maternal care behavior and does not appear to assist the embryons in any fashion. The two prerequisites of maternal assistance in the birth process include only the maintenance of a sitting position and relaxation of the pouch.

The migration of the neonate to the pouch appears to be guided by olfactory cues rather than by gravity or tactile responses to a trail licked in the abdominal hair by the mother. In the process of pouch and abdominal cleaning, the female saturates the area with saliva. Since the incisive ducts of Stensen are open at birth (McCrady, 1938) this provides a direct route of access to the Jacobson's organs, which are well innervated by olfactory nerves and connected to the ducts of Stensen at birth (Langworthy, 1928). The chemoreceptive ability of the Jacobson's organ could be sufficient to enable the embryon to orient toward areas of greater concentration of pheromones or other chemical cues of the mother's saliva or pouch secretions to achieve directional success.

McCrady (1938) considered that the female had to lick the hair between the vulva and the pouch for the neonate to successfully find the pouch by tactile orientation. Reynolds (1952) and the authors have observed females give birth without first licking the area between the pouch and vulva. It is a common pattern but not a prerequisite to successful migration of the embryon. Migrating embryons have been observed by Reynolds (1952) and the authors to make distinctive turns when they reach the area of the pouch, which further substantiates olfactory orientation rather than Hartman's (1928) suggestion of negative geotropism.

During actual birth, the female appears to undergo some physical stress and pain, however, not to any great extent. The early stages of labor are apparently the only high levels of pain to which the female reacts; at other times she is too busy to notice much pain and one is not able to observe any behavioral patterns associated with pain.

*Marmosa robinsoni* birth has been observed by few investigators. Barnes and Barthold (1969) reported their females assumed the typical posture for delivery with the shoulders propped against the cage and the tail extended between the legs. Prebirth activities include grooming, which becomes very frequent before the actual birth process. During the exit of the newborn and their journey to the nipples, grooming ceases. The process lasts about 5 min.

Female *Marmosa* do not appear to evidence pain and make no effort to retrieve embryons that go the wrong direction or fall off during birth. The maternal drive to retrieve the young apparently develops later. Some female *Marmosa* do not assume the typical birth posture. Births have been recorded in both Hunsaker's and Barnes' laboratories when the female was sitting with the tail directed backward and was simply grooming herself. Apparently the typical posture is not a requirement for successful birth as it appears to be in other marsupials.

In *Marmosa* there is a definite postparturition behavior shift in walking posture. Once the embryons are attached, the female raises the lumbar section of her back to compensate for the added girth caused by the young. This kyphotic posture protects the embryons from contact with the substrate and from mechanical abrasion. *Didelphis* embryons are well encased by the pouch and this postural compensation is not utilized. Although most of the other tropical genera have not been observed with attached young, we would assume that the species with reduced pouches utilize the arched back walking posture during the developmental period of the young.

## D. Maternal Behavior

The care and attention of the offspring of *Didelphis* lasts approximately 100 days. This period encompasses the time when the female first begins cleaning of the pouch, the birth, migration, and attachment of the baby to the teat, to the time it leaves the family to live alone. Figure 1 represents the sequence of the behavioral development of the young.

The mechanical swelling of the nipple in the buccal cavity and the structural changes of the developing opossum's mouth preclude the possibility of the neonate falling off, so maternal care in the first few weeks of life in the pouch is restricted to keeping the pouch environment clean and maintaining some control over thermoregulation by contraction and relaxation of the pouch muscles. This maternal care is closely correlated with the self-grooming activities and normal thermoregulation of the female, and is probably not directed specifically towards the well-being of the young. During this stage, the needs of the young opossums are fulfilled by adequate food supplies, ventilation, humidity, and thermal conditions.

For many years there was a controversy as to whether the mother actually pumped milk into the digestive system of the baby or whether the

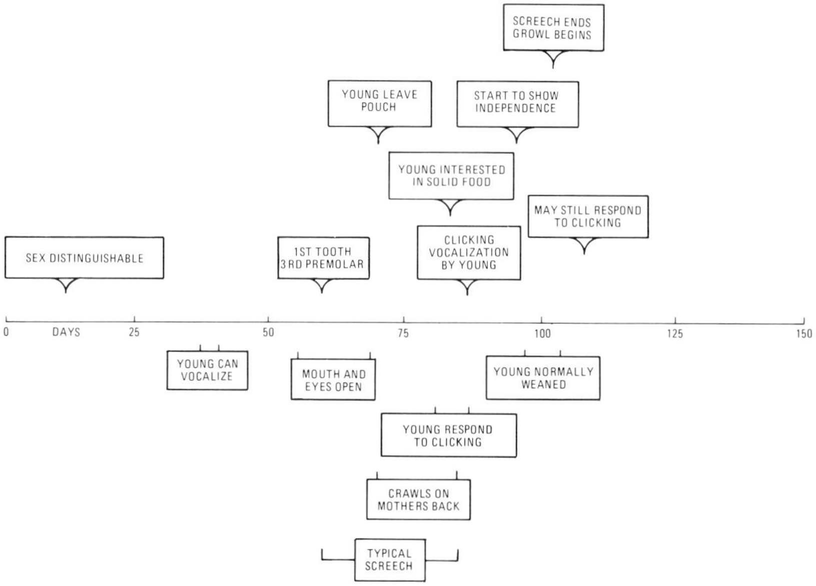


Fig. 1. Developmental behavior patterns and maturation times of responses of the Virginia opossum, *Didelphis virginiana*.

young opossum sucked the milk out. Only 5 years after Barton first witnessed and described the birth of opossums, Seiler (1828) stated that milk was pumped into the baby opossums by the contractions of the muscles surrounding the mammary gland of the female. Many investigators considered the iliomarsupialis muscle to be responsible for the expulsion of milk, and Owen (1868) reiterated that this muscle could compress the mammary gland and expel milk through the greatly elongated nipple. He considered the elongation of the nipples to be a normal process of pregnancy, a fact which Coues (1872) disputed. Coues apparently observed the suckling activity of the babies but considered the main function of this to be the elongation of the nipples, and after the baby was attached, the mother expelled milk by compression of the muscles over the mammary glands. McCrady (1938) made close observations on the suckling behavior of the newborn opossum and noted that after the fetus migrated to the pouch, the head was moved in lateral arcs until the snout encountered a nipple. In response to this tactile stimulus, the baby placed its rudimentary mouth next to the nipple and began to suck it into its mouth. McCrady was convinced that the entire nutritional process of the embryon was by sucking and that there were no mechanisms for pumping the milk to the fetus. Enders (1966) conducted some interesting research on the possibility that the iliomarsupialis muscle and other ventromusculature could cause ejection of milk from the mammary gland. Utilizing lactating *Didelphis,* he stimulated the iliomarsupialis and other muscles but did not observe the expulsion of any milk from the nipples. He was able to manually expel milk from the glands to establish that they were functional, and so concluded that the fetus actively sucked milk from the nipples. Enders also noted that *Marmosa* did not discharge milk from the nipples when a fetus was removed. No milk was ejected from the nipple while thus vacated, but upon reattachment the baby began suckling and continued to grow.

Maternal behavior during the first few weeks of pouch life is limited to an active and sometimes violent reaction toward any attempt to molest the pouch area. This has been noted in our captive animals and Reynolds (1952) stated that his animals reacted by kicking and growling when he attempted to examine the pouch. Jurgelski and Porter (1974) also reported violent snapping by females when the pouch was investigated. This presumably would hold true for wild opossums and would generally serve to protect the pouch young from external influences. Suckling of the young apparently maintains the lactation of the female and more than one pouch young must be attached for lactation to continue. Reynolds (1952) stated that if two or more young reach the pouch, lactation continues and estrus does not reoccur until the young are weaned. The suckling of a single pouch young is insufficient stimulation to maintain lactation of the female, and so a single baby will die. The authors know of no case where a female Virginia opossum has raised a single embryon.

The oral attachment of the embryon to the nipple of the female has been a point of controversy for many years. The old stories of the tissues of the embryon actually growing to the nipple have long since been disproved. Hartman (1952) noted that the young opossums did not bleed when carefully removed from the nipple, indicating that there were no intercellular and certainly no intervascular connections between the young and the mother. He suggested that the dried milk around the nipple and mouth formed a seal. This, in addition to other secretions, in fact, does contribute to the suction. Barker *et al.* (1967) described a layer of mucopolysaccharide material at the oral cavity–nipple interface, which probably assists in creating a positive seal.

Modern evidence has established that the embryons have a series of small projections, microbarbs, on the lips and surface of the tongue. There are also small microgrooves on the palate. These structural modifications, including small grooves in the nipple to receive the microbarbs, in conjunction with the swelling of the nipple in the buccal cavity, firmly affix the neonate to the mother. Barker *et al.* (1967) sectioned neonates attached to the nipples of *Didelphis virginiana* and determined several adaptations of the oral membranes that were important in the attachment. They observed that the nipple of the mother and buccal cavity of the young were reciprocally modified to assure a positive attachment.

The microbarbs begin to form a few days after the embryon is attached to the nipple. Barker *et al.* observed them in young that ranged from 6 to 60 days of age. They are exceedingly small, only a few microns in length, and fit tightly into the juxtaposed grooves of the maternal nipple. These grooves are approximately the same size. By the time the neonates are 17 days old, they have also developed larger barbs (macrobarbs) which in cross section are triangular, hollow, and can be best described as Barker did, as appearing like small shark's teeth. These are directed posteriorly, occur in horizontal and lateral rows, and probably represent a major factor in the attachment of the young to the nipple of the mother. They appear to be particularly concentrated near the tip of the tongue on the upper surface.

There is some variability in the amount of surface covered by the microbarbs and larger keratinized macrobarbs, but these in combination with distinct convolutions of the mouth cavity of the older pouch young produce a very secure attachment. These particularly complex modifications of the squamous epithelium in the oral cavity and on the nipple of the mother represent highly specialized adaptations to facilitate the protection and growth of the marsupial young outside of the body cavity. Neonate marsupials suck with a pumping action of the tongue, with the jaws held motionless. All the required musculature is present for this pump-sucking nourishment at birth. Müller (1968a,b; 1973) and Lillegraven (1975) pointed out that the wide gape of the neonate's mouth is closed at the corners by secretions to reduce the size of the opening, facilitate secure attachment to the nipple, and hold the jaws motionless. Both discussed the evolution of the marsupials and placentals in some detail.

For the first 60 days of the embryon's life, it is attached to the nipple which gradually elongates to compensate for the increased size of the fetus. When the young are 50 days old, the nipples exceed 20 mm in length and the embryons will entwine the elongated nipples while still attached. It is not until they are 60 days old that the young can leave the pouch, although some are still attached. They are occasionally seen nursing outside of the pouch while the mother sleeps. About this time, the eyes and mouth open so the young can detach and reattach themselves to the nipple, and nursing becomes intermittent. The earliest time an embryon can be removed from the nipple and reattach is 29 days of age. Cadzow (1974) was able to get two embryons to reattach that were 29 days old. They required about an hour to reattach to the nipple after being replaced directly above the teat that they were removed from.

At about 70 days of age, significant changes occur in the behavioral and physical maturation of the pouch young. Langworthy (1925) stated that prior to this time, the young were unable to support themselves and move about, but that at 70 days they could move to the extent that they could leave and reenter the mother's pouch. This corresponds to our observations and Reynolds' (1952). Motor control over the eyes and mouth also occurs simultaneously with this physical ability so that the young can be observed to begin directed movements, to detach and reattach to the nipple and leave the pouch. If the female leaves the young at this time, they begin their screeching calls which will sometimes bring the female back to join them. This vocalization occurs before the mouth can be opened. Langworthy (1925) reported it in opossums at 41 days of age and Reynolds (1952) reported it to occur at 37 days. This is an interesting case whereby the behavioral capabilities of vocalization occur long before the physical capabilities are present to utilize this behavior. It definitely serves as an orientation call for the female and presumably for the other siblings.

As the emergence of the young from the pouch begins, the female's behavior becomes more oriented toward the care of the young. Previous to this time, when the young were firmly attached to the nipples, pouch sanitation was her prime concern. Little or no effort was expended to pay individual attention to the young themselves. However, after the young begin to enter and leave the pouch and clamber around her, she begins a clicking vocalization to assist the young to identify her and definitely orients toward the young when they cry for assistance or care.

During the time that the young opossums are leaving the pouch, they become increasingly responsive to the clicking of the female. This appears to be more a learned association with the female than an instinctive response. It has definite survival value in that the accidental approach by the young to a mature male opossum could lead to the death and probable cannibalization of the youngster. McManus (1970) reported that interactions of adult male opossums with the pouch young males, which were approximately 100 days of age, varied from casual investigation to aggressive following of the young. Reactions of the young opossums to the approach of either a male or female was one of fear. Behavioral patterns of the young involved hissing, growling, or rapid retreat depending upon the circumstances. McManus reported that a female's actions toward the pouch young were of immediate interest by the adult, even if the young were not her own. All the females tested attempted to sniff and nuzzle the anal regions of the young opossums, and one female out of five began the clicking vocalizations.

The behavioral and physical development during the period of weaning is very rapid and marked from 100 to 110 days. By this time the young opossums will not follow the mother when she leaves the nest. They are becoming more and more independent of maternal influences and sibling activities, and begin the process which will result in the independent and solitary life that follows. This is a critical time for the young because if they leave the nest too early they will not have the physical coordination and muscular strength necessary to seek food and escape predators. On the other hand, evidence suggests that to prolong the family group for more than 150 days reduces the probability of a successful existence. Reynolds (1952) and other authors have observed that captive family groups, when maintained in confined cages for more than 150 days, tend to increase mortality and decrease the probability of living by a significant level. Increased mortality results from aggression, infection due to lesions from fighting, stress, and presumably other factors. Although aggression is rarely observed in siblings, when one animal becomes incapacitated or dies, it is consumed by the rest of the group. Cannibalism has been mentioned by several authors and observed in our laboratories. Even when there is an adequate supply of food, the adult and young opossums will eat their deceased cage mates.

There is a difference in the range of maternal behavior exhibited by the didelphids toward their young. *Didelphis* will not retrieve very young embryons when they become detached from the nipples but *Marmosa* will.

Various authors have reported *Marmosa* does not retrieve, but in general after the young have nursed for a short time, the female will actively retrieve them. The nonretrieval by some females could be due to a variety of reasons, one of the most probable is that the mother is under stress in the observational situation. It also could be due to the individual variation in the female maternal behavior. Second litter mothers could exhibit more retrieval drive than first litter mothers, for example. There appears to be a period of maturation that is necessary for this behavior to be activated. In *Didelphis*, the young are about 70 days old before they detach from the nipples and crawl around in the cage or den area. At this time, the mother will respond to their distress cries and return to them. This is in marked contrast to the newborn that fall to the ground at birth. The mother will not render any aid at this time and as a result, approximately 35% of the newborn never attach to the nipples and are lost. It is easy to see how, in nature, this retrieval behavior would not be too functional at birth since the fallen embryons would be almost impossible to locate in the nesting material. One would theoretically expect more attention and care for the embryons as they are being born and are searching for a nipple, rather than for the ones that fall out of the vagina or pouch before they can make sufficient noise to be heard by the mother. Arboreal species would certainly be more susceptible to loss of the fallen young than those in close association to the ground.

*Didelphis* exhibits little maternal behavior until the young are detached from the teats and are almost self-sufficient. In numerous attempts to get *Didelphis* mothers to retrieve their young, none were successful. The distress calls of the young are plainly discernible after the fifty-fifth day, about when the laryngeal development is complete. The distress call can be a reaction to tactile response, internal pain, hunger, or other stress. The mother does not even investigate the pouch area when the entire group of embryons are uttering their stress calls. Since we (in Hunsaker's lab) often disattached and otherwise molested the young, it could be a unique condition of our lab animals that they were not too responsive to the young. Cadzow (1974) observed the same lack of maternal response in the colony at Long Beach, California. Most of the other authors who have worked with embryons and young opossums have failed to mention any retrieval behavior or active responses by the female to the vocalizations of the young animals. In general, *Didelphis* mothers will accept young from other litters when transferred into their pouches. They also rest comfortably while their young are crying piteously and will walk over a fallen embryon on the way to a food dish with no more than a shameless glance. It is interesting to note that the well-developed self-grooming of the young at early stages, active nipple seeking and reattachment behavior, and staying in close proximity to the mother tend to compensate for the lack of maternal behavior.

*Marmosa* will retrieve the young after a few days of suckling. The mother will pick them up, as with any object they wish to investigate, with both front feet, and place them on the pouch area. This is done while in a sitting position. The embryon will then climb to the fur and begin the characteristic lateral and dorsoventral head movements to find the teat and attach. There appears to be a period of time from first attachment to about 10 days when it is difficult for *Marmosa* young to reattach.

Thrasher *et al.* (1971) tested the retrieval behavior of five female *Marmosa robinsoni* with litters between 22 and 28 days of age. They removed a young opossum, put it in a separate mouse pan which contained sawdust, and introduced the mother. She would approach the young, pick it up with her forefeet, and place it on the pouch area. It would cling to the fur and by both grasping and biting the hair move to a nipple and reattach.

One female would retrieve brown lab mice pups ($DBA_2/C_2gl^6$) but she would not retrieve white mice pups (MM:BRVS SW). Four other females would retrieve only their own young, even when they were mixed with baby mice. All would retrieve the young of other *Marmosa.*

*Marmosa robinsoni* are actively climbing over the mother at 40 days of age and are weaned by 70 days. In the laboratory, there is increased mortality if the family groups are maintained for periods longer than 70 days because of the aggressive nature of the species.

The conclusions of these observations of *Marmosa robinsoni* support Beach's (1939) study on the maternal behavior of *Marmosa cinerea.* His observations were made on a single female with pouch young. She would actively retrieve her babies when they were removed and would also retrieve the young of laboratory mice. If the one specimen is representative of *M. cinerea,* then that species is less selective than *M. robinsoni* in retrieval of other species young since the latter would retrieve only certain strains of mice. However, in all probability these results reflect as much individual variation as the variation between species. It would appear, therefore, that the genus does exhibit a higher level of development of maternal behavior than *Didelphis.*

Merchant and Sharman (1965) were able to transfer the neonates of several species of macropods into the pouches of other species. In general, there were no problems with the physical attachment as long as the female was in the correct stage of lactation and reproductive cycles. One swamp wallaby young showed accelerated growth when raised in the pouch of a foster red kangaroo female. There were indications that the adult females would respond to the calls of their new charges even if the species vocalizations were different.

These observations suggest that maternal care in the marsupials is not too specific and that learning by both the mother and young is important in their interactions. The self-sufficient behavior of the young offsets the lack of a high level of maternal behavior directed to the neonates.

### E. Nest Building

Various authors have described the nest building behavior of *Didelphis* (Pray, 1921; Smith, 1941; Reynolds, 1945; Layne, 1951; McManus, 1970). Smith (1941) described a field observation of a young animal carrying bundles of leaves with his tail into a nesting hole in the ground. He was less than 15 feet away and observed it carrying four loads in 10 min, each load consisting of six to eight mouthfuls of material. Reynolds (1945) recorded eight den boxes being filled with leaves in one night by five *Didelphis virginiana*.

An opossum will take a small amount of leaves or other material in its mouth and deposit it between its forelegs. While supporting the body on the hind feet, the material is grasped by the forefeet and passed back to the hind legs. Shifting the support of the body to the forelegs and tail, the animal moves the material with the hind legs back to the tail area and pushes the litter into the coiled tail. By this time, another mouthful of litter has been passed to the forelegs, and the process continues until up to 10 mouthfuls have been accumulated. The animal works rapidly and when the tail is full to carrying capacity, the animal proceeds to the nest.

Most of the tail is involved in the coil, and the bundle of material is carried with the base of the tail extended parallel to the ground. This position apparently reduces the amount of material lost due to friction with the ground. Layne (1951) observed a young male with half his tail lost by self-mutilation that, although "somewhat handicapped," managed to carry litter consisting of dead leaves, feathers, and other material.

Once the material is brought to the nest site, the process of nest building is carried out with the mouth and forefeet. The litter is desposited in the nestbox or container, picked up with the mouth, and deposited on the floor between the forefeet. The material is then moved into position by the forefeet. During this phase of litter movement, the tail remains tightly coiled. The size of the nest is a function of the volume of the cavity to be filled. One opossum brought in a bushel basket full of leaves to make a nest under the hood of a car (McManus, 1970), but usually a cubic foot of material will be adequate in a nesting box. If the nest is constructed on an open surface, the animal will add the material to its nest while pivoting in the center. This results in a circular nest which is slightly hollowed in the middle.

In the northern limits of its range, nest building principally functions to protect *Didelphis virginiana* from cold weather. The nest also affords the more tropical species shelter from rain, predators, and cold. Nests are probably important to the maintenance of body heat of the mother, enabling her to warm the young animals before they can effectively thermoregulate. McManus (1970) pointed out that in particularly cold weather his captive animals were seen to construct large nests that were covered on top with litter, and were hollow balls of straw that measured up to 25 cm high and 35 cm thick.

Nests appear to be constructed by most of the other didelphids as temporary sleeping quarters and as a place in which to raise their young. The temporary ones are fashioned out of leaves and other material after a night's hunting is completed. The most permanent nests are made by *Chironectes* which digs a hole in vertical stream banks. The hole is from 100 to 150 mm in diameter with the nesting cavity located close to the waterline. A nest excavated by Zetek (1930) had a 45° tunnel leading downward from a 4-inch hole above the waterline of a small stream. The hollowed-out cavity at the end of the 60 cm tunnel was below the waterline and was filled with water. Beebe (1923) described the nesting tunnel of *Chironectes* to be 6 inches in diameter with the tunnel either level or rising slightly as it went into the bank. Well-worn trails to the nesting site suggested they were used for considerable periods of time.

*Lutreolina crassicaudata* also occupies burrows near streams, but usually they have been dug by other animals. It has been reported to use armadillo and viscacha burrows in the Argentine pampas and tree holes in forests. It also constructs globular nests of leaves in wet areas.

*Marmosa* either builds nests for shelter, occupies abandoned bird nests, or takes up residence in a sequestered tree hole. Apparently they move from nest to nest in their nomadic life. Enders (1935) used nest boxes in his studies on Barro Colorado Island and found *Marmosa* would bring nest materials to the boxes. One box was used consecutively by four separate animals. His boxes were wooden and located about 2 to 3 m above the ground. Simon Daza in Baranquilla, Colombia constructed 1-m-high cement shelters on the ground, but was unable to entice *Marmosa* to use them. *Caluromys* and most of the other genera appear to use leaves for nesting material, either on the ground or in trees. All will probably inhabit other species' nests if unoccupied.

Tail coiling has been observed to facilitate carrying nesting material in *Didelphis virginiana, Monodelphis domestica, Caluromys derbianus,* and *Marmosa robinsoni*. The fact that more species have not been observed to do so is probably a function of inadequate observations. *Monodelphis* has a tail which, shorter than the body, is not very prehensile, yet they have been reported carrying paper with their tails, curling it downward rather than laterally (Walker, 1964). This behavior is probably a common trait within the family, but more observations should be recorded.

The tight coiling of the tail also occurs in *Didelphis marsupialis, Philander opossum, Marmosa,* and *Caluromys* as a function of stress. It occurs when an animal is frightened and can be considered a submissive posture. One of the authors (Hunsaker) has observed this behavior in *Didelphis marsupialis* of Central America, but not in *D. virginiana*. The latter species is not as aggressive as *D. marsupialis*, so this may be a reflection of their difference.

An interesting aspect of this nest material transporting behavior is that *Didelphis* seldom coils its tail unless climbing, but will immediately coil it once he orients toward, or grasps material in his mouth. Smith (1941) noted that the tail coils as the opossum picks up the first mouthful. Layne (1951) mentioned that a young male opossum he had observed would stop to feed or drink while carrying litter with the tail, but kept the tail tightly coiled. This suggests that this directed behavior is a complex combination of patterns and responses. Shupe (1967) has filmed tail carrying of nesting material in tamed *Didelphis virginiana*. Bright movie lights did not disrupt the ongoing carrying behavior, which often persisted for several hours between midnight and 3 AM.

### F. Vocalization

In general, the didelphids are primarily visually and olfactorily oriented, but vocalizations do play an important role in their lives. As is the case in many vertebrates, their vocalizations are utilized primarily in their intraspecific behavior. Several authors have described some vocalizations of *Didelphis virginiana*, including Hartman (1923), Reynolds (1952, 1953), McManus (1967, 1970), and others. From our observations in the laboratory and in the field, as well as other authors' descriptions of the noises that opossums make, it appears that the entire functional vocal repertoire of mature *Didelphis virginiana* consists of four distinct vocalizations. One of these, a lip clicking noise, is utilized in a variety of behavioral situations, while the other three, a hiss, a growl, and a screech, are utilized in agonistic interactions.

When the animals are nervous, they will sometimes quiver their lower jaw and their teeth will make contact, producing a slow tooth chattering sound. We have not observed any social significance or response to this sound.

Doutt (1954) referred to a purring noise that *Didelphis* made while swimming. Although they may have this distinct vocalization, other authors including ourselves have not observed this. In all probability, the purring is a growl made while the mouth is closed. McManus (1970) heard a growling type vocalization while one of his animals was swimming.

The clicking noises are uttered by males, females, and young opossums. It was originally described as a metallic click by Hartman (1923); it appears to be made by the opossum elevating the posterior portion of the upper lips and, with a sudden relaxation of the muscles involved, allowing the lips to smack down in place. The resulting noise is a very definite click that is audible several meters away. Apparently only the posterior portions of the lips are involved since the anterior part of the mouth remains closed and only the cheek teeth are exposed in the process of sound production.

Clicking noises are well defined and are spaced at regular intervals, 0.4 sec apart. It serves several functions for the Virginia opossum. It is utilized while the animals are investigating, and probably serves to announce the presence of a foraging opossum to other members of its species. In addition to the utilization of this vocalization during foraging behavior, it is an important part of courtship. Most of the investigators who have done research on the behavior of opossums (Hartman, Reynolds, and McManus) have described the utilization of this behavior by the male in the courtship sequence. During male/female interactions this vocalization is not used by the female. When a male encounters a female, he immediately begins a courtship behavior pattern. It is during this phase of the interaction that the male utilizes his clicking vocalizations. The clicking continues as he explores her genital region and tries to mount. Neither the authors nor other investigators have ever observed a female to utilize this vocalization during the courtship period. Usually the clicking stops as the male mounts and copulation occurs. Apparently the clicking noise serves to identify the male as a courting animal rather than as an aggressive one and can thereby serve to reduce the aggressive responses of the female.

The female does utilize the clicking noise when she has pouch young. As the youngsters begin to reach the age where they leave the pouch, at about 90 to 100 days, the female will utter clicks. In some instances the young do not appear to pay any attention to the female's vocalizations, but we think that this is probably due to the young being in a stable situation in which they are not fearful of loosing contact with the mother. McManus (1970) pointed out that the incidence of clicking by females increases during the periods of time when the young wander away from the mother for brief periods. When we have observed young opossums in a stressful situation, they will orient to the female when she utters the clicking noise, especially if she is out of sight.

A hissing noise is uttered by *Didelphis* and by most other members of the family during agonistic encounters. It is a low amplitude sound that lasts only for 1 or 2 sec. The hiss represents a very low level aggressive threat or is associated with the defensive postures of a threatened animal. It appears to be a startle reaction and in some individuals is a very common vocalization, especially in wide caught animals that are in laboratory situations. If the encounter continues to increase in severity, the hiss is shifted into a growl. The growling noise of opossums is probably the most common vocalization that one hears in captive animals. This medium level threat behavior pattern is used as an aggressive threat. It is basically a low amplitude sound which has the main components between 0.1 and 1.5 and between 2 and 4 kHz. The growls are of relatively short duration, lasting from 1 to 2 sec or slightly longer. Figure 2 shows a spectrographic analysis of the sound, and as can be seen, it appears to be white sound from below 1 to 4 kHz with some components up to 8 kHz. The growl of an adult opossum is very similar to that of the young, but the young animals do not appear to

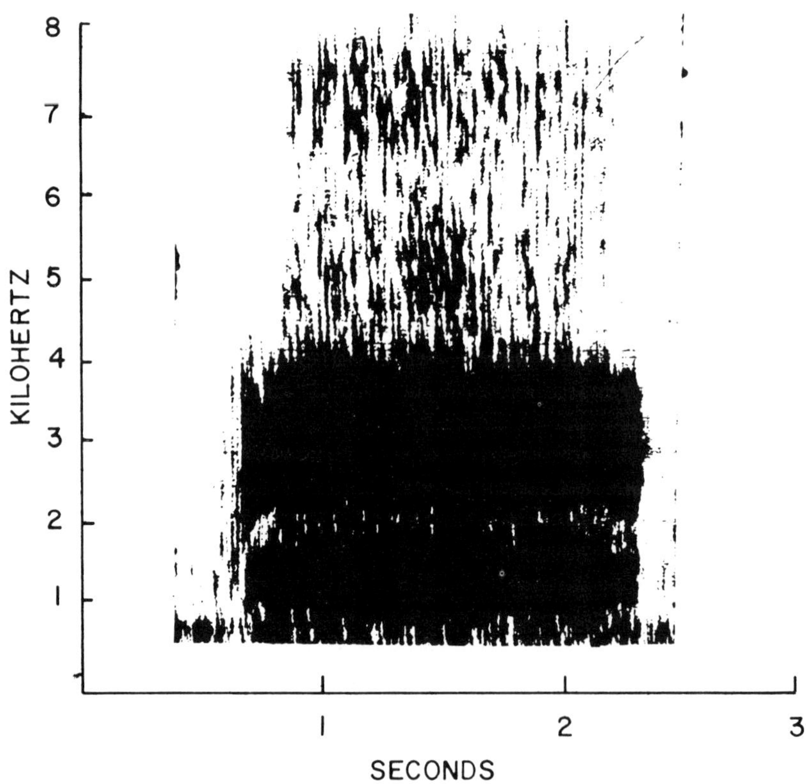


Fig. 2. Sound spectrogram of the growl vocalization of an adult Virginia opossum, *Didelphis virginiana*. From McManus (1970).

produce as much evenly distributed white noise as the adults do. As would be expected, the amplitude of the smaller animals is somewhat less than the mature, large-sized adults.

Under high levels of stress and aggression, opossums utter a screeching noise which is basically a tower of nonstructured white noise ranging from less than 1 to above 16 kHz. Figure 3 illustrates the frequency components of an intermediate growl–screech and a full screech of an adult Virginia opossum. Screeching vocalization is the highest level of aggression that an animal can communicate through vocalizations. The screeching vocalizations of two battling males afford the listener with an impressive array of noises. At this intensity of activity, the males are usually fighting, slashing, and in a full aggressive engagement. Figure 4 shows the tongue in a characteristic raised configuration during growling and screeching. Screeching noises are also utilized by young animals, but again as with the growl, there is not as much white noise involvement and the call is somewhat more structured with dominant frequencies in 5 to 5 kHz range with harmonic bands at 12, 18, and 25 kHz (Fig. 5). The screeching noise serves to alert other community members of the encounter. The screeching can be elicited in some individuals when roughly handled or grabbed by a predator, so there apparently is some selective advantage provided by the alarm reaction for the population. In this interpretation then, the screech also functions to warn conspecifics of emminent danger.

Vocalizations of other genera of didelphids are similar to *Didelphis* but observations are scanty. *Caluromys* has a very distinctive agonistic vocalization, a squealing noise which has a dominant frequency of 4 to 6 kHz which lasts approximately 0.5 sec (Fig. 6). This vocalization is used by both adult males, females, and young animals. The structure of the call differs significantly from the four types observed in *Didelphis*. Spectrographic analyses are not available for other species at this time.

Thrasher *et al.* (1971) observed some interesting vocalizations of pouch young during his work on the retrieval behavior of *Marmosa robinsoni*. They found that the females of this species are apparently oriented to the high-pitched clicking or squeeking sound of the newborn young. A female, when separated from the young, would retrieve them and could apparently hear their vocalizations up to 2 m away. When close enough to visually observe the young animals, she would also retrieve newborn laboratory mice mixed in with her own young. This is an interesting observation since newborn mice have been observed to emit high frequency calls to which their mothers orient. There is a possibility that newborn *Marmosa* utilize ultrasonic vocalizations to attract the attention of the female.

McManus (1970) recorded the characteristic screeches of young *Didelphis* while they were still in the pouch and pointed out that these vocaliza-

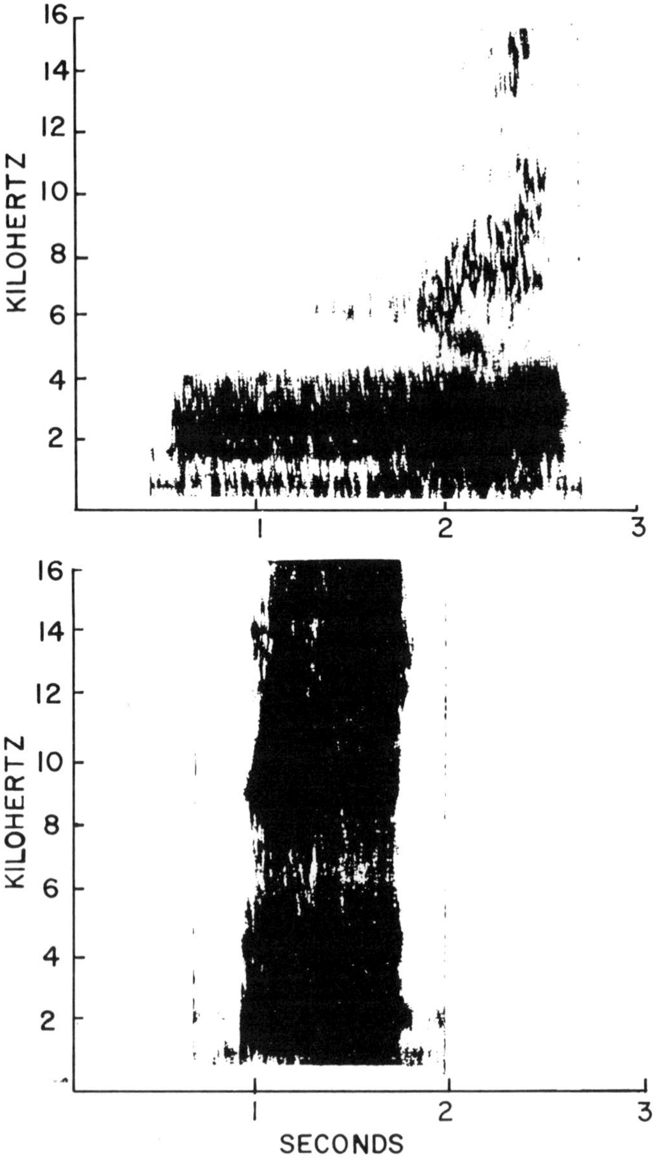


Fig. 3. Sound spectrograms of a transitional vocalization of a growl to a screech (upper section) and of a typical screech vocalization (lower section) of an adult Virginia opossum, *Didelphis virginiana*. From McManus (1970).

Fig. 4. Typical aggressive posture of an adult Virginia opossum during a screech vocalization. Note the characteristic inverted V shape of the tongue during this vocalization.

tions had components of 26 kHz with some as high as 36 kHz. This suggests that ultrasound is produced by young didelphids during periods of stress. Noirot (1966, 1972) and others have recorded ultrasounds in several species of immature rodents. These ultrasonic vocalizations have been reported to be as high as 80 kHz with a call rate of 250 per minute. *Mus musculus domesticus*, *Microtus arvalis*, *Apodemus flavicollis*, and other baby rodents utilize these sounds under stressful conditions (when they are cold, deprived of food, or removed from their mother). Since both *Marmosa robinsoni* and *Marmosa cinerea* females have been observed to orient toward the sounds of their young, the fact that vocalizations by the young are used for maternal detection has been established. This possible use of ultrasound, however, waits to be tested.

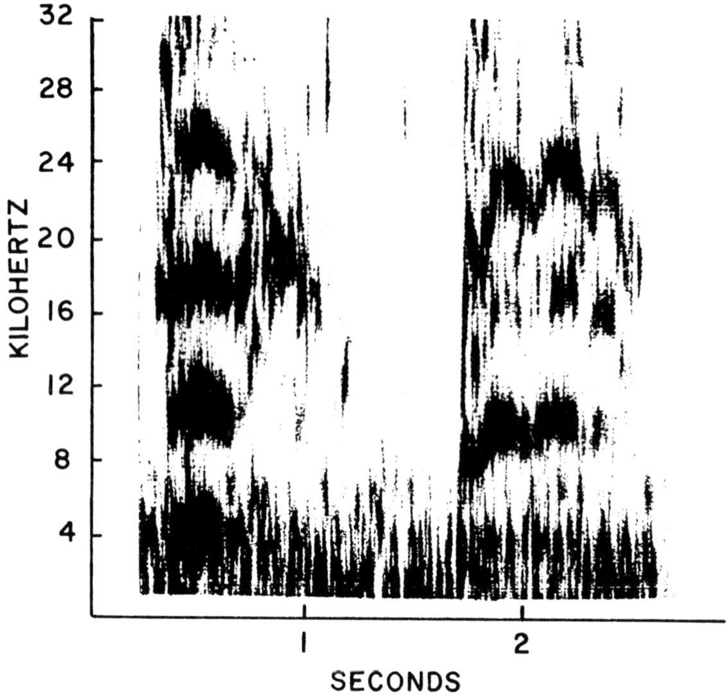


Fig. 5. Sound spectrogram of the screech vocalization of a young Virginia opossum. From McManus (1970).

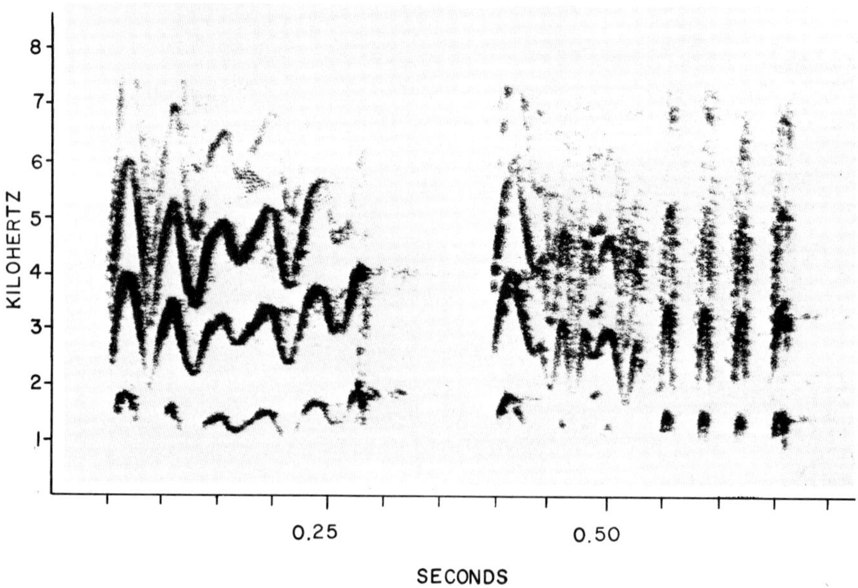


Fig. 6. Sound spectrogram of an agonistic vocalization of a female woolly opossum, *Caluromys derbianus*.

*Chironectes minimus* has been observed to produce another sound that is not utilized by *Didelphis, Marmosa,* or *Caluromys*; it is a loud, snapping sound made by rapid closing of the jaws during defensive or aggressive behavior. This animal commonly snaps its jaws together making a very audible sound. It may be a component of the threat behavior of opening and closing of the jaws. *Dromiciops australis* has a distinctive two-part vocalization, a chirring sound with a terminal cough.

### G. Evolution of Social Behavior

Raven and Gregory (1946) suggested that the earliest marsupials were probably descendants of the Phalangeridae. They considered the musk kangaroo (*Hypsiprymnodon moschatus*) to be the closest living approximation of a primitive marsupial. The musk kangaroos are diurnal, insectivorous animals and are restricted to distribution in the Queensland, Australia tropical forests. In all probability, the earliest ancestors of the family Didelphidae were also tropical forest dwellers. These animals were small to medium size omnivores which evolved in a tropical habitat. Increased sociality can usually be correlated with several ecological factors. These adaptations have been discussed by Eisenberg (1966), Fisler (1969), Crook (1970), and Estes (1974). In general, the more social animals are those which live in an open habitat, which increases their capability to develop complex visual and auditory communication systems. This enhances mutual dependency of the individuals in the protection from predators. Some of the most social groups of animals, for example, the bovids, are typically large sized, diurnal, and utilize food types which are more or less localized.

It is reasonable to assume that the more primitive mammals were smaller forms which tended to serve as a source of food for the more dominant reptilian faunas. Most of the small mammalian species that are living today are of a more solitary nature rather than a social one. It is easy to picture the early marsupials as small forest-dwelling animals that lived in isolated family groups which gradually disassociated as the young matured and became sexually independent. As long as individuals in the order Marsupialia remained small, nocturnal, and omnivorous animals, we would not expect that they would form complex social groups. It was not until the macropods began evolving into large-sized diurnal animals that interdependent group living began to occur. The individuals in these groups could rely on each other for mutual protection from predators, had easy access to mates, benefited in raising their young in a group, and would not overutilize their food resources. At this time they began to associate in semipermanent or permanent groups.

The variability in the ecological adaptations of the Didelphidae are extensive. It includes the aquatic habits of *Chironectes*, the partially aquatic *Lutreolina*, and the more arid adaptations of *Marmosa robinsoni* and some of the *Marmosa elegans* group. Although we have species which are well differentiated into arboreal, terrestrial, and aquatic adaptations, there have been no major evolutionary patterns of the specialization of social behavior in conjunction with these adaptations.

Early investigators such as Bensley (1903, 1906), Gregory (1910, 1920, 1921), and Winge (1923) considered *Marmosa* as the most primitive living genus of the Didelphidae. Tate (1933) suggested that the primitive didelphoid ancestor was an arboreal animal of moderate size with a well-developed pouch. The mammae were abdominal and the tail was prehensile. Early evolution of the ancestoral stock produced two main groups. One included an unspecialized didelphoid ancestor, the mustelid-like *Lutreolina*, and the aquatic *Chironectes*. The other major line was for the arboreal group of *Philander, Marmosa,* and *Monodelphis* which, in general, had reductions in size and weight coupled with the loss of the pouch. Some of these have remained terrestrial, while others have developed a seasonal incrassation of the tail, and some became terrestrial. Since there was no general tendency for an increase of size beyond the relatively small *Didelphis*, especially when compared to the evolutionary patterns of the more social species of mammals, the group has remained basically a solitary, tropical-adapted family that behaviorally is probably close to the ancestral stocks.

Kaufmann (1974) studied the social ecology of the whiptail wallaby (*Macropus parryi*) in northeastern New South Wales. In his excellent and extensive field study, he determined that the whiptails live in open group mobs which have developed one of the most complex social systems of the marsupials. These large-sized wallabies exhibit a reduced level of aggressiveness; they live in an open grassland area and actually become familiar with individuals in neighboring mobs. In contrast, the majority of the Macropodidae have remained in savanna and general forest environments. *Macropus, Onychogalea,* and *Lagorchestes* live in a grassland environment; *Petrogale* and *Peradorcas* have become adapted to living in a rockly habitat and one genus, *Caloprymnus*, is well adapted to a deserticulous existence. Kaufmann pointed out that *Macropus parryi, M. giganteus, M. fuliginosus,* and *M. rufus* have the highest levels, in general, of social development based on the size and permanence of their groups. There is a variety of group size, structure, and hierarchical organization down to *Setonix*, which is basically a solitary animal which will form loose aggregations at watering holes and particularly favorable feeding environments.

If we consider that there is a continuum of social interaction that is available to an individual from a solitary life style through territoriality, social hierarchy, and into a social flock, mob, or herd (Hunsaker, 1974), we can see that most of the marsupials have remained in the areas of solitary, territoriality, and social hierarchy. Certainly the didelphids have not evolved any of the ecological adaptations that are usually associated with the shift toward the more complex social interactions of group living.

Figure 7 represents a scale of social behavior with a representation of the approximate location of the didelphids, caenolestids, and some of the macropods. Other than some forms like *Caenolestes obscurus* and *Monodelphis brevicaudata* which reach remarkably large population levels, the social tolerance and cohesive aspects of their behavior patterns are about the same. The macropods represent a family which has evolved into a more social animal; it is also a family on which a considerable amount of field and social behavioral work has been done.

It is very difficult for an animal that is primarily nocturnal to establish any definite interrelationships on a continuous basis with other individuals. Problems associated with communication in the darkness obviously require vocalizations and exceedingly well-developed night vision or olfactory responses. Extensive vocalization also serves to attract predators and nocturnal vision and acute olfactory responses require a great deal of evolutionary selection. Most of the advanced nocturnal mammals, i.e., owl monkeys, lemurs, galagos, and other prosimians, tend to live in a semisolitary fashion with groups oriented around a family group relationship. In contrast to the nocturnal species, the diurnal lemurs have developed extensive behavioral complexities involving conspecifics.

In this respect, many of the didelphid marsupials occupy the same niche

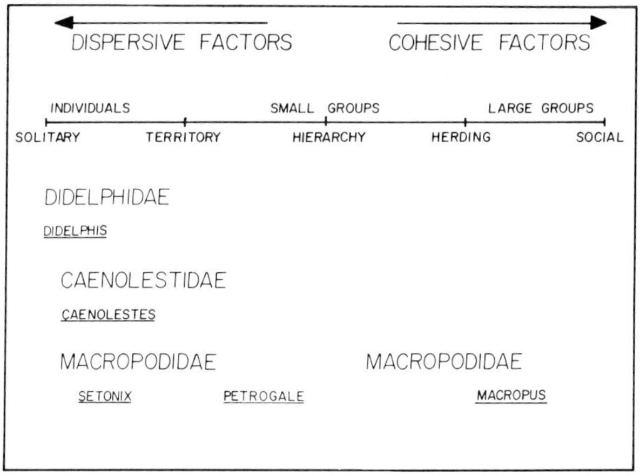


Fig. 7. Organization of social behavior of Didelphidae, Caenolestidae, and Macropodidae in relation to a continuum from solitary to social groups of mammals.

as the prosimians. Galagos, pigmy lemurs, and other small nocturnal genera represent groups which have maintained many of the ancestoral and primitive characteristics but are highly specialized in their own right. They also represent groups which have not evolved far from the ancestral social and behavioral stock which gave rise to the advanced and very successful primates. *Marmosa* and *Caluromys* are both genera which have the general appearance, activity levels, and behavioral sophistications which remind one very much of the prosimians, and they certainly occupy a similar niche. It appears that the same general isolation and selection which provided the prosimian evolution in Madagascar also occurred in the marsupial fauna of South America. In the absence of some of the early primates for competitors, marsupials began to occupy the niches that would be utilized by the prosimians in other parts of the world.

There is no indication that any of these factors which enhance the development of complex social systems with social hierarchies and intraspecies communication systems have evolved in the didelphids. All are reasonably small sized, omnivorous, and utilize food sources which are widely distributed in the tropical areas. Most have maintained the nocturnal habits which preclude complex social interactions involving mutual defense from predators, visual signal systems, etc. Another important factor which has been operative in the didelphids and has not changed through evolution is a high degree of agonistic behavior. One of the more basic factors in the development of a complex social system is that the individual distance has to be decreased with a similar reduction in the level of aggressive behavior patterns. Since all the species of didelphids that have been observed exhibit a high level of aggression, they have not changed the basic social structure of solitary territorial behavior that their ancestors presumably had.

The evolution of the social structure of a particular species is a complex thing; many factors relating to the ecology and behavioral patterns of parental stocks are involved. The didelphid marsupials can be characterized as solitary animals. Their interpersonal social interactions are limited to those necessary for the continuation of the species: reproduction and care of the offspring until weaning. Beyond this, agonistic behavior is characteristic of any interpersonal relationships. Social dominance and avoidance reactions are major constituents of their behavior. Male *Didelphis virginiana* probably spend less than 20% of their life span associating with other opossums. In all probability, the majority of this 20% is represented by the 100 or so days that the animal spends in close association with its mother and siblings. The remaining time would be spent in the company of a female in estrus. In the 1095 days of a 3-year-old male, 100 days with his parent and 4 weeks during each year with a female represent only 184 days spent associating with another animal of his species. In the vast majority of other social interactions, the first and primary response is an aggressive threat of high intensity which is persistent in males and to a lesser extent in females. The females, if given an opportunity to avoid the association with another male or female, will do so. Only under the influence of a behavioral hormonal set will the females become passive and accept the presence of a male for courtship behavior. In considering agonistic behavior as a very basic factor in the social structure, we find that fighting, injuries, and cannibalism, in addition to environmental limitations of food and weather, serve to keep the adult population numbers down and the individuals spaced throughout the habitat. This serves to maintain the population numbers well within the carrying capacity. In addition to the reproductive acceptance of the proximity of another individual, females will associate with other adult females only during exceedingly cold winters when communal nesting becomes beneficial for the survival of both individuals. The aggressive behavior levels of males are so high as to preclude even this association in dens. Thus, the species has evolved a social strategy of a diffuse population of solitary animals with a reactive society, one in which the individuals come together occasionally, react to the immediate set of social stimuli, and avoid any further interaction except courtship behavior. At reduced population levels, these reactive interactions probably are of a very low intensity, involving only sexual identification, a threat, and a parting of the ways. However, there have been observations of two sets of opossum tracks in the snow that coalesce, with only one track leaving the scene and a half-eaten carcass remaining.

## III. GENERAL BEHAVIOR

### A. Death Feigning

For a subject that has attracted so much interest, "playing possum" has had very little investigation. Francq (1969, 1970) has conducted more basic research into the physiological characteristics and behavior of death feigning than anyone else. The actual functional significance and survival value has not been concretely established, although many authors have discussed them.

When a *Didelphis virginiana* is pursued by a predator and captured, death feigning, or playing possum, occurs a relatively small percent of the time. To the casual observer and probably a predator, the animal appears dead for all intents and purposes. The eyes and mouth are open and there is copius salivation. The head is bent downward and the immobile body has a general ventral flex. The digits are partly closed and tend to grasp anything in contact with them. The animal invariably lies on its side with the tail usually curved ventrally, sometimes going completely between the legs. The body position, sequence of patterns, and general behavior of the animal while in this condition are very stereotyped. It appears that the animal has definite opinions as to what a dead opossum is supposed to look like, and will assume these positions if changed by an investigator. If the opossum's eyes are closed, he opens them; if he is shifted into a position where the tail is straight and the ventral surface is on the ground, he immediately falls over to the side and recurves the tail.

Francq (1969) compared the behavioral differences between sleep and feigned death. He listed the characteristics shown in the following tabulation for both behaviors.

| Postures while asleep | Death-feigned response |
|---|---|
| 1. Eyes and mouth closed | 1. Eyes and mouth open |
| 2. Dorsal aspect upward | 2. Lateral aspect upward |
| 3. Feet tucked under | 3. Feet flat and visible, toes usually flexed |
| 4. No response to sharp sounds | 4. Ears twitch at a sharp sound |
| 5. Arouses with a start and turns head toward source of prodding or blowing of air on head | 5. No response to prodding or blowing air on head; may retract lips slightly |

Although this state may superficially resemble sleep, Norton *et al.* (1964) found the central nervous system reacted more as if the animal were awake than as if in sleep. The recordings of an electroencephalogram (EEG) showed a slight increase in fast wave potentials with a reduction in the amplitude. This slight flattening of the record is characteristic of an animal in an excited state, that is, fully alert and active. All the components of the EEG that were observed in the catatonic state were also found in the recordings of an active, alert, awake animal. They concluded, based on the activity of the central nervous system, that *Didelphis virginiana* was doing nothing more than "playing possum."

Francq (1970) did a similar study on the electrocardiograms (ECG) of active and death feigning opossums. His normal electrocardiograms were of similar value to those found by Weber (1955). In a series of 14 records from four different opossums, Francq found no significant differences between those that were alert and active and those that were death feigning. His normal animals exhibited a mean of 166 heart beats per minute. The ECG wave components were measured and consisted of a 0.12 sec QT interval, 0.48 mV QRS deflections, and 0.05 mV T wave deflection. With the same animals death feigning, he observed mean heart beats per minute to be 178,

QT interval of 0.11, QRS deflection was 0.49, and the T wave deflection was 0.05.

The EEG and ECG studies did not show any significant differences between the electroencephalogram and electrocardiogram of opossums during this and a waking, active state. The basic role and involvement of the central nervous system and heart are unknown at the present time. Francq (1970) said that it would be premature to state generalities considering cardiac activity at that time.

Observations of the death feigned response in opossums indicate that in the majority of cases, tactile stimulation and grabbing by the predators are necessary to elicit the response. Francq (1969) subjected six adult opossums that had been in captivity for a short while to a variety of visual and auditory stimuli which included yelling, hand clapping, and poking with objects toward the head for 10 min. Only one of the six specimens had a positive response for death feigning that occurred after 5 min of this treatment. More than just capturing and casual handling are necessary for a response. In the process of trapping numerous opossums over the years, Hunsaker has only rarely observed death feigning resulting in an animal that was removed from the trap and transferred to carrying cages. Francq also mentioned that no death feigning was observed in the trapping and collection of the series of 37 opossums for his studies. Several *Didelphis marsupialis* and *D. albiventris* have been observed and grabbed, even by dogs, while Hunsaker was observing them yet no death feigning was apparent. Dr. Jorge Hernandez C. has reported (personal communication) that *D. albiventris* will feign death, but it is rarely observed in the Colombian population of this species. Gardner (1973) stated he has never seen *D. marsupialis* "playing possum."

In the authors' opinion, the state is mediated by the central nervous system and interaction with the ventral ganglia. In any case, the phenomenon of death feigning is a behavioral response that is an adaptation to the predator and prey relationships of *Didelphis virginiana*. It is different from both the freezing reaction sometimes seen where a prey species becomes immobile in the presence of a predator, and the type of response elicited when an animal remains motionless when captured. Whether these differences are qualitative, and in fact have a distinct evolutionary history and physiological basis, or whether they are quantitatively different and are basically related but have different degrees of expression, is not known.

Several authors have published papers relating to immobility in prey species (Armstrong, 1942; Gilman *et al.*, 1950; Ratner and Thompson, 1960; Clark, 1962; Matthews, 1964; Hinde, 1970). These authors have pointed out a variety of other species and conditions in which immobility results as a response to fear, attack, or social threat situations. There is an obvious advantage to immobility that occurs before contact with the predator is made. This is clearly demonstrated in the freezing position assumed by young birds and mammals when an alien predator is sighted.

Many animals, such as deer, will momentarily freeze until a source of noise or the potential predator is located. The immobility coupled with camouflaging coloration can be a very effective means of avoiding a predator. The freezing response prepares the animal for flight, allows it to closely observe and analyze the environment, and makes it less conspicuous. The problem with this survival strategy in opossums is that the immobility occurs after the contact has been made. The sight of a fleeing animal will elicit a chase and attack response in most of the larger predators. Clark (1962) pointed out that predators are much more likely to attack a moving prey than one that is immobile. We have numerous observations that substantiate this fact. Many predators will not start an attack on a prey species until it moves and are especially prone to attack an animal if it is running away.

There is also an obvious advantage if a prey species becomes immobile when captured by a predator. Predators will treat a dead animal differently than a live one, and, if recovery is instantaneous, the prey will have an opportunity to escape. Armstrong (1942) pointed out that birds that become immobile in the process of capture by a predator have an advantage to escape when treated as dead by a captor. This process does not appear to work with opossums since they take some time to recover from death feigning and therefore could not escape rapidly. It could be to some advantage in that once it was apparently dead and covered with odiferous anal secretions, a predator might prefer not to eat it. This could contribute to the selection of this behavior since carnivores tend to act with reflexive actions when pursuing and capturing their prey, but tend to consume them in a more leisurely fashion after the kill. This is probably one of the most important factors in the survival benefits of "playing possum."

### B. Locomotion

Observations of opossums in an open area or climbing in trees give an observer the distinct impression that speed and agility are not distinctive characteristics of *Didelphis* locomotion. The rather methodical movements on land with the tail moving from side to side and the head bobbing up and down indicates that their movements are those of slow investigation of a varied habitat. Even when fleeing from a predator, they clumsily crash through the underbrush in undirected flight, often seeking refuge in holes under rocks and sometimes climbing trees or escaping into water. Although obviously adapted to an arboreal life with an opposable thumb and prehensile tail, climbing activities are slow and often involve falling.

Several authors have studied locomotion of *Didelphis*: Langworthy (1925); Magne de la Croix (1936); Layne and Benton (1954); and McManus (1970). Doutt (1954), Moore (1955), and McManus (1970) also observed their swimming patterns. In general, they utilize the same limb movement sequences whether they are engaged in slow speed walking or swimming patterns. A plantigrade, slow speed system of locomotion is used and a metherptic system for higher speed running. They can also utilize a pacing stroke in high speed swimming. Many animals shift to simultaneous use of their rear legs in a richochetal escape when suddenly frightened, but this escape reaction is not exhibited by the opossum. However, it is sometimes observed in the more arboreal genera, i.e., *Marmosa*, when they are frightened.

Walking speed is about 0.3 m/sec (0.7 mph) as measured by Langworthy (1925) and McManus (1970). The range of walking speeds was from 0.1 to 0.5 m/sec (0.3 to 1.0 mph). In all probability, an opossum seldom sustains a walking speed of more than 0.5 m/sec since their walk–pause investigating behavior when they are foraging precludes rapid walking for any significant distance.

In a slow moving animal there are three feet on the ground at the same time to provide a stable platform for support. In *Didelphis*, slow forward progression is maintained by moving the opposite limb sequentially in the following pattern: right forefoot, left hind foot, left forefoot, right hind foot (see Fig. 8). McManus (1970) analyzed movie films of walking *Didelphis* and stated that the trajectory of the forefoot was 2.5 cm off the ground, which is greater than the observed 5 cm trajectory of the hind foot. The digits and tarsus of both feet are flexed until the foot approaches the substrate.

During the forward progession of an opossum, the body sways from side to side and the head moves up and down. The lateral movement of the body appears to be somewhat greater than it actually is. The appearance of these movements is accentuated by the lateral movements of the tail. The tail is used as a balancing weight, alternately being shifted to the same side as that of the hind foot that is being raised. As the hind foot goes forward, the tail is lowered; as the foot begins its power stroke, the tail is raised and begins to shift toward the other side. McManus (1970) stated that the movement of the tail tip can be described as a figure 8 when viewed from the rear.

As an animal's forward speed is increased, it is possible to maintain only two feet on the ground simultaneously. Balance while running is attained in the metherptic gait on diagonally opposed feet. Thus, the right forefoot and left hind foot are on the ground while the left forefoot and right hind foot are in motion (see Fig. 8). The hind foot is raised slightly sooner than the opposing forefoot. Both the dorsoventral movements of the head and the tail are greater when *Didelphis* is running, giving the appearance of a less than graceful progression. Layne and Benton (1954) and McManus (1970) measured and stated an opossum could run at 2.0 m/sec (4.4 mph).

The typical walking foot progression sequence is followed in climbing activities. The opossum is able to climb small trees as well as large ones. On large diameter trees, the legs are extended laterally, and upward movement is slow. On smaller trees and limbs, the legs are held tighter to the body and climbing is more rapid. The tail is used for posterior purchase and contributes considerably to the stability of the animal when climbing. It is also used to wrap around small branches when *Didelphis* is foraging and grooming. Descent to the ground appears to be more difficult than going upward. Usually, Virginia opossums proceed head downward when they are leaving trees. Sometimes, especially in captivity, they will descend tail first down the trees in the enclosure or on the cage wire. Figure 9 illustrates climbing progression patterns in Virginia opossums.

*Didelphis* locomotion patterns are probably similar to the ancestral stock of marsupials since the quadrapedal plantigrade system is the most primitive. Although rare in the didelphids, bipedalism has evolved to a high level of specialization in the Macropodidae. The only time *Didelphis virginiana* is bipedal is when it stands on its hind legs and partially supports the body with the tail for short periods of time. *Marmosa*, and presumably other didelphids, are bipedal during agonistic behavior when they raise their forequarters off the ground to use the forelimbs in combat (see Fig. 10). During these encounters, several steps are sometimes taken using the hind feet. *Didelphis, Marmosa*, and *Caluromys* are also bipedal while feeding when they use their forefeet to manipulate or secure food. Both *Marmosa* and *Caluromys* use the forefeet to manipulate food and other objects to a greater extent than does *Didelphis*. In many instances the forefeet are used to hold a food item next to the ground while being torn apart by the teeth. At other times, especially with soft foods, the item is held by one or both forefeet and transported to the mouth.

Bipedalism usually evolves due to a differential length and movement of the front and hind legs. In most quadrupeds, the hind legs are longer than the forelegs. This causes the gait to become unsymmetrical at higher speeds. The forelegs are unable to equal the increased stride of the hind legs and therefore must be in contact with the substrate for a shorter period of time. This differential movement then acts as a speed-limiting handicap to the animal. At increased rates of locomotion, the center of gravity shifts toward the rear causing elevation of the anterior portion of the animal. This renders the posterior components increasingly important while decreasing the

Fig. 8. Footfall sequences and appendage movement patterns of a walking Virginia opossum (upper section) and those of a running Virginia opossum (lower section), *Didelphis virginiana.*

Fig. 9. Sequence of limb movements and tail usage of a climbing Virginia opossum, *Didelphis virginiana*.

Fig. 10. Bipedal aggressive posture of a male *Marmosa robinsoni*.

importance of the forelegs. This is probably the mechanism which has produced bipedalism in the macropod marsupials.

The obvious selective advantage of the increased manipulative ability of the forefeet was also important in the evolution of bipedalism in the arboreal primates and other vertebrates. The reptilian Iguanidae with the basilisk (*Basiliscus*) and collard lizards (*Crotaphytus*) are good examples of living reptiles which are bipedal. Of the marsupials and the placentals, the macropods and the primates are the most typically bipedal groups of living vertebrates.

The opposable big toe, the hallux, of the hind foot in the didelphids is an adaptation for an arboreal or semiarboreal life. Romer (1959) pointed out that the primitive mammals probably had divergent thumbs and big toes that were opposable to some extent. The grasping ability that this affords is of greater advantage to a scansorial or arboreal species than to a terrestrial one. This suggests that the most primitive ancestors of the New World marsupials were basically arboreal, and the Virginia opossum and other terrestrial species retained the primitive hind foot condition after becoming terrestrial. Figure 11 shows the footprint of *Didelphis virginiana*. The opposable hallux does appear to afford the species some degree of lateral balance control in addition to its grasping ability. This is probably one of the reasons it was retained, since it affords both balance and stability when foraging as well as increased ability to escape terrestrial predators by climbing trees. The prehensile tail is probably a similar characteristic that is retained from the arboreal ancestral stock. These specialized modifications are surely of survival value and greatly expand the food searching and predator avoidance capabilities of these animals. Basically, however, it remains a terrestrial animal with a very distinct capability to climb in trees.

*Didelphis* are good swimmers and will often take to water to escape. They do not tire rapidly, are able to float for a period of time to rest, and can keep their nose and eyes out of the water with ease. They will sometimes dive completely under the water and can stay submerged for several minutes.

Doutt (1954), Moore (1955), McManus (1970), and Hunsaker (unpublished) have observed the swimming techniques of *Didelphis*. When swimming slowly, they use alternate, contralateral leg movements. The opossums

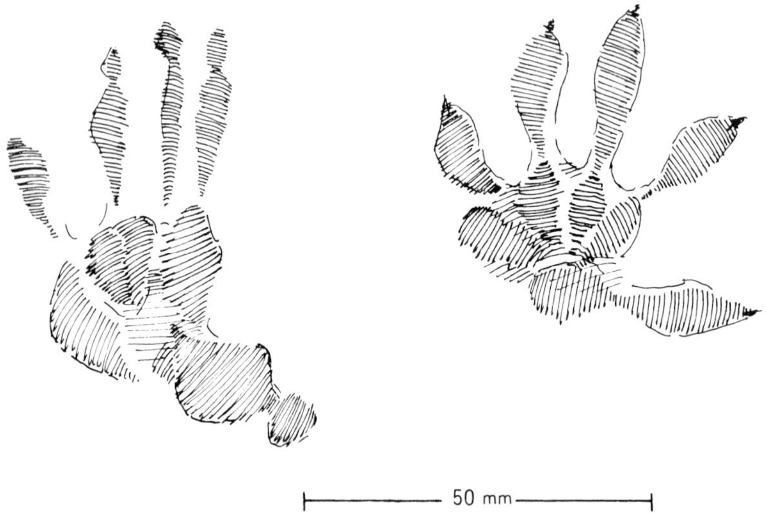


Fig. 11. Footprint patterns of left hind foot (left side) and left forefoot (right side) of a Virginia opossum, *Didelphis virginiana*.

Fig. 12. The swimming posture of *Chironectes minimus*. Note the extended forelegs which do not appear to be used in the power strokes, but are used to search for food items. The New York Zoological Society photo.

appear as though they are just walking underwater when they are swimming slowly. When exerting more energy for faster swimming, they use a pacing stroke, one in which both legs on one side are used at the same time. The toes are flexed on the upstroke and spread on the downward or power movement. At the completion of the downward stroke of one pair of legs, the contralateral ones begin theirs. The sculling motion of the tail probably provides more lateral control than forward thrust. It moves laterally to the same side on which the legs are in the backward stroke position. *Marmosa* utilize the same swimming strokes as *Didelphis*. Hunsaker has observed them swimming and found them to be able to swim above and below the water, but they seemed to tire easily and were not too agile while swimming.

The most curious swimming behavior of any marsupial is that of *Chironectes*. Unique among the swimming patterns of semiaquatic animals, observations by people who have maintained them in captivity indicate that they swim by alternate strokes of the webbed hind feet while keeping their forefeet extended in front of them (Fig. 12). The majority of other aquatic and semiaquatic quadrupeds either fold the forefeet posteriorly to their sides or use them for power strokes while swimming. This curious habit of swimming probably evolved as an adaptation to carrying food items in their forefeet rather than in the jaws, as well as feeling for prey along the stream beds as they swam. The forefeet of *Chironectes* are reputably well innervated and sensitive to tactile stimulation. Piles of crustacean shells are often observed on stream banks where yapoks (*Chironectes*) have taken them to eat after holding them in their forepaws while swimming.

In general, the didelphids have maintained the primitive plantigrade gait and utilize methods other than speed to avoid predators. A slow, methodical foraging pattern, freezing and immobility when a predator is sensed, and a mouthful of teeth have been sufficient adaptations to keep the species viable. The arboreal species utilize secrecy, agility, and speed to escape predators.

### C. Activity Periods

The activity rhythms of the didelphids are, like most of their other behavior patterns, generalized and similar to other nocturnal and crepuscular vertebrates. Unless they are hungry or disturbed, they do not have a significant amount of activity during daylight hours but become active soon after sunset and continue activity intermittently throughout the night until just before sunrise. From observations in laboratory colonies, trapping data, and other investigations, the didelphids appear to have either a bimodal or broadly unimodal periodicity in their diurnal activity pattern. In the laboratory they become active in the early hours of the evening or shortly after the lights are turned off. They remain active for approximately 3 or 4 hr, after which there is a period of rest with activity resuming in the early morning hours. Light intensities higher than about 0.1 lux definitely inhibit their activities. Indications are that it is the inhibitory effect of bright lights, rather than a circardian rhythmicity, that maintains their nocturnal activity cycle. Variation in the activity levels of didelphids are probably caused by either environmental factors (temperature) or physiological factors involved in mate seeking during the reproductive season, increased caloric needs during winter, and when raising the young.

Observations in the laboratory and trapping data both suggest there is an increase in activity periods during the reproductive season. Enders (1935) and other authors (Tate, 1933; Fleming, 1972) indicate that *Marmosa, Philander,* and *Caluromys* are much more commonly trapped during the reproductive season than at other times. Enders (1935) stated that as many *Marmosa* were trapped in Panama during the first 2 weeks of the breeding season as the number trapped during the remainder of the year.

In Hunsaker's lab, activity levels of *Caluromys derbianus* and *Marmosa robinsoni* were monitored at various times by the author, J. Boggs, J. Lambert, and M. Brimm. We were not only interested in the activity rhythms per se, but also in their ability to shift their activity periods to facilitate behavioral observations. It is very easy to shift the activity cycle of these animals by reversing the period of illumination. Bright light is especially functional in causing an almost immediate shift in activity periods in both *Marmosa* and *Caluromys*. Under reversed illumination they appear to be reasonably well adjusted (behaviorally) within 48 hr and within a week their sleep cycles, food consumption, and general behavior indicate that they are under little or no stress due to the reversal.

*Didelphis* in Shupe's laboratory were monitored with contact microphones and recorders to determine their activity period. The animals were active from about 2200 to 0400 hours with a mean activity period of 6 hr. Most investigators have noted maximum activity at about 2000 to 2100 hours. This corresponds well to McManus' (1971) observations on the activity periods of the Virginia opossum. He observed activity from 1700 to 0700 hours eastern standard time in a group of 23 opossums. Activity levels were higher in spring and summer. The activity curve flattened and became slightly bimodal in his fall observations. Peaks of activity during this period were at 2200 and 0400 hours. Stationary postures represented about 57% of the total activity; locomotion was 20%; comfort activities, including grooming and yawning, were about 7%; ingestive, 10%; and nest building was 3%. In Hunsaker's laboratory, when *Caluromys derbianus* was put on an illumination cycle of 13.5 hr light/10.5 hr darkness, we found that the animals spent 86% of their time inactive and 14% of their time active. These observations of *Didelphis* and *Caluromys* both correspond well with those of Van Twyver and Allison (1970), who reported that *Didelphis* slept 85% of the time. Using white rat activity cages with digital counters recording the activity wheel revolutions, we found that *Caluromys* had an activity period similar to *Marmosa*, with the greatest period of activity just after the lights were extinguished and just before they were turned back on. The greatest activity levels occurred 2 to 3 hr after darkness and 1½ to 2 hours before light. The time spent in the first hours of activity was primarily oriented towards cage exploration, grooming, searching for food, feeding, and drinking water.

Food deprivation periods lasting for 24 to 48 hr caused the *Caluromys* to double their activity levels in our lab, but they were active only during the periods of the darkness cycle. As the hunger levels we utilized in the experiments did not overcome the inhibition of the illumination, the animals were not active during the complete 24-hr period. On particularly cold nights, *Didelphis* will not forage, but this behavior is often modified after 2 or 3 days of cold weather. This suggests that the 48-hr period of fasting does not stress them too much. This could possibly explain one of the reasons for increased activity during the reproductive season; the increased metabolic requirements of raising young and of maintaining the animals in a reproductive state would require a considerable increase in food intake. The obvious increase in activity while searching for a mate during the courtship and copulation phase of the reproductive cycle would also cause an increase in activity levels.

Forty-eight hours of constant illumination resulted in a 40% decrease in activity by the *Caluromys*. We considered two factors to be operable in the reduction in the levels of activity; one was the inhibitory effects of high levels of illumination and the other was the stress due to the constant illumination. Maxwell (1960) observed that many animals will increase the time they spend sleeping under stressful conditions. Although he observed that otters, foxes, house mice, and other vertebrates increased their sleep levels as a function of stress, it appeared to us that the major factor in the reduction of activity was light levels rather than stressful conditions.

A rough estimate of the distances that *Didelphis* or *Caluromys* travel during these periods of activity can be arrived at by considering the number of revolutions and the circumference of the activity wheel. The circumference of our apparatus was 127 cm from which we could estimate the linear distance traveled to be 1200 m when the wheel was turned 1000 revolutions. The average number of revolutions traveled during a 24-hr period by the four *Caluromys* observed was 1688 revolutions or 1977.6 m (6592 feet). Although this could be considered to be the average linear distance traveled under laboratory conditions, one should be extremely careful about deriving what should be considered normal as far as behavior in the field is concerned when making deductions from observations in the laboratory. However, these data are remarkably close to the 1.5 to 1.8 miles that Fitch and Sandidge (1953) observed in *Didelphis*, but was more than twice the distance traveled as recorded by Davis (1946) in recapture data by *Caluromys* in their natural habitat. One male traveled only 4.0 m in 2 days between captures. Since *Caluromys* is more arboreal than *Didelphis*, they probably travel as much vertical distance as linear distance in the trees. This distance would be recorded in our activity cages and would not be reflected in trap line data.

### D. Grooming, Exploration, and Play

Opossums are not as fastidious about the condition of their bodies as some other animals; they spend a relatively small amount of time with grooming procedures. During the 6 or so hours they are active during the night, they spend less than an hour in this activity. Observations by the authors suggest that the Virginia opossum spends less than 10% of its activity time in grooming. McManus (1971) found that captive opossums spend about 7% of their time in grooming or comfort activities. As with most other mammals, the facial area is cared for much more than the rest of the body and receives the majority of the attention. After sleeping most of the day, the animals will normally wake up and begin moving in their nest enclosures. After a few minutes of observing and testing the environment, they will begin to groom themselves. There is a good deal of individual variation in the amount of time spent grooming as well as the attention given to various parts of the body. Both males and females will spend about the same amount of time grooming the facial area, and females use significantly more time grooming the abdominal area than the males do. Mutual grooming is rare and most often only involves the female cleaning and licking the young.

On a dry day, the animals will allocate approximately 10 to 15 min to grooming themselves when they first begin their activity period. They begin by licking their forefeet, and then using the moistened palmar area, rub their feet over their face and snout. McManus (1970) observed that the forefeet are moved in a circle while grooming the face, with the lower half of the circle involving the cleaning of the feet and upper half of the circle involving the cleaning of the nose and facial area. The normal procedure is for the animal to use both feet at the same time in rubbing the facial area from the eyes to the snout, while the cleaning of the paws occurs as they are being drawn posteriorly in the lower half of the circular movement. In a typical 10-min cleaning period, the grooming of the facial area will involve approximately 40 to 50% of the total time. Facial grooming and forepaw cleaning is also frequently utilized during feeding activities. The animals clean the sides of the face, forepaws, and snout intermittently during the eating process.

Most of the grooming of the back of the head, the ears, and dorsal parts of the body are by the use of the hind feet. During the process, the hind foot with the four digits extended is drawn through the fur for 5 to 10 sec in a series of 10 to 20 repeated scratching movements. It is during this cleaning procedure that most of the hair observed in stomach analysis and fecal material is probably ingested. The hind feet are carefully cleaned by licking after each series of strokes. When the hind foot is used to clean the back of the head and ears, it is elevated dorsally and laterally over the forearm and the head is bent partially to the side to facilitate the combing action.

The abdominal area, genital, and tail regions are cleaned from either a sitting or a lying position. In the sitting position during grooming, the tail is extended backward from the body or laterally curved around while the animal licks the ventral region including the thorax and genital areas. Females typically clean the pouch thoroughly at least once or twice a day. Except when the tail has a sore on it or has been frostbitten, little attention is given to it. Grooming is more directed toward the base than the tip. If for some reason the tip has been irritated or injured, the grooming sometimes takes the form of self-cannibalization of that part. Injuries to the tail sometimes appear to receive more attention than injuries to the other parts of the body which are only casually cleaned during the grooming process.

Cadzow (1974) observed three distinct types of head grooming behavior by embryons ranging in age from 52 to 84 pouch days. The juvenile grooming acts were only observed when the embryons were under maximum maternal protection, either in the marsupium or lying on or next to the mother, and usually during the hottest part of the day. It has therefore been hypothesized that juvenile grooming may be a thermoregulatory activity. McManus (1970) described in detail adult cranial grooming which is believed closely related to embryon head grooming, one possibly being a mature form of the other. Grooming behavior in the adult opossum involves many regions of the total body. However, juvenile head grooming was the only type of grooming activity observed in embryons.

*Marmosa, Philander,* and *Caluromys* groom in much the same fashion but for longer periods of time than *Didelphis.* The facial and snout areas of both males and females receive the special attention, and the abdominal or pouch area of the females is also cleaned regularly. During rainy or particularly damp weather, all species appear to groom more often and spend not only longer periods of time, but the grooming activities are also more numerous.

The percentage of time that the South American species groom themselves more nearly approximates the somewhat more extended periods that the Macropodidae utilize. Stodart (1966), Russell (1970), and Kaufmann (1974) commented on the grooming activities of macropods. Apparently the facial cleaning with the licked forepaws occurs most commonly and has been reported in several species: the whiptail wallabys by Kaufmann (1974), bettongs by Stodart (1966), and a variety of captive macropods that Hunsaker has observed at the San Diego Zoo. This appears to be a relatively basic marsupial grooming pattern, and one which commonly occurs in most placentals. The macropods scratch themselves more with the hind feet than the didelphids do. In some, the hind feet have been modified by two toes forming a comb for grooming. In many instances the opossums, kangaroos, and wallabys scratch themselves in spontaneous fashion which appears to be a response to some localized stimulation of the skin.

After the grooming procedures are completed, the opossums set out to explore their environment. The exploratory behavior consists of slow progression through the area with frequent stops to listen and monitor the environment visually and olfactorily. The primary sensory input is by olfaction, secondarily by hearing. Visual orientation during exploration appears to be minimal, although any slight movement which catches their eye will cause them to stop and peer intently in the direction of the movement. Olfactory perception is well developed, and the animals will frequently stop and sniff the air. In the process, the head is usually moved dorsoventrally in a series of nodding movements which coincide with the rapid inhalation of air. They can easily locate food items concealed below leaf litter and loose soil, and spend a considerable amount of time rooting around in leaves and at the bases of trees. The slightest rustling of any prey species or predator will cause an immediate freezing of the animal, and many times they can be seen with their ears up, snout waving in the air, appearing to be in suspended animation with one foot completely off the ground. If a slight rustling continues they will approach cautiously until visual or olfactory contact is made. In captivity they spend approximately 80% of their active period moving around the cage, investigating the litter, and observing activities outside of the cage. While walking, the animal will usually keep his head low and the nose in close approximation to the ground or litter at the bottom of the cage. Exploration in captivity also includes climbing the sides of the cage, inspecting corners, and traversing the ceiling, especially if it is made of wire or other material to which the animal can cling. Exploratory behavior also involves climbing small bushes and trees, although *Didelphis virginiana* is much better adapted for a terrestrial existence than an arboreal one. In the process of climbing and exploring bushes and trees, the prehensile tail is used to provide stability and to cling to one branch while moving toward another one. During their climbing activities, olfaction appears to be their primary receptor. In arboreal exploration, the animals move slowly and cautiously, even more so than when on the ground. They have been known to remain suspended from the roof or precariously perched on small limbs and branches of trees in an almost motionless fashion for up to 15 min while industriously sniffing the air and intently listening for movements. The authors have never observed the animals in a sitting position on branches smaller than 50 mm in diameter. Occasionally they will slip and fall from one limb to another or to the ground.

Behavior which can be called play is rarely seen in marsupials. There is some playful activity with inanimate objects by young didelphids but very little social play. They will generally exhibit playlike behavior by manipulating objects, moving around in an erratic fashion, and curiously pulling at their mother's fur. Evans (1959) reported on a series of studies on young Virginia opossums and stated that play behavior was limited in the species. He has seen young opossums that would play tug of war with a brush used to clean cages.

Kaufmann (1974) reported that the mothers and young whiptail wallabies (*Macropus parryi*) play while the young were still in the pouch and that the behavior persisted while the young were at foot. The young batted their mother's nose or paws and the mother would playfully bat back. By the eleventh month, social play becomes almost nonexistent in whiptails. They evidenced solitary play by running back and forth and by object manipulation. Wünschmann (1966) described play in wombats (*Lasiorhinus latifrons*).

There have been few observational data which include mention of play and no extensive studies specifically on play in the marsupials. In addition to the remarkably small amount of play these animals perform there is the old argument about just what constitutes play behavior (Wilson, 1975). This, as well as the value of play to a species (the functionalists vs the structuralists), tends to discourage some investigators from even mentioning such behavior patterns. Many consider play as an activity that is performed which does not have survival value, or is "useless," while others state that play occurs when a young animal's basic needs are satisfied. The work of Loizos (1966) and other authors (Leyhausen, 1965; Ewer, 1968; Schaller, 1972; Bekoff, 1976) suggest that the origin and causes of play are still incompletely understood and agreed upon by the scientific community. Basically, play does not appear to serve many immediate functions but does serve to familiarize the young animal with the environment and enhance motor skills. Fagen (1974) provides a good evaluation of the selective and evolutionary aspects of play behavior and the various schools of thought on the subject.

The low level of development of play in marsupials is probably due to their simplistic behavioral activities. With limited rapid running ability, agility in the trees, or complex motor activities, one would expect play behavior to also be limited. This would be especially true if play behavior was utilized by an animal species to develop complex motor skills through repetition of various play activities. Kaufmann (1974) suggests the limited learning capacity of the macropods may partly account for the paucity of play behavior. Ewer (1968) suggested that since the main advantage of play is learning, then those species or groups with limited learning would be expected to have limited play activities. Although the correlation can be made, this author (Hunsaker) does not think that the level of development of play behavior is entirely dependent upon the learning ability of an animal. It is this author's opinion that there are probably several factors which determine the amount of play and that it is not completely known as to how they relate to the learning ability or phylogenetic position of animals. Certainly more quantitative data are needed to arrive at definitive conclusions.

### E. Sleep

Didelphids have a variety of postures during sleep periods; usually they sleep curled up on their side with their head, limbs, and tail tucked closely into the fur. They sleep deeply and require a great deal of stimulation to awaken. They can even sleep with their eyes open.

The body posturing during sleep is a function of ambient temperature and serves to control the amount of heat loss that is required to maintain the optimum body temperature of the animal. During particularly hot weather, it is not uncommon to see *Didelphis* sleeping on its back with all four feet extending up into the air. At colder ambient temperature levels, the animal will tend to curl tightly into a ball with all the extremities very well protected by the insulation of the fur. Tight curling sleeping postures with the nose well covered by the abdominal fur and the limbs tightly tucked in the thoracic region of the opossum are found at temperatures below 15°C. McManus (1970) considered 20°C as that temperature above which the sleep posture becomes more variable. Below 20°C, the sleeping posture would tend to be curled on the side to reduce heat loss. Above 25°C, the animal exposes as much of its body to the environment as possible. The marsupium of the female is relaxed, the scrotum is flaccid in the male, and the animals' extremities are extended from the body. Snyder (1965) has observed penile erection during the sleep of *Didelphis virginiana,* but it is not known whether this functions in a thermoregulatory manner or whether it is a response to erotic dreams.

Virginia opossums do not have much variability in the total amount of sleep time that they spend during the day. Van Twyver and Allison (1970) established that the total sleep time (TST) of five opossums represented 80.0% of the 24-hr period, which agrees with Snyder's (1965) study in which *Didelphis* spent from 75 to 85% of the 24-hr period sleeping. Van Twyver and Allison's studies showed that 29.1% of the total sleep was paradoxical sleep (PS), with sleep epics at a mean of 22.6 per day. Their subjects were maintained under continuous light for 3 days prior to the recordings to adapt to the chamber. During the sleep recording period they were fed at 11 AM and maintained in continuous light. Under these circumstances, the sleep periods were clearly established in a nocturnal cycle. Major waking periods were between 2100 and 2300 hours with some observed activity up to 0600 hours. The rest of the 24-hr period was spent sleeping, alternating between slow wave sleep and paradoxical sleep with no tendency to be associated with either state.

*Didelphis virginiana* displays the electrophysiological characteristics of sleep similar to that recorded in higher animals. Electrocorticograms (ECoG) of large amplitude 8- to 11-Hz spindles were the most prominent features of slow wave sleep (SWS) in the opossums observed by Van Twyver and Allison (1970). During the sleep phase the animals maintained a regular heart rate, but exhibited a decreased respiration rate from 38.1 breaths/min to 21.8 in SWS. Brain temperature was closely correlated with wakefulness (36°C) and was reduced to 32°C during SWS. In PS, the electromyogram displayed low voltage, and there was a desynchronized electroencephalogram with an irregular heart and respiration rate and a higher brain temperature than during SWS. Rapid eye movements (REMs) in the opossum are slower and less frequent than in most mammals. Many PS periods showed no clear evidence of REMs (Van Twyver and Allison, 1970). The arousal threshold during SWS was much higher than during waking and higher still during PS, which occupied 29.1% of the TST.

Van Twyver and Allison (1970) concluded that cortical and hippocampal activity during waking and both the SWS and PS stages of sleep in the opossum were typically mammalian. However, they maintained that PS is no more different between marsupials and placentals than between orders of placentals. They concluded that the central nervous system and physiological manifestations of sleep in marsupials and placentals are identical. From these results, Van Twyver and Allison (1970) postulated that SWS and PS were fully developed in the early ancestors which gave rise to the marsupials and placentals.

The high percentage of PS in the opossum was attributed to the animal's particular physiological and behavioral adaptations. The difficulty one experiences waking an opossum from PS substantiates the fact that they probably do not have to escape predators often when sleeping. Since these animals have sleeping arrangements which are relatively secure from predators, they are afforded extended sleep periods with high amounts of deep or REM sleep.

## IV. EXPERIMENTAL STUDIES

Studies on marsupials by experimental psychologists have been limited, with their research conducted almost exclusively on the Virginia opossum (James, 1937, 1953, 1955, 1958–1960; Dember and Fowler, 1958; James and Turner, 1963; Tilley *et al.*, 1966; Friedman and Marshall, 1965). In general, these studies have been concerned with conditioned responses, spontaneous alternation, maze learning, and learning set in the opossums. The primary interest in these studies is generated by the fact that the opossum represents a primitive animal that occupies a taxon which relates it to a "prototype placental mammal" and has a characteristic marsupial brain which is of great interest (Herring *et al.*, 1966). Thusly considered, the experimental data then would provide information on the evolution of more complex behavior patterns that are observed in advanced mammals and, to some investigators, would shed light on the problems of comparative behavior of phyla, developmental behavior patterns, and the evolution of intelligence quotient in humans. Wiedorn (1954) suggested that the Virginia opossum would be of great value in psychiatric research.

James (1955, 1958, 1959) conducted a series of experiments on the conditioned responses of *Didelphis virginiana*. His early work, using wild caught animals, consisted of attempting to elicit a conditioned response of a foreleg to avoid shock stimulus. The opossums did not acquire a conditioned avoidance response in a series of 300 trials, although they did form a generalized body reaction and showed changes in breathing rate in the process of the study. The subjects were also unable to develop a conditioned running response when the investigator shocked their feet and sounded a signal. The response observed in this series of experiments was a passive defense in which the animal would lie on its side rather than run to escape the shock. The defense posture of lying on its side was a stimulus–avoidance response, since the hair insulated the animal from the shock and enabled it to avoid the stimulus. In the usual experimental situation, the wild caught animals were very defensive when disturbed and did not adapt well to the standard experimental conditions. They would respond to the investigators with the passive defense reflex and lie down on their sides thereby causing obvious problems in the experiment. Later James shifted to laboratory raised opossums since they proved easier to work with, and found that they could discriminate visual signals on the doors of feeding apparatus, operate the Guthrie–Horton puzzle box, and give operant responses by pushing levers in other experimental conditions.

He considered that the opossum was unable to form specific conditioned motor responses, e.g., moving its legs, because the animal exhibited a spreading of the excitation to all four legs and would have a generalized, rather than a specific response. Since the rear legs do not have motor representation in the cortex of the brain, James considered the species to be a spinal type animal. Specific foreleg actions cannot be conditioned either, yet the foreleg does have cortical representation. He found that the escape behavior as a general body reaction could easily be conditioned as well as a defense reaction to an approaching object.

James' testing apparatus confined the animal while at the same time allowing freedom of movement to the desired leg in which the response was to be elicited. A buzzer was sounded for a duration of from 1 to 5 sec and a brief shock was applied to the leg. In his first study, he utilized 300 signals and in the second study 550 signals were given without the specific leg response occurring. Although a definite breathing and a total body response were conditioned in the last 25 signals, in no case was the isolated leg response observed.

In his 1955 report, James also tested the conditioned escape response of three *Didelphis* in a double box apparatus. At the signal, a shock was applied to the feet of the subject in one box, and a door was simultaneously opened which led to the second box to which he could escape and avoid the response. The buzzer was again sounded from 1 to 5 sec. The shock was applied via an electrified grid and the restraining door was opened to enable the animal to escape into the second compartment. One of the subjects developed a conditioned response after only five signals and completed a series of 10 consecutive responses after 17 trials, with an average delay of 2.72 sec. The slowest animals to develop the conditioned escape response completed 10 consecutive responses after 53 signals, with an average delay of 4.08 sec. These animals were then tested for the extinction of the response. The subject which most rapidly acquired the conditioned response had the longest extinction period; it gave eight responses to the first 20 signals without shock and exhibited no responses in the subsequent 10 signals without shock.

These data suggested to James that the opossum is a generally inhibited animal and would therefore tend to extinguish rapidly. This is in contrast to excitable dogs which sometimes never completely lose an avoidance response to a signal that has been accompanied by shock (James, 1953). More inhibited dogs did form complete extinction of a response, which would lead one to suspect that there is a positive correlation between the level of inhibition in an animal species and the ability of that species to rapidly extinguish responses.

The ability of opossums to solve complex problems and work in the Guthrie–Horton puzzle box was studied by James in 1955. In this test, a subject was placed in a box in which it could escape by pushing on a post. As the door opened, a photo was taken which recorded the time of day and the position of the animal relative to the post and escape door. The opossum learned to operate the puzzle box in a relatively short time, although there was no measurement of the types of movements or activity time involved in the learning process. The majority of the time (42 times in 78 observations), the subjects would face the door and push the pole with their tail to open the door. In 7 of the 78 observations, they were not facing the door when they pushed the post. Fourteen times they were facing the door and pushed the post with their sides.

Fink (1954) studied the behavior of a series of animal species in a maze learning experiment utilizing the Fink–Arrow maze. The test apparatus gave the subjects a choice of four parallel alleys leading off to one side of a runway. He was able to rate the performance of his subjects based on a series of calculated scores: the performance quotient (*P* value) which is the sum of all trials required for learning all four alleys; trial quotient (*T* value) which excludes the correct trials when learning the maze; and alley quotient (*A* value) which was a measure of the number of trials an animal partially entered the wrong alley and withdrew. These entrance errors are included in the *A* value.

Fink considered man to have a 100% level of the factors referred to as *T* quotient, *P* quotient, and *A* quotient when the subject would learn the maze by finger movements through the maze while blindfolded. By a comparison of the *PQ*, *TQ*, and *AQ* of the test subjects and man, Fink could relate the ability of other animals to solve this maze to the ability of man. He derived values for these three factors for man, pig, dog, goat, chick, rat, cat, and rabbit. The average of these three values for the three most competent species were man, 100%; pig, 47.5%; and dog, 45.3%. James (1959) utilized five adult *Didelphis*, laboratory born and reared, to test their ability in the Fink–Arrow maze. Under the same experimental conditions as utilized by Fink, James determined that the average of *PQ*, *TQ*, and *AQ* for opossums was 58.9, which indicated that opossum under these experimental conditions ranked between the pig and man in its ability to solve this maze. Table II reflects the values for the animals tested in the Fink–Arrow maze. Since the opossum exhibits a relatively simple brain cortex, most people expected that it would have a considerably lower ability to solve problems, but these experimental results suggest that in regard to spatial location, *Didelphis* surpasses many animal species which exhibit a higher level of intelligence.

Tilley *et al.* (1966) studied spontaneous alternation in the maze behavior of *Didelphis*. They were primarily interested in establishing the universality of the spontaneous alternation phenomenon and utilized the opossum as a comparative species for the work that had been done on a variety of other animals. They used five adult subjects and ran them in a standard cross maze which had a T-choice point. The subjects were each given four masked trials in which the maze was run with no reinforcement, and the doors in the goal box automatically closed when the opossum entered the box. After a 15-sec delay, the subjects were returned to the start box to begin another trial with an intertrial interval of approximately 30 sec. In a total of 20 trials, the subjects made 14 left and 6 right turns, assuming that either arm of the maze had an equal probability of being chosen. Out of a

**TABLE II**

Values and Ranking of the Scores of a Variety of Mammalian Species Tested in the Fink–Arrow Maze[a]

| Animal | *P* | *T* | *A* | *PQ* | *TQ* | *AQ* | *Aq* $\bar{X}$ | Species ranking |
|---|---|---|---|---|---|---|---|---|
| Man | 14.8 | 9.2 | 25.1 | 100 | 100 | 100 | 100 | 1 |
| Opossum (2)[b] | 24 | 22 | 26 | 61 | 51.5 | 96 | 69.5 | 2 |
| Opossum (1) | 27 | 20 | 29 | 54 | 46 | 84 | 61 | |
| Opossum (5) | 27 | 19 | 35 | 54 | 48 | 71.7 | 57.9 | |
| Opossum (3) | 29 | 28 | 39 | 47 | 32 | 64 | 47.6 | |
| Opossum (4) | 38 | 28 | 26 | 38.9 | 48.4 | 53 | 46.7 | |
| Pig | 31.8 | 21.8 | 46.7 | 46.5 | 42.2 | 53.7 | 47.5 | 2 |
| Dog | 38.4 | 26.1 | 40.4 | 38.5 | 35.2 | 62.1 | 45.3 | 3 |
| Goat | 54 | 37 | 56 | 27.4 | 24.9 | 44.8 | 32.2 | 4 |
| Chick | 46.4 | 34.2 | 68.1 | 31.9 | 26.9 | 36.9 | 31.9 | 4.5 |
| Rat | 53.7 | 32.8 | 62.4 | 27.6 | 28 | 40.2 | 31.9 | 4.5 |
| Cat | 61.1 | 43.1 | 89.5 | 24.2 | 21.3 | 28 | 24.5 | 5 |
| Rabbit | 67 | 37 | 67 | 22.1 | 24.9 | 37.5 | 28.2 | 6 |

[a] From James (1959).
[b] Numbers were assigned to each of the five experimental opossums.

total of 15 opportunities of alternation, the subjects alternated 10 times, with *P* less than 0.05. Apparently these wild caught opossums, and one female, the least tame of the group, did not alternate. From these findings it was concluded that opossums do alternate in a T-maze situation, but the level of alternation is somewhat less than the 80% which Dember (1965) determined in rats. Some opossums, as in other species, do not alternate but choose the same arm of the maze on every trial.

Learning set is a characteristic of animals to improve their ability to solve a problem in the process of learning a series of similar problems. The transfer of the ability to solve similar problems, learning set, was suggested by Harlow (1959) to be a function of brain complexity. Friedman and Marshall (1965) determined that *Didelphis* could acquire a learning set. They selected a two choice discrimination box which would provide a test for position discrimination that would be independent of perceptual motor capacities. The testing apparatus was similar to that used by Dufort *et al.* (1954). Doors to the food boxes were marked with black and white stripes so that the subject could discriminate between the two. A sliding panel was utilized to cover both doors until it was raised; this panel was also lowered to prohibit correction of any choice the opossum made.

The project was set up in four distinct parts. Each part was dependent upon reversal of the previously incorrect side to become the correct side in the next series of trials. Criterion of learning was 11 correct responses out of 12, with the last eight responses being correct. The reinforcement for a correct choice was 12 g of ground horsemeat. Part 1, reversals 0 through 15, consisted of four trials each day. After criterion was reached, the problem was reversed on the first trial of the following date. Part 2, reversals 16 through 25, consisted of daily sessions in which the tests were run for as many trials as were required for the animal to reach criterion. Each day was begun with a new reversal. In this section, reinforcement was 6 g of ground horsemeat. Part 3, reversals 26 through 30, utilized the same conditions as Part 2, which consisted of daily sessions with as many trials as necessary. The tests were begun each day with the opossum running the test apparatus until it made four correct responses to the side that it had been trained to choose the previous day. After the subject had achieved four correct responses, the next trial consisted of a reversal of the goal boxes and it was required to continue on until criterion was reached. The fourth part, reversals 31 through 33, were the same as those used in Part 1, i.e., four trials were given each day. Figure 13 reflects the errors to reach criterion for each of the reversals. The scores of one female ($F_2$) are shown separately from the scores of the group. It can be noted that $F_2$ exhibited very rapid learning of $R_0$ although, in general, she had very poor performance in Part 1 of the experimental program. A significant difference in the learning was exhibited between Part 1 and Part 2 of the experimental program. Figure 13 reflects a very rapid decrease in the number of errors required to reach criterion when the number of trials was significantly extended.

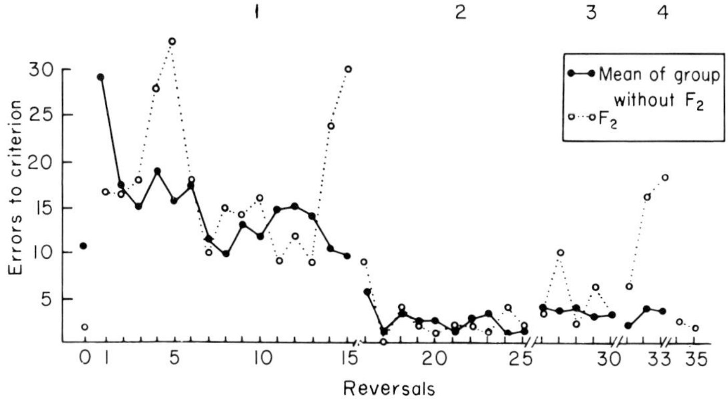


Fig. 13. The number of errors to reach criterion for each of the reversals of the Virginia opossum. Data for $F_2$ (a female) are shown separately. 1, reversals 0–15 (4 trials per day); 2, reversals 16–25 (as many trials per day as required to reach criterion); 3, reversals 26–30 (as many trials per day as required to reach criterion); 4, reversals 31–33 (4 trials per day). From Friedman and Marshall (1965).

Consideration of these results leads one to believe that the disruptive effects of early training were relatively difficult to extinguish during Part 1. There is also no evidence of formation of the learning set; no facilitation of learning occurred in the first part with only four trials to criterion. However, in Part 2 there was marked evidence of learning set in the rapid reduction in errors. Friedman and Marshall considered the 24-hr separation between blocks of four trials in Part 1 to be an important factor. It appeared that intraproblem learning was impeded due to this time span. Comparison of their results with those of Dufort *et al.* (1954) and Stretch *et al.* (1964) on rats indicated that opossums acquire learning set much more slowly than rats and also that spacing of trials is much more important in opossums than in rats. Friedman and Marshall (1965) pointed out that their findings, i.e., long delays between blocks of trials tended to impair learning, are in agreement with Harlow's (1959) statement that interproblem learning is more difficult than intraproblem. They considered the superiority of the rat to the opossum to be a function of the relative phyletic level between the marsupials and rodents.

The importance of time as a function of learning in the learning set experiments relates back to James' work on conditioned responses. In the previous experiments, it was found that the opossum forms extinction very rapidly. These data coupled with the learning set information and the wandering, nonfixed foraging behavior of *Didelphis* suggest that opossums tend to live a very opportunistic life which is not complicated by a high level of memory, negative responses, and complex avoidance reactions of places or things. These factors are probably very basic to the ability of the animal to exist in a variety of environments.

## V. BEHAVIOR IN CAPTIVITY AND COLONY MANAGEMENT

The opossum possesses a unique instinctual behavior pattern and, as Krupp and Quillin (1964) pointed out, "In contradiction to its apparent primitive development is a demonstrated capacity to learn in psychological research studies." Dr. James, one of the most active of researchers in the behavior of opossums in psychological experiments, in a personal communication with Hunsaker, agrees with the authors that *Didelphis* is a much more intelligent and useful animal in behavioral research than many investigators think.

To a great extent, the interest in utilization of the exotic marsupials (*Marmosa* and *Caluromys*) for research has been due to the inability of researchers to get *Didelphis* to breed and do well in captivity. Thrasher *et al.* (1971), Wolf (1970), Barnes (1968a,b), Eisenberg and Maliniak (1967), Barnes and Barthold (1969), Barnes and Wolf (1969), and Collins (1973) have all written about care and behavior of *Marmosa* or *Caluromys* in captivity. In addition, the Virginia opossums have also developed lesions and other behavioral and medical problems in captivity which tend to discourage people from utilizing the species. Krupp and Quillin (1964) described the problem of foot and tail sores developed from wire bottom cages, and other researchers have been unable to get *Didelphis* to reproduce in small indoor enclosures. Sherwood *et al.* (1969) reported on a variety of problems experienced with the Virginia opossum in captivity. Because of this, many have resorted to outdoor cages utilizing extra large enclosures for *Didelphis,* or they have shifted to another species altogether.

Since Hartman's (1923, 1928) early work with *Didelphis virginiana,* researchers have for the most part relied on wild-trapped pregnant females for their subjects. Hartman and others (Farris, 1952; Fritz, 1971) have also maintained young and adults in outdoor fenced enclosures in order to have animals readily available for their experiments. Some breeding has been recorded in these large enclosures but, in general, experimenters consider *Didelphis* a less than perfect colony species. Control of mating pairs and weaning of young have been haphazard. Since successful prophylactic treatment of parasites and diseases has been developed (Krupp and Quillin, 1964; Jurgelski, 1974), it seems foolish to continue to expose the subjects to the obvious reinfection potential of keeping many animals of random age and sex outdoors on the ground where adequate sanitation is nearly impossible. Fighting often results in cuts and abrasions which can become medical problems requiring time and money to treat. Jurgelski has had excellent success with a semioutdoor, individualized enclosure system in North Carolina. His reports (Jurgelski, 1974; Jurgelski *et al.*, 1974; Jurgelski and Porter, 1974) are of great help to investigators who keep colonies of *Didelphis*.

While the authors share Krupp and Quillin's (1964) and other investigators' observations that wire floors cause lesions on feet and tails of *Didelphis,* the wire floor is unsurpassed for enabling the animals to keep themselves clean and odor free when drop pans have paper, sawdust, or other litter material which is changed daily. A plastic covered hardware cloth will help alleviate the irritation to the feet of the animals. In Shupe's laboratory, where single or paired animals were maintained for up to 4 years in (40 × 81 cm) cages constructed on (25 × 50 mm) welded galvanized wire, *Didelphis virginiana* were always in good health and appeared to be very well adjusted. Medical problems were alleviated since they never contacted their feces. These animals were clean, odor free, and very pleasant to handle.

Wire cages are preferred as long as the animals do not develop skin lesions, though they rarely develop on young healthy animals. One can always return the animals to solid floors should this become necessary. The best treatment for skin lesions of *Didelphis* is a light antibiotic for infection followed by placing the animal in a rich environment where the increased attention to other things, and activity, reduces the licking and biting of the lesions. Boredom and extensive cleaning and stimulation of the area appears to us to be of greater importance than bacterial complications. Wire shelves two-thirds of the way up the ends of these cages provide the animals with the incentive to do a great deal of climbing in their cages at night. We consider an environment with numerous stimulus factors to be an important consideration if *Didelphis* is to be well adjusted and have a high reproductive rate in the laboratory. The natural habitat of this species is far more complex than the average cage environment, and the animals' adjustment to captivity will suffer from a cage system with too little stimulation. Increased activity levels will avoid the ascending paralysis observed in some animals and provide a healthier, better adjusted animal. Two of the most successful early researchers in *Didelphis* reproduction (Reynolds and Farris) both used a seminatural caging system for their subjects. Many investigators consider 10 square feet per animal about the ideal space for breeding in outdoor facilities. Recently, Jurgelski (1974) has published the plans of an excellent enclosure system for breeding opossums in 3.75 square feet per individual. He has had 113% birth rate in closely monitored breeding programs (Jurgelski and Porter, 1974).

Five-gallon square tin cans or similar sized plastic containers make excellent nesting boxes and are easy to clean. The animals use them more readily if they are sprayed with flat black acrylic paint. The dark paint reduces the incident light inside the boxes which encourages the animals to use them. An ideal bedding or nesting material (if any is desired for *Didelphis*) is paper hand towels. In our laboratories we consider bedding to be a positive factor in the comfort and general adaptation to cage life. When the animals are tame, they will exhibit their curious tail carrying of nesting material if fresh paper towels are supplied after cage adaptation has occurred (Shupe, 1967).

When *Didelphis virginiana* is made reasonably docile by handling and taming, it makes an extremely amenable experimental animal. The fact that opossums have never shown any affective responses in the years that investigators have worked with them does not detract from their usefulness as experimental subjects. In fact, the total devotion of their periods of activity toward food seeking, eating, nest building, grooming, and sexual behavior makes them ideal for some forms of research. They do not contaminate experiments by approach behavior, and they are not easily distracted once a goal oriented activity is begun.

If time for taming is a limiting factor, it is much more feasible to tame young animals rather than adults. Taming of the pups should begin as soon as they are seen emerging from the pouch (usually about 2 months after birth). The critical period for socialization begins prior to 60 days of age (the age that their eyes open), and extends to 70 to 85 days. After this, the young are noticeably more difficult to tame. Friedman (1964) and others agree with the present authors that this period of social development is the optimum one to train the animals. To tame an animal, one must get them used to handling so that one does not cause aggressive or fear responses upon approach. To obtain this objective, they must be exposed to human sounds, smells, movements, and handling techniques. This may be accomplished by having an assistant talk or read aloud while near them for several periods during their activity period (Friedman, 1964). Talking to and handling them for extensive periods of time is the system that the authors prefer. Another successful method employs imprinting as utilized by Fritz (1971). There is some controversy about mammalian imprinting, but the phenomenon does appear if early experience (around age 25 to 35 days) exists between humans and young opossums. Fritz (1971) reported that a definite preference for humans exists in imprinted opossums; they will leave a group of their own species to follow an investigator who has raised them. It will be interesting to see the results of preference testing during estrus of imprinted subjects when this line of research is pursued.

To facilitate handling embryons from a wild-trapped mother, one should place an 8 or 10 cm (3–4 inches) diameter can or container about 15 cm deep over her head and hold it there. The adult females adapt very quickly to this procedure, and thus do not require lengthy taming procedures that require extensive time which can better be spent with the young. With the female so immobilized, the pups may be removed from the pouch and handled extensively.

The key to taming young opossums is to provide as much contact with humans as possible with no teasing, cage rattling, or other factors that will tend to elicit fear or aggressive responses in the animals. On numerous occasions, Shupe has assigned pregnant or pouch-filled females to grade school classes for taming. Lengthy quarantine and prophylactic procedures for parasites are required if this is considered. Being raised by a grade school class and handled extensively by children is without a doubt one of the most effective taming procedures available. The only rule that must be enforced is that no one may tease the animals in any way. After 2 months of handling, one has a litter of healthy, tame, experimental animals that are excellent subjects. Each animal may be handled at will, hand fed, and taken about on a leash if desired. Injections may be administered without restraint and they are extremely resistant to stress. The only drawback to this technique is that the children or assistants want to keep the subjects. Children are not very good about security measures and their parents are on the side of the opossums!

Taming of wild-trapped adult *Didelphis virginiana* may be accomplished by extensive gloved handling at the risk of uncomfortable puncture wounds at first, or by food adaptation (Fritz, 1971). Fritz uses successive approximations to reduce fear responses in the animal by feeding bits of banana on the end of a short (20 cm) stick until food can be offered by hand. Then the animal may eventually be petted and fed, and after a while, picked up. Most of the investigators find that talking to the animals, numerous visits to the subjects, and patience will eventually succeed in taming wild opossums. Working during the night cycle, when they are alert, appears to help. The taming process of any wild animal is slow, but it usually can be done. It should be remembered that a mature, large specimen, especially a male with long canines, can inflict very serious injury to a careless or naive assistant or animal keeper. There is some individual variability of the tamability of animals, and the few that are so aggressive as to require maximum security and a great deal of time should be discarded.

*Marmosa* may be picked up by the tail and placed in the hand of the investigator from the first day of arrival at the laboratory. For its size, *Marmosa* has a fearful array of teeth but it seldom bites after the first few weeks of handling, especially when it is tended by an experienced person. The rare animal that appears untractable after a few short weeks work should be discarded as he poses a threat to safety and tends to disrupt the entire colony and staff.

Pair and like-sex compatibility may be tested in *Didelphis* and *Marmosa* (after Egoscue, 1957 and Thrasher, 1969) by placing subjects together and observing their behavior in a moderate size cage, 60 × 90 cm (25 × 35 inches), for single or paired housing of *Didelphis*. We found, as did Egoscue, that compatibility shows up quickly. In Shupe's laboratory, same sex and opposite sex pairs of *Didelphis* lived together with no difficulty for 4 years, and Hunsaker has kept *Marmosa* and *Caluromys* paired with very few losses. In fact, when *Didelphis* were placed in large screened runs with other individuals, they continued to sleep with partners of prior caging rather than any of the other animals available.

As reported by Shupe (1967) and James and Turner (1963), there is every reason to believe that wild-trapped adult animals do not show anything approaching normal behavior in captivity, especially in the first weeks after capture. The reasons for atypical behavior in captivity can be supported from a physiological health point of view by the studies of Barenfus. He has examined numerous *Marmosa robinsoni* and *Didelphis virginiana* brought into the UCLA vivarium for research purposes. He found that many animals undergo a period of time when they contract a variety of pathological problems during their first days of confinement. Cats undergo at least two complete sickness cycles before being used for research and wild marsupials should undergo even more isolation and treatment before being used.

Krupp and Quillin (1964) reported routine methods for removing parasites from *Didelphis virginiana,* and Barnes (1968a) discussed some measures taken in his laboratory. Hunsaker has seen numerous animal collecting and holding facilities in tropical countries where 50% mortality of wild-trapped animals was not uncommon. The work of Selye and others suggests that the effects of stress compounded with endemic disease or parasitism interacting on an animal pushes them beyond the limits of their systems, and biased results are obtained in a research project (Hunsaker, 1974).

Two suggestions are offered to enable researchers to study behavior of wild species without the cloud of uncertainty caused by the poor health and stress of the subjects. Behavioral research should be conducted only on laboratory-born subjects which have undergone extensive handling for taming purposes, or on wild-trapped young adults which have been quarantined for a minimum of 2 months. During this 2-month period, the wild-trapped individuals should be receiving extensive antiparasitic purgatives, and should also be handled routinely to facilitate taming.

The authors insist that behavioral research done on animals receiving less than the proposed minimal treatment will provide data containing complex interactions of experimental treatment effects and health stress effects, which cannot be sorted out of the data.

It has also become obvious over the past 10 years that conditions of housing or caging are ultimately crucial to collecting sound experimental data on captive animals. *Didelphis virginiana* seem to prefer large outdoor enclosures for breeding, but they are not absolutely necessary. *Marmosa robinsoni* wild-trapped adults will breed in small laboratory cages. The authors believe that solitary indoor caging is essential for experimental animals. Rectangular section 1 × 2-inch welded wire (plastic coated) caging is recommended. This size encourages feces to drop through the wire and makes cage cleaning easy. Cages are easily constructed with the material, and if the floor is suspended 3 inches above a galvanized pan, the animals are less likely to reach down through the floor and retrieve fallen material. One of the best criteria for cage size is the size of folded newspapers. If cage pans are constructed of galvanized iron in multiples of standard newspaper size, then cleaning cages is greatly facilitated. A very useful cage size is 35 × 51 cm (14 × 20 inches); this provides a 4-inch overlap of paper in the center of the cage. Paper soaks up liquid well and does not produce as much odor from urine as does sawdust. Single animals have room to move back and forth and to climb. As previously mentioned, we have installed 15 cm (6 inches) deep wire shelves two-thirds of the way up each of the 69 cm (27 inches) high cages and the animals spend a great deal of time crawling up to these shelves. These cages, thus equipped, lend themselves well to activity monitoring and the placement of recording devices.

All this discussion about cage size is based in part on the necessity of caging and handling experimental animals consistent with an accepted experimental standard. Cage size, early experience, and richness of the environment are, in all probability, ultimately crucial to adult behavior. The real value of the white rat to experimental psychology and physiology is that years of selection have produced a very standardized research subject which has a minimum of genetic and behavioral variability. When dealing with these varieties, it is nice to have such a standardization of subjects. Laboratory rats used in behavioral studies have for years been selected as experimentally naive male and female rats, 90–120 days of age. All are usually tamed by 5-min daily handling for 2 weeks at the beginning of the experiment. Researchers with marsupials have not been fortunate enough to have sufficient numbers of subjects available for such selection.

It is always amazing to the authors to see the way many biologists select experimental animals. They frequently choose animals that have been used in previous experiments. These animals, often in widely divergent age classes, are sometimes even mixed with wild caught and lab-reared individuals. This practice tends only to confuse the picture of what really is "normal" behavior for an opossum or other vertebrate. When at all possible, this should be avoided. When raising *Didelphis* for behavioral research, it is wise to raise several litters of similar ages, and then randomly assign the subjects from each litter to experimental groups. In any event, we should not fool ourselves that we really understand the normal behavior of a species until a very large body of well-conducted research is available to us.

## REFERENCES


Armstrong, E. A. (1942). "Bird Display." Cambridge Univ. Press, London and New York.

Audubon, J. J., and Bachman, J. (1851). "Quadrupeds in North America," Vol. 2. New York.

Barker, P. R., Bergman, H. C., Hunter, W. H., Lawrence, W. H., Steers, C. W., Wright, L. N., Willis, R. A., and Barrow, D. M. (1967). "Marsupial Biomodule Evaluation Study," Report no. SID66-1647. U.S. Defense Documentation Center, Cameron Station, Alexandria, Virginia.

Barnes, R. D. (1968a). *Lab. Anim. Care* **18,** 251–257.

Barnes, R. D. (1968b). *N.A.S.—N.R.C., Publ.* **1594,** 88–97.

Barnes, R. D., and Barthold, S. W. (1969). *J. Reprod. Fertil.* **6,** 477–482.

Barnes, R. D., and Wolf, H. G. (1969). *Marsupial Newsl.* **1,** 11–12.

Barton, B. S. (1823). *Ann. Philos.* [N. S.] **6,** 349–354.

Beach, F. A. (1939). *J. Mammal.* **20,** 315–322.
Beebe, W. (1923). *Zool. Soc. Bull.* **26,** 150–156.
Bekoff, M. (1976). *In* "Perspectives in Ethology" (P. P. G. Bateson and P. H. Klopfer, eds.), Vol. 2. Plenum, New York.
Bensley, B. A. (1903). *Trans. Linn. Soc. London* (*Zool.*) **9,** 179–185.
Bensley, B. A. (1906). *Univ. Toronto Stud., Biol. Ser.* **5,** 1–13.
Biggers, J. D. (1966). *Symp. Zool. Soc. London* **15,** 251–280.
Biggers, J. D. (1967). *J. Mammal.* **48,** 678–680.
Boggs, J. (1969). M.S. Thesis, San Diego State University, San Diego, California (unpublished).
Burns, R. K., and Burns, L. M. (1957). *Rev. Suisse Zool.* **64,** 595–605.
Cabrera, A., and Yepes, J. (1940). "Mamiferos de Sud Americanos." Compania Argentina de Editores, Buenos Aires.
Cadzow, G. (1974). M.S. Thesis, pp. 1–90. California State University, Long Beach.
Clark, L. D. (1962). *In* "Roots of Behavior" (E. L. Bliss, ed.), p. 179–186. Harper, New York.
Collins, L. R. (1973). *Smithson. Inst. Publ.* **4888.**
Coues, E. (1872). *Mem. Boston Soc. Nat. Hist.* **2,** 41–149.
Crook, J. H. (1970). *Anim. Behav.* **18,** 197–209.
Davis, D. E. (1945a). *Ecol. Monogr.* **15,** 243–295.
Davis, D. E. (1945b). *J. Mammal.* **26,** 119–127.
Davis, D. E. (1946). *Ecology* **27,** 168–181.
Dember, W. N. (1965). *Am. Sci.* **4,** 409–427.
Dember, W. N., and Fowler, H. (1958). *Psychol. Bull.* **55,** 412–428.
Doutt, K. (1954). *J. Mammal.* **35,** 581–583.
Dufort, R. H., Guttman, N., and Kimble, G. A. (1954). *J. Comp. Physiol. Psychol.* **47,** 248–249.
Egoscue, H. J. (1957). *J. Mammal.* **38,** 472–481.
Eisenberg, J. F. (1966). *Handb. Zool.* **8** (10), 1–92.
Eisenberg, J. F., and Maliniak, E. (1967). *Int. Zoo Yearb.* **7,** 78–79.
Enders, R. K. (1935). *Bull. Mus. Comp. Zool.* **78,** 385–497.
Enders, R. K. (1966). *Symp. Zool. Soc. London* **15,** 195–203.
Estes, R. D. (1974). *In* "Ungulate Behavior and its Relation to Management" (V. Geist, ed.), Proc. Calgary Symp. 1971. I.U.C.N. Morges.
Evans, L. T. (1959). *Proc. Int. Congr. Zool., 19th, 1958* Sect. XI.
Ewer, R. F. (1968). "Ethology of Mammals." Plenum, New York.
Fagen, R. (1974). *Am. Nat.* **108,** 850–858.
Farris, E. J. (1952). *In* "The Care and Breeding of Laboratory Animals" (E. J. Farris, ed.), pp. 210–267. Wiley, New York.
Fink, H. K. (1954). "Mind and Performance." Vantage Press, New York.
Fisler, G. F. (1969). *Los Angeles City Mus. Contrib. Sci.* No. 167, pp. 1–32.
Fitch, H. S., and Sandidge, L. L. (1953). *Univ. Kans. Publ. Mus. Nat. Hist.* **7,** 305–338.
Fleming, T. H. (1970). *J. Mammal.* **51,** 473–490.
Fleming, T. H. (1972). *J. Mammal.* **53,** 619–623.
Fleming, T. H. (1973). *J. Mammal.* **54,** 439–455.
Francq, E. N. (1969). *Am. Midl. Nat.* **81,** 566–568.
Francq, E. N. (1970). *J. Mammal.* **51,** 395.
Friedman, H. (1964). *Nature* (*London*) **201,** 323–324.
Friedman, H., and Marshall, D. (1965). *Q. J. Exp. Psychol.* **17,** 250–254.
Fritz, H. I. (1971). *Int. Zoo Yearb.* **11,** 46–49.
Gardner, A. L. (1973). "Special Publications of the Museum." Texas Tech. University, Lubbock.

Gilman, T. T., Marcuse, F. L., and Moore, A. V. (1950). *J. Comp. Physiol.* **43,** 99–111.
Goldman, E. A. (1920). *Smithson. Misc. Collect.* **69,** No. 5.
Gregory, W. K. (1910). *Bull. Am. Mus. Nat. Hist.* **27,** 217–221.
Gregory, W. K. (1920–1921). *J. Dent. Res.* **2, 3,** 151–158.
Grote, J. C., and Dalby, P. (1973). *Ohio J. Sci.* **73,** 240–241.
Hall, E. R., and Dalquest, W. W. (1963). *Aust. J. Sci.* **30,** 105–106.
Hall, E. R., and Kelson, K. R. (1959). "The Mammals of North America," Vol. 1. Ronald Press, New York.
Harlow, H. F. (1959). "Psychology: A Study of a Science," Vol. 2, pp. 492–537. McGraw-Hill, New York.
Hartman, C. (1923). *Smithson. Inst., Annu. Rep.* (1921) 347–364.
Hartman, C. (1928). *J. Morphol. Physiol.* **46,** 143–216.
Hartman, C. (1952). "Possums," Univ. of Texas Press, Austin.
Herring, F. H., Mason, D. J., Doolittle, J. H., and Starrett, D. E. (1966). *Psychol. Rep.* **19,** 755–757.
Hinde, R. A. (1970). "Animal Behavior," 2nd ed. McGraw-Hill, New York.
Hunsaker, D. (1970). *Marsupial Newsl.* **2,** 5–7.
Hunsaker, D. (1974). *In* "Centennial Symposium on Science and Health" (R. Snyder, ed.), pp. 141–150. Philadelphia Zoological Society, Philadelphia.
James, W. T. (1937). *J. Genet. Psychol.* **51,** 95–100.
James, W. T. (1953). *Ann. N.Y. Acad. Sci.* **56,** 171–183.
James, W. T. (1955). *J. Genet. Psychol.* **87,** 203–206.
James, W. T. (1958). *J. Genet. Psychol.* **93,** 179–183.
James, W. T. (1959). *J. Genet. Psychol.* **94,** 199–203.
James, W. T. (1960). *J. Genet. Psychol.* **97,** 127–130.
James, W. T., and Turner, W. W. (1963). *Psychol. Rep.* **13,** 921–922.
Jurgelski, W., Jr. (1974). *Lab. Anim. Sci.* **24,** 376–403.
Jurgelski, W., Jr., and Porter, M. (1974). *Lab. Anim. Sci.* **24,** 411–425.
Jurgelski, W., Jr., Forsythe, W., Dahl, D., Thomas, L., Moore, J., Kotin, P., Folk, H., and Vogel, S. (1974). *Lab. Anim. Sci.* **24,** 404–211.
Kaufmann, J. (1974). *Anim. Behav.* **22,** 281–369.
Krupp, J. H., and Quillin, R. A. (1964). *Lab. Anim. Care* **14,** 189–194.
Langworthy, O. R. (1925). *Am. J. Physiol.* **74,** 1–13.
Langworthy, O. R. (1928). *J. Comp. Neurol.* **46,** 201–247.
Lay, D. W. (1942). *J. Mammal.* **23,** 147–159.
Layne, J. M. (1951). *J. Mammal.* **32,** 464–465.
Layne, J. M., and Benton, A. H. (1954). *J. Mammal.* **35,** 103–104.
Leyhausen, P. (1965). *Z. Tierpsychol.* **22,** 412–494.
Lillegraven, J. A. (1975). *Evolution* **29,** 707–722.
Llewellyn, L. M., and Dale, F. H. (1964). *J. Mammal.* **45,** 113–122.
Loizos, C. (1966). *Symp. Zool. Soc. London* **18,** 1–9.
McCrady, E., Jr. (1938). *Am. Anat. Mem.* **16,** 1–114.
McManus, J. J. (1967). *J. Mammal.* **48,** 486–487.
McManus, J. J. (1970). *Am. Midl. Nat.* 144–169.
McManus, J. J. (1971). *J. Mammal.* **52,** 846–848.
Magne de la Croix, P. (1936). *J. Mammal.* **17,** 51–54.
Matthews, L. N. (1964). *In* "The Natural History of Aggression" (J. D. Carthy and F. J. Ebling, eds.). Academic Press, New York.
Maxwell, G. (1960). "Ring of Bright Water." Longmans, Green, New York.
Merchant, J. C., and Sharman, G. (1965). *Aust. J. Zool.* 593–609.

Moore, J. C. (1955). *J. Mammal.* **36,** 559–561.
Müller, F. (1968a). *Acta Anat.* **71,** 581–624.
Müller, F. (1968b). *Rev. Suisse Zool.* **75,** 630–643.
Müller, F. (1973). *Rev. Suisse Zool.* **79,** 1599–1685.
Noirot, E. (1966). *Ann. Soc. R. Zool. Belg.* **95,** 47–56.
Noirot, E. (1972). *Dev. Psychobiol.* **5,** 371–387.
Norton, A. C., Berne, A. V., and Misrahy, G. A. (1964). *Nature* (*London*) **204,** 162–163.
Owen, R. (1868). "On the Anatomy of Vertebrates." Longmans, Green, New York.
Phillips, C. J., and Jones, J. K. (1968). *J. Mammal.* **49,** 320–321.
Pray, L. L. (1921). *J. Mammal.* **2,** 109–110.
Ratner, S. C., and Thompson, R. W. (1960). *Anim. Behav.* **8,** 186–191.
Raven, H. C., and Gregory, W. K. (1946). *Am. Mus. Novit.* **1309,** 1–14.
Reynolds, H. C. (1945). *J. Mammal.* **26,** 361–379.
Reynolds, H. C. (1952). *Univ. Calif., Berkeley, Publ. Zool.* **52,** 223–284.
Reynolds, H. C. (1953). *Sci. Am.* **188,** 88–94.
Romer, A. S. (1959). "The Vertebrate Story." Univ. of Chicago Press, Chicago, Illinois.
Russell, E. M. (1970). *Z. Tierpsychol.* **27,** 385–404.
Sanderson, I. T. (1955). "Living Mammals of the World." Hanover House, Garden City, New York.
Schaller, G. B. (1972). "The Serengeti Lion." Univ. of Chicago Press, Chicago, Illinois.
Seiler, B. W. (1828). *Isis von Oken* **13,** 475–477.
Sherwood, B. F., Rowlands, D. T., Jr., Hackel, D. B., and Lemay, J. C. (1969). *Lab. Anim. Care* **19,** 494–499.
Shirer, H. W., and Fitch, H. S. (1970). *J. Mammal.* **51,** 491–503.
Shupe, D. (1967). Paper presented at Western Psychological Association Convention, San Francisco.
Smith, L. (1941). *J. Mammal.* **22,** 201–202.
Snyder, F. (1965). "Psychoanalysis and Current Biological Thought," pp. 275–315. Univ. of Wisconsin Press, Madison.
Stodart, E. (1966). *CSIRO Wildl. Res.* **11,** 91–99.
Stretch, R. G. A., McGonigle, B., and Morton, A. (1964). *J. Comp. Physiol. Psychol.* **57,** 461–463.
Tate, G. H. H. (1933). *Bull. Am. Mus. Nat. Hist.* **66,** 6–236.
Thrasher, J. D. (1969). *Lab. Anim. Care* **19,** 67–70.
Thrasher, J. D., Barenfus, M., Rich, S. T., and Shupe, D. V. (1971). *Lab. Anim. Sci.* **21,** 526–536.
Tilley, M. W., Doolittle, J. H., and Mason, D. J. (1966). *Psychol. Rep.* **19,** 593–594.
Tyndale-Biscoe, C. H., and Mackenzie, R. B. (1976). *J. Mammal.,* **57,** 249–265.
Van Twyver, H., and Allison, T. (1970). *Electroencephalogr. Clin. Neurophysiol.* **29,** 181–189.
Walker, E. P. (1964). "Mammals of the World," Vol. 1. Johns Hopkins Press, Baltimore, Maryland.
Weber, C. G. (1955). *J. Mammal.* **36,** 284–286.
Wiedorn, W. S. (1954). *Science* **119,** 360–361.
Wilson, E. O. (1975). "Sociobiology." Harvard Univ. Press, Cambridge, Massachusetts.
Winge, H. (1923). *Pattedyr. Slaegter* **1,** 56–86.
Wiseman, G. L., and Hendrickson, G. D. (1950). *J. Mammal.* **31,** 331–337.
Wolf, H. G. (1970). *Marsupial Newsl.* **2,** 1.
Wünschmann, A. (1966). *Z. Tierpsychol.* **23,** 56–71.
Zetek, J. (1930). *J. Mammal.* **11,** 470–471.

# 6

# The Development of the Lymphatic and Immunohematopoietic Systems

B. J. BRYANT



## I. THE LYMPHATIC SYSTEM

Systematic study of the marsupial lymphatic system has been limited to American opossums. McClure and Silvester (1909), Wood (1924), Zimmermann (1940), and Block (1964) examined *Didelphis virginiana.* Azzali and DiDio (1965) examined the Brazilian species, *Didelphis albiventris* and *Didelphis marsupialis,* trapped in Minas Gerais. Of these studies, only Zimmermann's (1940) dealt comprehensively with the histogenesis and adult form of the entire system. Azzali and DiDio (1965) and Wood (1924) examined only the adult system, McClure and Silvester (1909) only the adult thoracic duct, and Block (1964) dealt only with lymph node histogenesis. Observations other than these have been incidental. Virtually nothing is known of the system in other marsupials.

## A. Lymph Vascular System

The marsupial lymph vascular system, as represented by the opossum, displays arrangements that differentiate it from both the typical eutherian and reptilian plans. It nonetheless shares definitive aspects of each, giving the impression of intermediacy and of overlapping them. The main lymph trunks of adult and pouch young opossums as visualized by injection or dissection of gross specimens or by microscopic examination of serial whole-body sections are shown in Figs. 1 to 5. To obviate repetition, statements not referenced in the following are based on Zimmermann's (1940) and Azzali and DiDio's (1965) findings, where they are mutually confirmatory.

### 1. Adult Form of the System

**a. Deep Lymphatic Vessels.** The deep collecting lymph vessels of the posterior extremities and gonads follow the blood vessels and empty into the paraaortic lymph nodes situated at the caudal pole of each kidney. The lumbar lymphatics emerge from these nodes as paired channels running forward along the inferior vena cava and the abdominal aorta to the level of the kidney hilum, where they anastomose into a plexus corresponding to the eutherian cysterna chyli. This plexus consists of two sacciform dilatations of 1 cm length, located dorsolaterally to the aorta between the medial crura of the diaphragm, and joined below the aorta by one to four large vessels. The mesenteric lymph trunk contributes to the left dilatation and the hepatoduodenopancreatic lymph trunk contributes to the right. The thoracic ducts emerge bilaterally from these dilatations to run cranially in close relation to lateral aspects of the aorta and esophagus. The duct, frequently paired with multiple interconnections, admits visceral thoracic lymphatic tributaries in the anterior thorax. These tributaries are complex, differing considerably from the eutherian arrangements, and are detailed in relation to the regional nodal system in Section I, B, 1.

Considerable variability of form characterizes the thoracic duct of didelphids. Azzali and DiDio (1965) documented this variability in a large series of Brazilian opossums, finding regular, but variable, partial bilateral duplication of the thoracic duct. The efferents of the cisterna chyli, in their more typical arrangements, fused variably between the 12th and 6th thoracic vertebra levels, crossed to the left at the 4th to 3rd thoracic vertebra levels, and joined the left venous jugulosubclavian confluence at the supraclavicular fossa (Fig. 1). Other specimens revealed slender rami of the main anterior thoracic duct passing through the posterior cranial mediastinal lymph node into the right jugulosubclavian confluence (Fig. 1). Less frequently, the main terminal portion of the thoracic duct was found to empty into the right jugulosubclavian confluence. However, Zimmermann's (1940) examinations of Virginia opossums revealed a regular and equal bifurcation of the thoracic duct in the anterior thorax in a significant minority of specimens. The bilateral rami followed the dorsomedial aspect of the precava to the small, posterior cranial mediastinal lymph nodes on the respective sides. The nodal efferents opened bilaterally as cervical segments of the thoracic duct into the blood near the junction of external jugular and cephalic veins.

This diversity of observation was presaged in the early work of McClure and Silvester (1909), who found in one of five mature *D. virginiana* individuals a bilateral termination of the thoracic duct resembling Zimmermann's (1940) later observations. However, the thoracic duct had nonpaired terminations in the remaining four animals, one of them in the right venous jugulosubclavian confluence, three in the left (Fig. 2). Results thus show the latter lymphaticovenous arrangement as the more regular one for *Didelphis*, as well as being the typical one in Eutheria (Yoffey and Courtice, 1970). Bilateral thoracic duct terminations are also found in eutherian species, including man, but at frequencies that are considerably lower than are apparent in opossums (Kampmeier, 1970). Other lymphaticovenous communications than these have not been found in the thorax or abdomen of *Didelphis*, although Huntington (1910) reported a lymphatic outlet into the posterior vena cava in one specimen of kangaroo, *Macropus rufus*. Partial bilateral duplication of the thoracic duct, the regular arrangement in *Didelphis*, has also been observed in *Macropus*. Hodgkin, examining a specimen of adult kangaroo (Owen, 1847), found two thoracic ducts extending into the thorax from the plexiform cisterna chyli. The right duct crossed to the left at the 7th thoracic vertebra level. The combined trunk, after further division and reunion, terminated at the left jugulosubclavian confluence.

Asymmetry of the thoracic duct and its terminations is a late acquisition in evolution, as evidenced by the regular presence of paired right and left thoracic ducts from myxinoid Agnatha to Aves and by the "memory" of

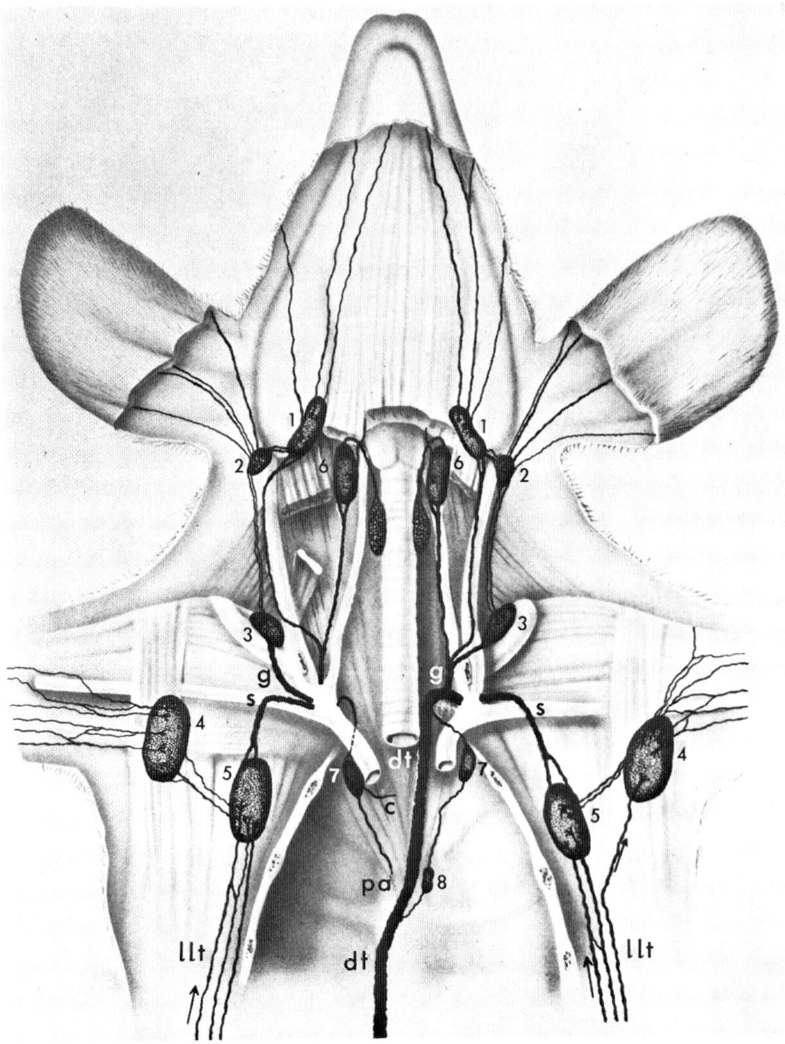


Fig. 1. *Didelphis albiventris*. Adult, anterior ventral view. Lymph nodes: 1, submandibular; 2, auriculoparotid; 3, superficial cervical; 4, superficial axillary; 5, deep axillary; 6, deep cervical; 7, posterior cranial mediastinal; 8, posterior middle mediastinal. Lymphatic vessels: llt, lateral lymph trunk; c, anterior mediastinal mode efferent vessel; dt, thoracic duct; g, jugular trunk; pa, paraesophageal node efferent vessel; s, subclavian trunk. Modified from Azzali and DiDio (1965).

this paired structure in eutherian, including human, embryos (reviewed by Kampmeier, 1970). Marsupial thoracic duct structuring thus seems transitional between the complete symmetry of premammalian vertebrates and

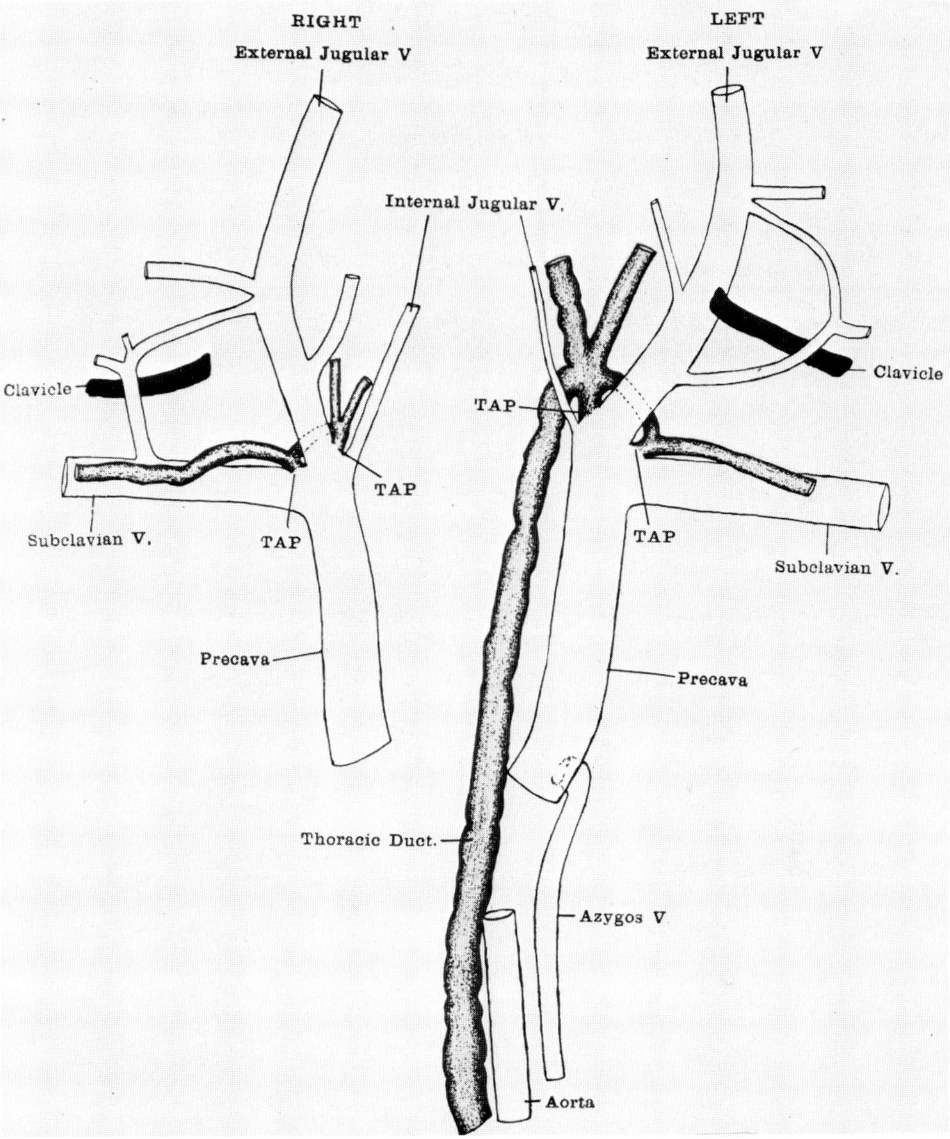


Fig. 2. *Didelphis virginiana.* Adult, region of lymphaticovenous communications, ventral view. TAP, lymphaticovenous juncture point. Modified from McClure and Silvester (1909).

the typical asymmetry of Eutheria. However, it should be noted that certain eutherian orders, e.g., Rodentia, regularly display partially duplicated thoracic ducts (Kampmeier, 1970).

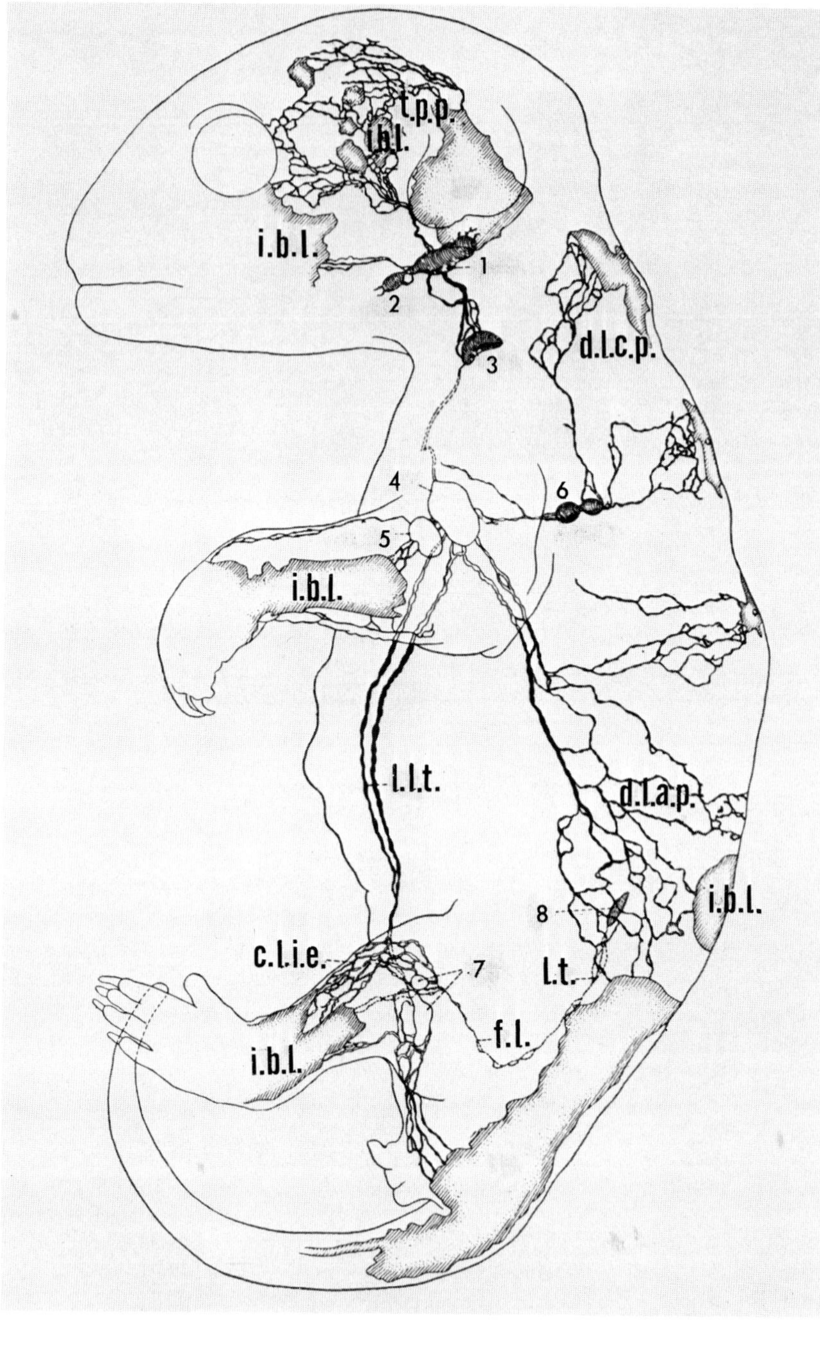
t.p.p.
i.b.l.
i.b.l.
1
2
3
d.l.c.p.
4
6
5
i.b.l.
l.l.t.
d.l.a.p.
i.b.l.
8
c.l.i.e.
7
l.t.
f.l.
i.b.l.

Other deep lymphatic trunks include the visceral thoracic system (cf. Section I,B,1), the subclavian, and the jugular lymphatics. The subclavian lymphatic fuses from efferents of the deep axillary lymph nodes, associates with the subclavian veins, and empties into the blood jointly or separately with the thoracic duct, bilaterally at the jugulosubclavian angle. The jugular lymphatics traverse the series of nodes in the head and neck, draining lymph from these regions to the blood by a common outlet with the subclavian and thoracic ducts or by an independent one in the jugulosubclavian region (Fig. 1).

**b. Superficial or Peripheral Lymphatic Vessels.** Among the most conspicuous and, phylogenetically, most interesting of peripheral lymphatics in the opossum are the paired vessels paralleling the body axis on each side and extending from the inguinal to the deep axillary lymph nodes (Figs. 3 and 4). One of the two vessels on each side may be smaller than the other and may unite with it. Tributaries to either of them from the abdominal wall are relatively scarce. They are abundantly valved. These lateral lymph trunks, like the thoracic duct, are very ancient channels and are fully homologous with the lateral line lymphatics found in lower vertebrates from myxinoid cyclostomes to reptiles (Kampmeier, 1970). These vessels serve in lower species as the chief somatic collectors of lymph in the trunk and tail. They flourish for a time in embryonic life of birds and mammals but have not been found postnatally in any of them, excepting opossums. Their survival in adult opossums, along with the structuring of the marsupial thoracic duct, evidences the somewhat more reptilian character of the marsupial lymphatic system compared to that of Eutheria.

Wood (1924) described the lateral lymphatics of the adult opossum as "large thoraco-epigastric collecting trunks extending parallel to the thoraco-epigastric veins." Azzali and DiDio (1965) described them as "subcutaneous abdomino-thoracic efferent collecting vessels." Zimmermann (1940) noted that no other lymphatics of the pouch young could be demonstrated by injection as easily or constantly, which attests to their probable function as collaterals for the thoracic duct in the conveyance of lymph from the posterior regions of the trunk and tail. Injections at the tip of the opossum tail bring into view lymph vessels on both sides of the tail

Fig. 3. *Didelphis virginiana.* Pouch young male, 5.5-cm crown-rump length. Cleared specimen with injection blisters (i.b.l.) and extravasates near points of injection. Lymph nodes: 1, parotid; 2, mandibular; 3, superficial cervical; 4, deep axillary; 5, ventral superficial axillary; 6, dorsal superficial axillary; 7, superficial inguinal; 8, lumbar. Lymphatic vessels: l.l.t., lateral lymph trunk; d.l.a.p., doroslateral abdominal plexus; d.l.c.p., dorsolateral cervical plexus; t.p.p., temporoparietal plexus; f.l., femoral lymphatic; l.t., lumbar trunk; c.l.i.e., cutaneous lymphatics of inferior extremity. Modified from Zimmermann (1940).

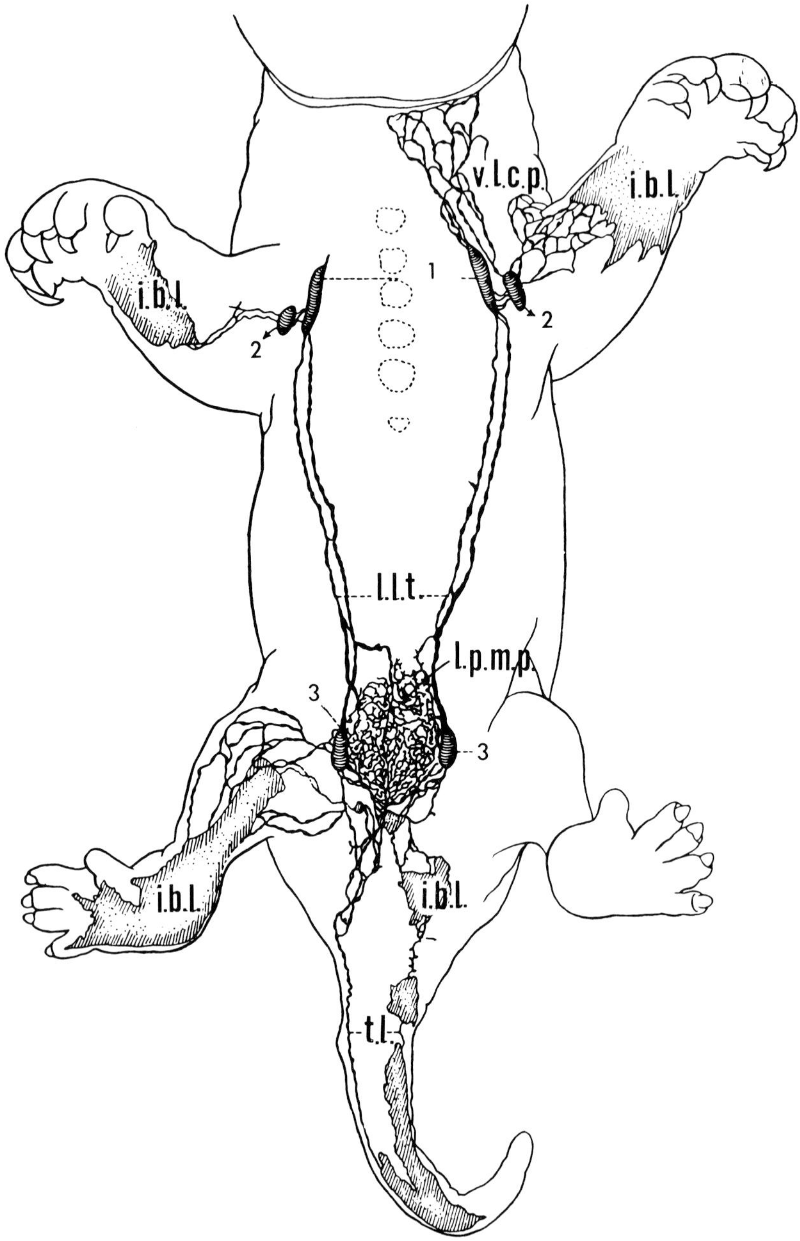


Fig. 4. *Didelphis virginiana.* Pouch young female, 4.5-cm crown-rump length. Cleared specimen with injection blisters (i.b.l.) and extravasates near points of injection. Lymph nodes: 1, deep axillary; 2, ventral superficial axillary; 3, superficial inguinal. Lymphatic vessels: v.l.c.p., ventrolateral cervical plexus; l.l.t., lateral lymph trunks; l.p.m.p., lymphatic plexus of the marsupial pouch; t.l., tail lymphatics. Modified from Zimmermann (1940).

that may in effect be caudal segments of the lateral lymph trunks (Zimmermann, 1940). They unite with each other in a plexus at the ventral root of the tail and end in the inguinal lymph nodes, the efferents of which emerge cranially as the lateral lymph trunks.

The presence of the marsupial pouch in the female causes the lymphatics in this region to have a much different pattern in the two sexes. The skin covering the pouch floor in the female is provided with a more intricate lymphatic network than is found in any other cutaneous region (Fig. 4). Afferents to this plexus arrive from the one ventral to the tail root and its efferents go directly to the inguinal lymph nodes and the lateral lymph trunks. The pouch plexus drains lymph from the mammary field, which in pouched marsupials is limited to the pouch confines. These observations are consonant with the supposition that equates the reptilian embryo lateral line with the mammalian embryo "milk ridge" (Broman, 1921). The appreciably more ventral position of the marsupial lateral lymph trunks compared with their lateral line antecedents is understandable on this basis.

## 2. Developmental Aspects

Ontogenetic mechanisms in the mammalian lymph vascular system provoked heated controversy among anatomists for several decades. The main issue, which is still not completely settled, was whether lymphatic endothelium is derived from the undifferentiated mesenchyme—the local origin theory—or whether it is derived from embryonic veins—the angioblast theory of His (cf. Sabin, 1901). However, Kampmeier (1970) has impressively marshaled the evidence substantiating the local theory, which must now be regarded as the more favored of the two views. According to this theory, lymphatic endothelium derives from periintimal islets of mesenchymal cells lying in the near vicinity or in contact with venular bordering cells of a blood vascular plexus. These islets are the cellular anlagen of periintimal mesenchymal clefts or open spaces which form by progressive extravascular infiltration of lymph, a process especially pronounced in, but not limited to, the vicinity of venous plexuses. This lymph at first collects in discontinuous pools (i.e., the mesenchymal clefts), but these become confluent by endothelial proliferation and extension as the lymph volume increases, and an outlet is sought. The latter is provided by degeneration of the plexiform venules. The incipient lymph stream follows the path of least resistance, usurping the decadent venules. The phenomenon of an orderly system of vessels being destroyed to make way for another one in precisely the same site, both of them functionally related, has precedents in other organ systems in development. The skeletal system provides the classic example of this in its radical demolition and rebuilding processes as it proceeds from the chondral to the osseous stage. This mechanism also accounts for the atypical lymphatic communications found at times in all mammalian species with such veins as the azygos, renal, and inferior vena cava (Yoffey and Courtice, 1970).

Onset of lymph vascular system development in all amniotes (i.e., mammals, birds, and reptiles) is marked by genesis of the jugular lymph sacs. These are transient structures disposed bilaterally in the embryonic cervical region at a position corresponding to the confluence of lymphatic vessels discharging at vent III of fishes, a region destined to support the typical lymphaticovenous communications of higher vertebrates (Kampmeier, 1970). Four other lymph sacs form shortly after this. These are the paired iliac sacs and two unpaired sacs, the retroperitoneal sac and the cysterna chyli.

The jugular lymph sacs are foreshadowed in early embryonic life by development of an evanescent blood vascular plexus arising dorsolaterally from the cardinocuverian venous confluence on both sides. This transient plexus recapitulates the venolymphatic plexus presaging the anterior lymph hearts of amphibians, but gives way in amniotes merely to enlarging mesenchymal lymph spaces (Kampmeier, 1970). These coalesce around and between the degenerating veins forming a large lymph space. Valved openings form between this space and the jugulosubclavian veins. The valves prevent reflux of blood and also limit the exit of lymph such that the sacs become disproportionately large, looking like vast collecting reservoirs. The sacs are joined at this time by the major anterior lymph trunks through their independent, centripetal growth. The sacs regress soon thereafter leaving only their terminal portions to serve as the main lymphatic passages to the veins, as their larger, circumferential portions give rise to the cervical and axillary lymph nodes (Yoffey and Courtice, 1970).

Zimmermann's (1940) paper on the opossum embryo and pouch young was one of the more influential of the seminal papers establishing the validity of these amniote developmental processes (Kampmeier, 1970). It also remains the only available study of marsupial lymphatic system ontogeny. The following account therefore summarizes the essentials of Zimmermann's monograph as it relates to lymph vascular development.

The opossum lymphatic system has its inception on the eleventh gestational day, 3 days before term, with the formation of the jugular lymph sac anlagen. These arise bilaterally as periintimal mesenchymal clefts circumjacent to the vessels of a retrogressing venous plexus lying dorsal to the precardinal veins. Coalescence and expansion of the perivascular clefts during the eleventh and twelfth gestational days lead to the formation of progressively larger lymphatic spaces, culminating at the close of the twelfth prenatal day in the establishment of the definitive jugular lymph sacs. The defunct venous plexuses separate from the parent cardinal veins during the early part of the twelfth gestational day, but a valved connection between the jugular lymph sac and the cardinal vein forms within an additional half-day. This valve is the only one to appear in prenatal life in either the lymphatic or venous systems.

The jugular lymph sacs at the end of the twelfth gestational day consist of two widely divergent axillary and cephalic portions connected by a more attenuated intermediate portion. The axillary portion diffuses into the dorsolateral region of the shoulder by coalescing from a perivascular mesenchymal plexus. The cephalic portion, on the other hand, presents a medial diverticulum destined to receive the developing thoracic duct. The thoracic ducts and lateral lymph trunks originate at this time. The former arise independently along the aorta from discontinuous mesenchymal clefts associated with retrogressive venules derived from the cardinal veins. The latter arise in similar fashion by fusion of mesenchymal clefts along the external mammary blood vessels. These events result in joining of the jugular lymph sacs with the thoracic ducts and the lateral lymph trunks on both sides during the first half of the thirteenth gestational day. At birth on the following day, the jugular lymph sacs are maximally developed as vast cavities superseding in size all other structures of the cervical and axillary regions. The communication of the major lymphatic trunks to the jugular lymph sacs, and the valved communications of the latter to the blood, signify that the anterior lymphatic system of the opossum is functional at birth, even if structurally immature.

The thoracic ducts at birth receive as tributaries the superior and inferior pulmonary lymphatics and the esophageal lymphatic vessel. The latter is significant in that it establishes an early communication through the diaphragm between the upper thoracic duct and the gastric lymphatics arising at the cardia. However, somewhat more complex processes of lymph vascular genesis obtain with respect to the major abdominal lymphatics. Primordia of two separate lymphatic plexuses are found in this region at birth. Zimmermann (1940) designated these as the mesenteric and subcardinal and observed them to originate directly from mesenchymal clefts not associated with evanescent venules. Confluence and expansion of these plexuses proceed rapidly on the first postnatal day, producing the retroperitoneal lymph sac. Continuity of this major abdominal lymphatic structure with the thoracic ducts is brought about by caudal extensions of thoracic portions of the latter, completed by the third postnatal day (13-mm pouch young). The mandibular, retropharyngeal, and cutaneous lymphatics, all developing independently in the anterior half of the newborn in association with retrograde venous plexuses, also join the jugular lymph sacs by centripetal growth within the first 3 days of postnatal life. Involution of the opossum jugular lymph sacs encompasses the third and fourth postnatal days. Two fat bodies form and usurp most of the sac spaces, leaving only a few narrowed spaces serving as the permanent lymphatic trunk connections.

The earliest valves of the lymphatic system, other than those terminal to the jugular lymph sacs, are formed within the cranial subcutaneous lymph plexuses of 3- to 4-day-old (14-mm) pouch young. They develop in the deep anterior lymphatics of 4-day-old (15-mm) pouch young and in peripheral lymphatics of the anterior trunk and forelimbs of 5-day-old (16-mm) pouch young. Much later, they are found in the deep abdominal lymph vessels (12-day-old pouch young) and in the lymphatics of the tail and the medial aspects of the hind limbs (20-day-old pouch young).

### B. Lymph Nodal System

Data on the marsupial lymph nodal system are limited to studies of various American opossum species by Zimmermann (1940), Block (1964), and Azzali and DiDio (1965). A striking general finding of these studies is the smaller number of lymph nodes in didelphids as opposed to comparably sized Eutheria. Zimmermann (1940) found a total of 46 lymph nodes in 22 stations in adult *D. virginiana* individuals. Azzali and DiDio (1965) found a total of 45 lymph nodes in 19 stations in adult *D. albiventris* individuals. The major determining factor is the regular singularity of lymph nodes in the regional nodal stations of the opossum and their variable multiplicity in Eutheria. A second factor is the absence in opossums of any nodes in certain of the regional stations, e.g., the popliteal, iliac, ischial, subpubic, or epitrochlear, common to Eutheria. Despite this, lymph nodes in the opossum are always interposed between individual organs or drainage regions and the blood, a situation not true in Eutheria (Azzali and DiDio, 1965).

Quantitatively, the opossum lymph nodal system, if it is not quite eutherian, is nonetheless definitively mammalian. The relatively primitive lymphoid nodules characteristic of amphibians and reptiles and the presence of four, or at most, six nodes of macroscopic size in birds indicate the impressive quantitative advance represented by the opossum lymph nodal system. We shall see in Section II,B,2 that lymph nodal organizational quality also conforms closely with the advanced mammalian type.

The arrangement and development of opossum lymph nodes are described according to anatomic region in Sections I,B,1 and I,B,2. Again, statements not referenced are based on mutually confirmatory findings of Zimmermann (1940) and Azzali and DiDio (1965).

### 1. Adult Form of the System

Lymph nodes of the cephalic, cervical, and superficial thoracic regions of the opossum range from 6 to 15 mm in length or diameter and are regularly bilaterally paired (Fig. 1). On the other hand, nodes of the deep thoracic and abdominal regions are usually smaller and unpaired (Fig. 5).

**a. Cephalic and Cervical Nodes.** An auriculoparotid node located below the ear at the caudal pole of the salivary gland collects lymph vessels draining the auricular, parietal, temporal, and occipital cutaneous regions. A submandibular node situated at the angle of the jaw receives lymphatics draining the nose, lips, oral mucosa, and related structures. The nodal efferents follow the external jugular vein, are joined by lymph vessels draining the superficial lateral aspect of the neck, and terminate in the superficial cervical node lying in the supraclavicular region. This node is variably present, sometimes replaced by an ampulla of the thoracic duct (Zimmermann, 1940), but receives in either instance efferents of the nearby, large deep cervical node. The latter node in turn drains lymph from the thyroid gland, pharynx, and root of the tongue.

**b. Superficial Thoracic and Abdominal Nodes.** The cutaneous lymphatics of the dorsum of the neck, the forelimb, and part of the thorax lead to the superficial axillary node, embedded in the axillary adipose tissue. Multiple efferents radiate from this node to the deep axillary node. The latter, a frequently doubled (Zimmermann, 1940) large node, receives lymph from the deep forelimb structures and also serves as the anterior terminus of the phylogenetically significant lateral lymph trunks (cf. Section I,A,1) originating with the inguinal lymph node.

**c. Deep Thoracic Nodes.** The deep thoracic nodal system is only partially bilaterally duplicated. The symmetric components of it begin with peripheral lymphatics of the ventral sector of the diaphragm merging on each side and following the internal mammary arteries to the small paired sternal node that lies ventral to the thymus on the middle of the third rib. Its efferent runs cranially to the paired anterior mediastinal node, lying dorsal to the thymus and lateral to the trachea at the level of the first rib. This node also receives lymphatics from the thymus and heart. Its efferent runs to the paired posterior cranial mediastinal lymph node. The efferents of these nodes discharge into the venous system on each side near the jugulosubclavian angle.

Other nodes of the deep thorax are not bilaterally paired and their occur-

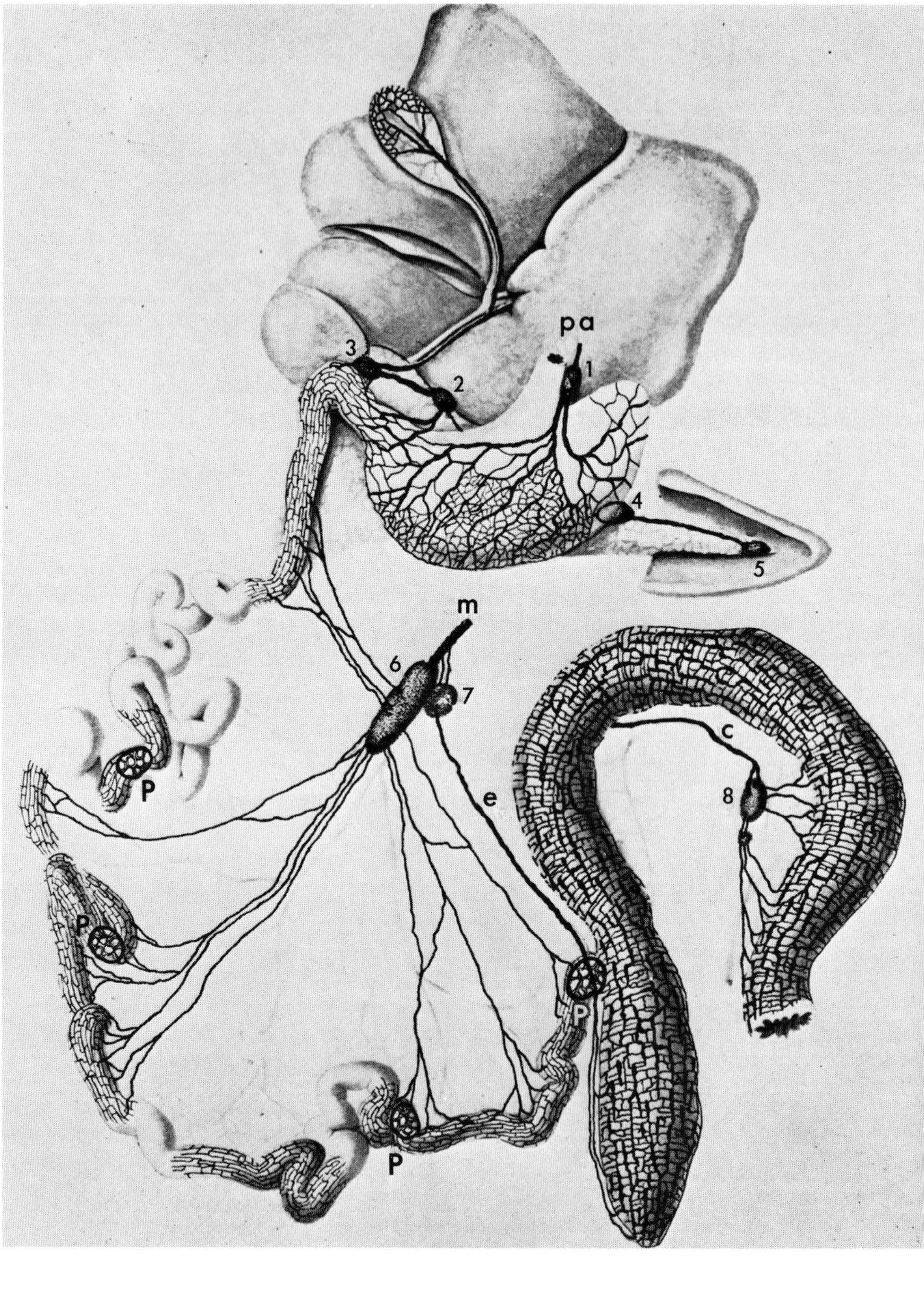
pa
3
2
1
4
5
m
6
7
c
8
e
P
P
P
P

rence varies with species and observers. The right posterior cranial mediastinal node is better developed than the left one and receives lymph from two additional nonpaired thoracic nodal systems. One is the system comprising the paraesophageal (Azzali and DiDio, 1965) or retroesophageal (Zimmermann, 1940) lymph node, lying on the left aspect of the esophagus near the diaphragm and draining dorsal sectors of the diaphragm and a node at the gastric cardia. The other is the nodal system comprising the posterior middle mediastinal node and those distal to it, the cranial and caudal bronchial nodes (Azzali and DiDio, 1965). The cranial bronchial node lying ventral to the tracheal bifurcation drains lymph from the trachea, most of the lungs, and the caudal bronchial node. The latter in turn drains lateral sectors of the diaphragm and portions of the left lung. On the other hand, Zimmermann (1940) found only one bronchial node at the tracheal bifurcation, sending slender short rami directly to the right and left thoracic ducts. Moreover, only the latter observer found five pairs of minute intercostal nodes, lying in the sixth to tenth intercostal spaces.

**d. Abdominal Nodes.** The paired anterior paraaortic nodes or lumbar nodes lying dorsal to the caudal pole of the kidney are the largest of the parietal series. Their efferent vessels transmit lymph to the cisterna chyli and thoracic ducts. Their afferents collect lymph from the colic lymph node, the iliac region, and the gonads. The latter lymph first traverses the nonpaired posterior paraaortic node, found only in the female usually on the right side at the ilial–vena cava confluence, draining ventral aspects of the gonads, the deep lymphatics of the posterior extremities, and the right vesicle node. The paired vesicle node drains dorsal aspects of the gonads and the bladder in both sexes.

A cardial node at the esophageal end of the stomach, a pyloric node at its pyloric end, two gastric nodes at its lesser curvature, two or three pancreatic nodes, and a splenic node at the splenic hilum are located in the anterior abdomen, draining the respective organs (Fig. 5). The node at the head of the pancreas additionally drains the liver and gallbladder. A colic node and the mesenteric lymph node are found in the posterior abdomen. The first, situated in the central mesocolon, drains the transverse, ascending, and sigmoid colon. The second node, the largest (2–3 cm long) in the opossum, drains the superficial and deep networks of the jejunum and ileum. A cecocolic node draining the cecum is associated with it (Fig. 5).

Fig. 5. *Didelphis albiventris*. Adult, abdominal viscera (displaced), ventral view. Lymph nodes: 1, cardial; 2, pyloric; 3, caput pancreatic; 4, corpus pancreatic; 5, cauda pancreatic; 6, mesenteric; 7, cecolic; 8, colic. P, Peyer's patch. Lymphatic vessels: pa, paraeosphageal node afferent vessel; m, mesenteric trunk; c, colic node efferent vessel; e, cecolic node afferent vessel. Modified from Azzali and DiDio (1965).

## 2. Developmental Aspects

The development of the opossum lymph nodal system has been studied in depth only by Zimmermann (1940). The first nodal primordia to develop are those of the paired deep cervical nodes on the third postnatal day (13-mm pouch young). Their early differentiation from cephalic portions of the involuting jugular lymph sacs (Section I,A,2), confirmed by Block (1964), is followed on the next day (14-mm pouch young) by the differentiation of more than a dozen other nodal primordia. Three pairs of them are the additional nodal anlagen generated by the jugular lymph sacs. These comprise the superficial cervical nodes arising, like the deep ones, from cephalic portions of the jugular lymph sacs, and the deep and superficial axillary nodes arising from their axillary portions. Other nodes differentiating at this time comprise all of the nodes of the cephalic region, the paired sternal nodes, the bronchial nodes, the mesenteric node, the paired anterior mediastinal nodes, the paraesophageal node, and the cranial pancreatic (coeliac) node. The last three nodal groups differentiate in intimate relation to the thoracic ducts at the relevant levels. Other nodal anlagen develop in similar relation to regional lymph plexuses. This entire group of nodes is designated as early lymph nodes by Zimmermann (1940). Other nodes differentiating between the fifth and twentieth days form a group of delayed lymph nodes. On the fifth postnatal day (15-mm pouch young), the inguinal and paraaortic (lumbar) nodal pairs begin development. On the sixth or seventh postnatal day (16-mm pouch young), the intercostal nodes develop. The remaining lymph nodal primordia all develop by the twelfth postnatal day (22-mm pouch young), with the exception of the mesocolic node, whose anlage does not become perceptible until the 30-mm pouch young stage. Opossum lymph nodes thus follow a generalized, if imperfect, anterioposterior developmental gradient. A similar but less pronounced gradient is found in eutherian development (Papermaster and Good, 1964).

# II. THE IMMUNOHEMATOPOIETIC COMPLEX

## A. General Characteristics

Organization and function in the immunohematopoietic organ system of adult marsupials parallel the eutherian plan, but as with the lymphatic system (Section I), a number of noneutherian peculiarities are evidenced which recall the precursory reptilian plan.

The thymus in Eutheria is a thoracic organ closely paired in the anterior mediastinum, although rarely, e.g., in the mole (*Talpa*) and guinea pig (*Cavia*), it is positioned subcutaneously in the neck. American marsupials may also be limited to one set of paired thoracic thymuses, if the various examined species of opossum (*Didelphis*) (Zimmermann, 1940; Block, 1964; Azzali and DiDio, 1965) and the pouchless marsupial mouse (*Marmosa robinsoni*) (Bryant and Shifrine, 1974) are representative. However, Australian marsupials variably possess one to three sets of paired thymuses. This was recognized early by Simon (1845), who first identified the thymus in marsupials. The koala bear (*Phascolarctos*), wombat (*Vombatus*), bandicoot (*Perameles*), and native cat (*Dasyurus*) have one set of thymuses, but they are located cervically in the former two species and thoracically in the latter two (Symington, 1898, 1900; Fraser and Hill, 1916). The kangaroo (*Macropus*) and quokka (*Setonix*) possess both cervical and thoracic paired thymuses (Symington, 1898; Stanley, 1966; Yadav and Papadimitriou, 1969). The brush-tailed possum (*Trichosurus vulpecula*) has, additionally to cervical thymuses, two pairs of thoracic thymuses (Fraser and Hill, 1916). The cervical thymuses ordinarily lie ventral to the origin of the sternomastoid muscle and seem generally to be absent in polyprotodont (omnivore) marsupials but present in diprotodont (herbivore) marsupials. This pattern probably expresses evolutionary affinities, as some reptiles also develop thymuses at multiple levels (Fraser and Hill, 1916). Despite this anatomic diversity and their radically different embryonic origins (Section II,B,1), these marsupial thymuses are unremarkable histologically, and possible functional differences between them are only beginning to be studied (Section III). They, like their eutherian equivalents, contain plasma cells and lymphoid follicles at much lower frequencies than are found in avian thymuses (Thorbecke *et al.*, 1957).

The structure of lymph nodes in adult opossums (*D. virginiana*) (Block, 1964) and *Marmosa* (*M. robinsoni*) (Miller, 1969; Bryant and Shifrine, 1974) does not differ from that found in adult Eutheria. Moreover, paracortical nodal regions in adult *Marmosa* abundantly possess cuboidal endothelium-lined postcapillary venules, which are specialized in rodents for transport of lymphocytes recirculating from blood to nodal parenchyma (Gowans and Knight, 1964). The surveillance function postulated for recirculating pool lymphocytes in Eutheria (Burnet, 1969) would thus seem operative in adult marsupials.

Gastrointestinal lymphoid tissues also seem unremarkable in marsupials. Tonsils and Peyer's patches have been observed in *D. albiventris* (Fig. 5) and in *M. robinsoni* (Bryant and Shifrine, 1974).

However, the spleen is remarkable in that it has in *D. virginiana* (Block, 1964) and *M. robinsoni* (Bryant and Shifrine, 1974) a triradiate or T-shape, like reptilian spleens (Jordan, 1938). Moreover, the postnatal attenuation of splenic hematopoiesis, found in most Eutheria, is not found in either of these marsupial species. Also, both species evidence in adulthood an exclusively intravascular mode of erythro- and megakaryocytopoiesis in bone marrow and spleen (Fig. 6), unlike Eutheria but like reptiles and birds (Jordan, 1938). This arrangement probably reflects the absence of mechanisms of delivery of relatively immobile types of blood cells from generative parenchyma to blood, for granulocytopoiesis is extravascular at all stages of development of these marsupial species.

We owe to Block (1964) the one available study of marsupial lymphoreticular and hematopoietic organ development. To avoid undue repetition, specific reference to his comprehensive study of opossum (*D. virginiana*) pouch young is made later only when studies dealing with individual organs are also described. In summary, current data (Sections II,A–D) show that marsupials share the salient developmental characteristics of Eutheria. These include the three overlapping stages of erythropoiesis, i.e., the yolk sac, hepatic, and myeloid stages; the two ordered stages of granulocytopoiesis, i.e., the mesenchymal and myeloid stages; and the thymus-first pattern of lymphoid organ development.

## B. Development of Lymphoreticular Organs

Unlike hematopoietic cells (cf. Section II,C), lymphocytes differentiate from mesodermal stem cells invading specialized inductive epithelia (Burnet, 1969). These epithelia, perhaps exclusively arising at points of early contact of ectoderm and endoderm and usually containing elements of both, give rise to the thymus, the avian bursa of Fabricius, and the mammalian tonsils and lymphoid tissues of Waldeyer's ring (Ruth *et al.*, 1964). The thymus in all vertebrates is the first of these epithelial organs to develop and it differentiates the first definitive lymphocytes in the body (Papermaster and Good, 1964). Thymectomy performed in Eutheria at sufficiently early developmental stages effects in later life a relative lymphopenia (especially of small lymphocytes) in peripheral lymphatic tissues, i.e., in lymph nodes, spleen, Peyer's patches, and blood. Thymectomy in opossum (*D. virginiana*) pouch young brings about very similar deficiencies (Miller *et al.*, 1965), suggesting that the thymus acts in both marsupials and eutherians as a mediator or progenitor of lymphoreticular organ lymphoid development.

### 1. Thymus

The epithelial primordia of the thymus arise at points of contact of branchial pouch ectoderm with pharyngeal pouch endoderm. The thymic

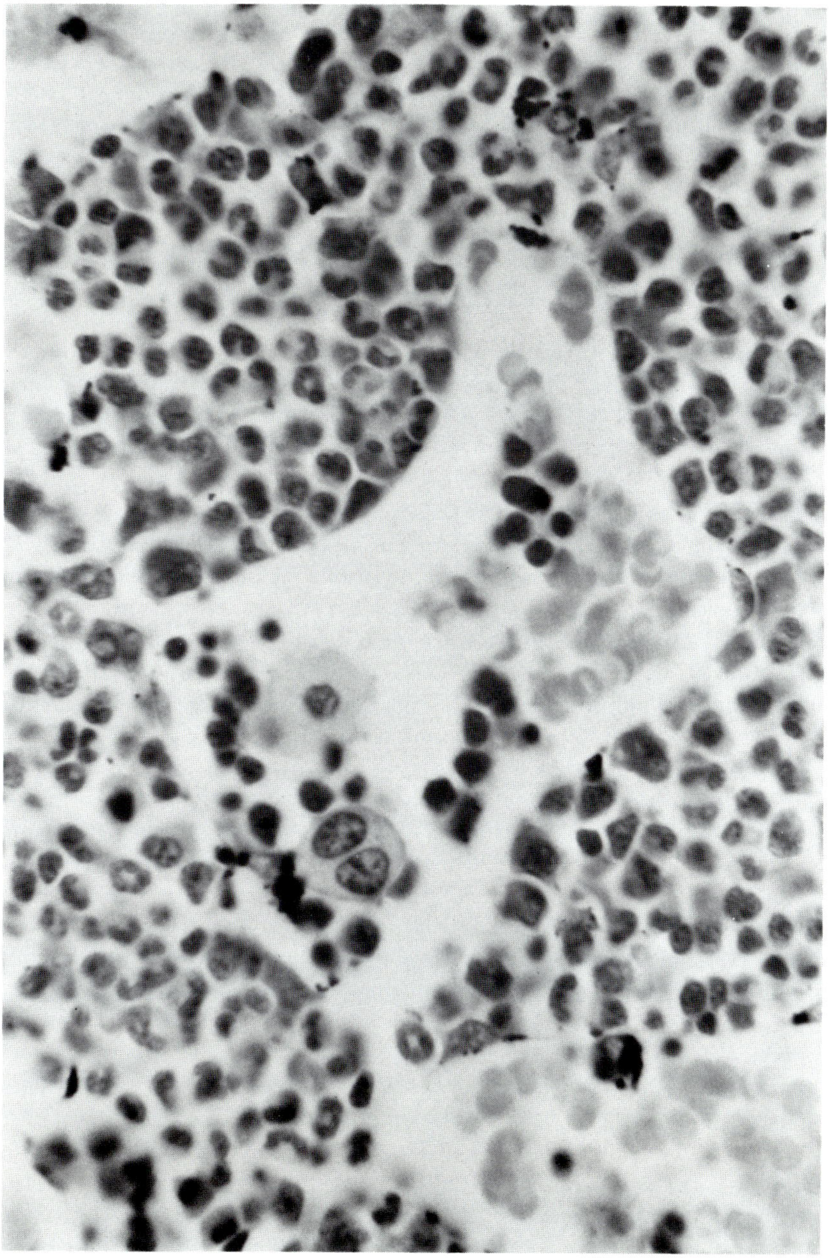

Fig. 6. *Marmosa robinsoni.* Adult bone marrow, H&E-stained 8-$\mu$m section. Typical pattern of intravascular erythro- and megakaryocytopoiesis and extravascular granulocytopoiesis. 1850×.

epithelial rudiments of reptiles arise as dorsal, mainly endodermal, derivatives of pouches I, II, III, and, in some, IV (Fraser and Hill, 1916). Those of birds contain mainly ectoderm, derive from both dorsal and ventral surfaces of pouches III and IV, and develop cervically (Romanoff, 1960). Those of Eutheria contain mainly endoderm, derive solely from ventral surfaces of pouch III, although IV also participates in some species, and migrate to the anterior mediastinum during development (Ruth *et al.*, 1964).

Marsupials show a diversity of thymus origins, recalling reptiles (Fraser and Hill, 1916). The thoracic thymuses arise, like their eutherian counterparts, as mainly endodermal derivates of pharyngeal pouch III (and IV in *Trichosurus*). However, as with birds, both dorsal and ventral walls of pouch III contribute. The cervical thymuses found in diprotodont marsupials (cf. Section II,A) arise as mainly ectodermal derivates of pouch II. Interestingly, the few Eutheria possessing a cervical thymus (e.g., *Cavia, Talpa, Sus*) derive it from pouch III; it contains mainly ectoderm and fails to descend (Fraser and Hill, 1916).

The thymus in all marsupials studied to date remains purely epithelial until birth, despite the earliness of development of the definitive thymic primordia (7-mm stage in *Trichosurus,* Fraser and Hill, 1916) and wide variation of gestational period. Thus, *Trichosurus vulpecula* after 20 days of gestation, *D. virginiana* after 12 days, *Setonix brachyurus* after 25 days (Yadav and Papadimitriou, 1969), and *Marmosa robinsoni* after 14 days (Bryant and Shifrine, 1974) all present at birth an epithelial thymus showing beginning capsularization by condensation of surrounding vascularized mesenchyme, beginning lobularization by connective tissue ingrowth, and beginning reticularization at the center of the epithelial mass. The thymus of Eutheria, excepting the hamster (*Cricetus*), always acquires lymphoid maturity well before birth, even if the gestational period is brief, e.g., the 21-day period of the mouse (*Mus*).

However, postnatal thymic development in marsupials proceeds rapidly. Significant accumulations of lymphocytes, the first anywhere in the body, are found on postnatal day 2 in the thymus of *Marmosa* (Bryant and Shifrine, 1974) and *D. virginiana* and on postnatal day 3 in the cervical thymus of *Setonix brachyurus* (Yadav and Papadimitriou, 1969). These lymphocytes, probably of hematopoietic origin (Burnet, 1969), first collect at the corticomedullary boundary in *Marmosa* and at the center of the lobule in *D. virginiana.* The subsequent rapid organ development is unremarkable in both species. However, the cervical and thoracic thymuses of *Setonix* have much different life histories (Stanley, 1966). The cervical thymus develops rapidly after birth, reaching maximum weight about halfway through pouch life (180 days). The main development of the thoracic thymus then ensues. Accordingly, a specific protective role for the cervical thymus against pouch flora has been proposed (Stanley, 1966).

## 2. Lymph Nodes

The lymph nodal system attains its greatest complexity of distribution (Section I,B,1), structure, and function in mammals. This contrasts with the associated lymph vascular system, which exhibits reduction and lessening of importance vis-à-vis the blood vascular system in the course of vertebrate evolution (Kampmeier, 1970). The lymphocytes initially populating lymph nodes and splenic follicles in Eutheria are thought to be cells peripheralized from lymphoepithelial organs, e.g., thymus, which are themselves not sites of immune reactivity (reviewed by Burnet, 1969). The lymphocytes in peripheral sites derive at later times mainly by homoplastic proliferative processes stimulated by antigen exposure. These populations increase gradually to adolescence when the contribution of newly peripheralized cells ceases by virtue of lymphoepithelial organ involution (reviewed by Papermaster and Good, 1964).

Marsupial lymph node histogenesis, as represented in *Marmosa robinsoni* (Bryant and Shifrine, 1974) and *Didelphis virginiana,* resembles the eutherian model in general terms, although differences exist that probably reflect exigencies of the relatively early assault of environmental pathogens on the developing marsupial. Chief among these is the very brief (2 to 3 days) interval elapsing in *Didelphis* and *Marmosa* between lymphocyte acquisition by the thymus and the first lymph nodes. The comparable interval in Eutheria ranges from 1 to 3 weeks (reviewed by Papermaster and Good, 1964). The brevity of this interval in marsupials probably expresses a need for rapid peripheralization of immunocompetent cells following lymphocyte formation in the thymus, which begins only postnatally in marsupials (Section II,A). The earliest stigmata of this peripheralization process are found on postnatal day 5 in *Marmosa* (Bryant and Shifrine, 1974) and *Didelphis,* when the jugular lymph sac-derived cervical and axillary lymph nodes, then in early stages of their condensation, begin to acquire greater concentrations of lymphocytes and granulocytic cells (at about a 1:1 ratio) than the surrounding mesenchyme. The lymphocyte:granulocyte ratio increases gradually over the following days and beginning corticomedullary differentiation becomes manifest on about postnatal day 10. The lymph sinuses then contain large numbers of hematopoietic cells, additionally to a few lymphocytes, reflecting the hemophoric function of the early lymph vascular system in transport of these cells from the diffuse mesenchyme to the blood (Kampmeier, 1970).

The lymph space content of hematopoietic cells and the parenchymal granulocytopoiesis begin to decrease after postnatal day 12 in parallel with decreasing myeloid metaplasia in the diffuse mesenchyme.

The cortex and medulla in the early group of lymph nodes (Section I,B,2) become clearly differentiated by postnatal day 20. Diffuse lymphocytes of varying sizes are then found in the cortex, but only small lymphocytes and some granulocytopoietic cells are found in the medullary cords. By postnatal days 45 to 65, the cortex forms a continuous zone interposed between the medulla and the subcapsular lymph space, and it evidences a continuing increase in density of large and medium lymphocytes. However, medullary cords evidence only small lymphocytes, very rare plasma cells, and residual granulocytopoiesis. Unlike other immunohematopoietic organs, the lymph nodes remain in a state of rapid evolution in the postnatal interval from 65 to 100 days, probably expressing necessary adjustments to varied environmental flora associated with the gaining of independence from the mother and the deprivation of protective colostrum (Section III). During this interval, as the nodes increase further in size, mainly by continued lymphocyte population expansion, primary and secondary (germinal center) nodules develop in the superficial cortex, and plasma cells increase greatly in the medullary cords. The more mature animals of this group evidence cessation of medullary granulocytopoiesis, but their germinal centers are smaller and lack the encompassing ring of small lymphocytes found in individuals older than 100 days.

### 3. Splenic Follicles

Splenic lymphatic tissue lags behind most lymph nodes in terms of early developmental events. The follicular anlagen are absent before postnatal day 10, as only unsheathed arterial capillaries are then found in the red pulp, except adjacent to the hilum. Arterioles lined by two or three layers of perivascular mesenchymal cells paralleling the vessel axis spring up throughout the splenic pulp in 10- to 12-day-old pouch young. A few lymphocytes invade the perivascular mesenchymal anlagen on about postnatal day 20, representing the beginning of follicular lymphoid tissue formation. Gradual increases in the lymphocyte: mesenchymal cell ratio follow, but remain below 1:1 up to postnatal day 32. However, the arteriolar tree consolidates with smooth muscle formation and thickening of the mesenchymal layer. The lymphocyte: mesenchymal cell ratio increases to 1:1 out to the terminations of the arteriolar tree in 45-day-old pouch young. These trends give way to a substantial increase in the ratio of splenic white to red pulp by postnatal day 65. The follicles then show a very densely lymphocytic periarteriolar core and a much less densely lymphocytic periphery. However, primary and secondary (germinal center) nodules are not evident in the follicles until after postnatal day 65 and are not prominent and well formed until after day 100. Germinal center and plasma cell formation thus develop synchronously in the splenic follicles and larger lymph nodes of *Didelphis virginiana.*

### C. Development of Hematopoietic Organs

Vascular endothelial and blood cells of all types derive from the extra- and intraembryonic mesenchyme in vertebrates and ultimately, therefore, from the mesoblast (reviewed by Romanoff, 1960). The earliest cellular primordia of the circulatory system arise extraembryonically in the area opaca from "blood islands" of syncytial mesoderm layered between endoderm and ectoderm. These mesodermal blood islets develop into a plexus of cellular cords which, by a cavitation process, generates angioblasts peripherally and hemangioblasts centrally to the cords, giving rise to the area vasculosa of the developing yolk sac. Hemangioblasts (megaloblasts) of the extraembryonic yolk sac thus constitute the first blood cells. Yolk sac hematopoiesis in Eutheria gradually subsides with the initiation of intraembryonic hematopoiesis following chorioallantoic placentation (reviewed by Bloom, 1938). However, yolk sac functions in marsupials are abruptly terminated at birth. Thus, nutrient supply and oxygenation are undertaken by the gastrointestinal tract and lungs, in the absence of a placenta, and hematopoiesis is undertaken by the embryonic mesenchyme, as in Eutheria. Hematopoiesis in mammalian embryonic mesenchyme localizes first in liver and undifferentiated mesenchyme and later in development in bone marrow and spleen.

#### 1. Liver

Hematopoiesis in the opossum (*D. virginiana*) begins prenatally in the liver mesenchyme. Block (1964) believed that these hematopoietic cells originate locally from hepatic mesenchyme, owing to their scarcity at birth. The hepatic epithelial syncytial cell seen at this stage of development has also been proposed as the genetic antecedent of both hepatocytes and hematopoietic cells (Stohlman, 1970). However, a metastatic origin from the yolk sac for the first hepatic hematopoietic cells cannot be ruled out, despite the fact that erythropoiesis is megaloblastic in the former site and normoblastic in the latter, for such divergences might reflect different inductive environmental factors acting on a common progenitor stem cell.

Approximately half of the liver space is occupied by hematopoietic cells at birth of the opossum. At this time, granulocytic cells are as numerous as erythroblasts, few of which are matured beyond the basophil stage. Hepatic hematopoiesis may therefore have its inception only 1 or 2 days before birth. However, the neutrophil lineage exhibits slightly greater maturity than the erythroblast. Occasional megakaryocytoblasts but few eosinophilic or basophilic precursor cells are found. Over the next few days, erythroblast production in the opossum outstrips all other types of hematopoiesis as the hepatic phase of hematopoiesis becomes predominately erythroblastic. The numbers of erythroblasts increase more rapidly than the size of the liver during the first 5 days of postnatal life and then parallel it to the seventh day of life. These erythroblasts are not evenly spread in the liver but comprise discrete intravascular foci within which are clustered cells of similar maturation. Block (1964) ascribes this pattern to the origin of erythroblastic foci from single primitive precursor cells. Individual foci progressively acquire greater morphologic maturity in the early 0 to 3 postnatal day interval, suggesting that maturation of erythroblasts from stem cell to mature red blood cell requires about 4 days.

On postnatal day 3, granulocytopoiesis increases and its restriction to the hepatic capsular and portal connective tissues becomes apparent. Giant myelocytes are common and neutrophils greatly predominate over eosinophils. On day 5, nonerythroid hematopoietic cells all show increases relative to erythroblasts, despite the massive predominance of the latter. Eosinophilic and megakaryocytic precursors show the greatest gains. On days 6 and 7, these gains are perpetuated and there are as many eosinophilic and basophilic as neutrophilic cells. On days 8 and 9, the beginning decrease in hematopoietic cell concentration and increase in the mature to immature cell ratio in all lines of hematopoiesis are first seen. Granulocytic precursors more immature than the metamyelocyte stage are rare. Basophils are rare. On days 10 to 12, the concentration of hematopoietic cells, especially neutrophilic precursors, continues to decrease. Eosinophils outnumber neutrophils. Nuclear debris is found in the Kupffer cells. Between days 13 and 30, liver and body weights increase in parallel. Hematopoietic sites continue to become more focalized and to gradually decrease in incidence. On days 17 to 22, intrasinusoidal erythroblastic and megakaryocytic foci become clearly separated from the periportal and subcapsular granulocytopoietic foci, which are predominately eosinophilic. Beginning on about the thirtieth day of life, the amount of hepatic hematopoietic tissue begins to decrease more rapidly. Only occasional small erythroblastic foci composed of cells of the same degree of maturation are found. The foci vary widely in degree of cellular maturation. Granulocytopoiesis is absent in the subcapsular region and is a minor component in the portal connective tissue. On days 45 to 65, the decrease of hematopoiesis is such that only one of every four lobules contains a small erythroblastic focus. There is practically no hematopoiesis in the liver after the sixty-fifth day of life.

### 2. Mesenchyme

Hematopoiesis in diffuse mesenchyme takes place in eutherian and avian embryos during a period prior to differentiation of the mesenchyme into connective tissue (reviewed by Bloom, 1938). Mesenchymal hematopoiesis begins with or slightly later than hepatic hematopoiesis and is equally expressive of the embryo's needs, for hematopoiesis is mainly granulocytopoietic in the mesenchyme, as it is mainly erythropoietic and megakaryocytopoietic in the liver. The alternatives of extrinsic or local origin of the involved hematopoietic stem cells are unresolved in mesenchyme as they are in liver (Section II,C,1), and accordingly, undifferentiated mesenchymal cells may either differentiate into hematopoietic cells or play some essential environmental role for hematopoietic stem cells originating in the yolk sac. Axial mesenchyme of the mediastinum and the perivertebral thoracic and lumbar regions is the primary, but not sole, focus of mesenchymal hematopoiesis, perhaps because these mesenchyma are replete with nascent lymph sacs and vessels, which serve to deliver the developing mesenchymal blood cells to the blood (Section I,A,2). Mesenchyme ceases hematopoiesis coincidentally with the deposition of extracellular collagenous fibers and seems unable, therefore, to simultaneously play both a structural and a hematopoietic role (Block, 1964). Hematopoiesis of lesser magnitude also takes place in a few epithelial sites besides liver, e.g., in adrenal gland and mesonephros, where it is predominately erythropoietic as it is in liver.

Hematopoiesis in diffuse mesenchyme does not begin before birth of the opossum (*D. virginiana*). Zimmermann (1940) reported that hematopoietic foci associated with the walls of the periaortic lymph vessels appear on postnatal day 4 in the thorax of 16-mm pouch young. He found these predominately granulocytopoietic foci to diffuse thereafter into the surrounding mesenchyme and to extend caudally on day 12 into the lumbar periaortic mesenchyme of 22-mm pouch young. Block (1964) confirmed and extended these observations of noninflammatory associated mesenchymal hematopoiesis. He found only rare histioid wandering cells, myelocytes, and mast cells in the mediastinum of the newborn opossum. Histioid wandering cells and granulocytic cells thereafter increase in the mediastinum. This region develops the first mesenchymal basophilic erythroblasts on postnatal day 1 and mature neutrophils on day 3. Other mesenchyma become hematopoietic on day 3, especially those of the mesonephric hilum and the perivertebral space between the ribs, where diffuse islands of primitive neutrophilic cells begin to form. One-third of the cells in the corium are histioid wandering cells and primitive mast cells in the anterior half of the animal, but these cells are less numerous in the posterior half. Mast cells gradually mature during the following period as the corium begins to deposit collagen fibers in an anterioposterior gradient.

On postnatal day 4, erythropoietic islets become evident, more frequently in parenchymatous organs such as the adrenal and mesonephros, than in the axial mesenchyme. The thoracic perivertebral granulocytic foci extend into the abdominal prevertebral mesenchyme and eosinopoiesis increases. On days 5 to 8, diffuse mediastinal hematopoiesis begins to decrease and eosinophil and neutrophil precursors become equally abundant. A few foci of erythroblasts are still evident and megakaryocytes remain very rare in all mesenchyma. Granulocytic cells, now predominately eosinophilic, and histioid wandering cells are still numerous near intercostal vessels, the hilum of the mesonephros, and in the posterior mediastinum and retroperitoneal tissues. Leukocytic infiltration is found near the end of this period in glomeruli and between the degenerating peripheral mesonephric tubules. Mast cells increase significantly in all mesenchyma. On postnatal days 9 to 11, diffuse mesenchymal hematopoiesis continues to decrease and erythroblasts increase relative to granulocytic cells in the mediastinum. The erythroblasts within individual foci are of the same stage of maturation as they are in the liver (Section II,C,1). Most of the granulocytic precursors are eosinophilic and they are noticeably decreased in the glomerular hilum and intercostal spaces. Mature mast cells in the corium peak out as collagen fiber deposition becomes prominent in both the anterior and posterior halves of the animal.

On days 12 to 15, only residual hematopoiesis persists in the mediastinum, but sizable islands of myelocytes, mainly eosinophilic, persist in the retroperitoneal tissues, the mesonephric hilum, and the intercostal spaces. On days 16 to 21, these myelocytes are greatly reduced and hematopoiesis persists thereafter as a scattering of eosinophilic and neutrophilic foci in the perivertebral axial mesenchyme and small erythroblastic islets in the submucosa of the small intestine. Hematopoiesis in connective tissue is very scattered after day 32 and is absent, as in adults, after postnatal day 45.

### 3. Splenic Red Pulp

The vertebrate spleen in development acquires hematopoietic tissue before lymphoid follicles and retains the former throughout life in lower vertebrates, marsupials (opossum and marmosa), and several eutherian species (Section II,A). The neonatal spleen of the opossum (*D. virginiana*) is a relatively avascular mesenchymal rudiment condensed from the dorsal mesogastrium. The main splenic developments in the early postnatal period are progressive increase of organ size and elaboration of the sinusoidal system so that, by postnatal day 3, the latter becomes prominent and begins to acquire small foci of basophilic erythroblasts. On day 4, the spleen is replete with foci of one to three hematopoietic cells of the same degree of maturation. These comprise basophilic or polychromatophilic erythroblasts in the numerous small sinusoids and neutrophilic myelocytes in the parenchyma. Occasional megakaryocytoblasts appear singly in the sinusoids. On days 5 to 8, the hematopoietic islets, although progressively larger and more numerous, are as yet discrete and of internally uniform maturation. Toward the end of this period, eosinophil myelocytes become as numerous as neutrophil myelocytes. The relative proportions of cells in the various myeloid lineages approximate that of the liver, but the ratio of proliferating to maturing cells in all lineages is greater than in the hematopoietically more mature liver (Section II,C,1). On days 9 to 15, the increase in erythropoiesis progressively outstrips granulocytopoiesis and is the primary factor accounting for a fourfold increase in spleen size during this period. Foci of immature granulocytes are fewer and smaller than those of immature erythroblasts. The largest hematopoietic foci consist of mature erythroblasts. Eosinophilic foci are more numerous than neutrophilic foci. Granulocytopoietic foci show a tendency toward the end of this period to localize near the developing capsule, arterioles, and connective tissue septa. On days 16 to 21, these trends give way to diffuse filling of the spleen with erythroblasts and to a lesser extent megakaryocytes, masking the sinuses and mesenchymal cells. Granulocytic, mainly eosinophilic, precursors are relatively infrequent but continue to focalize along trabeculae and arterioles. On days 22 to 31, the concentration of hematopoietic cells subsides somewhat so that the sinuses are not completely obscured, but on days 32 to 44, further subsidence is not seen. On days 45 to 64, spleen growth mainly expresses white pulp growth, but after this period, the ratio of red to white pulp does not alter significantly. At day 100, the spleen still differs from the mature spleen in having a higher concentration of erythroblasts and megakaryocytes and less clearly defined sinuses and mesenchymal cells.

### 4. Bone Marrow

Bone marrow differs from other hematopoietic tissues in originating only after the associated mesenchyme has reverted from a preceding development into cartilage cells in endochondral bone and into osteoblasts, osteocytes, and osteoclasts in membranous bone. However, as with other hematogenic mesenchyma (Sections II,C,2 and 3), and for similar reasons, it is not established whether bone medullary mesenchymal cells in their hematopoietic role actually generate stem cells, provide an inductive and perpetuating role for stem cells of extrinsic origin, or both. In the opossum (*D. virginiana*), localized reconversion of the specialized bone cells into mesenchymal cells in endochondral bone is brought about by a series of events characteristic for vertebrate myeloid tissue (reviewed by Yoffey and Courtice, 1970). These consist of conversion of the pericondrium into dense periosteum, vascularization of the mesenchyme outside the center of the periosteum, development of a periosteal gap through which the mesenchyme containing large, dilated sinusoids enters, and conversion of cartilage at this point into an early mesenchymal marrow. Block (1964) believed the vascular mesenchyme near the diaphysis to play an inductive role in these events.

At birth of the opossum, endochondral bones are cartilaginous, and membranous bones are composed of irregularly calcified tissue. Mesenchyme first appears in endochondral bones on postnatal day 4 and in membranous bones only after day 100. Medullary hematopoiesis in early life of the opossum is thus restricted to the axial and appendicular skeleton. Marrow mesenchyme differentiates according to an early anterioposterior gradient, a pattern resembling that of the lymphatic and other organ systems (Section I). Marrow hematopoietic tissue characteristically develops first in central bones and later in more peripheral appendicular bones. Block (1964), remembering the conversion of fatty to active marrow in a rat's tail after its intraabdominal insertion (Huggins and Blocksom, 1936), believed temperature gradients to play a determining role in this.

On postnatal days 5 and 6, the diaphyseal cartilage of endochondral bones is replaced by a myxoid hypocellular mesenchyme in the caudal half of the animal and by a mesenchyme containing many extravascular granulocytes, principally mature neutrophils and eosinophils, in the cranial half. On postnatal days 7 and 8, the marrow space increases at the expense of cartilage. On days 9 to 11, marrow cellularity and the eosinophil to neutrophil ratio are both increased. The marrow at this time continues to resemble a granulocyte-infiltrated tissue, although beginning hematopoiesis is clearly indicated by the appearance of a few extravascular myelocytes and metamyelocytes and intravascular basophilic erythroblasts. During days 12 to 21, hematopoietic cells progressively fill the marrow cavity of most of the long bones. The marrow sinusoids of older animals of this group become filled with sheets of erythroblasts and occasional megakaryocytes. Eosinophils in the extravascular spaces continue as the predominate type of granulocytic cell. The ratio of proliferating to maturing cells is greater among erythroblasts than among granulocytic precursors. These trends are perpetuated until, by about postnatal day 45, intrasinusoidal erythroblasts comprise about one-third of the hematopoietic tissue, eosinophilic cells comprise fewer but still a majority of granulocytes, and the ratio of proliferating to maturing cells of the latter lineages is considerably raised. Megakaryocytes are also increased in the sinusoids in terms of numbers and average degree of maturation. Continued development before postnatal day 65 shows intrasinusoidal erythroblasts and megakaryocytes occupying more of the marrow space than the parenchymal granulocytopoietic tissue. The ratio of proliferating to maturing erythroblasts decreases but remains above that of mature animals. The ratio of proliferating to maturing granulocytic cells increases to mature levels. The ratio of neutrophilic to eosinophilic granulocytes increases but remains below 1.0. Between postnatal days 65 and 100, endochondral bone marrow matures as neutrophils become as numerous as eosinophils.

### D. Development of Inflammatory and Blood Cells

Marsupials, after their brief gestation, face the vicissitudes of extrauterine life seriously deficient in hematologic factors ordinarily considered necessary for survival, i.e., leukocytes, platelets, and plasma factors for hemostasis, and possibly erythrocytes. The leukopenia, comprising both a total lymphocytopenia (Section II,B) and a total granulocytopenia (vide infra), seems independent of differences of gestational length and suggests a potential neonatal state of great infectious susceptibility. This has been confirmed in the opossum (*D. virginiana*) where a minor wound in the skin of the newborn produces an acute, massive, fatal bacterial invasion of the tissues without eliciting an inflammatory response (Block, 1960). Moreover, if the wound is prevented from becoming grossly infected, wound healing progresses slowly as long as the animal is leukopenic. Later in life, when the concentration of blood leukocytes begins to approximate adult levels, similar injury or infection elicits the ordinary rapid sequence of inflammatory cells, leading to protection from infection and prompt wound healing.

Survival of newborn marsupials is remarkable in view of the probability that the marsupial pouch, if present, has little shielding value and in fact carries potentially infectious flora (Stanley, 1966). The best explanation would seem to lie with innate mechanisms of host defense, e.g., the skin barrier and the integrity of the nipple attachment, which subserves the important function of passive transfer of protective maternal antibodies in the colostrum (Section III).

The similarity of sequence with which hematopoietic cells appear in the opossum and in Eutheria of various gestational periods stresses the importance of genetic control over environmental influence in the timing of these developments (Block, 1964). The sequence of blood cell development in the opossum and in Eutheria also runs parallel.

The first blood cells in all vertebrates are yolk sac hemangioblasts (Section II,C), which arise in the opossum (*D. virginiana*) embryo on the ninth day of gestation (Block, 1964). At birth, 3 days later, large numbers of eosinophil megaloblasts are found in the blood of the opossum, indicating that megaloblastic cytodifferentiation from hemangioblast through the intermediate basophil and polychromatophil stages to the eosinophil stage minimally requires 3 days. The neonatal count of nucleated blood cells, almost all of which are nondividing eosinophil megaloblasts, is very low, approximately 600,000 to 700,000 per $mm^3$. Reticulocytes comprise only 1 to 2% of erythrocytes, indicating a very low level of neonatal erythropoiesis.

The eosinophil megaloblasts thus tide the newborn opossum over as hepatic erythropoiesis gets underway. The absolute number of megaloblasts remains constant through postnatal days 8 or 9, although their count falls during this interval in accordance with the expanding blood volume. They disappear during the ensuing 3 to 4 days, probably by senescence and death rather than transformation into megalocytes, coincidentally with a significant rise of red blood cells. Hepatosplenic erythropoiesis is characterized by an early blood phase component brought about by an influx of erythroblasts into the blood beginning on postnatal days 3 and 4 and diminishing gradually after days 8 or 9. The blood erythroblasts are confined mainly to the more tortuous capillaries of the lungs and mesonephric glomeruli. Maturation of the early hepatosplenic and blood erythroblasts results in a peak reticulocyte percentage (90 to 100%) of blood erythrocytes on days 5 to 8. These reticulocytes mature forming substantial numbers of erythrocytes as absolute megaloblast numbers begin to decline on day 9. These events bring about a rapid escalation of the mean red blood cell count from 400,000 to 2,000,000 per $mm^3$ during the interval from postnatal days 5 to 10, which is impressive because of the concomitant increase in blood volume. After day 10, the red cell count increases much more slowly, but does so in linear fashion to day 90, when adult levels of 4 to 6 million erythrocytes per $mm^3$ are reached. However, unlike the adult, the 90-day-old opossum still has a few erythroblasts and an elevated reticulocyte count in its blood. Interestingly, the *Marmosa* (*M. robinsoni*), in the period from weaning to adulthood (70 to 210 days of age), shows a slight increase in mean red blood cells (from 5.7 to 6.6 million per $mm^3$) and a significant decrease in reticulocyte count (from about 10 to 2% of erythrocytes) (Wolf *et al.*, 1971).

Leukocyte counts at birth of the opossum range from 300 to 600 per

$mm^3$. Most (90%) of these cells are neutrophil myelocytes and metamyelocytes of hepatic origin (Section II,C,1), for granulocytopoiesis does not develop in the mammalian yolk sac as it does in the avian (Romanoff, 1960). The remaining cells comprise naked nuclei and large, unidentifiable cells. Thrombocytopenia is indicated. The platelet count is 25,000 to 50,000 per $mm^3$, compared to a normal adult count of 150,000 to 250,000 per $mm^3$. Platelet and possible plasma factor deficiencies might explain reports that intraabdominal surgery in opossums results in hemorrhagic death when attempted before the eighth day of postnatal life (Moore, 1952). However, the lack of understanding of hemostasis in the newborn opossum is indicated by the fact that hemorrhagic diathesis of birth trauma is unimpressive (Block, 1964). Platelet counts and morphology begin to resemble those of adult opossums only at 7 and 8 days after birth.

The leukocyte count on postnatal days 2 and 3 rises to 3,000 to 12,000 per $mm^3$, where it is to remain, although the differential count changes dramatically with progressive age. Most of the leukocytes at this time are neutrophil granulocytes ranging from metamyelocytes to mature forms. Basophilic granulocytes of corresponding maturation are also increased but remain a minority. Platelets are increased. Leukocytes, megakaryocytes, and erythroblasts appear marginated along the more tortuous capillaries. On postnatal days 4 to 6, eosinophilic granulocytes begin to appear and basophilic granulocytes decrease. On postnatal days 7 to 11, occasional medium and small lymphocytes appear for the first time and eosinophils rise sharply to account for 25 to 40% of leukocytes during this period. Neutrophils decrease correspondingly to 40 to 60% of leukocytes. On postnatal days 12 to 15, lymphocytes form 5 to 20% of leukocytes. Older pouch young of this age group have more small lymphocytes and a higher lymphocyte differential than younger animals. Neutrophils and eosinophils each comprise about 30% and basophils about 20% of leukocytes. These trends continue on days 16 to 21 as lymphocytes form 25 to 40% and basophils about 10% of leukocytes. The great majority of lymphocytes are small. Neutrophils and eosinophils are unchanged. During the postnatal day 22 to 100 interval, lymphocytes increase to about 50%, eosinophils increase to 25–45%, neutrophils decrease to 10–30% of leukocytes, and basophils decrease greatly.

Cursory examination of quokka (*Setonix brachyurus*) pouch young has shown that leukocytes develop in this species in much the same fashion as in the opossums (Yadav and Papadimitriou, 1969). Despite differences in gestational period, the quokka after 25 days and the opossum after 12 days are born without lymphocytes in the blood or elsewhere. However, the first quokka blood lymphocytes appear on postnatal days 4 and 5, slightly earlier than the postnatal interval of 7 to 8 days noted for definitive medium and small opossum blood lymphocytes. The *Marmosa* (*M. robinsoni*), between weaning and adulthood (70 to 210 days of age), shows a transient rise of the blood leukocyte count during the juvenile period (100 to 160 days of age) followed by a subsidence in the adult to 16,000–18,000 per $mm^3$. Lymphocytes decreased gradually from 80% at 70 days to about 50% at 210 days. Neutrophils increased from about 8 to 27% in the same period. Most of the remaining cells are eosinophils (Wolf *et al.*, 1971).

## III. ONTOGENY OF IMMUNE MECHANISMS

The special attractions that marsupials hold for biologists at large, viz. their essentiality for comparative phylogenetic study within Mammalia and the precocity of their neonatal state, have yielded only a small volume of immunologic literature to date. Marsupials nonetheless have significant potential for immunobiologic study, especially developmental aspects thereof. They are born before definitive lymphocytes develop in the thymus or elsewhere (Section II,A); they are equivalent at birth to immunologically incompetent eutherian embryos, e.g., to 8-week-old human embryos (Block, 1964); and their "external" embryos can be manipulated experimentally during development without the impediment of the intervening maternal membranes of Eutheria. The following account summarizes available immunologic data on this inadequately exploited resource.

### A. Cell-Mediated Responses

The cellular aspects of host resistance to infection in marsupials are poorly delineated, as suggested by the sparseness of data available on inflammatory cell development in opossum (*D. virginiana*) pouch young (Section II,D). Delayed hypersensitivity responses in adult opossums nonetheless are intact. They develop tuberculin hypersensitivity reactions, experimental allergic encephalomyelitis, and contact dermatitis to dinitrochlorobenzene, although these responses are claimed to be somewhat less vigorous than in Eutheria (Taylor and Burrell, 1968). However, opossums do reject skin allografts with eutherian-like vigor, i.e., within a 13.7-day mean period for first-set allografts and within a shorter period for second-set allografts (La Plante *et al.*, 1966).

When pouch young were grafted with maternal skin, those grafted before 13 days of age accepted them permanently, those grafted at 17 days of age rejected them in delayed fashion after 90 days, and those grafted at 90 days rejected them in adult first-set fashion. When the question of induced tolerance was tested in a 3½-month-old litter with permanently surviving allografts, two of five such animals rejected a second maternal graft in first-set fashion while the others accepted a second maternal graft for more than 90 days. Third-party grafts were rejected in first-set fashion and the original maternal grafts were unchanged in all cases. These results indicate that partial specific tolerance to histocompatibility antigens can be induced in some opossum pouch young in the early postnatal period (La Plante *et al.*, 1969).

### B. Serologic Responses

The earliest electrophoretic separation of marsupial serum proteins, i.e., the single opossum (*D. virginiana*) included in Gleason and Friedberg's (1953) study of small mammals, revealed the five distinct components which characterize eutherian serum. Fantl and Ward (1957), examining hemostasis mechanisms in Australian marsupials, made similar observations in the brush-tailed possum (*Trichosurus vulpecula*), the ring-tailed possum (*Pseudocheirus peregrinus*), the black-tailed wallaby (*Wallabia bicolor*), and the tiger cat (*Dasyurus maculatus*). They found $\alpha_1$ components greatly exceeding $\alpha_2$, total serum proteins ranging from 5.1 to 7.4 g/100 ml, and a low level of albumin. Rowlands and Dudley (1968, 1969) studied the development of serum proteins in opossum pouch young. Total serum proteins in 7-day-old pouch young averaged 2.9 g/100 ml. They remained at about this level to 50 days of age, after which they increased to 80 days of age, when adult levels of 6.4 $\pm$ 0.5 g/100 ml were attained. The albumin:globulin ratio was about 1.0 at 50 days of age and 0.6 at maturity. The electrophoretic patterns showed $\alpha_2$ components exceeding $\alpha_1$, bringing them more in line with eutherian patterns, except for the unusually low level of albumin. Timmons and Marques (1969) confirmed the adult data, finding an opossum serum albumin:globulin ratio of 0.4. Wolf *et al.* (1971), examining adult marmosa (*M. robinsoni*), found an albumin:globulin ratio of 0.8 and a mean total serum protein level of 8.3 g/100 ml. However, this low albumin:globulin ratio was not attained until adulthood and was due to gradual rises in $\beta$- and $\gamma$-globulins during development. $\alpha$-Globulins remained constant between weaning and adulthood with $\alpha_2$ components exceeding $\alpha_1$ by factors of 4 to 5. The serum albumin:globulin ratio of adult marsupials falls well below the 1.0 or greater ratio found in Eutheria, thereby recalling the sera of lower vertebrates (Marchalonis and Edelman, 1966).

The immunoglobulins, like other marsupial serum proteins, have received relatively little attention. The serum immunoglobulin complex of opossums nonetheless does appear to be simpler than that of Eutheria. Only two immunoglobulin classes have been isolated to date. They are analogous to the IgM and IgG of immunoglobulins of other vertebrates in terms of electrophoretic mobility, sedimentation velocity, and relative sensitivity to reducing agents (Rowlands and Dudley, 1968, 1969; Taylor and Burrell, 1968; Genco and Liebert, 1970; Rowlands, 1970). This relatively simple configuration, like the low albumin level, recalls lower vertebrate sera, which also contain only one or two immunoglobulin classes (Burnet, 1969). These, like those of the opossum, closely resemble the IgM and IgG immunoglobulins of Eutheria.

Two IgG subclasses have been identified in adult opossums in serum, colostrum, and intestinal extracts (Genco and Liebert, 1970). Both subclasses in serum were found to contain antibody activity after stimulation with dinitrophenylated bovine $\gamma$-globulin.

Antibody activity in adult opossums appears in a biphasic manner in IgM and IgG immunoglobulins after stimulation with nonbacterial antigens. However, there is a slower rate of conversion from IgM to IgG antibodies than is apparent in Eutheria, again recalling lower vertebrates (Taylor and Burrell, 1968; Rowlands and Dudley, 1968; Rowlands, 1970). IgG antibodies become predominant only later than 3 weeks after one or two injections of soluble or particulate antigens; however, they prevail at or slightly before 3 weeks, if antigen is given in hyperstimulating dosage (Taylor and Burrell, 1968). Rowlands (1970), using subcutaneous injections of a bacteriophage, found that primary response neutralizing antibodies began to appear at day 7 and reached peak titers at days 10 to 21. A second injection resulted in increased titers. A 100-fold increase in antigen dosage yielded higher titers over a longer period but did not affect the lateness of the conversion to lighter IgG-type antibodies. Titers in comparably treated rabbits reached higher peak levels at earlier times than any of these. The opossum thus has a less facile, slower progressing humoral immune response than is found in Eutheria. The reasons for this, however, are as yet unclear, for the lymphatic organs of adult opossums generally correspond to those of Eutheria in development, structure, and distribution (Sections I,B and II,A).

The opossum and the Rottnest Island quokka (*Setonix brachyurus*) are the only marsupials which have been used in studies of the development of humoral immune capacity. This is unfortunate as several marsupial species are now known to be totally agammaglobulinemic at birth (Section III,C), suggesting a general pattern. Also unfortunate is the fact that too few data have been published to date to indicate whether marsupials might share the sequential ontogeny of humoral reactivity to different antigens that has been found during development of the fetal lamb (Sterzl and Silverstein, 1967).

However, *Salmonella typhi* flagella injected into opossum pouch young on postnatal day 8 or later will elicit specific serum agglutinins 7 days later, i.e., as early as postnatal day 15 (La Via *et al.*, 1963; Rowlands *et al.*, 1964). Treatment of opossum pouch young with bacteriophage $\phi$X on postnatal day 11 or later will elicit serum antibodies by postnatal day 17, as indicated by accelerated phage clearance (Kalmutz, 1962). Treatment of opossum pouch young with bacteriophage f2 at 5 days of age or older or with dinitrophenylated bovine serum albumin at 12 days of age or older will elicit specific antibodies in serum 3 weeks later (Rowlands and Dudley, 1969). A few quokka pouch young treated with sheep erythrocytes on postnatal days 10–12 will subsequently form specific serum hemagglutinins; however, most pouch young will respond if treated on postnatal day 13 or later (Stanley, 1966).

The development of these reactivities does not correlate in any readily understandable fashion with the developmental histology of lymphoid organs. The first opossum lymphocytes appear in the epithelial thymus on postnatal day 2, in most lymph nodes on days 4 to 10, and in the spleen on days 17 to 20. Germinal centers and plasma cells make their first appearance around day 60. Tonsils and Peyer's patches appear only after day 100 (Section II,A). Hence, the humoral reactivities observed thus far do not manifest themselves until several days after definitive lymphoid tissue has appeared in thymus and lymph nodes. However, they do become manifest before lymphoid tissue appears in spleen, tonsils, or Peyer's patches and well before germinal centers and plasma cells appear anywhere in the body, and early antigen stimulation does not provoke such structures (Rowlands *et al.*, 1964). The antibodies elicited in early life of the opossum thus do not derive from cells of the germinal center–plasma cell complex.

Antigen treatment in the early period of opossum life apparently can promote not only a specifically reactive state but an enduring state of partial specific tolerance as well (Major and Burrell, 1971). Opossums injected multiply with cellular or soluble forms of human blood group B substance beginning at 5 to 14 days of pouch life yielded very low or null hemagglutinin titers on specific challenge at 10 months of age. These results and the long term acceptance of neonatal skin allografts by some opossums (Section III,A) parallel eutherian data (Burnet, 1969) and suggest that basic mechanisms of tolerance and immunity induction are similar in all mammals.

### C. Transmission of Passive Immunity

Passive transfer of maternal antibodies to the fetus or newborn has been observed in all Eutheria examined (Brambell, 1970). The transfer takes place variously by the yolk sac, placenta, or colostrum and may involve maternal IgM or IgG antibodies. Passive transfer also takes place in marsupials, and as shown by the quokka (*Setonix brachyurus*), brush-tailed possum (*Trichosurus vulpecula*), and tammar (*Macropus eugenii*), it involves only IgG-type antibodies and is effected entirely by colostrum (Yadav *et al.*, 1973; Yadav, 1971). The newborn of all of these species are agammaglobulinemic at birth but, within 48 hr of nipple attachment, maternal antibodies absorbed from the gut can be detected. The absorption continues for most of the pouch life. The capacity to absorb antibodies in the intestine is lost in the quokka at 170–190 days of age and in the possum at 98–145 days of age. The pouch young are able to leave the pouch for varying lengths of time at these ages. The rumen was not found to absorb immunoglobulin at any age.

The transmission of passive immunity in marsupials continues for a much longer duration than is found in any eutherian system. The significance of this is not known but seems best explained by the relative primitiveness of marsupials at birth.

## ACKNOWLEDGMENTS

I thank H. G. Wolf for providing marmosas from his colony; Dr. L. K. Bustad and Dr. M. Shifrine for their interest and support; and the U.S. Atomic Energy Commission (Contract no. AT[04-3]472) and the National Institutes of Health (Health Sciences Advancement Award RR-06138) for financial support.

## REFERENCES

Azzali, G., and DiDio, L. J. A. (1965). *Am. J. Anat.* **116,** 449.
Block, M. H. (1960). *Nature (London)* **187,** 340.
Block, M. H. (1964). *Ergeb. Anat. Entwicklungsgesch.* **37,** 237.
Bloom, W. (1938). *In* "Handbook of Hematology" (H. Downey, ed.), pp. 863–922. Harper (Hoeber), New York.
Brambell, F. W. R. (1970). "The Transmission of Passive Immunity from Mother to Young." North-Holland Publ., Amsterdam.
Broman, I. (1921). *Verh. Anat. Ges.* **53,** 27.
Bryant, B. J., and Shifrine, M. (1974). *J. Reticuloendothel. Soc.* **16,** 105.
Burnet, F. M. (1969). "Cellular Immunology." Melbourne Univ. Press, Melbourne.
Fantl, P., and Ward, H. A. (1957). *Aust. J. Exp. Biol.* **35,** 209.
Fraser, E. A., and Hill, J. P. (1916). *Philos. Trans. R. Soc. London, Ser. B* **207,** 1.
Fraser, E. A., and Hill, J. P. (1917). *Proc. R. Soc. London, Ser. B* **89,** 97.
Genco, R. J., and Liebert, B. (1970). *Fed. Proc., Fed. Am. Soc. Exp. Biol.* **29,** 704.
Gleason, T. L., and Friedberg, F. (1953). *Physiol. Zool.* **26,** 95.

Gowans, J. L., and Knight, E. J. (1964). *Proc. R. Soc. London, Ser. B* **195,** 257.
Huggins, C., and Blocksom, B. (1936). *J. Exp. Med.* **64,** 253.
Huntington, G. S. (1910). *Anat. Rec.* **4,** 1.
Jordan, H. E. (1938). *In* "Handbood of Hematology" (H. Downey, ed.), pp. 699–862. Harper (Hoeber), New York.
Kalmutz, S. E. (1962). *Nature (London)* **193,** 851.
Kampmeier, O. F. (1970). "Evolution and Comparative Morphology of the Lymphatic System." Thomas, Springfield, Illinois.
La Plante, E. S., Taylor, D. L., and Burrell, R. (1966). *Fed. Proc., Fed. Am. Soc. Exp. Biol.* **25,** 232.
La Plante, E. S., Burrell, R., Watne, A. L., Taylor, D. L., and Zimmermann, B. (1969). *Transplantation* **7,** 67.
La Via, M. F., Rowlands, D. T., and Block, M. H. (1963). *Science* **140,** 1219.
McClure, C. F. W., and Silvester, C. F. (1909). *Anat. Rec.* **3,** 535.
Major, P. C., and Burrell, R. (1971). *J. Immunol.* **106,** 1690.
Marchalonis, J., and Edelman, G. M. (1966). *J. Exp. Med.* **124,** 901.
Miller, J. F. A. P., Block, M. H., Rowlands, D. T., and Kind, P. (1965). *Proc. Soc. Exp. Biol. Med.* **118,** 916.
Miller, J. J., III. (1969). *Lab. Invest.* **21,** 484.
Moore, C. R. (1952). Cited by Block (1964).
Owen, R. (1847). *In* "Cyclopedia of Anatomy and Physiology" (R. B. Todd, ed.), Vol. III, pp. 257–330. Sherwood, Gilbert, & Piper, London.
Papermaster, B. W., and Good, R. A. (1964). *Adv. Immunol.* **4,** 1.
Romanoff, A. L. (1960). "The Avian Embryo." Macmillan, New York.
Rowlands, D. T. (1970). *Immunology* **18,** 149.
Rowlands, D. T., and Dudley, M. A. (1968). *J. Immunol.* **100,** 736.
Rowlands, D. T., and Dudley, M. A. (1969). *Immunology* **17,** 969.
Rowlands, D. T., La Via, M. F., and Block, M. H. (1964). *J. Immunol.* **93,** 157.
Ruth, R. F., Allen, C. P., and Wolfe, H. R. (1964). *In* "The Thymus in Immunobiology" (R. A. Good and A. E. Gabrielsen, eds.), pp. 183–206. Harper (Hoeber), New York.
Sabin, F. R. (1901). *Am. J. Anat.* **1,** 351.
Simon, J. (1845). "A Physiological Essay on the Thymus Gland." Hodder & Stoughton, London.
Stanley, N. F. (1966). *Thymus: Exp. Clin. Stud., Ciba Found. Symp., 1965,* pp. 207–211.
Sterzl, J., and Silverstein, A. M. (1967). *Adv. Immunol.* **6,** 337.
Stohlman, F. (1970). *In* "Regulation of Hematopoiesis" (A. S. Gordon, ed.), Vol. I, pp. 471–485. Appleton, New York.
Symington, J. A. (1898). *J. Anat. Physiol., Norm. Pathol. Homme Anim.* **32,** 278.
Symington, J. A. (1900). *J. Anat. Physiol.* **34,** 178.
Taylor, D. L., and Burrell, R. (1968). *J. Immunol.* **101,** 1207.
Thorbecke, G. J., Gordon, H. A., Wostman, B., Wagner, M., and Reyniers, J. A. (1957). *J. Infect. Dis.* **101,** 237.
Timmons, E. H., and Marques, P. A. (1969). *Lab. Anim. Care* **19,** 342.
Wolf, H. G., Shifrine, M., Klein, A. K., and Foin, A. (1971). *Lab. Anim. Sci.* **21,** 249.
Wood, G. N. (1924). *Anat. Rec.* **27,** 192.
Yadav, M. (1971). *Immunology* **21,** 839.
Yadav, M., and Papadimitriou, J. M. (1969). *Aust. J. Exp. Biol. Med. Sci.* **47,** 653.
Yadav, M., and Eadie, M. (1973). *Aust. J. Zool.* **21,** 171.
Yoffey, J. M., and Courtice, F. C., eds. (1970). "Lymphatics, Lymph, and Lymphomyeloid Complex." Academic Press, New York.
Zimmermann, A. A. (1940). *Ill. Med. Dent. Monogr.* **3,** 1.

# 7

# The Special Anatomy of *Marmosa robinsoni*

RAYMOND D. BARNES



## I. INTRODUCTION

The use of marsupials as experimental animals confers a variety of special advantages, some of which derive from anatomical differences between this group and the common laboratory mammals. Some of the ways in which metatherians differ from eutherians are of special usefulness; others are likely to be troublesome or disadvantageous.

The unique life cycle of marsupials that results in exteriorization of their late embryos is the most important single characteristic separating them from all other mammals, which they otherwise resemble in a truly remarkable number of ways. Although this precocious embryo has structurally modified organ systems vital to its early extrauterine life, the basic plan is still mammalian in every way, and the embryologist familiar with other mammalian material will be surprised only by the early development of the relevant features, not by their qualitative distinctness. For example, the lung of a marsupial is in every way a typical mammalian lung. It differs from that of the placental mammal in that it must function very much earlier in ontogeny. At birth its compartmentalization and arborization are hardly begun, and it resembles a reptilian lung superficially (Bremer, 1904).

Organ systems not vital to life support in the exteriorized embryo, such as the reproductive organs, eyes, and ears, are scarcely more than anlagen at birth. These stages may be studied more conveniently in the marsupial than in the eutherian. The entire endocrine system, as well as most of the adult nervous system, is undeveloped at birth and uniquely subject to external modification because of the absence of physiologic interference of maternal origin. For work of this kind any marsupial will provide the basic requirement of exteriorized embryos, so long as it can be either bred in captivity or obtained conveniently in the pregnant state from the wild.

In North America the Virginia opossum is the only marsupial available in the wild and has been the most widely studied because of this availability. Its anatomy is, with few exceptions, well documented although no description of general application has been widely available.

For reasons mentioned elsewhere in this volume and detailed further in Barnes and Wolf (1971), the small South American cousin of *Didelphis* known as *Marmosa robinsoni* (formerly *M. mitis*) has been selected for use as a laboratory marsupial. This chapter deals with the anatomy of *Marmosa robinsoni*; comparisons with *Didelphis* will provide literature references for the reader's use. No attempt is made to cover the Australian families of marsupials except where an occasional detail of a comparative nature seems relevant. The *Marmosa* descriptions are from original observations; histologic, embryologic, and gerontologic observations are included along with gross anatomy of adult specimens.

## II. GENERAL DESCRIPTION AND POSTURE

*Marmosa robinsoni*, the pouchless mouse opossum of north coastal Columbia, is cinnamon brown with buffy to yellow underparts and the black facial mask typical of all species in this genus. Males weigh 60 to 130 g as adults and females 40 to 70 g. The ratlike tail is from 1.0 to 1.3 times the head and body length and naked except for the first 1.0 to 1.5 cm. Typical body measurements for an adult male of 80 g are head and body, 160 mm; tail, 170 mm. The tail is prehensile and the digits of all four feet are well developed as grasping organs with digital pads and an opposable first toe.

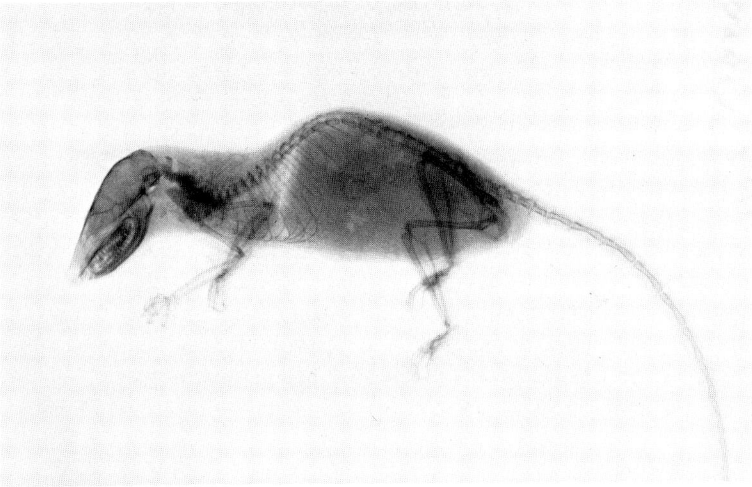

Fig. 1. Radiograph of adult female *Marmosa*; the kyphosis present between thoracic and lumbar regions is responsible for the hunchbacked posture normal to this species.

Although the general appearance of a mouse or small rat is strengthened by the pointed, whiskered snout and mouselike ears and tail, the dentition is unmistakably marsupial. The dental formula is I 5/4, C 1/1, P 3/3, M 4/4 × 2 = 50. The incisors are so tiny that they can hardly be counted without magnification. The canines are needle sharp. The premolars* are sectorial, whereas the molars are all multicuspidated. The sharp cusps of the molar teeth appear to be well adapted to handling insect prey. The bite of this animal is strong and will result in a tiny bleeding puncture if the canines catch the edge of a nailbed or soft fingertip. Their habit is to bite firmly and grip anything within reach, especially when grasped by the tail; release the tail and they will usually release their bite to scramble away.

While handling untamed animals one will notice that the mouselike ears crumple rather than lie flat. This is possible because the ear cartilage is in many small pieces joined together by striated muscle. The ears tend to flush red to purple when the animal is excited; this reaction is even more conspicuous in *Caluromys* than *Marmosa*.

The normal posture of *Marmosa* is recognizably different from the anus to ground sitting posture of rodents. *Marmosa* have a hunchbacked appearance owing to a strong kyphosis at the level of the thoracolumbar transition (see Fig. 1), and the pelvis is usually elevated. This posture serves

* Although the dental formula indicates three premolars, the only tooth replaced during the life of any didelphid marsupial is P3; in *M. robinsoni* P3 is replaced at about 140 days of age. The deciduous P3 is multicuspidated (molariform).

to protect a nursing litter or, in the case of males, the large and pendulous scrotum. The widely spread forelimbs look ungainly if the animal is on a flat surface, but on a branch or a wire cage wall everything looks normal; the animals' movements are deliberate but quick.

Determination of sex of adult *Marmosa* is very easy for two reasons. Males are much larger than females and their bright blue, glabrous scrotum contrasts sharply with the yellow underfur. The female is identified by smaller size and absence of a scrotum. There is no pouch on female *Marmosa* at any time. A female carrying a litter of less than 20 days of age walks with a stilt-legged gait. When the young are less than 10 days of age this posture will seem unnecessarily exaggerated. As the litter grows, it barely suffices to keep them from being rubbed off the nipples as she moves. From 22 days of age onward the young ride on the mother's back and flanks.

## III. INTEGUMENT

The skin of *M. robinsoni* is not loose over any part of the body in the manner that one expects from experience with mice or hamsters. The body contours are followed closely without looseness over the shoulders, inguinal, or flank areas. The fur is moderately dense but short and the hair is fine; it gives a soft, dry touch to the animal. This description does not pertain to all species of *Marmosa*; *M. elegans*, for instance, has a loose coat of fluffy, long hair, and a short tail used to store adipose tissue exactly like the fat-tailed marsupial mice of the dasyurid genus *Sminthopsis*.

The feet of *M. robinsoni* are all five toed with a well-developed grasping structure. The plantar surfaces are naked with ridged pads. The toe tips are also padded. All toes except the halluces are armed with strong claws; the halluces are unarmed.

The membranous ears are lightly pigmented and essentially naked. The tail is finely scaled and lightly pigmented. A downy pile of fine white hairs is present except at the basal 1 to 2 cm where the general body coat extends. There is species variation in tail color and hair coat; see Tate (1933) for further descriptions.

There are two areas that tend to be stained by cutaneous gland secretions. In males a prominent 10 to 15-mm strip of stained hair running along the midline of the throat onto the sternum will be observed. With age its color deepens from yellow to orange. This gland field is not dissectable from the skin. It consists of hypertrophied apocrine sudoriferous glands and sebaceous glands, both confined to the thickened dermis of the area. In females the sternal gland is recognizable histologically but usually not grossly. A field of integumentary glands surrounds each of the 15 mammary nipples. An orange to rust brown stain is produced around each nipple by grooming during pregnancy and later by the nursing young. The glands secrete most actively during the luteal phase of the estrous cycle (Fig. 9A). Other species of *Marmosa* may have slightly different colored stains, from purple to yellow.

Sudoriferous and sebaceous glands are found in the usual association with hair follicles over the entire body. The body hair follicles are simple, not compound.

The skin comprises 10 to 15 % of total body weight. Subcutaneous fat accumulates over the entire body uniformly. There is no thermogenic brown fat at any age. In old age (greater than 24 months) many females in the Davis colony developed pendulous folds of soft skin and fat paralleling the mammary area; this condition is not known to occur in nature. It has been, in this colony, a reliable sign of senescence in female animals.

## IV. SKELETON AND MUSCLES

Considerable literature exists regarding skeletal anatomy of *Didelphis* and other marsupials, much of it of interest primarily to taxonomists. The failure of epiphyses to ossify in marsupials is widely documented; Washburn (1946) concluded that chronological age bore less relation to epiphyseal union in the opossum than in any other mammal after a study of 30 skeletons in a collection of the Wistar Institute. However, there is no evidence in the Davis *Marmosa* colony for delayed epiphyseal closures; radiographs of sexually mature animals do not show open epiphyseal cartilages in any of the long bones. (No histologic examination of the skeleton has been conducted to confirm the radiologically negative finding.) One dried, defatted, beatle-cleaned *Marmosa* skeleton weighed 6.46% of live weight (41 g female).

Lowrance (1949) examined a series of 134 postcranial skeletons of wild opossums and confirmed the judgment of Washburn previously cited, indicating that nutrition during captivity had no major effect on the Wistar Institute animals. This work also established a greater variability between certain skull dimensions than is expected for eutherians. Hildebrand (1961) came to a similar conclusion with respect to the postcranial skeleton. These observations are relevant to future studies of nutrition and endocrine factors affecting growth, as well as to teratologic studies where inherent variability must be accounted for in experimental designs.

Coues (1872), Cheng (1955), and Kerr (1955) have reported on the locomotor system of *Didelphis*. This author has not made any specific study of this system in *Marmosa*. Muscle physiologists and pathologists should be aware that at birth marsupial skeletal muscle fibers are present as multinucleated tubes with centrally placed nuclei, the normal condition for embryonic skeletal muscle. Changes in function and various parameters of growth could conveniently be studied in a selected muscle by means of serial biopsies, if not in *Marmosa*, certainly in some larger species. Bridge (1970) has reported such a study for the quokka (*Setonix brachyurus*) in conjunction with studies of nutritional myopathy in that species. No other special feature of marsupial muscle has come to the attention of this author.

## V. CIRCULATORY SYSTEM

The heart and great vessels are described by Wade and Neely (1949) for the opossum as showing two variations from general eutherian structure: (1) the right ventricle is more crescentic in cross section than that of placental mammals, and (2) the right A-V valve consists of a single flap. These same conditions are seen in *Marmosa*.

Cardiac bleeding has been accomplished routinely in *Marmosa, Caluromys,* and *Didelphis*. A lateral approach is most successful, entering through the third intercostal space.

The lateral tail veins may be nicked for small quantities of blood, or threaded with a 27-gauge needle for injections. Adjacent arteries are sometimes cut but this has never caused any special problems. The ears are unsuitable for bleeding in didelphids except by razor cutting after the fashion of bleeding from guinea pig ears. The so-called retrobulbar venous plexus, used for bleeding hamsters and mice, will yield blood in *Marmosa* but not reliably. The convenience of tail vein bleeding makes this latter route unattractive.

Block (1964) and Rowlands *et al.* (1964) have described the hematopoietic system of *Didelphis* from birth to weaning. Circulating nucleated red blood cells are found at birth. These are gradually replaced by nonnucleated erythrocytes.

The differential white cell count in *Marmosa robinsoni* is characterized by a lymphocyte:neutrophil ratio between 2:1 and 3:1. Large and medium lymphocytes are always present in high proportion, and these grade into cells of the monocyte class which show kidney shaped to multilobed nuclei. The eosinophilic granulocyte is present at concentrations of 5 to 30% of white blood cells. Basophils average 0.5% but have been recorded as high as 10%. Rothstein and Hunsaker (1972) have reported on the hematology of *Caluromys*, the woolly opossum.

The *Marmosa* spleen is actively hematopoietic throughout life, although the level of activity varies markedly in response to hematopoietic stress. The splenic red pulp of *Didelphis* has been described by Hayes (1968) as lacking venous sinuses. *Marmosa* spleens also appear to lack sinuses unless one examines the spleen of an animal that has had an acute septicemic disease; this promotes pooling of blood in the portal system, and the venous sinuses are filled with red cells and are rendered prominent. The splenic white pulp is always scanty; some old animals, in which generalized lymphoid tissue atrophy has occurred, will appear to lack white pulp entirely.

## VI. RESPIRATORY SYSTEM

Bremer (1904) observed that respiration (gaseous exchange) in the neonatal opossum appeared to be carried out by modified bronchi and bronchioles before the infundibular and alveolar structures were developed. Selenka (1887, quoted by Bremer, 1904) reported that the lung of the newborn opossum had exactly the form of a reptilian lung. Selenka further assumed that development of the adult structure progressed by partitioning of the small number of air chambers (5 to 6 per lung) present at birth. Bremer agreed with Selenka regarding the reptilian nature of the neonatal lung but established that subsequent development was not as Selenka had assumed. "Also in both lungs there are bands of muscle fibers running circularly around the central air-chamber or bronchus, whose probable function is to contract the lung and force the air out during expiration." This further quote from Bremer is a novel explanation for respiratory capacity in the embryonic neonate; it remains to be validated by experiment. Bremer studied two age classes of pouch young, a neonate and an unaged specimen of 14 cm [which would have been about 11 weeks of age, using Reynolds (1952) growth curve for *Didelphis*] to establish that growth of the lung was not by subsequent partitioning of the chambers but by budding in the same manner observed in embryonic mammals.

Figures 2A and 2B are *Marmosa* lungs at 12 days and adult. It is probably not accidental that the air spaces have the same dimensions at birth and in the adult, although these specimens were not treated during preservation in any special manner to assure that "normal" volumes were retained.* The *Marmosa* neonatal lung has five to seven compartments to each lobe. Such a primitive lung offers an unusual model for study of the effects of various atmospheric variables such as oxygen partial pressures

* Neonates were fixed by immersion; some fixative was inhaled. Adult lungs were perfused intratracheally to a pressure sufficient to expand the collapsed lung to its anatomical volume in the thorax.

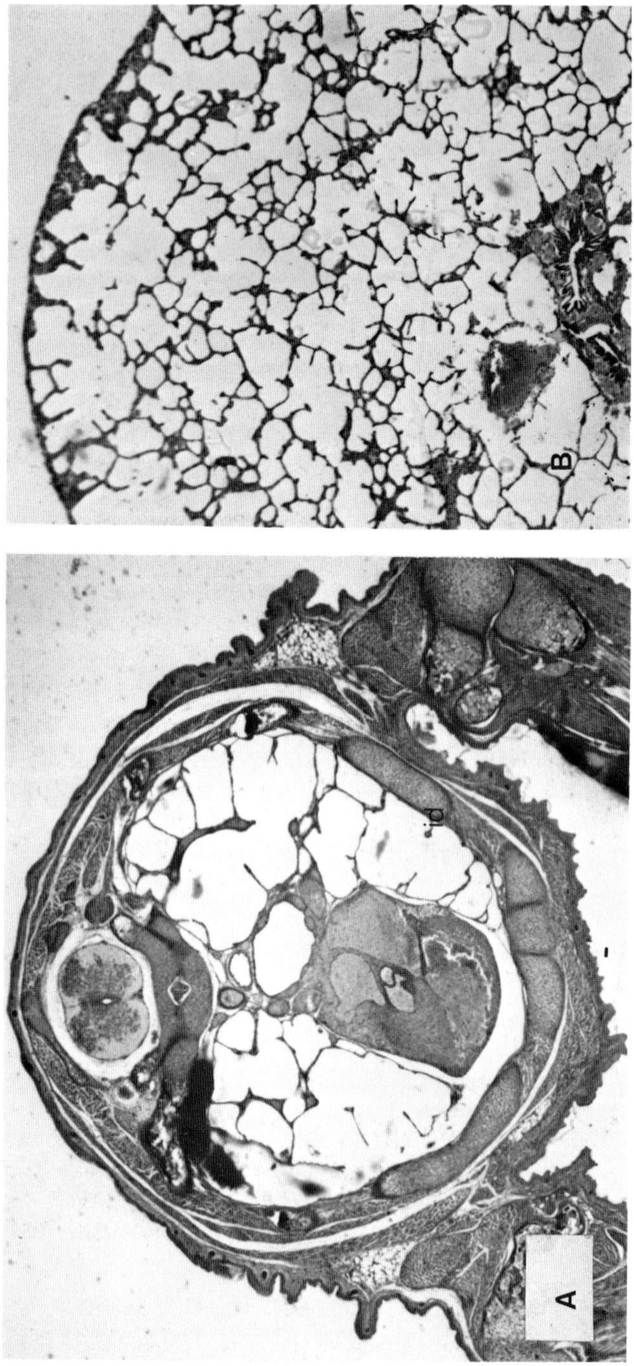


Fig. 2. (A) Lungs of 12-day old *Marmosa*. (B) Segment of adult *Marmosa* lung at same magnification. See text for discussion.

and air pollutants on pulmonary development and might be significant to the understanding of degenerative pulmonary diseases.

The adult upper respiratory tract is richly glandular; both mucous and serous glands are developed from the nasal walls. The conchae are elaborate scrolls. No paranasal sinuses are present in *Marmosa*; this author has not examined the skulls of larger marsupials for this feature.

Paired vomeronasal organs are present. These are lined by a sensory epithelium resembling that of the olfactory area. Patent foramina connect these with the oral cavity (the incisive foramina).

In nursing marsupials the tubular glottis is permanently grasped by the tight fitting nasopharyngeal orifice, forming a seal that allows respiration to continue uninterrupted by swallowing movements. During inhalation anesthesia of adults it is possible to ventilate the lungs by blowing gently into a soft rubber tube placed over the nostrils. The animal's mouth may be gaping widely and not effect the ventilation, so we assume that the glottis is at all times inserted into the nasopharynx. This adaptation is described for cetaceans also, where it presumably protects the nursing calf from being drowned by milk that is ejected under pressure from the mammary nipple. (A similar mechanism of milk ejection is postualted for marsupials; see Fig. 9C and Section IX.)

## VII. DIGESTIVE SYSTEM

The digestive system, like the respiratory system, must function very early in its ontogeny, and as is the case for the lungs, the alimentary canal at birth presents both cellular and tissue structures different from the adult morphology. Digestion does not occur during the first few days of pouch life; the milk proteins and lipids are absorbed unaltered and carbohydrates are lacking except for galactose, which is also absorbed unaltered (Barker *et al.*, 1967). The gastric mucosa is a simple cuboidal epithelium without glandular differentiation (Fig. 3). The villi of the small intestine are present but duodenal glands are undeveloped. The pancreas is recognizable although undifferentiated with respect to endocrine islets.

The adult didelphid alimentary canal is characterized as monogastric with a short, uncomplicated small intestine having two histologically definable regions (duodenal and ileal) and a short colon lacking haustra or other special features. A short cecum is present. Table I shows typical measurements for the alimentary canal of *Marmosa* compared with an herbivorous rodent, the gerbil.

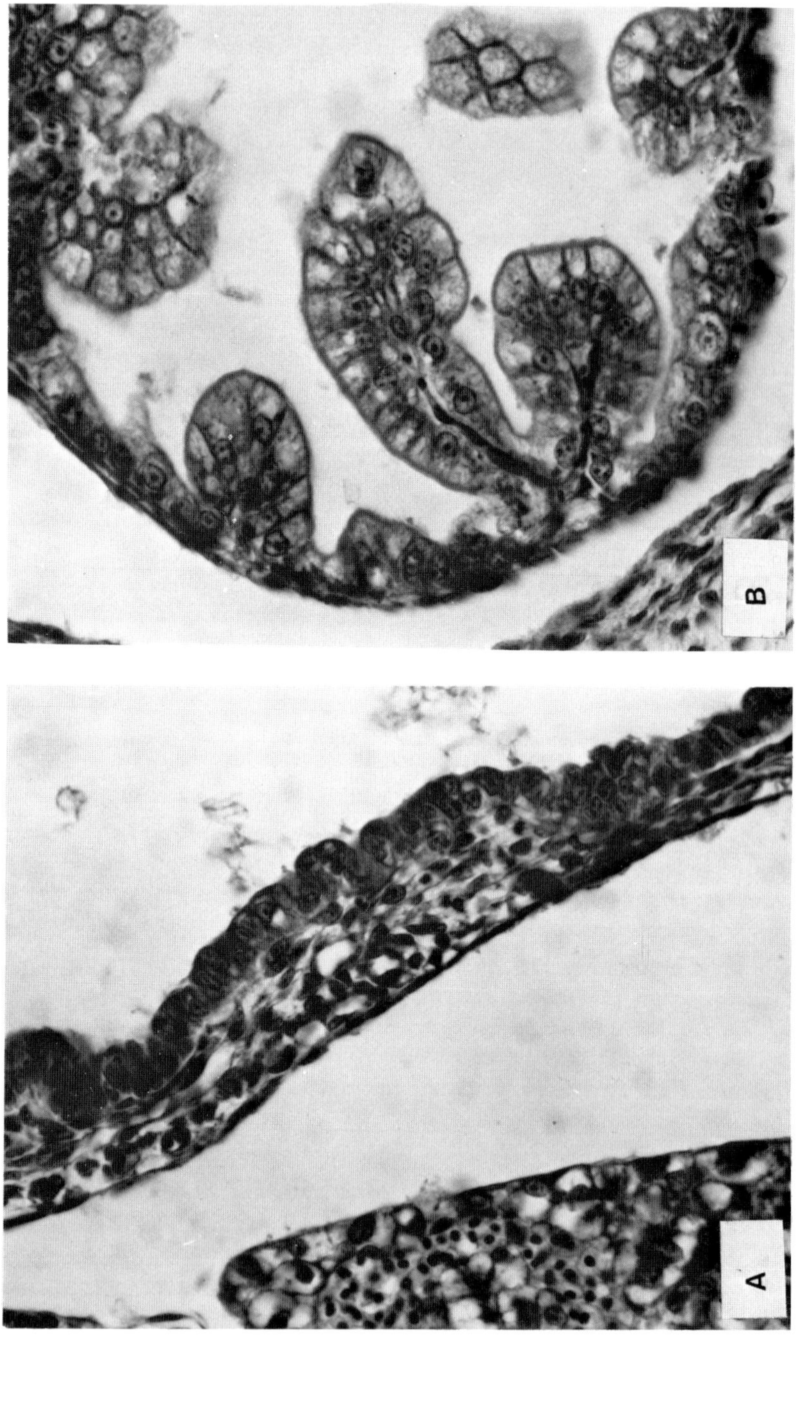


Fig. 3. (A) The gastric mucosa of the neonatal *Marmosa*; compare with Fig. 4B. (B) The duodenal mucosa of the neonate; compare to Fig. 6B.

**TABLE I**

Typical Alimentary Canal Measurements of *Marmosa robinsoni* and *Meriones unguiculata*

| | *Marmosa* | Gerbil[a] |
|---|---|---|
| Body weight | 46 g | 46 g |
| Body length | 122 mm | 115 mm |
| Small intestine | 136 mm | 320 mm |
| Cecum | 10 mm | 53 mm |
| Colon | 30 mm | 280 mm |

[a] Data of A. Kramer, Jr., Lederle Laboratories, Pearl River, New York.

Figures 4, 5, and 6 show the histology of the digestive canal of *Marmosa*. The important features may be summarized as follows:

A. The esophagus contains striated muscle in its initial segment but caudal to the trachea this is replaced by smooth muscle. Three muscle layers are present; mucous glands are absent.

B. The stomach is simple with glandular mucosa throughout; there is minimal differentiation of cardiac and fundic areas; the organization of zymogenic cells and parietal (oxyntic) cells resembles that seen in eutherian mammals. The pyloric region is mucous secreting (see Fig. 4).

C. The pyloris is clearly defined externally by the presence of a white collar of duodenal (Brunner's) glands (Fig. 6A).

D. The duodenum is characterized by large caliber and long, finger-like villi without duodenal crypts (Figs. 5 and 6B). Paneth cells are seen.

E. The ileum is of slightly smaller caliber but has a larger lumen than is seen in the duodenum owing to its shorter villi. Glandular crypts are short. Goblet cells gradually increase in number as the ileocecal valve is approached. Circular and longitudinal muscle layers are subequal in thickness. Peyer's patches are present (Fig. 4B).

F. The short cecum is histologically identical to the upper colon; villi are absent in both regions; crypts lined with abundant goblet cells form the mucosa. The circular *muscularis* increases in thickness gradually; Fig. 4D shows the lower colon just at the pelvic brim. There is no formation of haustra or taeniae.

The liver has a gallbladder and its organization and structure are both similar to other mammals. The liver "triads" show multiple bile ducts and are rarely seen as simple triads (bile duct, portal vein, and hepatic artery).

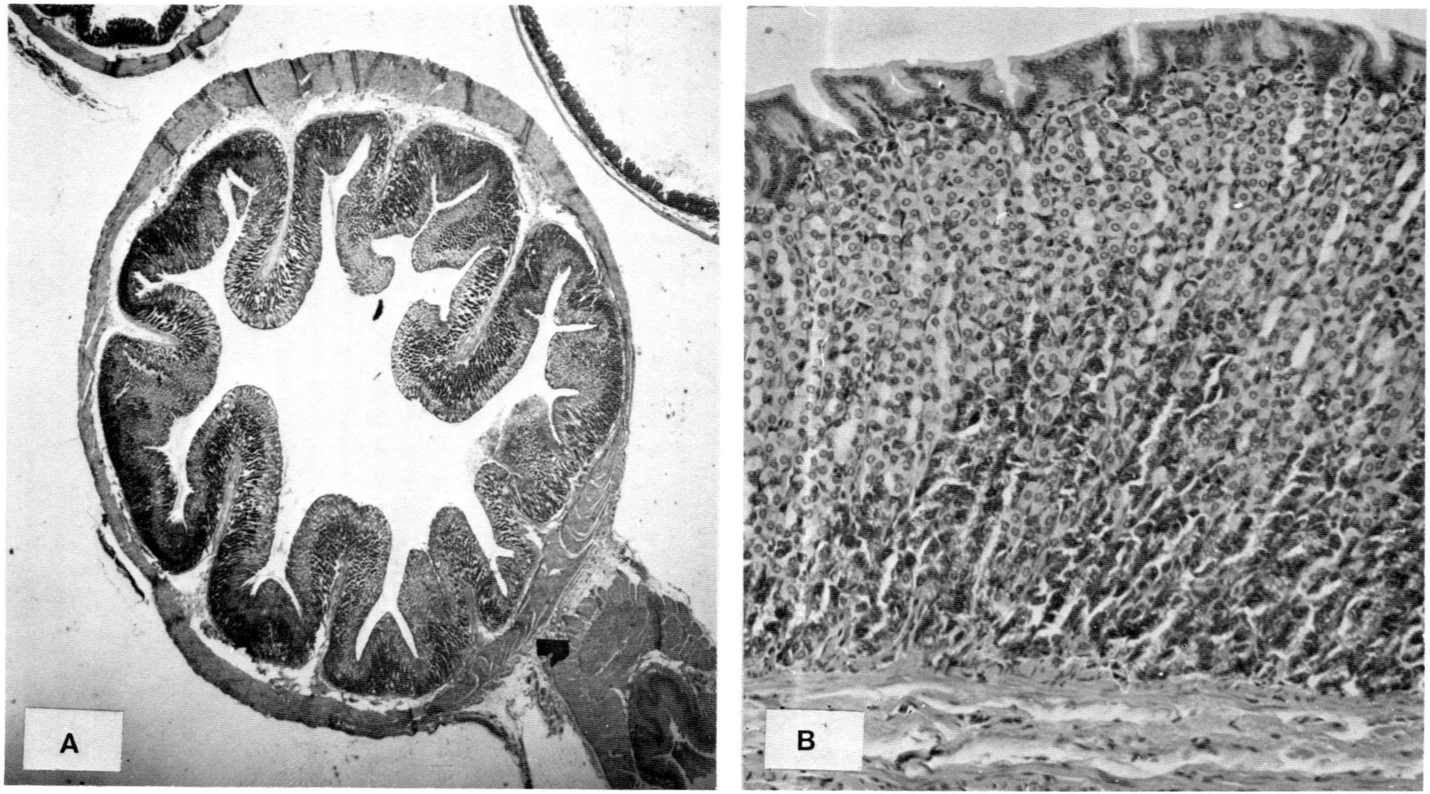


Fig. 4. (A) Cross section of stomach in fundic region. (B) The gastric mucosa; note the abundance of oxyntic cells.

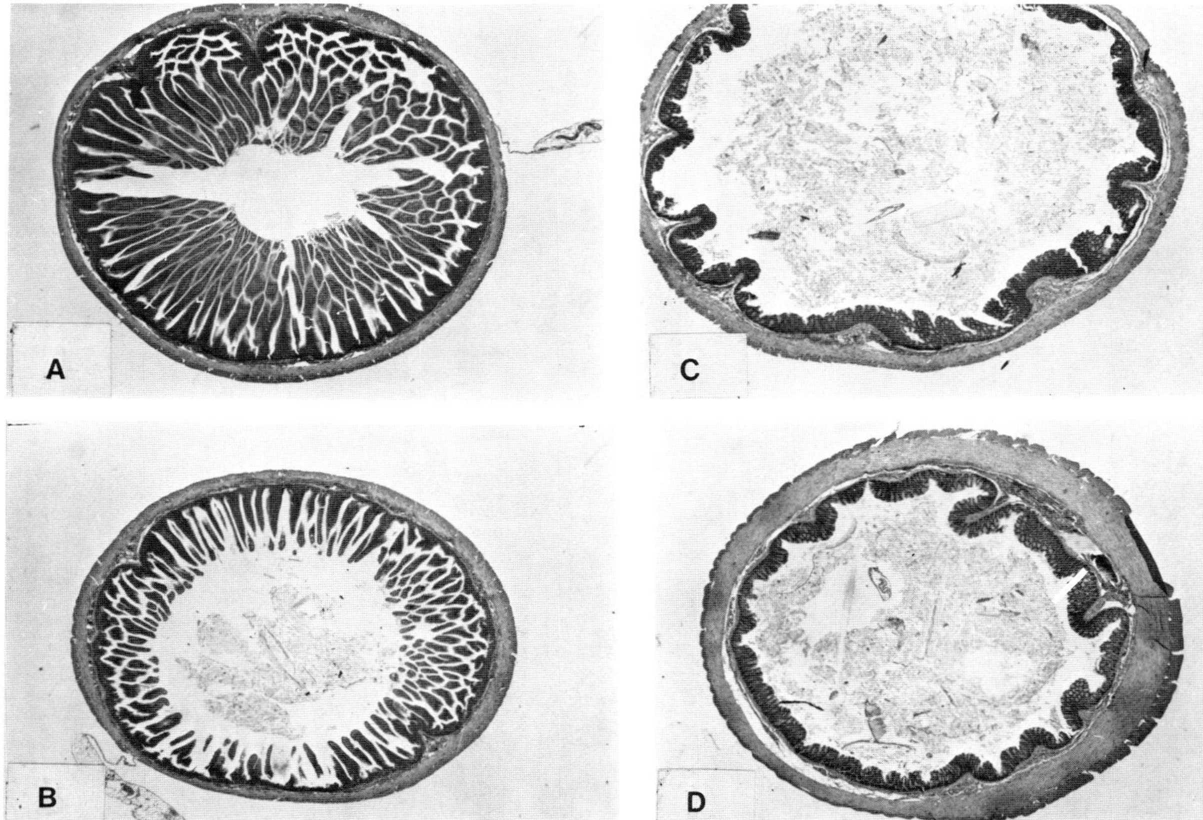


Fig. 5. *Marmosa* duodenum (A), ileum (B), cecum (C), and colon (D) at point just ahead of the pelvic brim. Note the large caliber of the "small intestine" and the changes in muscle layer thickness. See Fig. 6 for a longitudinal section of the unusual villi of *Marmosa* duodenum.

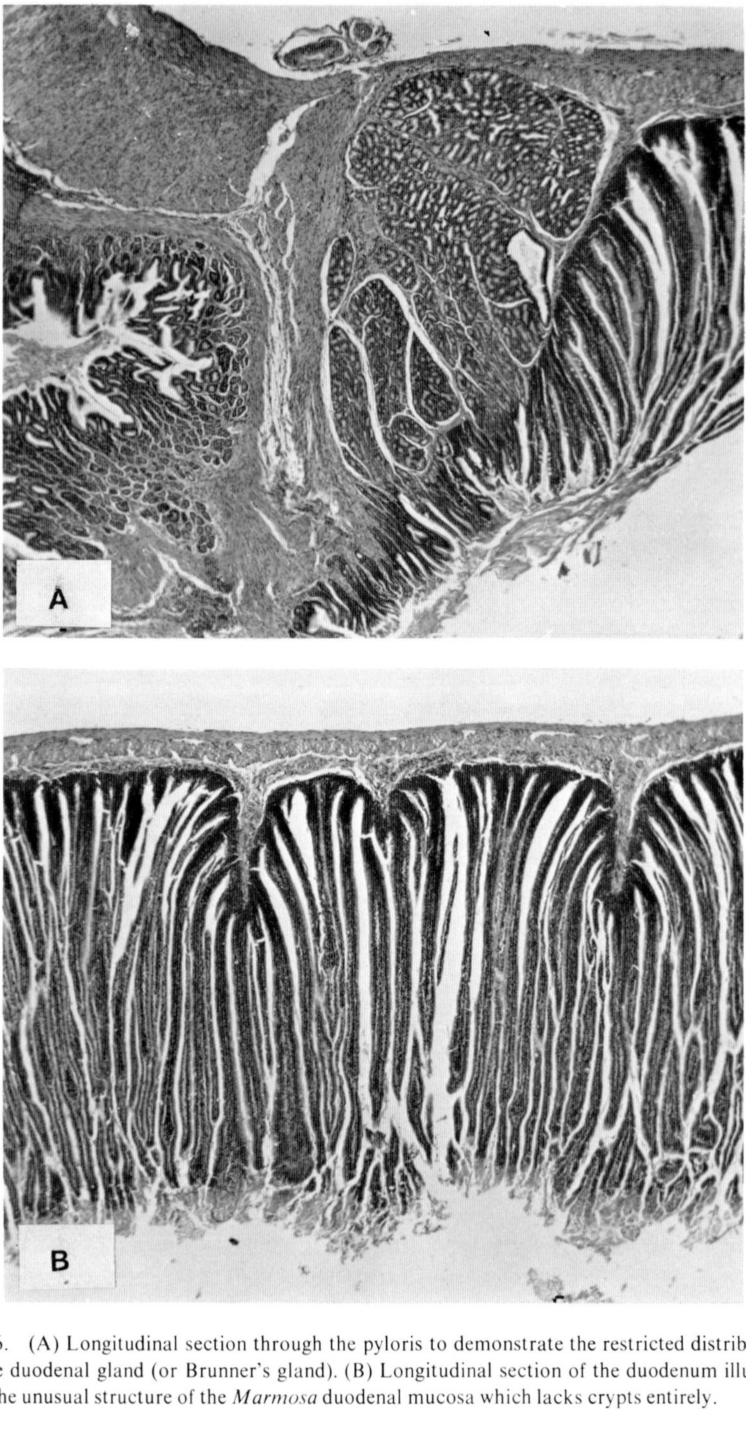


Fig. 6. (A) Longitudinal section through the pyloris to demonstrate the restricted distribution of the duodenal gland (or Brunner's gland). (B) Longitudinal section of the duodenum illustrating the unusual structure of the *Marmosa* duodenal mucosa which lacks crypts entirely.

Lymphatics are present in the "triads" though seldom visible unless distended by pathologic conditions. Polyploid giant hepatocytes are not uncommon. In 50 adult males ranging from 60 to 105 g body weight the mean liver weight (fresh) was 5.72% of body weight.

At birth the liver is strongly hematopoietic and reversion is observed in pathologic states in adults. The bile duct enters the duodenum in the first centimeter below the pyloris.

The pancreas is compact; in location and extent it resembles that of other small mammals. The organ is normally creamy white to pink in freshly killed animals. There is no region of special concentration of endocrine tissue. A small lymph node is seen at one edge of the body. The organ weighs about 300 mg in an 80-g animal. Only in pathologic states is adipose tissue intermixed with exocrine tissue. In the Davis colony of *M. robinsoni* there has been a high incidence of idiopathic pancreatic atrophy in colony-reared animals. This condition has not been seen before 150 days of age. It may escape detection until "natural" death at 2 years. Males were affected at a rate four times that of females. The atrophy shows no signs of being related to an inflammatory disease, obstruction of the pancreatic ducts, or any other morphologically apparent cause. Acinar tissue was seen in various states of degeneration and replacement by adipose cells. Marcrophages and inflammatory cells were absent. The terminal stages reveal as little as 20% of the original exocrine pancreas, but its original extent is clearly indicated by islands of endocrine tissue surrounded by fat. Figure 7A–D illustrates the normal pancreas and an atrophic organ.

## VIII. URINARY SYSTEM

The kidneys of marsupials are in every way comparable to those of placental mammals. Among the marsupials of Australia there are many species living in the arid central region that demonstrate water conservation mechanisms the equal of any eutherian. Urine with a calculated osmolality of 4000 mOsm was obtained from a mulgara (*Dasycercus cristicauda*) by Schmidt–Nielsen and Newsome (1962). Comparison of the kidney architecture of this small dasyurid and *M. robinsoni* indicates that the later species should also have a high concentrating ability; both species share the special feature of an elongated urinary papilla that has long been recognized as the essential feature of kidneys capable of producing urine of high osmolality. Figure 8 is a longitudinal section of a *Marmosa* kidney illustrating the extreme development of its single papilla.

Not only is the morphology of *Marmosa* kidneys of more than passing

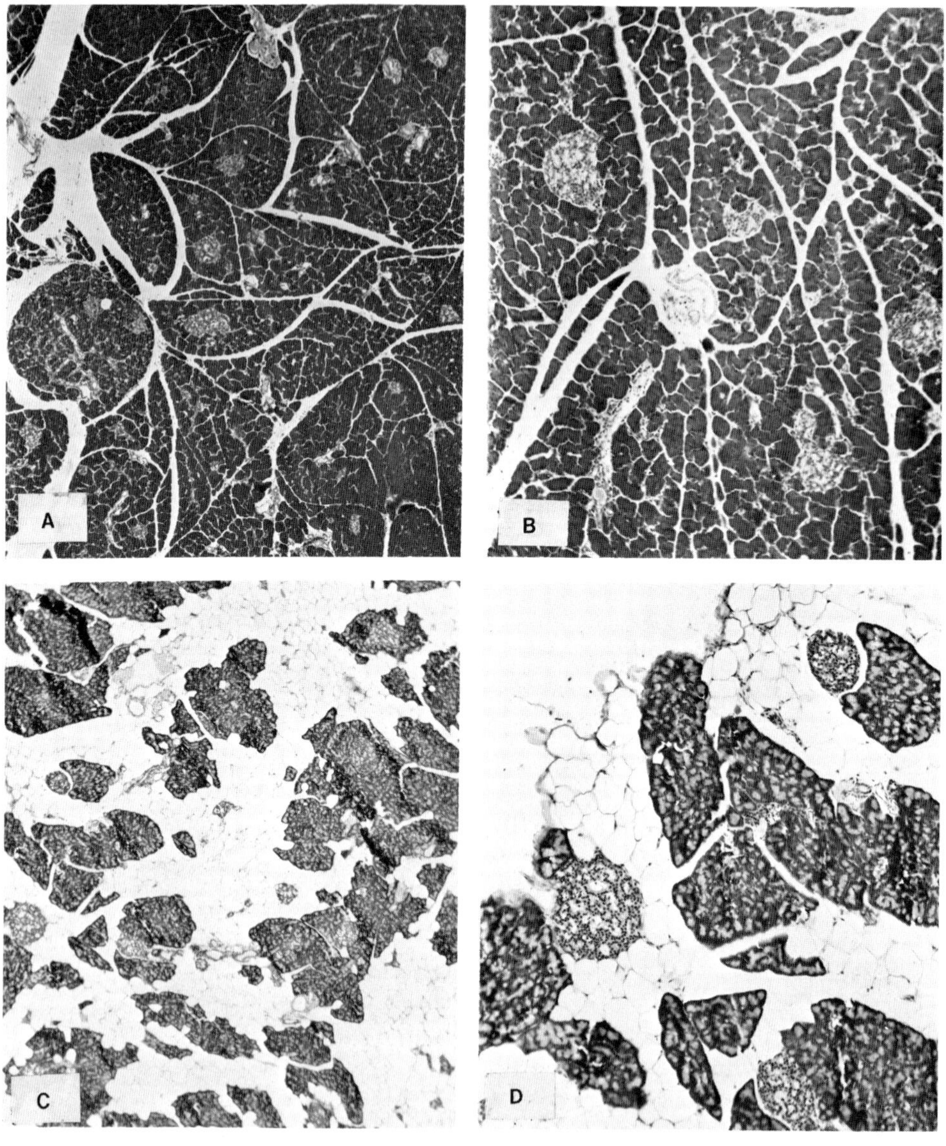


Fig. 7. Normal and atrophic pancreas. (A) and (B) are normal *Marmosa* pancreas; (C) and (D) are an atrophic specimen from 714-day-old male reared from birth in the Davis colony. The atrophy is idiopathic; a genetic basis has not been ruled out.

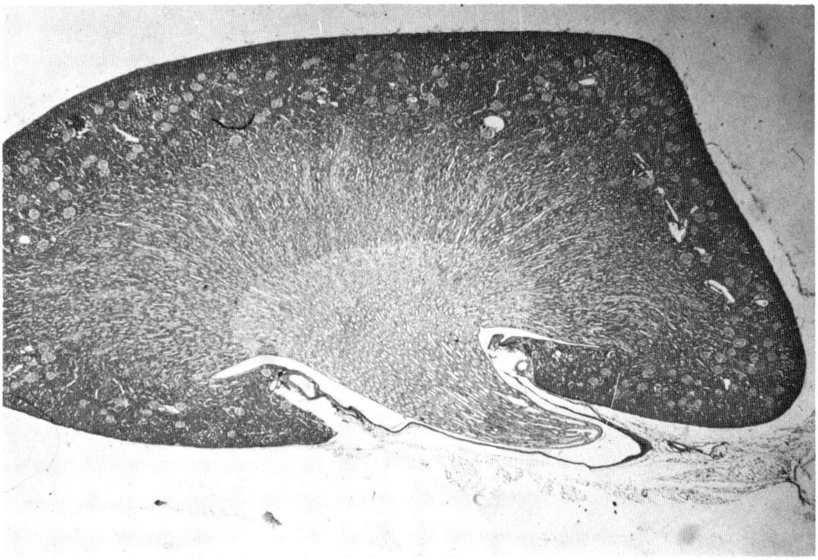

Fig. 8. Longitudinal section of *Marmosa* kidney; note unusually long urinary papilla and strongly developed inner and outer zones in the medulla.

interest, their size is also unexpectedly large. The formula of Rytand (1938)*

$$KW = 8.22BW^{0.85}$$

predicts a combined kidney weight of 682 mg for an 80-g animal. Observed values for 80-g *Marmosa*, whose kidneys were judged to be normal at histologic examination, ranged from 790 to 1150 mg. A further analysis has proven that over the entire range of body weights from weaning to sexual maturity (20 to 120 g), there is a consistent difference in kidney size between males and females with females' kidneys averaging 124% of males. Thirteen males had an average of 10.7 mg/g $BW$ (SEM $\pm$ 0.46) and ten females an average of 13.3 mg/g $BW$ (SEM $\pm$ 0.52). The weights used were from kidneys carefully screened microscopically for evidence of pathology; inflamed and fibrotic specimens were excluded.

It is somewhat of a mystery that *M. robinsoni* has a kidney which, on morphologic criteria, should belong to a desert animal when it originates in a tropical environment. The explanation may lie in the diet rather than with

* $KW$ equals weight of one kidney, $BW$ is body weight; formula derived from observed values for 13 species ranging from mouse to elephant.

the environmental scarcity of water. The carnivorous habit may demand both high efficiency and large capacity in the kidneys of such small mammals, in order to relieve them of the high water consumption rate that would otherwise be required to clear large amounts of urea (Schmidt-Nielsen *et al.*, 1961). The lower urinary tract is described in Section IX.

## IX. REPRODUCTIVE SYSTEMS

As stated in the Introduction, it is the reproductive systems of marsupials that most strikingly differ from the eutherian condition. Although some of the differences are clearly related to the exteriorized embryo habitus, others are simply derivatives of divergent evolution. Most female didelphid marsupials have the marsupium. This structure consists of a fold of skin that forms a protective pouch over the mammary glands. Slips of cutaneous muscle are oriented in such a fashion as to allow voluntary pursing of the opening in all the species; in *Chironectes*, the water opossum, this closure is said to be water tight, allowing females with young to swim without drowning their babies.

The genera lacking a pouch include *Marmosa* and *Lutreolina*. The mammary nipples of all marsupials are very small and those of *Marmosa* females are hidden from casual inspection by the abdominal fur. At the end of the first estrous cycle the female grooms some of this fur away leaving only the shortest pelage hairs in place. The nipples can then be seen in a circle in the posterior abdominal area. Nipple number varies widely in this genus, from a low of 7 to a high of 19. The uneven numbers are accomplished by the presence of single nipple in the center of two symmetrical half-circles or lines of nipples (Tate, 1933). *Marmosa robinsoni* has 13 to 15 nipples.

The microanatomy of the marsupial mammary gland does not differ substantially from eutherian histology except for the presence of slips of striated muscle coursing between the deep fascia and the base of each nipple (Fig. 9C). In *Marmosa*, each nipple receives three lactiferous ducts; no milk cistern is present.

The disposition of the striated muscle bands lends credence to the otherwise unsubstantiated idea that female marsupials pump milk into their babies. The well-developed tongue of neonates would appear to provide adequate sucking power, but it is conceivable that maternal reflexes acting via the neurohypophysis are supplemented by voluntary muscle contraction during the early period of obligate attachment of young to the nipples. No definitive experiment has been conducted on this point.

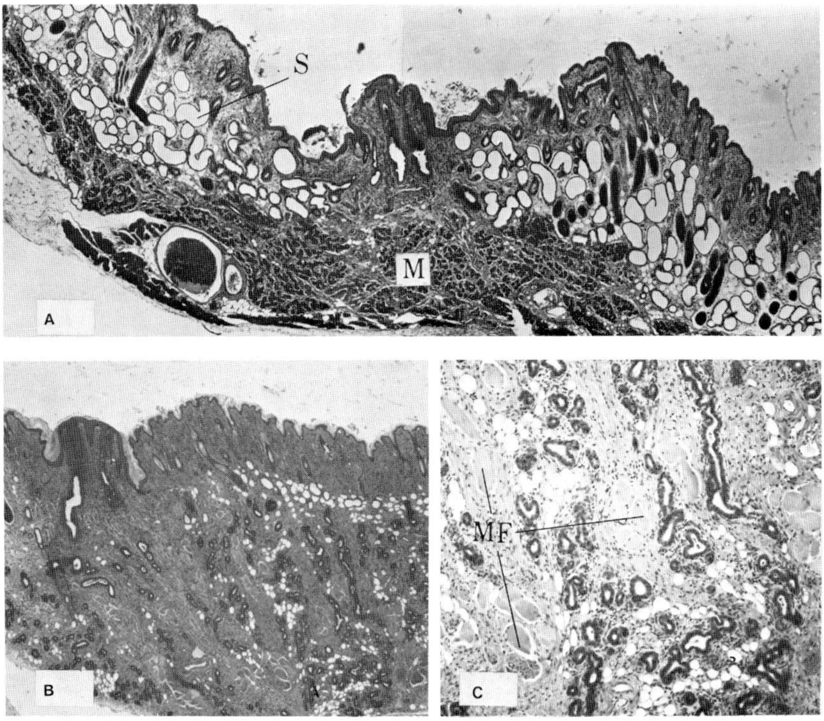


Fig. 9. (A) Mammary gland, nipple, and modified sudoriferous glands as seen in an old female on the eleventh day of pregnancy. Note that the mammary gland (M) is compact and nonsecretory and is comprised mainly of ducts. In contrast, the superficial layer of modified sweat glands is secreting actively (S). (B) Nipple and glands of young female *Marmosa* killed on fourth day of her first estrous cycle. Compare with (A), especially noting the undeveloped status of the modified sweat glands. (C) Higher magnification of (B) to demonstrate the striated muscle fibers present in *Marmosa* mammary gland (MF). These fibers run perpendicular to the skin surface or at shallow angles like poles in a teepee converging upon the dermis around each nipple.

The female urinary and genital systems share a common terminus. Internally the urethra joins the paired vaginal canals to form a urogenital sinus. Figure 10 is a ventral view of the female *Marmosa* genital tract. The didelphid name is derived from the paired vaginae which function to transport sperm upward but not to deliver young downward. The passage of young is via the shortest distance from the median vaginal cul-de-sac to the outside; because of the recurved pattern of the vaginal canals, this shortest distance proves to be through a median connective tissue bridge that is apparently rent apart by the descending neonate, which actively tunnels through to exit via this temporary pseudovaginal canal into the urogenital sinus.

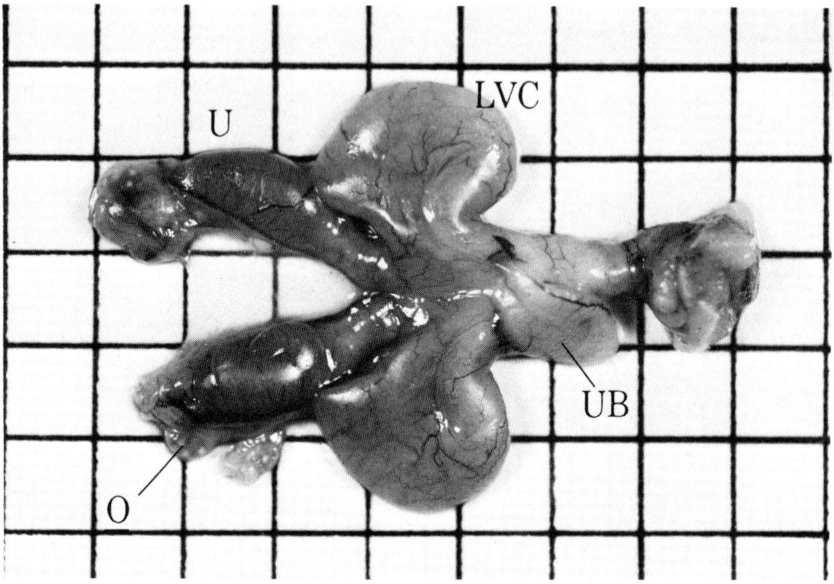


Fig. 10. Ventral view of freshly dissected female *Marmosa* genital tract; the urinary bladder has been reflected back. This specimen is from a female in the early luteal phase; at estrus the proportions of uteri and lateral vaginal canals change markedly with the latter structures hypertrophing to almost twice their size in this picture. The grid has ¼ inch spacing.

Each uterus is completely separate from the other tract. In *Marmosa* the median vaginal cul-de-sac is partitioned until birth of the first litter. At this time the median partition is ruptured and subsequently remains torn to leave the cul-de-sac as a median structure. The same situation has been observed in *Didelphis*. The cervical canals of *Marmosa* are long and coiled through approximately one and a half turns. Each uterus provides a single, large cavity that balloons during pregnancy to contain an average of seven young. The uterine myometrium is very thin, much thinner, in fact, than the muscle layers of the lateral vaginal canals (Figs. 11 and 12). The endometrium is villous with conspicuous and abundant uterine glands opening on every aspect of its complex surface. In females older than 700 days these glands become dilated and cystic. The original architecture of the uterine mucosa is obliterated.

The ovaries are only partially enclosed by bursal folds of the suspensory ligaments. The ovary of the opossum was described in 1942 by Martinez-Esteve. The follicular dynamics associated with estrous cycles of eutherian mammals are seen in every detail in marsupials. The ovarian events are the same insofar as morphologic events are concerned (Fig. 12C, D).

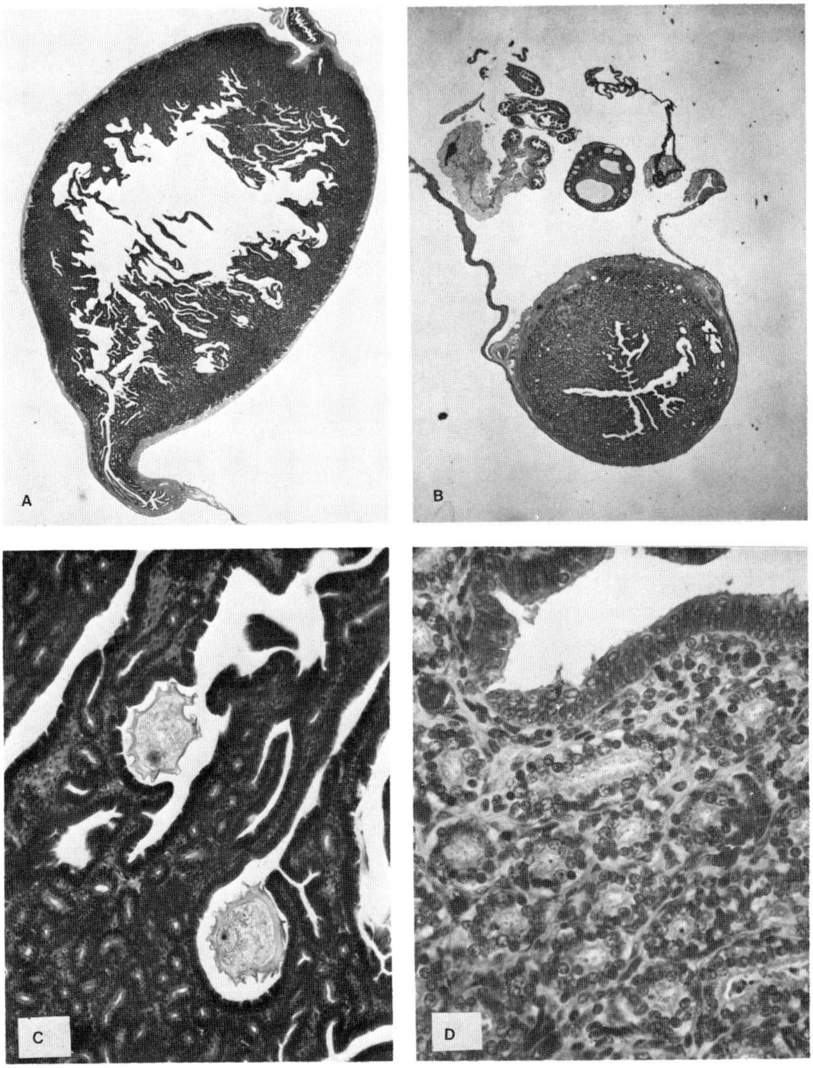


Fig. 11. (A) *Marmosa* uterus in longitudinal section; cervical canal is at bottom of picture. (B) Two unfertilized ova in folds of mucosa of uterus pictured in (A). (C) Cross section of uterus typical of follicular phase of cycle. Note in both (A) and (C) the very thin myometrium. (D) Uterine mucosa at time of estrus.

Likewise, the cyclic development of uterine glands and of exfoliating masses of cornified cells from the lateral vaginal canals follows the same patterns in marsupials as in eutherians. Only in the temporal aspects do we

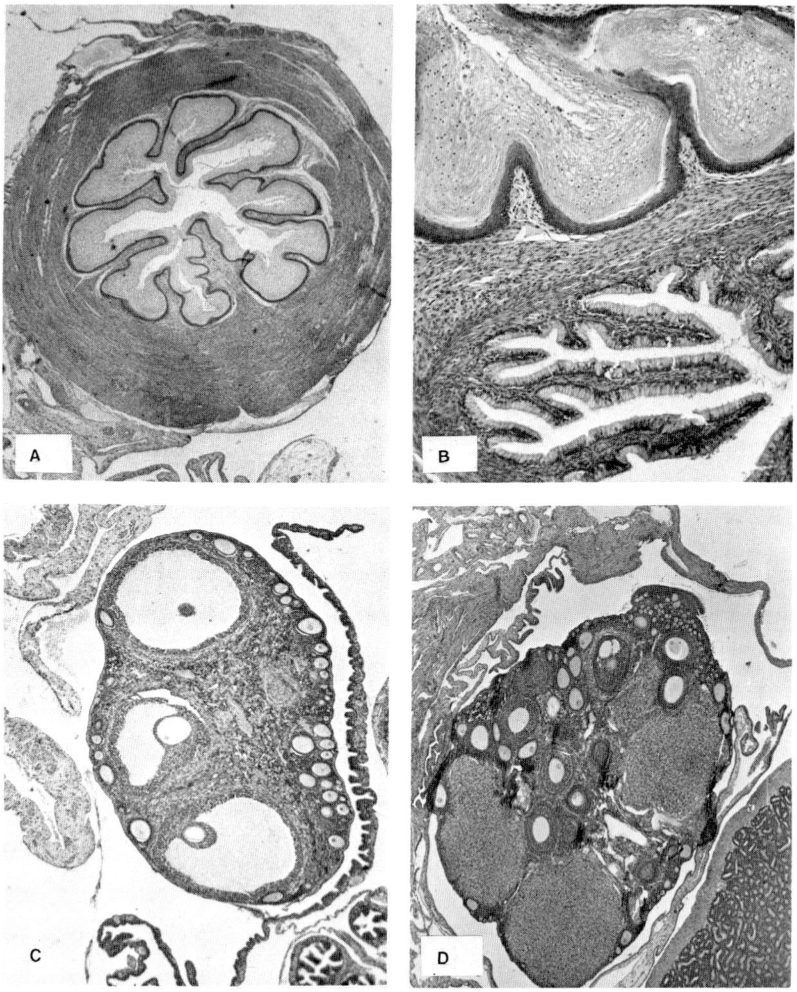


Fig. 12. (A) The lateral vaginal canal (LVC) in cross section; same magnification as the uterus of Fig. 11C; both of these illustrations and the ovary in Fig. 12C are from the same female. She was judged to be in the first day of estrus. (B) The LVC epithelium showing maximal cornification; adjacent mucous epithelium is the cervical canal. (C) Ovary of estrus female. (D) Ovary of female 3 to 4 days postovulation. Uterus with unfertilized ova is illustrated in Figs. 11A and 11B. Note that corpora lutea do not grow much larger than the Graafian follicles from which they develop.

recognize marked departures; the most essential of these is the failure of pregnancy to alter the ovarian cycle, or more specifically, the life span of the corpus luteum. This singular fact dictates that (a) either the young must be born at a very early stage of development or (b) the cycles must be very long. Marsupials seem to have arrived at a compromise where moderately long cycles of about 30 days allow for rebreeding in a particular season if a litter is lost early or an infertile mating occurs. The primitive yolk sac placenta seems to limit the rate of embryonic growth in this interval so that the young remain both very tiny and truly embryonic at birth.

Placentation in *Marmosa* is apparently identical to that described long ago for *Didelphis* (Selenka, 1887); An extensive trilaminar yolk sac placenta absorbs nutrients from the copious uterine secretions, and no adhesion of the embryonic membranes to the uterine epithelium takes place. This anatomical fact would seem to offer a preadaptation to *in vitro* culture of marsupial embryos, since the problems of implantation and true placentation are not present.

The male reproductive system of marsupials is sufficiently different from that of eutherians to have attracted many scientists over the years. Externally the system is always confusing because of the relations of penis to scrotum and the failure of the penis to present any externally visible profile. As noted in Section II, the scrotum is typically a conspicuous, pendulous type with a narrow neck that prevents the gonads of sexually mature animals from being withdrawn into the abdomen. Several small marsupials including *Marmosa, Caluromys,* and *Antechinus* show a glabrous scrotum that is either light blue or black. The pigmentation (presumably melanin) is located in the spermatic fascia or tunica dartos; with advancing age some migration of melanocytes into more superficial layers may cause a darkening of the scrotum from light blue to black. The scrotum is anterior to the penis and occupies the position of the pouch and mammary gland field of females; males never show external signs of mammary gland development.

In many marsupials (and all didelphids) the penis is bifid with the glans penis split for all or part of its length. A median groove in the glans area represents the urethra. It functions to conduct spermatozoa into both sides of the similarly bipartite vaginal tract. Although experimental proof of this functional arrangement is lacking, one can determine from measurements of the erect penis and the corresponding lengths of the female tract that such an arrangment is feasible. That it occasionally fails is likewise easy to prove as a semen plug may sometimes be palpated (in recently bred female *Marmosa*) unilaterally. This plug (Fig. 13) forms very high in the female tract at the most muscular area of the lateral vaginal canals; a liquid spermatozoa-rich ejaculate is trapped above it in the median vaginal cul-de-sac.

The male accessory sex glands, which produce the semen and vaginal plugs, are different from those of eutherians. It is doubtful that their borrowed nomenclature is entirely appropriate, but no attempt will be made to correct that situation here. Chase (1939) published a 24-page description of the opossum male reproductive system. More recently the histochemistry

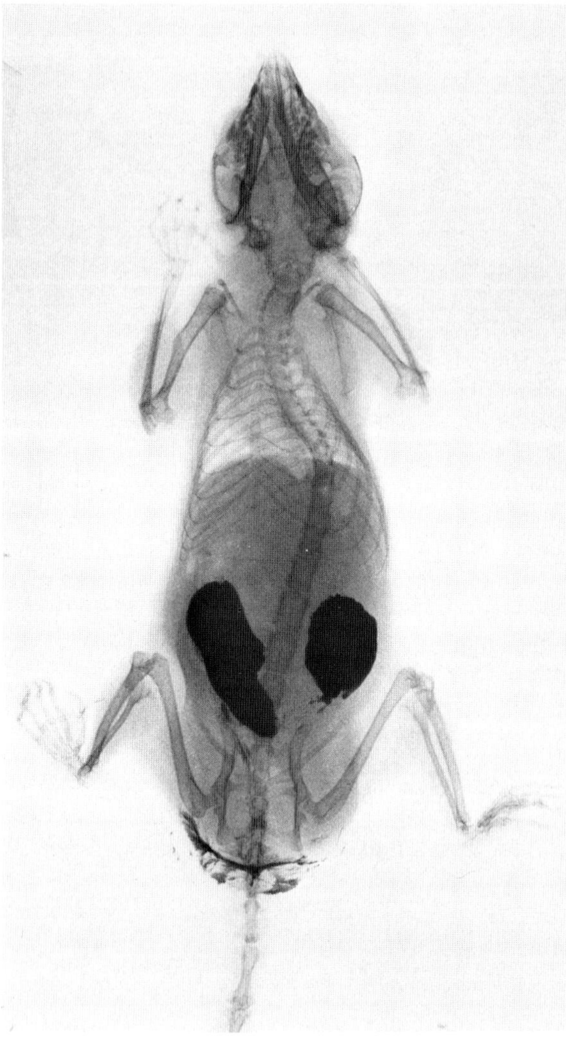

Fig. 13. Radiograph of adult female *Marmosa* illustrating the size and position of the semen plugs that occupy the lateral vaginal canals. For this demonstration the canals have been injected with radiopaque medium. Recently bred females can be palpated for the presence of these plugs.

and ultrastructure of the didelphid system have been described in papers by Martan and Allan (1965), Martan *et al.* (1967), and Ladman (1967). The structure of the penis and scrotum and the phenomenon of spermatozoan pairing in American marsupials are discussed at length by Biggers (1966).

The special anatomy of *Marmosa* is limited to the following observations:

1. The glans penis is conspicuously bifid.
2. The bulbourethral glands are represented by two pairs rather than three, which otherwise seems typical of most families and of other genera of didelphids.
3. The largest pair of bulbourethral glands gives rise to the semen plug when catalyzed by the secretions of prostate I (R. D. Barnes, unpublished).
4. The prostate gland shows the same three distinct zones as described for other marsupials; its size is unexpectedly large. At puberty (315 days of age) the gland weighs 1.7 g or 2.27% of body weight; in sexually active males 1 to 2½ years old it averages 2.8 g or 3.59% of body weight.
5. There is relatively little spermatozoa stored in any portion of the epididymis; spermatorrhea is demonstrable in adult males throughout the year.
6. The testes comprise slightly less than 1% of body weight in mature

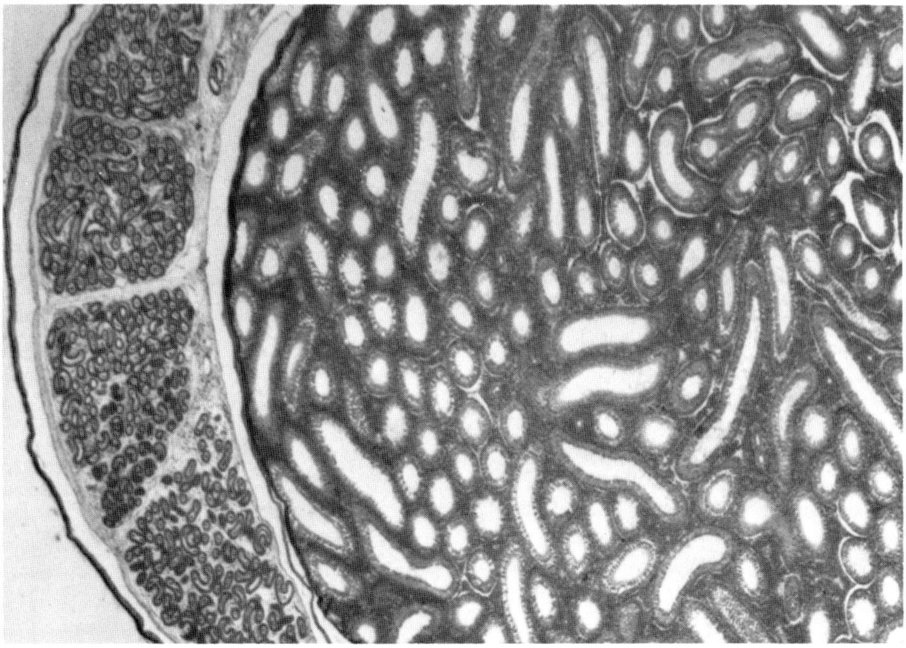

Fig. 14. *Marmosa* testis; note the abundant interstitial cells. The adjacent section of epididymis is from the head region (caput epididymis); note the small diameter of these tubules which never store large quantities of spermatozoa.

males (9.9 mg/g *BW*). Their microanatomy is remarkable for the quantity of interstitial cells (see Fig. 14).

7. The pairing of spermatozoa occurs in the cauda epididymis (about one-third of the way through this region).

8. In colony-reared males of known ages greater than 800 days degenerative changes leading to aspermia gradually appear. Seminiferous tubules comprised entirely of Sertoli cells increase from this age onward; the testes shrink and the prostate gland slowly involutes. Loss of turgor in the externally prominent bulbourethral glands is an early sign of senescence.

## ACKNOWLEDGMENTS

Original observations of *Marmosa robinsoni* reported in this chapter were made on animals supported by U.S. Public Health Service. National Institutes of Health Grant no. FR 00359 at the University of California, Davis, California, School of Veterinary Medicine, and the technical assistance of Laura Minning and Louis Mariani are gratefully acknowledged.

## REFERENCES

Barker, R. R., Bergman, H. C., Hunter, W. H., Lawrence, W. H., Steers, C. W., Wright, L. N., Willis, R. A., and Barrow, D. M. (1967). "Marsupial Biomodule Evaluation Study." Aerospace Medical Division, Air Force Systems Command, Brooks AFB, Texas. (Available from U.S. Defense Documentation Center, Cameron Station, Alexandria, Virginia.)

Barnes, R. D., and Wolf, H. G., (1971). *Int. Zoo Yearb.* **11,** 50–54.

Biggers, J. (1966). *Symp. Zool. Soc. London* **15,** 251–279.

Block, M. (1964). "Reviews of Anatomy, Embryology and Cell Biology." Springer-Verlag, Berlin and New York.

Bremer, J. L. (1904). *Am. J. Anat.* **3,** 67–73.

Bridge, T. D. (1970). *J. Anat.* **106,** 285–295.

Chase, E. B. (1939). *J. Morphol.* **65,** 215–239.

Cheng, C. (1955). *J. Morphol.* **97,** 415–472.

Coues, E. (1872). *Mem. Boston Soc. Nat. Hist.* **2,** 41–149.

Hayes, T. G. (1968). *J. Morphol.* **124,** 445–450.

Hildebrand, M. (1961). *Am. J. Anat.* **109,** 239–249.

Kerr, N. S. (1955). *Anat. Rec.* **121,** 481–494.

Ladman, A. J. (1967). *Anat. Rec.* **157,** 559–576.

Lowrance, E. W. (1949). *J. Morphol.* **85,** 569–593.

Martan, J., and Allen, J. M. (1965). *J. Exp. Zool.* **159,** 209–230.

Martan, J., Hruban, Z., and Slesers, A. (1967). *J. Morphol.* **121,** 81–102.

Martinez-Esteve, P. (1942). *Contrib. Embryol. Carnegie Inst.* **541,** No. 189,19–26.

Reynolds, H. C. (1952). *Univ. Calif. Berkeley Publ. Zool.* **52,** 223.

Rothstein, R., and Hunsaker, R., II. (1972). *Lab. Anim. Sci.* **22,** 227–232.

Rowlands, D., La Via, M. F., and Block, M. (1964). *J. Immunol.* **93,** 157–164.
Rytand, D. (1938), *Am. J. Anat.* **62,** 507–520.
Schmidt-Nielsen, B., O'Dell, B., and Osaki, H. (1961). *Am. J. Physiol.* **200,** 1125–1132.
Schmidt-Nielsen, K., and Newsome, A. E. (1962). *Aust. J. Biol. Sci.* **15;** 683–689.
Selenka, E. (1887). "Studien über Entwickelungsgeschichte. Viertes Hoft. Das Opossum." C. W. Kreidels Verlag, Wiesbaden.
Tate, G. H. H. (1933). *Bull. Am. Mus. Nat. Hist.* **66,** 6–236.
Wade, D., and Neeley, P. (1949). *J. Mammal.* **30,** 111–116.
Washburn, S. L. (1946). *Anat. Rec.* **95,** 353–363.

# 8

# Diseases of Marsupials

STEPHEN POTKAY



## I. INTRODUCTION

Marsupials are part of the native fauna of three continents. The American species have been known to the Western world since the early sixteenth century and those of Australasia since the mid-eighteenth century.

Historically, man has used certain marsupial species as food, hunted them for sport, and kept them as pets. General awareness of these animals has been largely sustained by their great anatomic diversity, the presence of a marsupium in most species, and folklore. Interest in marsupials as research animals appears to date from the 1920's when the unique characteristic of this order of mammals—birth at an early stage of development—was investigated. Subsequently, biomedical interest in marsupials has broadened and currently includes developmental anatomy, immunology, neurology, and many aspects of comparative medicine and behavior. It is imperative, therefore, to further our knowledge of the naturally occurring diseases and the disease susceptibilities of captive marsupials so that healthier specimens will be available for research studies, and the influence of natural diseases upon experimental results can be better evaluated. This knowledge may also shed light on the role of these animals in the spread of disease-producing agents to man, domestic animals, and wildlife.

Source material for this chapter came from zoological, wildlife, parasitological, and veterinary medical societies. Although many of these studies were well conducted, some were of a fragmentary nature, while others appeared to be casual observations. It is hoped that the incomplete nature of this material will stimulate the reader and point out the need for further study.

## II. DISEASES OF AMERICAN MARSUPIALS

### A. Viral Diseases

#### 1. Myxovirus Infection

**Rabies.** The incidence of rabies among wild *Didelphis virginiana* appears to be low. Sixteen confirmed cases occurred in the United States between 1955 and 1960 (Barr, 1963) and 13 between 1964 and 1967 (Potkay, 1970). Experimental studies indicated that *D. virginiana* is highly resistant to rabies infection (Barr, 1961, 1963; Beamer *et al.*, 1960; Constantine, 1966; Constantine and Woodall, 1966). Clinical signs of the experimental disease in this species are anorexia and paralysis. Young opossums appear to be more susceptible than adults.

Evidence presented by Barr (1963) suggested that a carrier state may be induced by experimental rabies infection. He isolated rabies virus from a healthy, young opossum in which hippocampal Negri bodies were demonstrated.

## 2. Herpesvirus Infection

**Pseudorabies.** Pseudorabies has been induced experimentally in two *D. virginiana* by Trainer and Karstad (1963). Clinical signs were pruritis, anorexia, excessive salivation, clonic spasms, and convulsions. They appeared 4 days after intramuscular injection of the Aujeszky strain and 7 days after administration of Strain A of pseudorabies virus. Death occurred 6 days after inoculation with the Aujeszky strain; the other opossum became moribund on the tenth day and was killed. Gross lesions were limited primarily to the injection sites and included alopecia, lacerations of the skin, and edema and hemorrhage in adjacent muscles. Naturally occurring pseudorabies has not been reported from American marsupials.

## 3. Arbovirus Infections

Table I lists the American marsupials from which naturally occurring arboviruses have been isolated and on which antibody testing was performed. The virus classification follows that of Wilner (1969).

As indicated in Table I, nine arboviruses have been isolated from five genera of wild Didelphidae (*Didelphis, Marmosa, Caluromys, Philander,* and *Metachirus*). Antibody to arboviruses was found in six genera. The frequency of both virus isolation and antibody detection was low, but antibody was more commonly encountered than viruses. Cross immunity phenomena should be considered in evaluating antibody responses because of the occurrence of group antigens among arboviruses. Although isolation is the most important consideration in determining the presence of virus in a given locale, isolation per se does not necessarily incriminate a vertebrate or arthropod in the cycle of virus dissemination (Taylor, 1967). The animal from which the agent is obtained may be a dead end or incidental host which is not involved in the transmission cycle which maintains the virus in nature. Marsupials probably do not play important roles as reservoir or amplifying hosts for arboviruses according to classical criteria. The duration or level of viremia or both are generally considered insufficient to provide optimal conditions for host—vector interaction. A less orthodox approach suggests that opossums, or other animals, in which immune responses to infection are infrequent, equivocal, or absent, may be suitable hosts for arboviruses. According to this view, infection may occur in immature animals which does not produce disease and is associated with suppression or paralysis of immune responses. It is conceivable that such animals could serve as important hosts during epizootic or interepizootic periods.

**TABLE I**

Summary of Arboviruses Associated with American Marsupials (Didelphidae)

| Virus | Marsupial | Virus Isolation[a] | Antibody[b,c] | Location | References |
|---|---|---|---|---|---|
| Group A (mosquito-borne) | | | | | |
| Pixuna | *Didelphis marsupialis* | 0/883 | 1/80 HI | Para, Brazil | Shope (1967e) |
| | Other marsupials | 0/1067 | 1/143 HI | Para, Brazil | Shope (1967e) |
| Mucambo | *Metachirus nudicaudatus* | 1/82 | HI and SN | Para, Brazil | Shope (1967e) |
| | *Philander opossum* | 1/237 | antibody common | Para, Brazil | Shope (1967e) |
| Venezuelan equine encephalitis (VEE) | *Marmosa robinsoni* | 1/28 | * | Trinidad, West Indies | Jonkers *et al*. (1968) |
| | *Marmosa* sp. | +[d] | + | Almirante, Panama | Grayson and Galindo (1968, 1969) |
| | *D. marsupialis* | – | 16/48 HI | Almirante, Panama | Grayson and Galindo (1968, 1969) |
| | *D. virginiana* | * | 1/184 HI | Florida | Bigler *et al*. (1975) |
| | *D. virginiana* | –[e] | 1/110 HI | Florida | Bigler (1971) |
| | *Philander opossum* | – | 26/84 HI | Almirante, Panama | Grayson and Galindo (1968) |
| Western equine encephalitis (WEE) | *D. virginiana* | 1/1 | * | California | Emmons and Lennette (1969) |
| | *D. virginiana* | 0/2 | * | Texas | Hayes *et al*. (1967) |
| | *D. virginiana* | *[f] | 0/159 HI | Florida | Bigler *et al*. (1975) |
| Eastern equine encephalitis (EEE) | *D. virginiana* | * | 1/3 HI | Iowa | Wong *et al*. (1970) |
| | *D. marsupialis* | * | 0/50 HI | Bahia, Brazil | Shope *et al*. (1966) |
| | *Caluromys* sp. | * | 0/40 HI | Bahia, Brazil | Shope *et al*. (1966) |
| | *Marmosa murina* | * | 0/55 HI | Bahia, Brazil | Shope *et al*. (1966) |
| | *M. cinerea* | * | 1/36 HI | Bahia, Brazil | Shope *et al*. (1966) |
| | *Philander opossum* | * | 0/12 HI | Bahia, Brazil | Shope *et al*. (1966) |
| | *Metachirus nudicaudatus* | * | 0/4 HI | Bahia, Brazil | Shope *et al*. (1966) |
| | *D. virginiana* | * | 1/199 HI | Florida | Bigler *et al*. (1975) |
| St. Louis encephalitis (SLE) and WEE | *Caluromys* sp. | * | 0/11 HI | Bahia, Brazil | Shope *et al*. (1966) |
| | *D. marsupialis* | * | 0/12 HI | Bahia, Brazil | Shope *et al*. (1966) |
| | *Marmosa murina* | * | 0/15 HI | Bahia, Brazil | Shope *et al*. (1966) |

| | | | | | |
|---|---|---|---|---|---|
| | *M. cinerea* | * | 0/10 HI | Bahia, Brazil | Shope *et al.* (1966) |
| | *Philander opossum* | * | 0/1 HI | Bahia, Brazil | Shope *et al.* (1966) |
| | *D. virginiana* | * | 2/199 HI | Florida | Bigler *et al.* (1975) |
| Group B (mosquito-borne) | | | | | |
| Ilheus | *Didelphis* sp. | * | 18/134 HI | Bahia, Brazil | Laemmert (1967) |
| | *Philander opossum* | * | 9/88 HI | Bahia, Brazil | Laemmert (1967) |
| | *Marmosa* sp. | * | 13/54 HI | Bahia, Brazil | Laemmert (1967) |
| Yellow fever | *D. marsupialis* | * | 0/227 SN | Bahia, Brazil | Laemmert *et al.* (1946) |
| | *Philander opossum* | * | 0/1 SN | Bahia, Brazil | Laemmert *et al.* (1946) |
| | *Metachirus nudicaudatus* | * | 1/246 SN | Bahia, Brazil | Laemmert *et al.* (1946) |
| | *Monodelphis americana* | * | 0/5 SN | Bahia, Brazil | Laemmert *et al.* (1946) |
| | *Marmosa cinerea* | * | 10/114 SN | Bahia, Brazil | Laemmert *et al.* (1946) |
| | *M. agilis* | * | 2/11 SN | Bahia, Brazil | Laemmert *et al.* (1946) |
| | *M. murina* | * | 5/57 SN | Bahia, Brazil | Laemmert *et al.* (1946) |
| | *M. incana* | * | 1/78 SN | Bahia, Brazil | Laemmert *et al.* (1946) |
| Bunyamwera supergroup (mosquito-borne) | | | | | |
| Bunyamwera group | | | | | |
| Maguari | Marsupials | 0/1950 | 0/98 HI | Para, Brazil | Shope and Whitman (1967) |
| Cache Valley | *D. virginiana* | * | 0/44 SN | Kentucky and Illinois | Kokernot *et al.* (1969b) |
| Group C | | | | | |
| Murutucu | *Caluromys* sp. | * | 1/33 HI | Bahia, Brazil | Shope (1967b) |
| | *Didelphis* sp. | * | 6/80 HI | Bahia, Brazil | Shope (1967b) |
| | *Marmosa* sp. | * | 5/81 HI | Bahia, Brazil | Shope (1967b) |
| | *Metachirus nudicaudatus* | * | 1/9 HI | Bahia, Brazil | Shope (1967b) |
| Itaqui | *Philander opossum* | 1/237 | HI antibody rare in marsupials | Bahia, Brazil | Shope (1967c) |
| Oriboca | *D. marsupialis* | 1/883 | * | Bahia, Brazil | Shope (1967a) |
| California encephalitis group | | | | | |
| California encephalitis | *D. virginiana* | * | 113/113 HI | Florida | Jennings *et al.* (1968) |
| San Angelo | Opossum | * | 3/7 SN | Texas | Grimes (1967) |

**TABLE I (Cont.)**

| Virus | Marsupial | Virus Isolation[a] | Antibody[b,c] | Location | References |
|---|---|---|---|---|---|
| Capim group | | | | | |
| Acara | *Philander opossum* | * | 1/3 SN | Bahia, Brazil | Belem Virus Laboratory (1967g) |
| | *Metachirus nudicaudatus* | * | 0/4 SN | Bahia, Brazil | Belem Virus Laboratory (1967g) |
| | *Didelphis* sp. | * | 1/6 SN | Bahia, Brazil | Belem Virus Laboratory (1967g) |
| | *Marmosa* sp. | * | 1/4 SN | Bahia, Brazil | Belem Virus Laboratory (1967g) |
| Capim | *Caluromys philander* | 1/183 | * | Bahia, Brazil | Belem Virus Laboratory (1967b) |
| Guama group | | | | | |
| Guama | *Caluromys* sp. | 1/183 | 10/33 HI | Bahia, Brazil | Shope (1967d) |
| | *Marmosa* sp. | 1/556 | 4/54 HI | Bahia, Brazil | Shope (1967d) |
| | *D. marsupialis* | 1/883 | 15/80 HI | Bahia, Brazil | Shope (1967d) |
| Unassigned in the Bunyamwera supergroup | | | | | |
| Mirim | *Marmosa* sp. | * | 1/20 HI | Bahia, Brazil | Belem Virus Laboratory (1967a) |
| Other small groups (mosquito-borne) | | | | | |
| Turlock | *D. marsupialis* | * | 2/45 HI | Para, Brazil | Viral and Rickettsial Disease Laboratory (1967) |
| | Other marsupials | * | 0/38 HI | Para, Brazil | Viral and Rickettsial Disease Laboratory (1967) |
| | *Caluromys* sp. | * | 0/11 HI | Bahia, Brazil | Shope *et al.* (1966) |

| | | | | | |
|---|---|---|---|---|---|
| | *D. marsupialis* | * | 0/12 HI | Bahia, Brazil | Shope *et al.* (1966) |
| | *Marmosa murina* | * | 0/15 HI | Bahia, Brazil | Shope *et al.* (1966) |
| | *M. cinerea* | * | 0/10 HI | Bahia, Brazil | Shope *et al.* (1966) |
| | *Philander opossum* | * | 0/1 HI | Bahia, Brazil | Shope *et al.* (1966) |
| Other small groups (phlebotomus-borne) | | | | | |
| Candiru | Marsupials | 0/1950 | * | Para, Brazil | Belem Virus Laboratory (1967c) |
| Icoaraci | Marsupials | 0/1950 | 6/223 | Para, Brazil | Belem Virus Laboratory (1967e) |
| Itaporanga | *Caluromys* sp. | 1/129 | 11/23 HI | Bahia, Brazil | Trapp and Shope (1967) |
| | *Marmosa* sp. | * | 2/11 HI | Bahia, Brazil | Trapp and Shope (1967) |
| | *Didelphis* sp. | * | 1/12 HI | Bahia, Brazil | Trapp and Shope (1967) |
| | *Marmosa murina* | * | 0/8 HI | Bahia, Brazil | Trapp and Shope (1967) |
| | *Philander opossum* | * | 0/3 HI | Bahia, Brazil | Trapp and Shope (1967) |
| Other small groups (miscellaneous vectors) | | | | | |
| Vesicular stomatitis group[g] | | | | | |
| Vesciular stomatitis New Jersey strain | *D. marsupialis* | * | 8/160 SN | Panama | Tesh *et al.* (1969) |
| | *Philander opossum* | * | 14/126 SN | Panama | Tesh *et al.* (1969) |
| | *Caluromys derbianus* | * | 8/86 SN | Panama | Tesh *et al.* (1969) |
| | *Marmosa robinsoni* | * | 0/25 SN | Panama | Tesh *et al.* (1969) |
| | *Metachirus nudicaudatus* | * | 2/33 SN | Panama | Tesh *et al.* (1969) |
| | *Chironectes minimus* | * | 1/2 SN | Panama | Tesh *et al.* (1969) |
| Indiana strain | *D. marsupialis* | * | 18/165 SN | Panama | Srihongse (1969) |
| | *D. marsupialis* | * | 3/16 CF | Darien Prov., Panama | Tesh *et al.* (1969) |
| | *Philander opossum* | * | 9/128 SN | Panama | Tesh *et al.* (1969) |
| | *Caluromys derbianus* | * | 3/87 SN | Panama | Srihongse (1969) |
| | *Caluromys derbianus* | * | 1/3 CF | Darien Prov., Panama | Tesh *et al.* (1969) |
| | *Marmosa robinsoni* | * | 0/25 SN | Panama | Tesh *et al.* (1969) |
| | *Metachirus nudicaudatus* | * | 1/34 SN | Panama | Srihongse (1969) |
| | *Metachirus nudicaudatus* | * | 1/7 CF | Darien Prov., Panama | |

**TABLE I (Cont.)**

| Virus | Marsupial | Virus Isolation[a] | Antibody[b,c] | Location | References |
|---|---|---|---|---|---|
| Other small groups (unknown vectors) | | | | | |
| Piry group | | | | | |
| Piry | *Philander opossum* | 1/237 | * | Bahia, Brazil | Belem Virus Laboratory (1967a) |
| | Other marsupials | 0/1713 | * | Bahia, Brazil | Belem Virus Laboratory (1967a) |
| Timbo group | | | | | |
| Chaco | Marsupials | 0/1950 | * | Para, Brazil | Shope (1967f) |
| Timbo | Marsupials | 0/1950 | * | Para, Brazil | Shope (1967h) |
| Marco | Marsupials | 0/1950 | * | Para, Brazil | Shope (1967g) |
| Ungrouped (tick-borne) | | | | | |
| Lone star | *D. virginiana* | * | 0/34 SN | Kentucky | Kokernot *et al*. (1969c) |
| Ungrouped (unknown vector) | | | | | |
| Pacui | *Philander opossum* | * | 1/3 SN | Bahia, Brazil | Belem Virus Laboratory (1967f) |
| | *Didelphis* sp. | * | 3/17 SN | Bahia, Brazil | Belem Virus Laboratory (1967f) |
| | *Marmosa* sp. | * | 0/2 SN | Bahia, Brazil | Belem Virus Laboratory (1967f) |
| Sawgrass | *D. virginiana* | * | 0/14 SN | Florida | Sather *et al*. (1970) |

[a] Number of isolations/number tested.
[b] Number with antibody/number tested.
[c] HI, hemagglutination-inhibition test; SN, serum neutralization test; CF, complement fixation text.
[d] +, positive (number of animals not given).
[e] −, negative (number of animals not given).
[f] Asterisk indicates isolation not attempted.
[g] Classification as arbovirus uncertain.

Attempts were made to infect opossums experimentally with several arboviruses. The results of these studies are summarized below.

**a. Yellow Fever.** Numerous studies were undertaken to elucidate the possible role of marsupials as reservoirs for yellow fever in South America (Bugher *et al.*, 1941; Bates, 1944; Laemmert, 1946; Anderson and Roca-Garcia, 1947). No virus isolations were obtained from wild Didelphidae, although antibody was present in *Metachirus nudicaudatus* and four species of *Marmosa* examined by Laemmert *et al.* (1946). In their review of experimental yellow fever infection, Bates and Roca-Garcia (1946) indicated that *Philander* and *Didelphis* rarely had detectable viremias following artificial administration of the virus, and that *Marmosa* was extremely difficult to infect. Further studies by Bates and Roca-Garcia (1946) showed that *Metachirus* and *Caluromys* were susceptible to infection by mosquitoes (*Haemogogus capricornii*) and, in turn, could transmit infection to mosquitoes. Waddell and Taylor (1948) reported a similar situation in their transmission experiments using the opossums, *M. nudicaudatus* and *Marmosa cinerea,* and mosquitoes (*Aedes aegypti*). However, *A. aegypti* were not readily or consistently infected by feeding on viremic opossums. Because of this and the aversion of *H. capricornii* for marsupials, it was considered unlikely that Didelphidae played a major or effective role in the sylvan cycle of yellow fever. The results of a survey of 1040 marsupials and their ectoparasites in an endemic yellow fever area of Brazil were consonant with this opinion. Only 19 opossums had antibodies against yellow fever and none of the ectoparasites (ticks, lice, and mites) carried the virus (Laemmert *et al.*, 1946).

**b. Vesicular Stomatitis.** The natural occurrence of neutralizing and complement-fixing antibody against vesicular stomatitis virus (VSV) in six species of opossums is shown in Table I. Although natural infection does occur, the studies of Tesh *et al.* (1970) and Srihongse (1969) suggest that the Didelphidae probably play a minor role as reservoir hosts of VSV.

Opossums appear to be highly susceptible to experimental VSV infection (Tesh *et al.*, 1970). Two *Caluromys derbianus* and five *Didelphis marsupialis* inoculated subcutaneously with the New Jersey strain and one *C. derbianus*, five *D. marsupialis*, two *Philander opossum*, and one *Metachirus nudicaudatus* given the Indiana strain of VSV developed neutralizing antibody within 12 days. Both strains of VSV were isolated from the blood, throat, and skin of two opossums from 1 to 7 days after infection. Virus clearance studies showed that VSV was eliminated from the blood within 10 min after intravenous inoculation of *D. marsupialis*. Although no clinical signs were observed in experimentally infected adults, paralysis of the hind limbs, spasms, convulsions, and death occurred following administration of VSV to three species of suckling opossums. However, opossums delivered by immune mothers had neutralizing antibody titers equal to that of the mothers, indicating that transfer of maternal antibody to the fetuses occurs.

**c. Eastern and Western Equine Encephalitis.** The role of *D. marsupialis* in the epidemiology of eastern (EEE) and western (WEE) equine encephalomyelitis is not known. While some investigators (Giltner and Shahan, 1963; Syverton and Berry, 1940) found *D. virginiana* refractory to infection with these viruses, WEE has been isolated from a wild *D. virginiana* (Emmons and Lennette, 1969) and antibody to EEE found in *Marmosa cinerea* (Shope *et al.*, 1966) and *D. virginiana* (Wong *et al.*, 1970).

**d. St. Louis Encephalitis.** No clinical signs of infection were observed following subcutaneous administration of St. Louis encephalitis (SLE) virus to five *D. virginiana*. Trace viremias occurred for 1 to 7 days after inoculation, and hemagglutination-inhibition antibody responses to SLE virus were detected in each animal (Kokernot *et al.*, 1969a).

**e. Cache Valley.** No detectable viremias occurred following infection of seven *D. virginiana* with Cache Valley virus (Kokernot *et al.*, 1969d). Neutralizing antibody was found in two of two opossums tested 1 to 2 months after infection. The authors presumed that the opossums had been exposed earlier to a related virus or that the experimentally administered virus produced an anamnestic response.

**f. Powassan.** Trace viremias were detected following subcutaneous inoculation of Powassan virus in four *D. virginiana* (Kokernot *et al.*, 1969d). Virus was isolated beginning on the second and continuing for 6 to 10 days after inoculation.

**g. Tensaw.** Sudia *et al.* (1969) were unable to detect viremias following subcutaneous administration of Tensaw virus to five *D. virginiana*. Hemagglutination-inhibition antibody was found 1 to 2 months after inoculation in two of three opossums tested.

**h. Wyeomyia.** Roca-Garcia (1967) could not demonstrate immune responses in 10 *D. marsupialis* inoculated intracerebrally or subcutaneously with Wyeomyia virus.

**i. B virus.** Syverton and Berry (1940) infected one *D. virginiana* with what was apparently an unspecified Group B arbovirus which caused progressive paralysis and death in 9 days.

## B. Bacterial Diseases

### 1. Mycobacterial Infections

Fox (1927–1928) and Hamerton (1935a) considered marsupials relatively resistant to tuberculosis. Their conclusions were derived principally from observations of zoological specimens. However, four cases were reported in captive *Didelphis virginiana* from Pennsylvania, two of which were pets (Fox, 1926, 1927–1928, 1929, 1937). Lesions were found in the spleen and mesenteric lymph nodes. R. DiGiacomo (personal communication, 1970) observed mycotuberculosis of the liver in one *D. virginiana* from Louisiana. The pale yellow or white tubercles measured up to 1 cm in diameter and contained acid-fast bacilli. Causative agents were not isolated in any of these cases.

Moore *et al.* (1971) observed fatal infection with mycobacteria (*Mycobacterium intracellulare*) of the Runyon Group III in a captive *D. virginiana* obtained from Pennsylvania or Virginia. Lesions containing large numbers of acid-fast bacilli occurred in the intestine, spleen, liver, and mesenteric lymph nodes. Experimental infection of mice with the organism caused slowly progressive disease and death within 10 to 14 months. Guinea pigs and rabbits failed to develop progressive disease unless overwhelmed by large doses of the organism.

Clark and Shepard (1963) induced tuberculous lesions in *D. virginiana* by inoculation with *Mycobacterium marinum*, the cause of spontaneous tuberculosis in fishes. In no instance did systemic tuberculosis occur.

Tuberculin testing of American marsupials with naturally occurring or experimentally induced tuberculosis has not been reported. However, a method for sensitizing *D. virginiana* to *Mycobacterium tuberculosis* by foot pad inoculation with killed bacteria and Freund's incomplete adjuvant has been described (Taylor and Burrell, 1968).

### 2. *Borrelia* Infections

Dunn and Clark (1933) found *Borrelia recurrentis* infection in six of 61 *Didelphis marsupialis caucae* collected from several locations in Panama. Larval forms of ticks (*Ornithodorus talaje*) were found on the opossums and were believed to be vectors. *Didelphis virginiana* was described as a natural reservoir for relapsing fever spirochetes in Texas (Bohls and Schuhardt, 1933) and *Ornithodorus turicata* was implicated as a vector. No clinical signs or lesions were seen in either subspecies of opossum associated with *B. recurrentis* infection.

### 3. *Pasteurella* Infections

**a. Tularemia.** Opossums must be considered as potential sources of *Pasteurella* (= *Francisella*) *tularensis* infection. Tularemia has been observed in human beings who were bitten by or who had killed, skinned, and dressed opossums (Mease, 1929; Francis, 1934, 1937).

Significant antibody titer to *P. tularensis* was found in 66 of 544 *D. virginiana* examined in Georgia and Florida and in seven of 24 opossums from Arkansas (Calhoun *et al.*, 1956; McKeever *et al.*, 1958d). *Pasteurella tularensis* was isolated from the spleen of one (McKeever *et al.*, 1958d). According to Volkmar (1932) the lesions of experimental tularemia infection in opossums included areas of necrosis in the liver, spleen, and lymph nodes. Clinical signs of *P. tularensis* infection have not been described.

**b. Other *Pasteurella* infections.** Pucak *et al.* (1969) described an acute epizootic of *Pasteurella multocida* in a colony of 28 *Marmosa robinsoni* in which 11 deaths occurred suddenly over a 3-day period. Polymyxin B and neomycin were administered to the remainder of the colony and no further cases occurred. *Pasteurella multocida* was also isolated from localized abscesses and adjacent lymph nodes of individually housed *M. robinsoni*, suggesting a carrier state. *Pasteurella multocida* and *P. pneumotropica* have also been recovered from *M. robinsoni* dying of septicemias by Thrasher *et al.* (1971). One human case of *P. multocida* infection which resulted from the bite of a *Didelphis virginiana* has been reported (Hubbert and Rosen, 1970).

### 4. *Salmonella* Infections

Salmonellosis has been reported in a number of wild and captive *D. virginiana*. In a 3-year survey, Schnurrenberger *et al.* (1968) isolated *Salmonella give* from 16 and *S. typhimurium* from one of 31 Illinois opossums. In Virginia, four of 18 opossums had *S. anatum* and one of 79 had *S. typhimurium* infections (Marx, 1969). *Salmonella typhimurium* was also cultured from an opossum by Moran (1942). *Salmonella give* and *S. typhimurium* were isolated from the urine of two wild opossums (Starr *et al.*, 1959). Beamer *et al.* (1960) reported fatal *Salmonella* infections in two captive opossums. *Salmonella* sp. and $\beta$-hemolytic streptocci were isolated from one and *S. bern* from another of 21 captive *D. virginiana* with vegetative valvulitis (Sherwood *et al.*, 1969).

Thigpen *et al.* (1975) isolated Salmonellae from 69 of 106 live opossums trapped in North Carolina. At necropsy, Salmonellae were found in 87 of them. Because *S. anatum, S. give, S. infantis,* and *S. rubislaw* were isolated from the feces of healthy as well as clinically ill opossums, it was suggested that these serotypes are normal inhabitants of the gastrointestinal tract and associated lymph nodes. Evidence of septicemia was found only in opossums infected with *S. typhimurium* and *S. bern,* and was indicative of the pathogenicity of these serotypes in the opossum.

According to Schnurrenberger *et al.* (1968), the presence of *Salmonella* in *D. virginiana* may be explained by their indiscriminate dietary habits, while among captive opossums, stress may predispose them to infection. *Didelphis virginiana* has not been directly incriminated in the transmission of salmonellosis to man (Marx, 1969).

### 5. *Streptococcus* Infections

*Streptococcus* sp. have been frequently isolated from captive *D. virginiana* with endocarditis. Fox (1923a) found *Streptococcus pyogenes* in three of 10 and La Plante and Burrell (1966) isolated *Streptococcus viridans* from one of three opossums. Thirty-four of 90 *D. virginiana* examined by Sherwood *et al.* (1969) had bacterial endocarditis. Alpha and $\beta$-hemolytic streptocci were obtained from 10 of the 15 animals in which cultures were attempted. Other bacteria (*Escherichia coli, Proteus mirabilis,* other *Proteus* sp., *Salmonella bern,* other *Salmonella* sp., *Aerobacter* sp., and *Pseudomonas* sp.) were also isolated but much less frequently.

Cardiac lesions observed in streptococcal infections most frequently involve the mitral and aortic valves and range in severity from small acute changes to large vegetations (Sherwood *et al.,* 1969). Other lesions observed in opossums with endocarditis were myocarditis; amyloidosis of the kidneys, liver, and spleen; microabscesses of the brain; focal necrosis of lymph nodes and adrenals; pyelonephritis; and acute renal infarcts; however, the relationship between these lesions and streptococcal infection is not clear.

Endocarditis was induced experimentally in *D. virginiana* by injections of *S. viridans* (Rowlands *et al.,* 1970; Vakilzadeh *et al.,* 1970). Endocarditis occurred in 12 of 21 test and two of 12 control opossums, usually within 2 weeks of inoculation. Infections were not associated with elevated antibody titers or increased white blood cell counts.

Studies by Sherwood *et al.* (1968) indicated that endocarditis was principally a disease of captive opossums. Its incidence among caged opossums was the same as among those maintained under simulated natural conditions. Endocarditis was prevented by sterilizing cages daily and eliminating contacts with other animal species.

### 6. *Bordetella* Infections

Switzer *et al.* (1966) reported the incidence of *Bordetella bronchiseptica* isolation from the tracheas of 105 *D. virginiana* from Iowa as less than 2%.

### 7. *Staphylococcus* Infections

Focal suppurative and disseminated lesions were listed by Thrasher *et al.* (1971) associated with *Staphylococcus aureas* infections in *Marmosa robinsoni.*

### 8. *Proteus* Infections

Endometritis in *Marmosa robinsoni* was caused by *Proteus vulgaris* and *Proteus mirabilis* (Thrasher, 1970).

### 9. *Leptospira* Infections

*Didelphis virginiana* appears to be an important host for leptospirosis based on bacterial isolation and serologic studies. In a survey of 995 opossums conducted by Gorman *et al.* (1962) in Georgia, 15.6% were positive for leptospirosis. Infections with *Leptospira ballum* appeared to be common (Evans *et al.*, 1962; Clark *et al.*, 1961; Galton *et al.*, 1958; Gorman *et al.*, 1962; Roth *et al.*, 1961; Starr *et al.*, 1959; Alexander *et al.*, 1972; Yager *et al.*, 1953; Shotts *et al.*, 1975) and Hyos strain infections were found frequently (Galton *et al.*, 1957; Gorman *et al.*, 1962; Roth *et al.*, 1961, 1963). *Leptospira hebdomonadia* (Galton *et al.*, 1960; Gorman *et al.*, 1962; Roth *et al.*, 1961), *Leptospira autumnalis* (Evans *et al.*, 1962; Gorman *et al.*, 1962; Alexander *et al.*, 1972; Roth *et al.*, 1961), *Leptospira australis* (Gorman *et al.*, 1962), *Leptospira canicola* (McKeever *et al.*, 1959), *Leptospira pomona* (Galton *et al.*, 1968; Roth and Knieriem, 1958), *Leptospira butembo, Leptospira grippotyphosa, Leptospira icterohemorrhagiae* (Evans *et al.*, 1962; Alexander *et al.*, 1972), and *Leptospira andamana* (Alexander *et al.*, 1972) have also been found in *D. virginiana.* Szyfres and Blood (1964) isolated *Leptospira paidjan* from the kidneys of two *Didelphis albiventris* in Corrientes Province, Argentina, and *Leptospira hallum* and *L. icterohemorrhagiae* were isolated from two *D. virginiana* in Maryland.

Santa Rosa *et al.* (1975) isolated *L. icterohemorrhagiae, L. grippotyphosa, L. szwajizak,* and *L. ballum* from *D. marsupialis* and concluded that this opossum was the most important reservoir of leptospirosis in Brazil. *Leptospira quaratuba* was also found in a *Philander opossum*.

Experimental trasmission of leptospirosis in *D. virginiana* was studied by Reilly *et al.* (1970). Two of 14 opossums that were fed mice infected with *L. grippotyphosa* developed leptospirosis. It was concluded that transmission may occur in nature as a result of eating infected animals. It was not determined whether opossums thus infected would serve as carriers. Clinical signs of natural or experimental leptospirosis were not reported.

### 10. Other Bacterial Infections

A host of bacteria isolated from six moribund, septicemic *Marmosa robinsoni* were tabulated by Thrasher *et al.* (1971). They included four species of *Proteus*, four of *Streptococcus*, three of *Paracolon*, two of *Pasteurella*, two of *Escherichia*, one of *Klebsiella*, one of *Lactobacillus*, and one of *Pseudomonas aeruginosa*.

## C. Rickettsial and Chlamydial Diseases

### 1. *Coxiella* Infection

Although Q fever is commonly found among wild Peramelidae in Australia (Derrick, 1944), evidence of *Coxiella burneti* infection could not be demonstrated by Irons *et al.* (1952) among feral *Didelphis virginiana* or a variety of their associated ectoparasites in an endemic Q fever area of Texas.

### 2. *Rickettsia* Infections

**a. Spotted Fever.** *Didelphis marsupialis aurita, Didelphis albiventris paraguayensis*, and *D. marsupialis* are reported to be natural hosts for rickettsiae of the spotted fever group (*Rickettsia rickettsii*) which, in Brazil, produce Sao Paulo typhus and Minas Geraes fever in man (Diaz and Martins, 1939; Travassos, 1948). Rocky Mountain spotted fever rickettsiae have also been isolated from wild *D. virginiana* in the eastern United States (Bozeman *et al.*, 1967; Alexander *et al.*, 1972; Fuller, 1963). Vectors of *R. rickettsii* which infest didelphids include the ticks *Amblyoma cajennense, Ixodes loricatus, Dermacentor venustus* (= *andersoni*), *Dermacentor variabilis*, and *Boophilus microplus*.

Experimental infection of opossums with spotted fever rickettsiae produces no clinical signs. However, rickettsemias have been observed for at least 18 days in *D. marsupialis aurita* (Travassos, 1948) and for 3 to 4 weeks in *D. virginiana* (Bozeman *et al.*, 1967). Travassos (1948) demonstrated a decrease in virulence of rickettsiae during a series of passages through six *D. marsupialis aurita*. Experimental studies in *D. virginiana* indicate that maximal complement-fixing antibody responses occur between 18 and 28 days after infection (Bozeman *et al.*, 1967). Titers gradually declined until antibody was no longer detected 3 months after infection.

Experimental infections of *D. auritas* with Sao Paulo rickettsiae are associated with degenerative changes and small histiocytic nodules in the kidneys, livers, and lungs. Although 3 species of opossums have been cited as natural hosts for spotted fever rickettsiae, their importance in the epidemiology of the disease in man remains to be elucidated.

**b. Murine Endemic Typhus.** Brigham (1936) demonstrated that *Didelphis virginiana* was susceptible to experimental typhus infection, and Adams *et al.* (1970) isolated *Rickettsia typhi* (= *R. rickettsia*) from a wild *D. virginiana* in California. Clinical signs of infection were not observed. Although Keaton *et al.* (1953) reported that serums from two of 27 opossums examined in Texas were positive, they considered the findings doubtful because test animals could not be infected with suspensions of their brains. However, the results of other serologic (complement fixation) surveys revealed that three of 345 *D. virginiana* examined in Georgia (Morlan *et al.*, 1950) and eight of 75 tested in California (Adams *et al.*, 1970) were positive.

Chapman and Chapman (1935) and Adams *et al.* (1970) considered that *D. virginiana* was probably a reservoir of endemic typhus, and the fleas may transmit the infection. Typhus was reported in human beings who had direct contact with opossums or who lived in proximity to opossums. Fleas (*Ctenocephalides felis*) are common ectoparasites of *D. virginiana* and infected *C. felis* have been obtained from opossums. However, it is possible that the fleas were infected from a murine source (Keaton *et al.*, 1953).

### 3. *Miyagawanella* Infections

Roca-Garcia (1949) described a disease in *Didelphis albiventris paraguayensis* and *Caluromys lanatus* in Colombia which was characterized by hind quarter paralysis, convulsions, and death. The etiological agent, which he called opossum A virus, has been renamed *Miyagawanella opossumi* (Phillip, 1957) in the lymphogranuloma-psittacosis group.

Experimental studies indicate that *M. opossumi* is pathogenic for *C. lanatus, Metachirus nudicaudatus, Philander opossum,* and *D. albiventris paraguayensis. Didelphis marsupialis* is resistant to infection. Gross lesions associated with *M. opossumi* infections are hepatomegally, splenomegally, pneumonitis, and peritoneal exudation. Microscopically, focal necrosis and perivascular hyaline degeneration with leukocytic infiltration are observed in the liver. Other lesions are meningitis, hemorrhage and necrosis in the spleen, and capillary thrombosis and edema of the lungs.

Roca-Garcia (1949) isolated another agent of the lymphogranuloma-psittacosis group from a Colombian opossum which he called opossum B virus. Further descriptions of the agent and its pathogenic effects were not given.

## D. Protozoan Diseases

### 1. *Toxoplasma* Infection

Seventy-six *D. virginiana* from Georgia were tested for *Toxoplasma gondi* infection by Walton and Walls (1964). Five opossums (7%) were considered positive based on the results of the Sabin–Feldman dye test. De Roever-Bonnet (1967) also employed the dye test in a survey of 22 opossums in Surinam. Serum from one *D. marsupialis* and one *Caluromys philander* gave positive reactions. In a similar survey conducted in Paraguay, no reactors were detected among the 14 opossums (species not identified) tested (De Roever-Bonnet *et al.*, 1969). Likewise, 14 *D. virginiana* examined by Paine (1969) in Illinois gave negative reactions to the indirect hemagglutination test. Miller *et al.* (1972) were unable to isolate *Toxoplasma* from nine *D. virginiana* which became seropositive following ingestion of toxoplasma cysts. However, the wide distribution of *Toxoplasma* throughout the world (De Roever-Bonnet, 1967) and the results of the surveys conducted suggest that the prevalence of this parasite in wildlife, including opossums, is probably high (Walton and Walls, 1964).

### 2. *Sarcocystis* Infection

*Didelphis marsupialis* and *Philander opossum* were reported as hosts of *Sarcocystis garnhami* by Mandour (1965) in Brazil and Shaw and Lainson

(1969) in British Honduras. Scorza (1957) identified *Sarcocystis didelphis* infection in *D. marsupialis* from Venezuela and Shaw and Lainson (1969) described *Sarcocyctis marmosae* in *Marmosa murina* from Brazil. In the United States, *Sarcocystis* infections were reported in three of 19 *D. virginiana* by Seneviratna *et al.* (1975). *Sarcocystis* infections in *Didelphis* sp. (Darling, 1910), *D. marsupialis aurita* (Barretto, 1940), and *D. albiventris* (Lage, 1968) are considered under Section II,D,3, in accordance with Mandour's (1965) suggested reclassification of these parasites.

Grossly, sarcocyst infections are inapparent or, if the parasites are present in large numbers, gray-white streaks are visible in affected muscles. Sarcocysts vary considerably in size and are located within skeletal muscle cells. An unidentified *Sarcocystis* was reported as common among *M. robinsoni* obtained from Colombia. Cysts were observed in the muscles of its tongue, neck, extremities, heart, and diaphram (Thrasher *et al.*, 1971). Specific diagnosis, according to Mandour (1965), is based on the host in which the parasite is found, the cyst wall structure, and spore sizes. Clinical signs have not been reported in marsupials with sarcosporidiosis.

### 3. *Besnoitia* Infection

*Besnoitia* infection has been described in *Didelphis* sp. (Darling, 1910) in Panama and in *D. marsupialis aurita* and *D. albiventris* in Brazil (Barretto, 1940; Lage, 1968). The original identification of these organisms as *Sarcocystis* was reviewed by Mandour (1965) and Schneider (1967), who concluded that *Besnoitia darlingi* was the correct name. *Besnoitia darlingi* infection was subsequently described in one of 18 *Didelphis marsupialis* examined in Panama (Schneider, 1967), in five *D. virginiana* from Kentucky (Conti-Diaz *et al.*, 1970), and in eight of 13 opossums from Missouri and Illinois (Flatt *et al.*, 1971). Stabler and Welch (1961) reported *Besnoitia jellisoni* infection in one *D. virginiana* from Texas.

Although the parasitic cysts of *B. darlingi* are most frequently observed in the heart, lungs, kidneys, tongues, skins, and skeletal muscles of the trunk and limbs of affected animals, they may also be distributed in numerous other tissues. The cysts are spherical or slightly oval, white, and measure about 1 mm in diameter. Spindle-shaped trophozoites are present within the cysts in enormous numbers. Although no inflammatory response is observed around intact cysts, those undergoing degeneration contain necrotic or mineralized trophozoites and are surrounded by granulomatous inflammation. Yellow-gray masses in the cranial dural sinuses have been associated with *B. jellisoni* infection.

*Besnoitia* isolated from *D. marsupialis* by Conti-Diaz *et al.* (1970) were found to be highly pathogenic for hamsters, but had little or no pathogenicity for white Swiss mice. They were nonpathogenic for adult guinea pigs. Darling (1910) produced an inapparent infection in one of two guinea pigs which he inoculated with *Besnoitia* sporozoites. Schneider (1967) inoculated *Caluromys derbianus* and *Philander opossum* intraperitoneally with a strain of *Besnoitia* established in mice. Both species developed acute and rapidly fatal infections. Schneider (1967) also infected two of eight *D. marsupialis* by the same method. One died 46 days after exposure and the other was killed on the seventh day. Lesions involving the hearts, lungs, spleen, and pancreas were similar to those reported in naturally infected opossums.

### 4. *Coccidia* Infection

Volk (1938) described *Isospora boughtoni* infection in *Didelphis virginiana.* Cells of the epithelium and subepithelium of the upper small intestine were parasitized, and the infection was characterized by hemorrhagic enteritis. *Isospora boughtoni* and other *Isospora* sp. were obtained from another Georgian opossum by Ernst *et al.* (1969). It was postulated that the unidentified species may have been derived from a bird which the opossum had eaten.

Thrasher *et al.* (1971) observed *Eimeria* sp. infections in approximately 20% of the *Marmosa robinsoni* which they obtained from Colombia. Oocysts of one organism were large, brown, and had a rough ectocyst. The stages of this *Coccidia* were found in the duodenal mucosa and lamina propria around the crypts of Lieberkühn. Histologically, affected areas were characterized by ulceration and hemorrhage. Smaller, clear, thin-walled oocysts of another species of *Eimeria* were found only in epithelial cells of the villi and crypts of the ileum. No pathology was associated with their presence.

Coccidiosis is considered an important disease of *M. robinsoni.* The slow proliferation and renewal rate of intestinal epithelium in this species may leave it particularly susceptible to bacterial invasion and malabsorption following extensive epithelial losses associated with *Eimeria* infections. Treatment of coccidiosis was successful using sulfamerazine administered orally at a rate of 1 mg/g body weight/day for 7 to 10 days. Daily sterilization of cages by autoclaving was also recommended (Thrasher *et al.*, 1971).

*Eimeria didelphydis* and *Eimeria gambai* have been reported from *D. marsupialis aurita* in South America according to Ernst *et al.* (1969).

### 5. *Trichomonas* Infection

Henger and Ratcliffe (1927) observed *Trichomonas didelphis* inhabiting the intestine of one *Didelphis virginiana* from Ohio. In 1965, *Tetratrichomonas didelphis* infection was reported in 18 of 20 *D. virginiana* examined by Andersen and Reily (1965), who considered that the parasites were probably the same as those described by Henger and Ratcliffe. *Trichomonas* sp. infections of the colon and urogenital sinus were observed in *Marmosa robinsoni* from Colombia (Thrasher *et al.*, 1971). Neither clinical signs nor pathology were reported in association with trichomoniasis in these species. Treatment with anisomycin (550 mg/kg body weight/day) for 3 to 4 days was found to reduce trichomonad populations.

### 6. *Babesia* Infection

Regendanz and Kikuth (1928) described natural *Babesia* (*Nuttallia*) *braziliensis* infection in two *Metachirus nudicaudatus* in Brazil. Although the opossums were anemic, other clinical signs of infection were not evident. Splenectomy was followed by increases in the numbers of circulating parasites. Thirty-five percent of the 110 *Didelphis albiventris paraguayensis* examined by Roca-Garcia (1945) in Colombia had *Nuttallia* (*Babesia*) *equi*-like organisms in their peripheral blood. Experimental infections in *D. albiventris paraguayensis* were associated with severe anemia. At the height of the induced parasitemia (10 to 16 days after inoculation) as many as 30% of the erythrocytes were parasitized. After the acute phase, parasites were present in reduced numbers for up to 11 months. Mild infections were experimentally induced in *D. marsupialis* and *M. nudicaudatus*; guinea pigs, mice, and a horse were not susceptible. Roca-Garcia (1945) considered the tick *Ixodes loricatus*, a common parasite of *D. albiventris paraguayensis*, as a vector of *Babesia equi*.

Frerichs and Holbrook (1970) examined 15 *D. virginiana* in Maryland for evidence of *Babesia* infection. Peripheral blood smears were negative, and splenectomy of the opossums failed to exacerbate latent infections.

### 7. *Trypanosoma* Infection

**a. Chagas' Disease.** *Trypanosoma cruzi*, the agent of Chagas' disease, has been reported from three genera of Didelphidae. Naturally occurring infections with *T. cruzi*-like organisms were described in *Didelphis marsupialis* by Clark and Dunn (1931, 1932), Herrer *et al.* (1966), and Pipkin

(1968) in Panama; Zeledon *et al.* (1970) in Costa Rica; Robertson (1930) and Petana (1969) in Honduras and British Honduras; and Deane (1961, 1964) and Deane *et al.* (1963) in Brazil. These infections were described in *Didelphis virginiana* by Packchanian (1942) in Texas; Brooke *et al.* (1957) in Georgia; McKeever *et al.* (1958b) and Norman *et al.* (1959) in Georgia and Florida; and Olsen *et al.* (1964) in Alabama. *Marmosa cinerea, Marmosa agilis* and *Marmosa alstoni,* and *Philander opossum* were hosts for *T. cruzi* in Brazil and Costa Rica (Wood and Wood, 1941; Deane, 1961, 1964; Correa and Barretto, 1964; Zeledon *et al.*, 1970). Wood and Wood (1941) also list *T. cruzi* isolations from *D. albiventris paraguayensis, D. albiventris,* and *Lutreolina crassicauda paranalis* in Argentina. Survey results showed that up to 67.3% (202 of 300) of *D. marsupialis* examined harbored hemoflagellates; smaller populations of other didelphids were sampled and the incidence of infection among them was lower (Zeledon *et al.*, 1970).

*Trypanosoma cruzi* may be found in the blood, skeletal and heart muscle, kidney, intestinal walls, peritoneal fluid, urine, and skin of opossums. Xenodiagnostic techniques are generally considered superior to examining blood films to establish diagnoses. Pathologic findings in opossums with *T. cruzi* infections include subacute or chronic inflammation of the myocardium and muscular layers of the digestive tract (Packchanian, 1942; de Brito and Deane, 1966). Involvement of the intestinal nerve plexuses of the digestive tract could lead to megaesophagus and megacolon in affected animals according to de Brito and Deane (1966). Clinical signs of *T. cruzi* infection, however, have not been reported in opossums.

Triatomid bugs are the natural vectors of *T. cruzi.* Those which infest opossums or their nests are listed in Table II under Hemiptera. The most important vectors of Chagas' disease which are associated with opossums include *Panstrongylus megistus, P. geniculatus, Triatoma infestans, Triatoma sordida, Triatoma dimidata,* and *Rhophalius neglectus* (Correa, 1968; Barretto, 1968; Zeledon *et al.*, 1970).

Both opossums and triatomids live in close association with man in many areas. According to Zeledon *et al.* (1970), there is a relationship between the cycles of Chagas' disease in man and wildlife, principally opossums. Opossums are considered the most likely vertebrate reservoir of *T. cruzi* in Panama and Costa Rica (Pipkin, 1968; Zeledon *et al.*, 1970); in Brazil, they reportedly share this role with dogs, cats, and rats (Correa, 1968).

**b. Other *Trypanosoma* Infections.** Deane (1964) reported *Trypanosoma rangeli*-like hemoflagellates from *D. marsupialis* and *P. opossum,* and *Trypanosoma freitasi* from *D. marsupialis,* in Para, Brazil.

Clark and Dunn (1933) consider that *Didelphis marsupialis caucae* does not sustain natural *Trypanosoma hippicum* infections longer than a few

**TABLE II**

Arthropod Parasites of American Marsupials (Didelphidae)

| Parasite | Host | Location[a] | References |
|---|---|---|---|
| Acarina | | | |
| Ticks | | | |
| *Amblyomma americanus* | *Didelphis virginiana* | * | Krupp and Quillin (1964) |
| *Amblyomma cajennense* | *Didelphis marsupialis aurita* | Brazil | Travassos (1948) |
| *Amblyomma striatum* | *Didelphis marsupialis* | Brazil | Travassos (1948) |
| | *D. albiventris* | Brazil | Travassos (1948) |
| *Dermacentor variabilis* | *Didelphis virginiana* | Virginia and Maryland | Irons *et al*. (1952); Travis (1941); Bozeman *et al*. (1967); Alexander *et al*. (1972) |
| *Ixodes loricatus* | *Didelphis albiventris* | Brazil | Travassos (1948) |
| *Ixodes richinus scapularis* | *Didelphis* sp. | * | Travis (1941) |
| *Ornithodoros talaje* (larvae) | *Didelphis marsupialis caucae* | Panama | Dunn and Clark (1933) |
| Mites | | | |
| *Bdellonyssus (Liponyssus?) bacoti* | *Didelphis virginiana* | Texas | Morlan (1949); Irons *et al*. (1952) |
| *Kymocta chironectes* | *Chironectes minimus* | Panama | Nadchatram (1970) |
| *Neojchoronyssus wernecki* | *Didelphis marsupialis* | * | Morlan (1949) |
| *Sarcoptes* sp. | *Didelphis virginiana* | New York Zoological Gardens | Blair (1921) |
| Mites (unspecified) | *Didelphis virginiana* | California | Adams *et al*. (1970) |
| Hemiptera | | | |
| *Cimex rotundatus* | *Didelphis marsupialis* | Tela, Honduras | Robertson (1930) |
| *Acrocomis sclerocarpa* | Opossum | Goias, Brazil | Barretto (1967) |
| *Lutzomyia flaviscutellata* | Opossum | Para, Brazil | Lainson and Shaw (1969b) |
| *Panstrongylus geniculatus* | Opossum | Sao Paulo, Brazil | Barretto (1968) |
| *Panstrongylus megistus* | Opossum | Sao Paulo, Goias, Brazil | Barretto (1967, 1968); Correa (1968) |
| *Psammolestes tertius* | *Didelphis marsupialis* | Sao Paulo, Brazil | Barretto and Albuquerque (1969) |

| | | | |
|---|---|---|---|
| *Rhodnius neglectus* | *Marmosa agilis* | Sao Paulo, Brazil | Correa and Barretto (1964) |
| | Opossum | Sao Paulo, Brazil | Barretto (1968) |
| *Rhodnius pallescens* | *Didelphis marsupialis* | Panama | Pipkin (1968) |
| *Rhodnius prolixus* | *Didelphis marsupialis* | Sao Paulo, Brazil | Clark and Dunn (1932) |
| | *Marmosa agilis* | Panama | Correa and Barretto (1964) |
| *Scheelea phaleratus* | Opossum | Goias, Brazil | Barretto (1967) |
| *Triatoma brasiliensis* | Opossum | Sao Paulo, Brazil | Correa (1968) |
| *Triatoma dimidiata* | *Didelphis marsupialis* | Panama; Costa Rica | Pipkin (1968); Zeledon *et al*. (1970) |
| *Triatoma geniculata* | *Didelphis marsupialis* | Panama | Clark and Dunn (1932) |
| *Triatoma infestans* | *Marmosa agilis* | Sao Paulo, Brazil | Correa and Barretto (1964) |
| | Opossum | Sao Paulo, Brazil | Barretto (1968); Correa (1968) |
| *Triatoma protracta* | *Didelphis virginiana* | California | Kofoid and Donat (1933) |
| *Triatoma pseudomaculata* | Opossum | Sao Paulo and Goias, Brazil | Barretto (1967); Correa (1968) |
| *Triatoma rufrofasciata* | Opossum | Brazil | Correa (1968) |
| *Triatoma sordida* | *Marmosa agilis* | Sao Paulo, Brazil | Correa and Barretto (1964); Correa (1968); Barretto (1968) |
| Mallophaga | | | |
| Lice | | | |
| *Cummingsia maculata* | *Lesteros inca* | * | Hopkins (1949) |
| *Cummingsia peramydis* | *Marmosa incana* | * | Hopkins (1949) |
| | *Monodelphis domestica* | * | Hopkins (1949) |
| *Gliricola porcelli* | *Philander opossum* | * | Hopkins (1949) |
| *Gnyopus ovalis* | *Philander opossum* | * | Hopkins (1949) |
| *Pulex similans* | *Didelphis virginiana* | California | Volkmar (1932); Adams *et al*. (1970) |
| *Trimenopon hispidium* | *Philander opossum* | * | Hopkins (1949) |
| Siphonaptera | | | |
| Fleas | | | |
| *Cediopsylla simplex* | *Didelphis virginiana* | Georgia | Mohr and Morlan (1959) |
| *Ctenocephalides felis* | *Didelphis virginiana* | California, Georgia | I. Fox (1940b); Irons *et al*. (1946); Keaton *et al*. (1953); Mohr and Morlan (1959); Adams *et al*. (1970) |
| *Ctenocephalides* sp. | *Didelphis virginiana* | Illinois | Leigh (1940) |

**TABLE II (Cont.)**

| Parasite | Host | Location[a] | References |
|---|---|---|---|
| *Echidnophaga gallinacea* | *Didelphis virginiana* | California, Texas | Irons *et al.* (1952); Adams *et al.* (1970) |
| *Leptopsylla segnis* | *Didelphis virginiana* | Georgia | Mohr and Morlan (1959) |
| *Odontopsyllus multispinosus* | *Didelphis virginiana* | * | Mohr and Morlan (1959) |
| *Orchopeas leucopus* | *Didelphis virginiana* | Iowa | I. Fox (1940a); Joyce and Eddy (1944) |
| *Orchopeas howardi* | *Didelphis virginiana* | Georgia | Mohr and Morlan (1959) |
| *Orchopeas wickhami* | *Didelphis virginiana* | * | I. Fox (1940a) |
| *Oropsylla arctomys* | *Didelphis virginiana* | * | I. Fox (1940a) |
| *Peromyscopsylla catatina* | *Didelphis virginiana* | * | Ewing and Fox (1943) |
| *Polygenis gwyni* | *Didelphis virginiana* | * | Dikmans (1931); I. Fox (1940b) |
| *Xenopsylla cheopis* | *Didelphis virginiana* | Georgia | Mohr and Morlan (1959) |
| Pentastomida | | | |
| Tongue worms | | | |
| *Porocephalus clavatus* (larvae) | *Marmosa murina* | * | von Haffner (1913) |
| | *Caluromys philander* | * | von Haffner (1913) |
| | *Didelphis marsupialis* | * | von Haffner (1913) |

[a] Asterisk indicates location not given.

weeks. This is in contrast to previous reports that this species is susceptible to and dies from infection within several weeks. *Trypanosoma hippicum* infections produce murina or derrengadera in horses and in man.

### 8. *Leishmania* Infection

Cutaneous leishmaniasis has been described in three *Marmosa murina* from Mato Grosso and Para, Brazil (Lainson and Shaw, 1969a,b, 1970). Lesions were also described in two *M. robinsoni* from eastern Panama (Herrer *et al.*, 1971; Telford *et al.*, 1972). Lesions consisted of ulcers at the base of the tail on the dorsal surface. *Leishmania* were observed in stained exudate from lesions. Organisms isolated following intradermal inoculation of ulcerated tissue into hamsters were similar to those associated with a local murine leishmaniasis. The vector of this disease among rodents was *Lutzomyia* (= *Phletotomus*) *plaviscutellata*. Other opossums (36 *Didelphis*, 125 *Marmosa*, 52 *Metachirus*, 18 *Philander*, and 9 *Caluromys*) examined for cutaneous lesions were negative. The low incidence of leishmaniasis among opossums indicates that they are probably incidental hosts of this zoonosis.

The causative agent of mucocutaneous leishmaniasis, *Leishmania brasiliensis*, when inoculated intradermally into *D. marsupialis* and *Caluromys derbianus*, failed to produce disease (Thatcher *et al.*, 1965).

The agent of enzootic rodent leishmania (*Leishmania mexicana amazonensis*) was isolated from five of 30 *Marmosa robinsonilone*, one of 9 *M. fuscata*, one of 23 *Caluromys philander*, and none of 25 *Didelphis marsupialis* in Trinidad, West Indies (Tikasingh, 1974). Occurrence of this disease in humans is reportedly rare in contrast to cutaneous leishmaniasis.

### 9. *Hemobartonella* Infection

One survey was performed to ascertain the incidence of hemobartonellosis among *Didelphis virginiana*. Peripheral blood smears taken before and after splenectomy of 15 wild Maryland opossums failed to demonstrate the presence of *Hemobartonella* sp. (Frerichs and Holbrook, 1970).

### 10. *Hemogregarina* Infection

Erythrocytes of peripheral blood from one of 50 Brazilian *Didelphis marsupialis aurita* examined by d'Utra e Silva and Arantes (1917) were found to contain hemogregarines (*Hemogregarina didelphydis*).

### 11. *Plasmodium* Infection

Wellde *et al.* (1966) inoculated two *D. marsupialis* with *Plasmodium berghei* (NYU-2 strain). No parasitemia was observed during the 3-week post-inoculation period. Naturally occurring malaria has not been reported among marsupials.

## E. Mycotic Diseases

### 1. Dermatomycoses

*Trichophyton mentagrophytes* was isolated from one of 59 *Didelphis virginiana* surveyed by Menges and Georg (1957) and six of 379 examined by McKeever *et al.* (1958c) in Georgia and Florida. Eleven opossums from the latter group carried "red" *Microsporum*. The absence of visible lesions may indicate that the fungi were saprophytes or contaminants. *Microsporum cookei* was reported from an "opossum" by Rieth (1966).

### 2. Systemic Mycoses

Histoplasmosis may be common among American marsupials. *Histoplasma capsulatum* was isolated from three of 95 *Didelphis virginiana* in Virginia (Emmons *et al.*, 1955) and seven of 57 in Panama (Taylor and Shacklette, 1962). Four of 18 *Philander opossum* and one *Caluromys derbianus* in the Canal Zone, Panama were also infected.

*Candida albicans* was cultured from ulcers of the gastroduodenal region of several *Marmosa robinsoni* (Thrasher *et al.*, 1971). The mycotic invasion was considered to be secondary in these debilitated animals.

## F. Nutritional Diseases

### Rickets and Osteomalacia

Rickets and the adult form of this disease, osteomalacia, are common in captive opossums. Fatal disease was reported in *Didelphis marsupialis* (Hamerton, 1931), *D. albiventris* (Hamerton, 1938), and *D. albiventris paraguayensis* (Hill, 1951). Corson-White (1922) mentioned eight cases in marsupials. Grossly, there is a softening of the bones, particularly those of the extremities and the ribs. The limbs of affected animals become distorted and they show signs of "cage paralysis." Barnes (1968) suggested adding bones to the diet to supplement calcium intake and prevent manifestations of the disease. Others, however, have successfully reared opossums (*D. virginiana*) on well-balanced commercial diets formulated for domestic carnivores.

### G. Neoplastic Diseases

A variety of neoplasms have been reported in zoological and laboratory specimens of *Didelphis virginiana*. The urogenital system was commonly affected: single instances of renal adenocarcinoma (Fox, 1923h) and papillary adenoma and carcinoma of the bladder were reported (Sherwood *et al.,* 1969; Fox, 1933). One mammary adenocarcinoma was observed by Fox (1923i).

Tumors of the respiratory system are common. Pulmonary adenomas of bronchial epithelial origins were found in 33 of 158 *D. virginiana* (Sherwood *et al.,* 1968, 1969). They were visible grossly as gray-white nodules beneath the visceral pleurae in the peripheral lung fields. Microscopically, the lesions were circumscribed, papillary, and consisted of nonciliated cuboidal to columnar epithelium. Nuclei were enlarged, centrally located, and vesicular; no mitoses were evident. Clinical signs of pulmonary adenomatosis were observed in a few opossums and consisted of sneezing, coughing, and a clear discharge from the eyes and nose. Animals with tumors frequently had pneumonia. Metastases were not observed. Vakilzadeh *et al.* (1971) found an increased incidence of pulmonary adenomas following installation of graphite particles intratracheally in opossums. A number of the tumors were atypical in that they were large, multiple, and not well circumscribed. Single occurrences of pulmonary squamous cell carcinoma and adenocarcinoma were reported by Snyder and Ratcliffe (1966) and Lombard and Witte (1959), respectively.

Osteosarcoma of the skull was described by Fox (1923j) in four opossums; in two instances the ribs were also affected. Hamerton (1930) reported the occurrence of hard, symmetrical osteoid tumors which involved the alveolar processes of the mandibles of one *D. virginiana.* He indicated that the condition may have been associated with a nutritional deficiency.

Generalized lymphosarcoma was reported in *D. virginiana* by Lombard and Witte (1959), and R. DiGiacomo (personal communication, 1970) described a reticulum cell sarcoma of the liver and spleen in this species. Affected organs in the latter animal were enlarged and their visceral and cut surfaces contained multiple tan-yellow lesions. Microscopically, the tissues were diffusely infiltrated with pleomorphic, malignant reticuloendothelial cells. A large round cell sarcoma arising from the left ventricle was observed by Hamerton (1937) in *D. albiventris*; metastases to the intestine, kidney, and urinary bladder were present.

Toft *et al.* (1973) reported a basal cell tumor of the skin of a female *D. virginiana.* The lesion was confined dermis and consisted of branching cords and nests of neoplastic columnar cells; there were no metastases.

In contrast to the high incidence of tumors of the gastrointestinal tract among the Macropodidae, no primary neoplasms involving this system were reported in *D. virginiana.*

### H. Diseases of Specific Organ Systems

The lesions of specific organs or systems reported in Didelphidae are listed in Table III. The affected marsupials were usually zoological display specimens or subjects of experimental studies. In most instances, the cause of death was not reported; however, the nature of many of the lesions listed indicates such a role.

#### Gastroduodenal Ulcers

Thrasher *et al.* (1971) reported a high incidence of gastroduodenal ulceration and hemorrhage in newly captured and laboratory reared *Marmosa robinsoni.* The lesions were acute and varied from solitary or multiple areas of superficial necrosis to perforated ulcers. Septicemia and mycotic enteritis were frequently observed in affected animals. Clinical signs included dehydration, rough hair coat, hunched posture, melena, tremors, and ataxia. Affected animals generally died within 12 to 36 hours. It was believed that the occurrence of gastroduodenal ulcers in *M. robinsoni* was related to some condition of captivity.

### I. Ectoparasites

The arthropod parasites recovered from American marsupials are listed in Table II. Ectoparasitism appears to be common. Krupp and Quillin (1964) recognized flea (*Ctenocephalides* sp.) infestation frequently in newly captured *Didelphis virginiana* and outlined a control program. The role of ticks (*Acarina*), fleas (*Siphonaptera*), and triatomids (*Hemiptera*) as vectors of spotted fever and endemic typhus rickettsiae and of trypanosome infec-

**TABLE III**

Specific Anatomical Lesions Observed in American Marsupials

| Lesion | Marsupial | Incidence[a] | References |
|---|---|---|---|
| Digestive System | | | |
| Dental caries | *Didelphis virginiana* | Common | Farris (1950) |
| Gastritis | *Didelphis virginiana* | Common | Apgar (1934) |
| Gastric granulomas | *Didelphis virginiana* | 1(90) | Sherwood *et al.* (1969) |
| Gastric and duodenal ulcers | *Didelphis virginiana* | 2(90) | Sherwood *et al.* (1968, 1969) |
| Gastric torsion | *Didelphis virginiana* | 1 | Fox (1923b) |
| Gastroenteritis | *Didelphis virginiana* | 1 | Goss (1940) |
| Enteritis | *Didelphis virginiana* | 2 | Fox (1923c) |
| Enteritis (chronic) | *Didelphis virginiana* | Common | Farris (1950) |
| Impaction of the colon | *Didelphis virginiana* | 1 | Goss (1940) |
| | *Caluromys philander* | 1 | Hamerton (1939) |
| Rectal abscesses | *Didelphis virginiana* | 1(90) | Sherwood *et al.* (1969) |
| Pancreatitis (acute) | *Didelphis virginiana* | 2 | Fox (1923c) |
| Hepatitis | *Didelphis virginiana* | 2 | Apgar (1934) |
| Abscess | *Didelphis virginiana* | 1(90) | Sherwood *et al.* (1969) |
| Amyloidosis | *Didelphis virginiana* | 6(143) | Aherwood *et al.* (1968, 1969) |
| Diffuse acute hepatitis | *Didelphis virginiana* | 5(90) | Sherwood *et al.* (1969) |
| Fatty metamorphosis | *Didelphis virginiana* | 4(90) | Sherwood *et al.* (1969) |
| Focal necrosis | *Didelphis virginiana* | 44(153) | Sherwood *et al.* (1968, 1969) |
| Passive congestion | *Didelphis virginiana* | 10(90) | Sherwood *et al.* (1969) |
| Periportal inflammation | *Didelphis virginiana* | 26(90) | Sherwood *et al.* (1969) |
| Cholecystitis | *Didelphis virginiana* | 1(90) | Sherwood *et al.* (1969) |
| Fat necrosis | *Didelphis virginiana* | 2 | Fox (1923c) |
| Peritonitis | *Didelphis virginiana* | 3(10) | La Plante and Burrell (1966) |
| Urogenital System | | | |
| Renal agenesis (unilateral) | *Didelphis virginiana* | 2(500) | Gupta and Feldman (1975) |
| Nephritis | *Didelphis virginiana* | 48(68) | Apgar (1934); Sherwood *et al.* (1969) |

**TABLE III (Cont.)**

| Lesion | Marsupial | Incidence[a] | References |
|---|---|---|---|
| Glomerulonephritis | *Didelphis virginiana* | 23(143) | Sherwood *et al.* (1968, 1969) |
| Proliferative glomerulonephritis | *Didelphis virginiana* | 24(33) | Rowlands *et al.* (1970) |
| Pyelonephritis | *Didelphis virginiana* | 33(153) | Fox (1923d); LaRue and Bosma (1958); Sherwood *et al.* (1968, 1969) |
| Interstitial nephritis | *Didelphis virginiana* | 53(143) | Sherwood *et al.* (1968, 1969) |
| Infarction | *Didelphis virginiana* | 26(143) | Sherwood *et al.* (1968, 1969) |
| Amyloidosis | *Didelphis virginiana* | 5(90) | Sherwood *et al.* (1969) |
| Cystitis | *Didelphis virginiana* | 10(90) | Sherwood *et al.* (1969) |
| | *D. albiventris* | 1 | Hamerton (1938) |
| Cystic calculi | *Didelphis virginiana* | 1(90) | Sherwood *et al.* (1969) |
| Urethritis (traumatic) | *Didelphis virginiana* | 1 | Fox (1923e) |
| Prostatitis | *Didelphis virginiana* | 6(90) | Fox (1923e); Sherwood *et al.* (1969) |
| Prostatic hyperplasia | *Didelphis virginiana* | 1 | Fox (1923f) |
| Prostatic arteritis (acute) | *Didelphis virginiana* | 1(90) | Sherwood *et al.* (1969) |
| Eversion of the uterus | *Didelphis virginiana* | 1(90) | Fox (1923g) |
| Endometritis | *Didelphis virginiana* | 2(90) | Sherwood *et al.* (1969) |
| Hyperplasia of the mammary gland | *Didelphis virginiana* | Common | Farris (1950) |
| Respiratory System | | | |
| Pneumonia | *Didelphis virginiana* | 68(161) | Hamerton (1938); Beamer *et al.* (1960); Lombard and Witte (1959); Sherwood *et al.* (1968, 1969) |
| | *Didelphis virginiana* | 1 | Hamerton (1938) |
| Bronchopneumonia | *Didelphis virginiana* | 1 | Hamerton (1939) |
| Pulmonary edema and congestion | *Didelphis virginiana* | 2(90) | Sherwood *et al.* (1969) |

| | | | |
|---|---|---|---|
| Focal pulmonary hemorrhage | *Didelphis virginiana* | 7(90) | Sherwood *et al.* (1969) |
| Alveolar microlithiasis | *Didelphis virginiana* | 90(107) | Sherwood *et al.* (1968, 1969) |
| Cardiovascular System | | | |
| Endocarditis, bacterial | *Didelphis virginiana* | 68(168) | Apgar (1934); Fox (1923a); Goss (1940); La Plante and Burrell (1966); Sherwood *et al.* (1968, 1969) |
| Myocarditis | *Didelphis virginiana* | 129(221) | Sherwood *et al.* (1968, 1969) |
| Myocardial infarction | *Didelphis virginiana* | 1(90) | Sherwood *et al.* (1969) |
| Fibrinous pericarditis | *Didelphis virginiana* | 2(90) | Sherwood *et al.* (1969) |
| Musculoskeletal System | | | |
| Degenerative bone disease | *Didelphis virginiana* | Common | Apgar (1934) |
| Hematopoietic System | | | |
| Acute splenitis | *Didelphis virginiana* | 2(90) | Apgar (1934); Sherwood *et al.* (1969) |
| Splenic infarction | *Didelphis virginiana* | 1(90) | Sherwood *et al.* (1969) |
| Splenic amyloidosis | *Didelphis virginiana* | 19(143) | Sherwood *et al.* (1968, 1969) |
| Asplenia | *Didelphis virginiana* | 1(90) | Sherwood *et al.* (1969) |
| Focal necrosis of lymph nodes | *Didelphis virginiana* | 7(90) | Sherwood *et al.* (1969) |
| Central Nervous System | | | |
| Meningitis | | 36(135) | Sherwood *et al.* (1968, 1969) |
| Microabscesses | | 13(90) | Sherwood *et al.* (1969) |
| Endocrine System | | | |
| Focal necrosis of adrenal | *Didelphis virginiana* | 3(90) | Sherwood *et al.* (1969) |
| Thyroiditis (acute) | *Didelphis virginiana* | 2(90) | Sherwood *et al.* (1969) |
| Miscellaneous | | | |
| Tail infections | *Didelphis virginiana* | *[b] | Fox (1923a); Krupp and Quillin (1964); Barnes (1968) |
| Foot and toe infections | *Didelphis virginiana* | * | Krupp and Quillin (1964) |
| Cannibalism | *Didelphis virginiana* | * | Pray (1921) |
| Autocannibalism | *Didelphis virginiana* | * | Barnes (1968) |

[a] The figures in parentheses indicate the total number of marsupials examined.
[b] Asterisk indicates incidence not specified.

tions was discussed in Sections II,C,2, and II,D,7. The only clinical disease associated with ectoparasite infestation was reported by Blair (1921); sarcoptic mange was characterized by pruritis, alopecia, and skin lesions in *D. virginiana*.

## J. Helminth Parasites

Internal parasitic infections are common in *Didelphis* and *Marmosa* (Table IV). Feldman *et al.* (1972) have provided good keys and visual aids for identifying helminth parasites of *D. virginiana*. There is a paucity of information concerning helminth parasites in other Didelphidae. The following discussion is restricted to those infections which were particularly common or injurious to their hosts.

### 1. Nematodes

*Physaloptera turgida, Cruzia americana,* and *Cruzia tentaculata* are commonly observed in North American opossums (*D. virginiana*). Although these parasites are frequently present in large numbers, only *P. turgida* produces noticeable pathology which includes hyperemia, erosions, and ulcers in the gastrointestinal tract, and in some instances, perforation of the stomach. Similar gastric lesions are associated with adult *Gnathostoma* infections. Clinical signs associated with *Physaloptera* infections include weight loss, anorexia, profuse diarrhea, poor hair coat, and death. Krupp (1962) reported that carbon disulfide was effective in treating infected opossums. Adequate sanitation and elimination of intermediate hosts of *Physaloptera* (beetles, earwigs, cockroaches, and crickets) are necessary to prevent reinfections.

Miller and Harkema (1970) found that about 70% of the 54 *D. virginiana* which they examined had *Viannia viannai* and *Longistriata didelphis* infections. Clinical signs and pathology were not reported. *Capillaria* infections were also observed frequently.

Thrasher *et al.* (1971) listed high incidences of *Syphacia, Trichostrongylus, Trichuris,* and *Capillaria* among recently captured *Marmosa robinsoni* from Colombia. Trichostrongyle infections were associated with intestinal mucosal ulcerations. Thiabenzadole (100 mg/kg body weight) administered orally for five consecutive days was used to eliminate *Syphacia* and *Trichostrongylus*. The treatment employed for *Trichuris* infections consisted of a single injection of phthalofyne (250 mg/kg body weight) given intraperitoneally. The adverse systemic effects (severe enteritis and ill-

**TABLE IV**

Helminth Parasites of American Marsupials

| Name | Host | Habitat | Location | References |
|---|---|---|---|---|
| Nematodes | | | | |
| Oxyuroidea | | | | |
| *Oxyuris* sp. | *Didelphis virginiana* | Intestine | Philadelphia Zoological Gardens | Barnes (1968); Canavan (1929); Fox (1923k) |
| *Aspidodera harwoodi* | *Didelphis virginiana* | Cecum | Texas | Chandler (1932) |
| *Aspidodera railliet* | *D. m. caucae, Caluromys lanatus* | Large intestine, cecum | Panama and Brazil | Imperial Bureau (1933); Foster (1939) |
| *Aspidodera scoleciformis* | *Marmosa murina, Monodelphis domestica* | Intestine, cecum | Brazil | Imperial Bureau (1933) |
| *Aspidodera subulata* | *Metachirus nudicaudatus* | Stomach, intestine | Panama | Imperial Bureau (1933) |
| *Cruzia americana* | *Didelphis virginiana* | Large intestine, cecum | Illinois, Virginia, North Carolina, Georgia | Babero (1957, 1960); Holloway and Dewler (1963); Holloway (1966); Miller and Harkema (1970); Feldman *et al.* (1972); Nettles *et al.* (1975) |
| *Cruzia tentaculata* | *Didelphis virginiana, D. m. aurita, D. m. caucae, D. quoaiquiqua, Monodelphis domestica, Marmosa murina, Metachirus nudicaudatus, Philander opossum, Caluromys philander* | Small intestine, cecum | Louisiana, Texas, North Carolina, Pennsylvania, Tennessee, Illinois, and Brazil | Dikmans (1931); Canayan (1929, 1931); Chandler (1932); Imperial Bureau (1933); Leigh (1940); Rieber and Byrd (1942); Rausch and Tiner (1949); Sherwood *et al.* (1968, 1969) |
| *Heterakis paradoxa* | *Marmosa murina* | Intestine | *[a] | Imperial Bureau (1933) |
| Ascaroidea | | | | |
| *Lagochilascaris turgida* | *Didelphis marsupialis, Lutreolina crassicaudata* | Stomach | Buenos Aires, Argentina | Canavan (1931); Imperial Bureau (1933) |

**TABLE IV (Cont.)**

| Name | Host | Habitat | Location | References |
|---|---|---|---|---|
| Rhabdiasoidea | | | | |
| *Strongylidae* | | | | |
| *Strongyloides* sp. | *Didelphis virginiana* | Intestinal tract | Louisiana | Little (1960) |
| *Globocephalus marsupialis* | *Philander opossum* | Small intestine | Rio, Brazil | Teixeira de Freitas and Lent (1936) |
| Trichostrongylidae | | | | |
| *Oesophagostomum* sp. | *Didelphis virginiana* | *[b] | Louisiana | Dikmans (1931) |
| *Maciella macieli* | *Caluromys derbianus*, *Didelphis marsupialis caucae* | * | Panama | Foster (1939) |
| *Trichostrongylus* sp. | *Didelphis virginiana* | * | Louisiana | Dikmans (1931) |
| *Ostertagia callis* | *Didelphis marsupialis aurita* | Small intestine | Brazil | Imperial Bureau (1933) |
| *Heligmosomum didelphae* | *Didelphis marsupialis aurita* | Small intestine | Brazil | Imperial Bureau (1933) |
| *Travassostrongylus guatuor*, *T. quintus* and *T. sextus* | *Metachirus nudicaudatus* | Small intestine | Rio, Brazil | Teixeira de Freitas (1937) |
| *Viannaia burobscura* | *Didelphis virginiana* | Small intestine | Louisiana, Georgia, North Carolina, Illinois, and Tennessee | Dikmans (1931); Leigh (1940); Rieber and Byrd (1942); Babero (1957) |
| *Viannaia conspicua* | *Philander opossum* | Small intestine | Brazil | Imperial Bureau (1933) |
| *Viannaia hamata* | *Didelphis virginiana*, *D. m. aurita* | Small intestine | North Carolina and Brazil | Imperial Bureau (1933); Miller and Harkema (1970); Feldman *et al*. (1972) |
| *Viannaia pusilla* | *Didelphis marsupialis aurita* | Small intestine | Brazil | Imperial Bureau (1933) |
| *Viannaia viannai* | *Didelphis virginiana*, *D. m. aurita* | * | Louisiana and Brazil | Dikmans (1943) |
| *Longistriata didelphis* | *Didelphis virginiana* | Small intestine | Louisiana and North Carolina | Dikmans (1943); Miller and Harkema (1970); Feldman *et al*. (1972) |

| | | | | |
|---|---|---|---|---|
| Metastrongylidae | | | | |
| *Heterostrongylus heterostrongylus* | *Didelphis marsupialis aurita* | Bronchi | Sao Paulo, Brazil | B. Schwartz (communicated by Vaz and Perier, 1934) |
| Metastrongylids (unidentified) | *Didelphis marsupialis aurita* | Nasal cavities | Rio and Espirito Santo, Brazil | Travassos (1946) |
| Spiruroidea | | | | |
| *Gnathostoma didelphis* | *Didelphis virginiana, D. m. aurita* | Stomach, liver | Texas and Georgia | Chandler (1932); Foster (1939); Babero (1960) |
| *Gnathostoma spinigerum* | *Didelphis marsupialis* | Stomach | Philadelphia Zoological Gardens | Canavan (1931); Ash (1962) |
| *Gnathostoma turgidum* | *Didelphis virginiana, D. m. aurita, D. albiventris* | Stomach | Louisiana; Argentina and Brazil | Dikmans (1931); Canavan (1931); Imperial Bureau (1933) |
| *Gnathostoma* sp. | *Didelphis virginiana* | Stomach | Georgia and North Carolina | Babero (1960); Feldman *et al.* (1972) |
| *Gongylonema longispiculum* | *Didelphis virginiana* | Esophagus | | |
| *Gongylonema marsupialis* | *Didelphis marsupialis aurita* | Esophagus | Sao Paulo, Brazil | B. Schwartz (communicated by Vaz and Perier, 1934) |
| *Physaloptera ackerti* | *Didelphis virginiana* | Stomach | New York | Stoner (1945) |
| *Physaloptera mirandai* | *Metachirus nudicaudatus* | Stomach | Rio, Brazil | Lent and Teixeira de Freitas (1937) |
| *Physaloptera turgida* | *Didelphis virginiana, D. albiventris, Didelphis marsupialis caucae, Caluromys derbianus* | Mouth, esophagus, stomach, duodenum, intestine, liver | Pennsylvania (Philadelphia Zoological Gardens), Louisiana, Illinois, Tennessee, New York, Georgia, and North Carolina; London Zoological Gardens, England; Chiapas, Mexico; and Panama | Cobbold (1874); Fox (1923e); Dikmans (1931); Leigh (1940); Rieber and Byrd (1942); Rausch and Tiner (1949); Krupp (1962); Sherwood *et al.* (1968, 1969); Miller and Harkema (1970); Feldman *et al.* (1972); Nettles *et al.* (1975); Caballero (1951) |

**TABLE IV (Cont.)**

| Name | Host | Habitat | Location | References |
|---|---|---|---|---|
| *Physaloptera* sp. | *Didelphis virginiana* | | Pennsylvania (Philadelphia Zoological Gardens) and Iowa | Fox (1923); Haley (1938); Joyce and Eddy (1944); Krupp and Quillin (1964) |
| *Spirocerca longispiculata* | *Didelphis virginiana* | Stomach | Oklahoma | Hill (1939) |
| *Dipetalonema* sp. | *Didelphis virginiana*, *Caluromys derbianus* | Lung, esophagus, heart, blood | Georgia, North Carolina, and Panama | Foster (1939); Babero (1960); Feldman *et al*. (1972) |
| *Dirofilaria* sp. | *Didelphis virginiana* | Heart, blood | Georgia | Babero (1960) |
| *Skrjabinofilaria pricei* | *Didelphis marsupialis aurita* | Subcutaneous tissue | Sao Paulo, Brazil | Schwartz (communicated by Vaz and Perier, 1934) |
| *Skrjabinofilaria philanderi* | *Caluromys derbianus* | Body cavity | Panama | Foster (1939) |
| *Agamofilaria* (group name) (larvae) | *Caluromys philander* | Peritoneal cavity | Philadelphia Zoological Gardens | Canavan (1929) |
| Dracunculoidea | | | | |
| *Dracunculus fuelleborni* | *Didelphis virginiana* | Subcutaneous tissue | Eastern Maryland | Alexander *et al*. (1972) |
| *Trichuroidea* | | | | |
| *Trichuris didelphis* | *Didelphis virginiana* | Intestine | Georgia | Babero (1960) |
| *Trichuris marsupialis* | *Caluromys derbianus* | Large intestine | Panama | Foster (1939) |
| *Trichuris minuta* | *Didelphis virginiana*, *D. albiventris*, *D. m. caucae*, *Marmosa murina*, *Metachirus nudicaudatus* | Intestine | Georgia and Brazil | Imperial Bureau (1933); Babero (1960) |
| *Trichuris* sp. | *Didelphis virginiana* | Intestine | Louisiana and North Carolina | Dikmans (1931); Imperial Bureau (1933); Krupp and Quillin |

| | | | | |
|---|---|---|---|---|
| | | | | (1964); Miller and Harkema (1970); Feldman *et al*. (1972) |
| *Capillaria aerophilia* | *Didelphis virginiana* | Lung | North Carolina and Georgia | Miller and Harkema (1970); Feldman *et al*. (1972); Nettles *et al*. (1975) |
| *Capillaria auritae* | *Didelphis marsupialis aurita* | Small intestine | Brazil | Imperial Bureau (1933) |
| *Capillaria longicauda* | *Didelphis virginiana* | Esophagus | Georgia and North Carolina | Babero (1960); Feldman *et al*. (1972) |
| *Capillaria* sp. | *Didelphis virginiana* | Bronchi | North Carolina | Sherwood *et al*. (1968, 1969) |
| *Trichinella spiralis* (larvae) | *Didelphis virginiana* | Muscle | Iowa | Zimmerman and Hubbard (1969) |
| Trematodes | | | | |
| Opisthorchidae | | | | |
| *Amphimerus pseudofelineus* | *Didelphis virginiana* | Biliary ducts, gall bladder | Illinois | Leigh (1940) |
| Echinonstomatidae | | | | |
| *Echinostoma coalitum* | *Didelphis virginiana* | Stomach | * | Byrd *et al*. (1942b) |
| *Echinostoma revolutum* | *Didelphis virginiana* | Intestine | Illinois | Leigh (1940); Rausch and Tiner (1949) |
| *Echinostoma* sp. | *Didelphis virginiana* | Small intestine | Louisiana and North Carolina | Dikmans (1931); Park (1936); Feldman *et al*. (1972) |
| Dicrocoelidae | | | | |
| *Eurytrema allentoshi* | *Didelphis virginiana* | * | Texas | Denton (1944) |
| *Opisthorchis pricei* | *Caluromys derbianus* | Biliary ducts | Panama | Foster (1939) |
| *Plagiorchis didelphis* | *Didelphis albiventris* | * | Paraguay | Imperial Bureau (1933) |
| *Platynosomum allentoshi* | *Caluromys derbianus* | Biliary ducts | Panama | Foster (1939) |
| Rhophaliadidae | | | | |
| *Rhophalias coronatus* | *Didelphis virginiana, D. m. aurita, D. marsupialis caucae, Chironectes minimus, Philander opossum, M. nudicaudatus* | Small intestine | Brazil and Chiapas, Mexico | Imperial Bureau (1933) |

**TABLE IV (Cont.)**

| Name | Host | Habitat | Location | References |
|---|---|---|---|---|
| *Rhopalias baculifera* | *Chironectes minimus* | Intestine | Brazil | Imperial Bureau (1933) |
| *Rhopalias horridus* | *Didelphis marsupialis aurita, Philander opossum, Metachirus nudicaudatus, Caluromys philander* | Stomach, small intestine | Angra dos Reis, Brazil | Imperial Bureau (1933) |
| *Rhopalias macracanthus* | *Didelphis virginiana, D. marsupialis, D. m. caucae* | Small intestine | Illinois, North Carolina, Georgia, Tennessee, and Texas; and Chiapas, Mexico | Chandler (1932); Leigh (1940); Byrd *et al.* (1942b); Babero (1957, 1960); Miller and Harkema (1970); and Feldman *et al.* (1972) |
| *Rhopalias* sp. | *Didelphis virginiana* | * | Louisiana | Dikmans (1931) |
| Troglotrematidae | | | | |
| *Achillurbainia recondita* | *Didelphis marsupialis* | Maxillary sinuses | Sao Paulo, Brazil | Travassos (1942) |
| *Paragonimus westermani* | *Didelphis virginiana* | Lung | Tennessee | Byrd *et al.* (1942a) |
| *Paragonimus kellicotti* | *Didelphis virginiana* | Lung | Georgia and North Carolina | McKeever (1958); Sherwood *et al.* (1969); Feldman *et al.* (1972) |
| *Paragonimus* sp. | *Didelphis virginiana* | Lung | North Carolina | Sherwood *et al.* (1969) |
| Strigeidae | | | | |
| *Diplostoma variabilis* | *Didelphis virginiana* | Small intestine | North Carolina, Tennessee, Louisiana, Georgia, and Illinois | Leigh (1940); Byrd *et al.* (1942b); Babero (1957, 1960); Miller and Harkema (1970); Feldman *et al.* (1972) |
| *Duboisiella proloba* | *Didelphis marsupialis aurita* | * | * | Baer (1938) |
| *Hemistomum pedatum* | *Didelphis marsupialis aurita, D. m. caucae, Metachirus nudicaudatus* | Small intestine | Brazil | Imperial Bureau (1933) |

| | | | | |
|---|---|---|---|---|
| *Levinseniella jagerskioldi* | *Didelphis marsupialis aurita* | Intestine | Manguinhos, Brazil | Imperial Bureau (1933) |
| *Maritreminoides nettae* | *Didelphis virginiana* | Small intestine | Illinois | Miller and Harkema (1970) |
| *Proalaria variabilis* | *Didelphis virginiana* | Small intestine | Texas | Chandler (1932) |
| *Fibricola cratera* | *Didelphis virginiana* | Small intestine | Tennessee and Michigan | Byrd *et al*. (1942b); Chandler and Rausch (1946); Rausch and Tiner (1949) |
| *Fibricola incida* | *Didelphis virginiana* | Duodenum | Louisiana | Kaplan (1964) |
| *Neodiplostomum lucidum* | *Didelphis virginiana* | Small intestine | Illinois, Louisiana, Georgia, Tennessee, and Texas | Dikmans (1931); Park (1936); Byrd *et al*. (1942b); Babero (1957); LaRue and Bosma (1958); McKeever (1958) |
| Schistosomatidae | | | | |
| *Heterobilharzia americana* | *Didelphis virginiana* | Mesenteric veins | Louisiana | Kaplan (1964); and Malek (1970) |
| Brachylaemidae | | | | |
| *Brachylaemus opisthotrias* | *Didelphis virginiana* | Small intestine | Tennessee, Georgia, and Illinois | Leigh (1940); Byrd *et al*. (1942b); Babero (1960) |
| *Brachylaemus spinosulum* | *Didelphis virginiana* | Intestine (experimental) | | Krull (1933) |
| *Brachylaemus virginiana* | *Didelphis virginiana* | Small intestine | Georgia, North Carolina, Louisiana, Tennessee, Illinois, Texas, and Virginia | Chandler (1932); Byrd *et al*. (1942b); Rausch and Tiner (1949); Babero (1957, 1960); Holloway and Dowler (1963); Holloway (1966); Kaplan (1964); Miller and Harkema (1970); Feldman *et al*. (1972); Nettles *et al*. (1975) |
| *Harmostomum opisthotrias* | *Didelphis virginiana, D. m. aurita* | Small intestine | Texas and Sao Paulo, Brazil | Chandler (1932); Imperial Bureau (1933) |
| *Harmostomum opisthotrias virginiana* | *Didelphis virginiana* | Ileum | North America | Dickerson (1930) |

**TABLE IV (Cont.)**

| Name | Host | Habitat | Location | References |
|---|---|---|---|---|
| *Harmostomum* sp. | *Didelphis virginiana* | * | Louisiana | Dikmans (1931) |
| Cestodes | | | | |
| Anoplocephalidae | | | | |
| *Anoplocephala* sp. | *Didelphis marsupialis* | * | | Krupp and Quillin (1966) |
| Linstowiidae | | | | |
| *Linstowia braziliensis* | *Didelphis bistriata, Monodelphis americana* | * | Brazil | Imperial Bureau (1933) |
| *Linstowia jheringi* | *Monodelphis americana* | * | South America | Meggitt (1926) |
| *Linstowia* sp. | *Marmosa robinsoni* | Small intestine | Colombia | Thrasher *et al*. (1971) |
| *Oochoristica bivittata* | *Metachirus nudicaudatus, Marmosa elegans, M. grisea, M. murina, Caluromys philander, Didelphis marsupialis* | Intestine | * | von Janicki (1904); Meggitt (1926); Baer (1927); Imperial Bureau (1933); Foster (1939) |
| *Oochoristica braziliensis* | *Philander opossum* | * | * | Baer (1927) |
| *Oochoristica didelphis* | *Marmosa elegans, M. murina* | Intestine | * | Imperial Bureau (1933) |
| *Oochoristica marmosae* | *Marmosa elegans* | * | South America | Imperial Bureau (1933) |
| *Oochoristica murina* | *Marmosa murina, M. elegans* | * | South America | Imperial Bureau (1933) |
| *Oochoristica surinamensis* | *Didelphis marsupialis aurita* | * | * | Baer (1927) |
| *Oochoristica* sp. | *Didelphis virginiana* | Small intestine | | Krupp and Quillin (1966) |
| Mesocestoididae | | | | |
| *Mesocestoides latus* | *Didelphis virginiana* | * | * | Rausch and Tiner (1949) |
| *Mesocestoides variabilis* | *Didelphis virginiana* | Small intestine | Illinois, Georgia, North Carolina, and Tennessee | Byrd and Ward (1928, 1943); Babero (1957, 1960); Miller and Harkema (1970); Feldman *et al*. (1972) |

| | | | | |
|---|---|---|---|---|
| *Mesocestoides* sp. | *Didelphis virginiana* | * | Louisiana | Dikmans (1931) |
| Hymenolepididae | | | | |
| *Hymenolepis (Weinlandia)* sp. | *Didelphis virginiana* | Small intestine | Illinois | Leigh (1940) |
| *Hymenolepis* sp. | *Didelphis virginiana* | * | Illinois | Leigh (1940) |
| Diphyllobothriidae | | | | |
| *Bothriocephalus didelphis* | *Didelphis marsupialis, D. albiventris* | Intestine | Sao Paulo, Brazil | Imperial Bureau (1933) |
| *Bothriocephalus* sp. *(plerocercoids)* | *Didelphis goagnia, D.* sp. | Connective tissue, musculature | Santa Cruz, Bolivia | Imperial Bureau (1933) |
| *Sparganum reptans (plerocercoids)* | *Monodelphis brevicaudata, Philander opossum* | Subcutaneous tissues | * | Imperial Bureau (1933) |
| *Sparganum* sp. *(plerocercoids)* | *Chironectes minimus* | Subcutaneous tissues | Sao Paulo, Brazil | Imperial Bureau (1933) |
| *Spirometra mansonoides (plerocercoids)* | *Didelphis virginiana* | * | Louisiana | Corkum (1966) |
| Acanthocephala | | | | |
| *Acanthocephala* sp. | *Didelphis virginiana* | Small intestine | Virginia and North Carolina | Holloway and Dowler (1963); Sherwood *et al*. (1968); Miller and Harkema (1970); Feldman *et al*. (1972) |
| *Centrorhynchus* sp. | *Didelphis virginiana* | Small intestine | Illinois | Miller and Harkema (1970) |
| *Echinorhynchus microcephalus* | *Didelphis virginiana* | Mesentery | * | Imperial Bureau (1933) |
| *Hamarniella microcephala* | *Didelphis virginiana, Caluromys derbianus* | Intestine | Illinois | Van Cleave (1953); Miller and Harkema (1970) |
| *Hamarniella tortuosa* | *Didelphis virginiana* | * | Illinois and Georgia | Babero (1957, 1960); Krupp and Quillin (1964) |
| *Hamarniella tumida* | *Didelphis virginiana* | * | * | Van Cleave (1953) |
| *Macracanthorhynchus ingens* | *Didelphis virginiana* | Ileum | North Carolina | Sherwood *et al*. (1969) |

[a] Asterisk indicates location not given.
[b] Asterisk indicates habitat not given.

ness for 2 to 4 days) limited the usefulness of this therapeutic regime. Piperazine citrate was employed by Barnes (1968) to treat *M. robinsoni* infected with *Syphacia* and other nematodes.

### 2. Trematodes

*Brachylaemus, Rhopalias, Diplostomum,* and *Neodiplostomum* are the most common trematode parasites of *Didelphis virginiana.* Their geographical ranges are coincident with that of the opossum in North America. Trematodes were also reported from Colombian *Marmosa robinsoni* by Thrasher *et al.* (1971). Unidentified flukes were observed in the biliary ducts, mesenteric blood vessels, and intestine. Trematode infections did not respond to anthelmintic therapy in this species (Barnes, 1968).

### 3. Cestodes

Tapeworms are the least frequently encountered helminths in the Didelphidae. However, a larger number of South American marsupials appeared to be hosts for these parasites than for other helminths. Thrasher *et al.* (1971) considered *Linstowia* infections injurious to captive *Marmosa robinsoni.* A single oral treatment with bunamidine hydrochloride (400 mg/kg body weight) eliminated these tapeworms.

## III. DISEASES OF AUSTRALASIAN MARSUPIALS

## A. Viral Diseases

### 1. Arbovirus Infection

Evidence of arbovirus infection among Australasian marsupials was based on serologic surveys (Table V). Arboviruses were not isolated.

### 2. Reovirus Infection

Stanley and Leak (1963) surveyed 72 *Setonix brachyurus* (Rottnest quokka) for hemagglutination-inhibition (HI) antibody to three serotypes of reoviruses. Presence of antibody was closely related to the degree of contact the animals had with human beings. Quokkas with positive titers ranged from 17% among those rarely exposed to man to 100% among those continuously exposed. Similar serologic evidence of reovirus infection in *S. brachyurus* was obtained by Stanley *et al.* (1964). Type 3 reovirus was recovered from one of six fecal specimens examined. Stanley and Leak (1963) were unable to induce reovirus epizootics among captive *S. brachyurus*. No clinical signs were reported and only one death was attributed to experimental reovirus infection; myocarditis was present in this animal.

Three of 11 wallabies examined by Stanley *et al.* (1964) had HI antibody against type 3 reovirus; the remainder were negative to all reovirus serotypes. Of 15 kangaroo serums tested by Stanley and Leak (1963), HI antibody was detected in nine against type 1, in 11 against type 2, and in six against type 3 reovirus. Clinical evidence of infections was not reported.

### 3. Picornavirus Infections

**Foot and Mouth Disease.** Snowdon (1968) demonstrated that a variety of Australian marsupials are susceptible to experimental foot and mouth disease (FMD) virus infection. Among the Macropodidae, viremias were detected up to 3 days after inoculation by the intradermal, intramuscular, intranasal, or oral routes. Virus replication occurred principally in the soft palate, pharynx, and base of the tongue. Serum neutralizing antibody responses to FMD virus administration reached a peak 14 days after exposure and then declined gradually.

Clinical signs of experimental FMD were observed in *Macropus rufus* and *Dendrogalus matschiei*. Lesions consisted of vesicles measuring up to 4 cm in diameter which ruptured leaving irregularly circular areas of superficial epithelial detachment on either the foot pads or the tongue. Lingual vesicles occurred 3 to 5 days after intramuscular inoculation, and foot lesions were observed within 7 to 14 days.

Other marsupials in which viremias were detected following experimental infection were *Vombatus ursinus, Trichosurus vulpecula, Potorous tridactylus,* and *Antechinus stuartii*. Viremic responses were not observed in *Perameles nasuta*.

FMD contact transmission studies were performed by exposing kangaroos, wombats, and cattle to one another for prolonged periods. Experimentally infected *M. rufus* were capable of transmitting FMD to susceptible cattle. Similarly, *M. rufus* and *V. ursinus* developed viremias and serum neutralizing antibodies during confinement with infected cattle for 24 days. However, *M. rufus* that were exposed to infected cattle for 24 hr did not

**TABLE V**

Summary of Arboviruses Associated with Australasian Marsupials

| Virus | Marsupial | Antibody[a] | Location | References |
|---|---|---|---|---|
| Group A (mosquito-borne) | | | | |
| Whataro | *Trichosurus vulpecula* | 12/76 SN | Westland, New Zealand | Miles (1967) |
| Group B (mosquito-borne) | | | | |
| Kokobera | *Macropus giganteus* | 14/68 HI | West Queensland, Australia | Doherty (1967a) |
| | *M. rufus* | 1/16 HI | West Queensland, Australia | Doherty (1967a) |
| Edge Hill | *M. rufogriseus* | 9/26 HI | Queensland, Australia | Dohherty (1967b) |
| | *M. parryi* | 2/4 HI | Queensland, Australia | Dohherty (1967b) |
| Murray Valley Encephalitis | Opossum | 2/3 CF | Victoria, Australia | French (1967) |
| | *Isoodon macrourus* | 1/75 HI | Queensland, Australia | Doherty *et al*. (1964) |
| | *Perameles nasuta* | 0/12 HI | Queensland, Australia | Doherty *et al*. (1964) |
| | *M. dorsalis* | 1/9 HI | Southeast Queensland, Australia | Doherty *et al*. (1964) |
| | *W. bicolor* | 0/1 HI | Southeast Queensland, Australia | Doherty *et al*. (1964) |
| | *Thylogale thetis* | 0/4 HI | Southeast Queensland, Australia | Doherty *et al*. (1964) |
| | *Trichosurus vulpecula* | 0/3 HI | Southeast Queensland, Australia | Doherty *et al*. (1964) |
| Ungrouped | | | | |
| Koongol or Wongal | Kangaroo | 2/12 HI | West Queensland and Mitchell River, Australia | Doherty *et al*. (1970) |
| | Bandicoot | 6/118 HI | Northeast and West Queensland, Australia | Doherty *et al*. (1970) |
| | Wallaby | 8/58 HI | Northeast Queensland, Australia | Doherty *et al*. (1970) |
| Mattuppa | Kangaroo | 16/74 SN | West Queensland and Mitchell River, Australia | Doherty *et al*. (1970) |
| | Wallaby | 21/70 SN | Southeast Queensland and Mitchell River, Australia | Doherty *et al*. (1970) |
| | Bandicoot | 0/51 SN | North Queensland, Australia | Doherty *et al*. (1970) |
| Trubanaman | Kangaroo | 21/53 SN | West Queensland and Mitchell River, Australia | Doherty *et al*. (1970) |
| | Wallaby | 34/70 SN | Southeast Queensland and Mitchell River, Australia | Doherty *et al*. (1970) |

| | | | | | |
|---|---|---|---|---|---|
| | Bandicoot | 0/41 | SN | North Queensland, Australia | Doherty *et al.* (1970) |
| Corripeta | Kangaroo | 7/49 | SN | West Queensland and Mitchell River, Australia | Doherty *et al.* (1970) |
| | Wallaby | 6/50 | SN | Southeast Queensland and Mitchell River, Australia | Doherty *et al.* (1970) |
| | Bandicoot | 1/41 | SN | North Queensland, Australia | Doherty *et al.* (1970) |
| Eubenagee | Kangaroo | 5/54 | SN | West Queensland and Mitchell River, Australia | Doherty *et al.* (1970) |
| | Wallaby | 12/69 | SN | Southeast Queensland and Mitchell River, Australia | Doherty *et al.* (1970) |
| | Bandicoot | 0/51 | SN | North Queensland, Australia | Doherty *et al.* (1970) |
| Kowanyama | Kangaroo | 8/15 | SN | West Queensland and Mitchell River, Australia | Doherty *et al.* (1970) |
| | Wallaby | 25/68 | SN | Southeast Queensland and Mitchell River, Australia | Doherty *et al.* (1970) |
| | Bandicoot | 3/50 | SN | North Queensland, Australia | Doherty *et al.* (1970) |
| MRM 4059 | Kangaroo | 4/49 | SN | West Queensland and Mitchell River, Australia | Doherty *et al.* (1970) |
| | Wallaby | 0/64 | SN | Southeast Queensland and Mitchell River, Australia | Doherty *et al.* (1970) |
| | Bandicoot | 1/39 | SN | North Queensland, Australia | Doherty *et al.* (1970) |
| C 5502 | Kangaroo | 4/52 | SN | West Queensland and Mitchell River, Australia | Doherty *et al.* (1970) |
| | Wallaby | 0/68 | SN | Southeast Queensland and Mitchell River, Australia | Doherty *et al.* (1970) |
| | Bandicoot | 0/39 | SN | North Queensland, Australia | Doherty *et al.* (1970) |
| Upolu | Kangaroo | 1/52 | SN | West Queensland and Mitchell River, Australia | Doherty *et al.* (1970) |
| | Wallaby | 0/69 | SN | Southeast Queensland and Mitchell River, Australia | Doherty *et al.* (1970) |
| | Bandicoot | 0/39 | SN | North Queensland, Australia | Doherty *et al.* (1970) |

[a] Number with antibody/number tested. SN, serum neutralization test; HI, hemagglutination-inhibition test; CF, complement fixation test.

subsequently transmit FMD to susceptible cattle. Snowdon (1968) concluded that the marsupials tested in these studies would participate in the spread of FMD under field conditions only under exceptional circumstances.

### 4. Other Virus Infection

**Mucosal Disease.** Antibody against mucosal disease (virus diarrhea) virus was found in two of 44 *Macropus rufogriseus frutica* (Munday, 1972). The significance of this finding and the possible role of the wallaby as reservoir host, however, is not clear. The susceptibility of marsupials to mucosal disease virus infection has not been investigated.

## B. Bacterial Diseases

### 1. *Nocardia* Infection

According to Fox (1923r) and Rewell (1948a) up to one-half of all captive wallabies and kangaroos throughout the world succumb to *Nocardia* infections. Affected species have included *Macropus rufus* (Hamerton, 1933; Fox, 1923), *Macropus robustus* (Hamerton, 1929, 1937, 1938, 1941), *Macropus agilis* (Hamerton, 1938), *Macropus parryi* (Hamerton, 1935a), *Macropus rufogriseus,* including *M. r. fructica* (Scott, 1927; Hamerton, 1932, 1937, 1938, 1941; Hill, 1951; Rewell, 1948b; Scott, 1926), *Macropus giganteus* (Fox, 1923r), *Macropus eugenii* (Scott, 1926), *Wallabia bicolor* (Rewell, 1948c), *Dendrogalus ursinus* (Fox, 1938), and a variety of other unspecified or hybrid Macropodidae (Hamerton, 1930, 1944; Plimmer, 1912; Boyd, 1929; Fox, 1924; Blair, 1916, 1918, 1924; Noback, 1929, 1930).

Nocardiosis has been reported as kangaroo disease, jaw disease, mycosis, actinomycosis, lympy jaw, and streptothricosis. The causative agent, *Nocardia macropodidarum* (Fox, 1923r), is associated with granulomatous and suppurative lesions which are frequently located near the nose and jaws and involve the lips, gums, teeth, tongue, palate, and cervical lymph nodes. The lungs, stomach, liver, and pancreas are also common sites of infection. Clinical signs are generally referable to the anatomic location of the lesions and include nasal and ocular discharge, dyspnea, anorexia, lethargy, depression, loss of teeth, weight loss, diarrhea, and subcutaneous swellings. The course of the disease is rapid after macroscopic lesions or clinical signs are observed; affected animals die within 4 days to 3 weeks (Fox, 1923r;

Tucker and Millar, 1953). A generalized form of nocardiosis has also been described (Fox, 1923r; Blair, 1924; Boyd, 1929).

The gross lesions of nocardiosis consist of purogranulomatous nodules which are small and firm in the early stages. Progressive central necrosis produces larger masses with fluctuant consistencies. Rupture of superficial suppurative lesions is frequent. Oral and gastrointestinal lesions are commonly ulcerated; necrotic osteitis of the palate and necrotic alveolar periostatis are observed. Necrosis and purogranulomatous nodules are seen in the liver, lungs, and other viscera. Histologically, the nodules appear as central areas which contain the filaments of *N. macropodidarum* and necrotic debris surrounded by a zone of cellular necrosis. Loose connective tissue infiltrated with mono- and polymorphonuclear leukocytes delimit the lesions from normal tissues.

Attempts to treat animals with nocardiosis have been unsuccessful (Boyd, 1929; Fox, 1923r; Tucker and Millar, 1953). The natural route of infection is through contamination of preexisting wounds or by those produced by sharp vegetable matter in the oral or gastrointestinal mucosa. The feeding of soft alfalfa or other hay, and rolled rather than crushed oats, are considered successful prophylactic measures (Fox, 1924; Goss, 1942). Isolation of infection animals and the elimination of contaminated food and bedding are also recommended control measures (Blair, 1916, 1918).

### 2. *Mycobacterium* Infections

Captive marsupials appear to be highly susceptible to tuberculosis despite evaluations to the contrary which were provided in earlier reports (Crisp, 1860; Hamerton, 1935b). One of two young Queensland opossums (= *Trichosurus*?) obtained by Moore (1904) in Calcutta, India died with pulmonary tuberculosis. Among zoological specimens, Crisp (1860) described tuberculosis of the liver in *Dendrogalus inustus,* Griffith (1928) observed *Mycobacterium* infection (bovine strain) of the lungs in *Aepyprymnus rufescens,* and Hamerton (1932) described pancreatic tuberculosis (avian strain) in *Macropus fuliginosus.* Widely disseminated tuberculous lesions were also found in three *Macropus fuliginosus* (Scott, 1928, 1930; Griffith, 1928), *Wallabia bicolor* (Hamerton, 1930), and *A. rufescens* (Lucas, 1925). An avian strain of mycobacterium was identified in two of the *M. fuliginosus* and *A. rufescens.* Stünzi (1950) described human *Mycobacterium* infection in two *M. rufogriseus frutica,* and Tilden and Williamson (1957) reported tuberculosis (avian strain) of the spinal canal in a *M. rufogriseus frutica.* Tuberculosis of the skeletal systems of four

*Macropus rufogriseus frutica* was diagnosed radiographically by Hime and Jones (1972). Human *Mycobacterium* was recovered from three and the avian strain was isolated from one of the wallabies.

An epizootic of tuberculosis (bovine strain) involving 14 captive *M. rufogriseus frutica* was described by Teuscher (1957). The reaction to tuberculin (0.1 ml of a 50% solution) administered intradermally in the eyelid consisted of thickening (0.8 to 3.0 mm) and erythema in nine of 10 tuberculous kangaroos. Evidence of infection obtained from dorsoventral chest radiographs was considered less diagnostic. Based on the anatomic location of the tubercles, it was suggested that the initial cases occurred through ingestion of infective material and that the disease was then spread by airborne transmission. Oral therapy with nicotinaldehyde thiosemicarbazon and isonicotinic acid hydrazide (INH) at the rate of 1 and 4 mg per kg body weight per day, respectively, arrested or caused regression of tuberculosis. Organisms could not be isolated from kangaroos that were treated for more than 6 weeks. The reaction to tuberculin was not altered by therapy.

Experimental studies by Bollinger and Bollinger (1948) indicate that *Trichosurus vulpecula* is highly susceptible to tuberculosis. Intramuscular and intraperitoneal administration of bovine and human strains of *Mycobacterium* to 10 phalangers was followed by generalized tuberculosis and death in 2 to 5 weeks. The disease was also passaged by oral administration of material from infected opossums. Further evidence of the susceptibility of this species was demonstrated by Ekdahl (1970), who isolated *M. tuberculosis* (bovine strain) from two of three wild *T. vulpecula* with tuberculous lesions in New Zealand. Twenty other phalangers which were not submitted for laboratory examination had generalized tuberculosis. All were trapped on a farm where tuberculosis in cattle was a problem.

Lesions are generally widespread in naturally and experimentally infected animals and frequently involve lymph nodes (axillary, cervical, iliac, lumbar, celiac, tracheobronchial, internal mammary, and submaxillary), lungs, and liver. Less commonly, the joints (scapulohumeral, sternoclavicular, and femorotibial), pancreas, and spleen are affected. The tubercles range from small milliary lesions to massive nodules. Their consistency varies from purulent to caseous. Microscopic lesions typically consist of central areas of necrosis surrounded by epithelioid and giant cells. Peripheral to these, lymphocytic aggregations and connective tissue may be observed. Clinical signs of tuberculosis, although usually not described, are presumably referable to the degree of impairment of the organ systems involved. Hime and Jones (1972) suggest that consideration of muscular disorders in macropod marsupials include tuberculosis and hypovitaminosis E among the differential diagnoses.

### 3. *Borrelia* Infections

At least two genera of Queensland marsupials are susceptible to *Borrelia* sp. infection (Mackerras, 1959b). In addition to *Isoodon obesulus*, two of 72 *Macropus giganteus* and two of 50 *M. rufus* harbored *Borrelia* spirochetes. Clinical signs of infection were not observed.

### 4. *Leptospira* Infections

*Perameles nasuta* and *Isoodon macrourus* are important reservoirs of leptospirosis in Queensland based upon the results of isolation and serologic studies of more than 600 bandicoots (Emanuel and Mackerras, 1964; Battey and Smith, 1964). *Leptospira broomi, Leptospira zanoni, Leptospira mini, L. grippotyphosa,* and *Leptospira celledoni* have been isolated from *Isoodon macrourus*; *Leptospira medanensis, Leptospira australis,* and *Leptospira kremastos* were recovered from both *I. macrourus* and *Perameles nasuta.* These species were also positive serologically for *Leptospira icterohemorrhagica, Leptospira canicola, Leptospira pyrogenes, Leptospira pomona,* and *Leptospira hyos. Leptospira hebdamonis* was isolated from one of 15 *Trichosurus vulpecula* and *L. grippityphosa* from one of 2 *Thylogale stigmatica.* Antibody was not found in small numbers of *Antechinus godmani, Aepyprymnus rufescens, Dendrogalus lumholtzi, Hypsiprymnodon moschatus,* and *Macropus agilis.*

Munday and Corbould (1973) found significant antibody to *L. pomona* in nine of 13 wild Tasmanian wombats (*Vombatus ursinus*). Affected animals had mild to severe interstitial nephritis. The status of this species as a reservior for leptospirosis in Tasmania was uncertain, however, because infected wombats were detected in areas where the disease was recently diagnosed in cattle. Experimental infection of two wombats with *L. pomona* resulted in severe disease. Clinical signs consisted of depression, anorexia, icterus, and death within 14 days. Interstitial nephritis and nephrosis were observed at necropsy.

Munday (1972) also reported antibody against *L. pomona* in a *Perameles gunnii* and *L. hyos* in three of six *Potorous tridactylus* in a survey of Tasmanian wildlife.

### 5. *Bacillus* Infections

**a. Anthrax.** *Bacillus anthracis* infection has been reported in two kangaroos (Blair, 1924) and may be an important disease of captive kangaroos and wallabies. *Bacillus anthracis* infections are rapidly fatal. Carcasses, bedding, and other contaminated items should be incinerated to prevent spread of the disease, and the surroundings should be disinfected.

**b. Other *Bacillus* Infection.** Several strains of *Bacillus* (*Sphaerophorus*) *necrophorus* have been isolated from the lesions of nocardiosis in *Macropus* sp. (Beveridge, 1934). Subcutaneous injections of these organisms into normal *Macropus* sp. produced large (8 cm diameter) ulcers within 16 days; the animals also became emaciated. *Bacillus necrophorus* of bovine origin produced massive necrosis and death 8 days after inoculation in *Macropus* sp. The significance of *B. necrophorus* in naturally occurring lesions, however, has not been determined.

### 6. *Bacteroides* Infections

*Bacteroides sp.* was isolated from abscesses occurring in five of a group of *Macropus rufus*, *M. fuliginosus*, *M. agilis*, and *M. robustus* by Watts and McLean (1956). *Fusobacterium* sp. was obtained from the lesions of a sixth animal. Abscesses were located in the soft tissues of the jaws, lips, and gums and resembled the lesions of nocardiosis. Treatment consisted of procaine penicillin administered parenterally, and aqueous penicillin instilled into the abscesses. Later, the lesions were drained surgically, curetted, and packed with sulfapyridine. Good results were obtained following therapy unless the lesions involved bony structures.

### 7. *Salmonella* Intections

Feral quokkas (*Setonix brachyurus*) and several other Australian marsupials constitute widespread reservoirs of salmonellosis. Diagnosis of salmonellosis (*Salmonella javiana*) in a child who had close contact with quokkas and their feces led Iveson and Bradshaw (1973) to perform a survey to determine the extent of the disease on Rottnest Island and mainland Western Australia. The results of quokka fecal pellet examinations showed that 50% of the samples were infected with *S. anatum*, *S. chester*, *S. havana*, *S. javiana*, *S. decatur*, *S. muenchen*, *S. newington*, *S. typhimurium*, and *Arizona* 26:26:25. A higher level of infection was found by direct sampling of 87 quokkas. In addition to the above serotypes, the following were isolated from 62 (71%) of the animals: *S. adelaide*, *S. bootle*, *S. newbrunswick*, *S. bahrenfeld*, *S. fremantle*, *S. give*, *S. oranienburg*, *S. orientalis*, *S. orion*, *S. wandsbek*, *S. waycross*, and *Arizona* 26:23:21 and

16:23:25. Multiple infections (up to four serotypes) were found in many quokkas. It was postulated that *Salmonella* infections contribute to the seasonal (summer) increase in morbibity and mortality among quokkas which is also coincident with the decline of good forage.

Salmonellae have also been isolated from a variety of other Australian marsupials including *Macropus robustus* (*S. typhimurium* and *S. decatur*), *M. fulginosus* (*S. adelaide, S. muenchen*, and *Arizona* 26:26:25), *M. eugenii* (*S. newington, A.* 26:26:25), *Macropus rufus* (*S. wandsbek*), *Petrogale brachyotis* (*S. jangwani*), *Dasyurus geoffroii* (*S. charity, S. chester*, and *S. muenchen*), *Sminthopsis murina* (*S. fremantle*), and *Isoodon obesulus* (*S. meleagridis*) (Iveson and Bradshaw, 1973). Other evidence of salmonellosis in kangaroos includes isolation of *S. birkenhead* in bile (Winter, 1958) and the demonstration of serum antibody against *S. typhimurium* and *S. paratyphi* in *Macropus rufus* (Schwab and Reeves, 1966).

### 8. Other Bacterial Infections

Several instances of bacterial isolation or serologic evidence of infection have been described. Their significance as pathogens in marsupials is not clear.

Markham and Markham (1966) isolated coagulase positive staphylococci from the nose of one of five *Trichosurus vulpecula*. Staphylocci, dipheroids, and *Micrococcus catarrhalis* were found in *Wallabia bicolor* by Hamerton (1938). Fox (1923) isolated pneumococci from pulmonary abscesses in *Macropus rufus* which also had nocardiosis. Urbain *et al.* (1947) isolated *Corynebacterium necrophorum* from the necrotic lesions on the head of *Macropus rufogriseus*. *Streptococcus equisimilis* was isolated from a purulent sequestrum in the center of a healed fracture of the right radius and ulna of a *M. giganteus (Keep, 1972)*.

## C. Rickettsial Diseases

### 1. *Coxiella* Infections

Susceptibility of *Isoodon macrourus, Trichosurus vulpecula,* and *Aepyprymnus rufescens* to experimental infection with Q fever rickettsiae (*Coxiella burnetti*) was demonstrated by Derrick *et al.* (1939, 1940). Clinical signs were not observed, and the only consistent pathologic changes associated with infection were splenomegaly and multiple areas of focal necrosis in the spleen. Other studies indicate that natural infections are common among wild bandicoots (Derrick *et al.*, 1939; Smith and Derrick, 1940; Derrick, 1944; Munday, 1972).

Smith and Derrick (1940) isolated six strains of *C. burnetti* from ticks (*Haemophysalis humerosa*) infesting *I. macrourus* on Moreton Island, Australia. Serums from all of the bandicoots tested contained agglutinins against *C. burnetti*. Derrick (1944) described *C. burnetti* as being confined to the epithelial lining and lumen of the intestine of *H. humerosa*; tick feces were highly infective. Ticks retained infectivity for as long as 68 days. The significance of *H. humerosa* as a vector consists principally of maintaining Q fever rickettsiae in the bandicoot population; this tick generally does not bite man (Smith and Derrick, 1940). Smith (1942) demonstrated that larval, nymphal, and adult *Ixodes holocyclus* could be infected with *C. burnetti* by feeding on laboratory animals and subsequently transmit the infection to bandicoots. *Ixodes holocyclus* frequently attacks man, and its most important wildlife hosts include *I. macrourus* and *Perameles nasuta*. Derrick *et al.* (1939), Seddon (1943), and Derrick (1944) considered that bandicoots, particularly *I. macrourus*, constituted an important reservoir of Q fever in Australia.

## 2. *Rickettsia* Infections

**a. Scrub Typhus (*Tsutsugamushi* Fever).** Heaslip (1941) demonstrated that *Isoodon macrourus* was capable of harboring *Rickettsia tsutsugamushi*, the causative agent of scrub typhus, after experimental infection. Serum from 24 of 82 wild *I. macrourus* surveyed in South Queensland contained agglutinins to *R. tsutsugamushi*. Mouse inoculation, blood culture, and agglutination tests indicated that of 52 *I. macrourus* examined, one harbored the rickettsia and seven were probably infected. Heaslip concluded that *I. macrourus* was a natural reservoir for scrub typhus.

**b. North Queensland Tick Typhus.** In a survey of wild marsupials in North Queensland for complement-fixing antibodies, Fenner (1946) found that two of six *Trichosurus vulpecula*, two of 11 *Aepyprymnus rufescens*, one of five *Perameles nasuta*, and two of 35 *Isoodon obesulus* had been previously infected with North Queensland tick typhus (*Rickettsia rickettsii*). *Ixodes holocyclus, I. tasmani,* and *Haemaphysalis humerosa* were found infesting many of the animals. However, attempts to isolate rickettsiae from the ticks were unsuccessful.

## D. Protozoan Diseases

### 1. *Toxoplasma* Infections

*Toxoplasma gondi* infection has been demonstrated in a variety of Australian marsupials. Natural infection is common in Queensland among the free-living bandicoots *Isoodon obesulus* and *Perameles nasuta* and the possum *Trichosurus vulpecula*, based on the results of complement fixation and mouse inoculation tests obtained in two surveys (Pope *et al.*, 1957a; Cook and Pope, 1959). Toxoplasmosis has also been reported in one *Phascogale tapoatafa* and one *Dasyurus viverrinus* (Seddon, 1952). Antibody was not found in the wild specimens of *Macropus littoralis, M. robustus, M. giganteus, M. rufus, Wallabia bicolor,* and *Macrotis lagotis* that were examined by Cook and Pope (1959). However, toxoplasmosis has been described in zoological specimens of *Macropus rufogriseus frutica* (Ratcliffe and Worth, 1951), *M. rufus* (Hackel *et al.*, 1953), and *Macropus eugenii* (Dobos-Kovacs *et al.*, 1975). Hamerton (1933, 1934b) and Coutelen (1932) reported infections in two captive wombats (*Vombatus ursinus*) which were attributed to *Toxoplasma wenyoni*. Serologic evidence of toxoplasma infection in seven of 13 captive wallaroos and kangaroos was provided by Rieman *et al.* (1974).

A wide variety of pathologic changes associated with *Toxoplasma* infection have been observed. Myocarditis, characterized by focal necrosis, was common in most of the species examined. Brain lesions were also common and varied from mild encephalitis associated with perivascular cuffing to extensive necrosis. The lungs and pleurae of affected animals sometimes contained small, firm nodules, and were edematous. Splenomegaly and splenic nodules were observed in the bandicoots and a wallaby. Nodules and ulceration of the stomach were reported in a kangaroo. Nodules and necrotic foci in the liver, pancreas, adrenal glands, and abdominal lymph nodes were infrequently observed. Microscopically, lesions usually contained *Toxoplasma* pseudocysts or individual organisms. In some instances, no tissue reaction was present; however, acute lesions were generally characterized by monocytic infiltration. Monocytes, plasma cells, neutrophiles, fibroblasts, and necrosis were associated with chronic lesions.

Clinical signs associated with toxoplasmosis were infrequently observed. When present they were clearly referable to the underlying pathology. Weakness, listlessness, loss of weight, nasal discharge, dyspnea, seizures, and death were observed in two immature, captive *Isoodon obesulus*. Ataxia occurred in the captive *Vombatus ursinus*. Anorexia, weight loss, incoordination, and paralysis were described in three *Thylogale eugenii*. Clinical signs were not observed in free-ranging marsupials with *Toxoplasma* infection.

The mode of transmission of *Toxoplasma* among marsupials was not studied. However, attempts by Pope *et al.* (1957a) to isolate *Toxoplasma* from pools of helminths and ectoparasites obtained from infected bandicoots were negative. The finding of toxoplasmosis in two bandicoots which had recently left the pouch led Cook and Pope (1959) to suggest that infection probably occurred by aerosol transmission of nasal exudate by the mother. Although Australian strains of *Toxoplasma* were sometimes highly pathogenic to bandicoots, the field serology performed by Cook and Pope (1959) indicated that the recovery rate was high. Nevertheless, they consider the bandicoot an important reservoir of toxoplasmosis in Queensland. Ratcliffe and Worth (1951) found that active toxoplasmosis was uncommon among zoological specimens. They postulated that *Toxoplasma* organisms, which may be widely distributed in animals, produce detectable disease when malnutrition, concurrent infection, and subsequent interference with the immune mechanisms weaken the host. Based on investigation of an outbreak of toxoplasmosis in kangaroos, however, Dobos-Kovacs *et al.* (1975) advised that feed supplies be protected against contamination by domestic cats whose feces sometimes contain *Toxoplasma* oocysts.

## 2. *Sarcocystis* Infection

Several species of *Sarcocystis* have been reported in captive marsupials. Gilruth and Bull (1911) observed *Sarcocystis macropidis* infection in three wallabies (*Petrogale* sp.), two of which died with acute enteritis. Gross pathology was limited to the small intestine. The serosal surface was mottled with irregular areas of congestion and contained groups of circular, raised lesions with pale centers and hyperemic peripheries. Dark red nodules projected from the mucosa which was characterized by catarrhal inflammation. Microscopically, spherical, translucent cysts surrounded by fibrous tissue were seen in the submucosa. Hyperemia, hemorrhage, loss of mucosal epithelium and villi, and atrophy of the glands of Lieberkühn were also evident. Triffitt (1926) described the occurrence of sporelike bodies in large vacuoles within the muscle fibers of the small intestine of a wallaby (*M. rufogriseus frutica*) which may have been *Sarcocystis*. *Sarcocystis* infection was also reported by Scott (1927) in *M. rufogriseus frutica*, and Triffitt (1927) described the pathology associated with *Sarcocystis tenella* infection in that wallaby. Small white nodules (cysts) were embedded in the wall of the small and large intestines; some projected from the serosal surface and others extended from the mucosa. Microscopically, the cysts were located between the muscle layers and produced distension, but no tissue response. The cysts contained many sickle-shaped, sporelike bodies. *Sarcocystis bettongiae* was described from the short-nosed rat kangaroo, *Bettongia lesueur* (Bourne, 1934). Inasmuch as identification was based only on the host, this parasite should probably be called *Sarcocystis* sp. (Mandour, 1965). *Sarcocystis mucosa* which was reported from *Petrogale penicillatus* was later shown to be *Globidium mucosa* (Triffitt, 1927).

### 3. *Ileocystis* and *Lymphocystis* Infections

Hemorrhagic enteritis with petechiae involving the mucosa of the small intestine and to a lesser degree the cecum occurs in *Macropus* sp. and *M. rufogriseus frutica* associated with *Ileocystis macropodis* and *Lymphocystis macropodis* infections (Gilruth and Bull, 1911; Wenyon and Scott, 1925; Triffitt, 1926). The parasitic cysts are visible as numerous, irregular, granular, grayish elevations on the mucosal surface. *Ileocystis macropidis* ranges in size from small, uninuclear, spherical stages to large, oval, multinuclear, mature forms. The small forms are located in the epithelial and submucosal layers, while the mature stages are found only in the submucosa, either singly or in groups of three to four. *Lymphocystis macropodis* is observed principally in the lymphocytes of the submucosa. Affected lymphocytes are greatly hypertrophied by the large size of the parasite. A few free-lying spores may be observed in the intercellular spaces. Intestinal infection with *Ileocystis wombati* was described in *Lasrorhinus latifrons* by Gilruth and Bull (1911). Cysts were present in large numbers, particularly around the periphery of the villi which were greatly hypertrophied. The intestinal mucosa appeared papillomatous. Barretto (1940) suggested that *Ileocystis* should be included in the genus *Sarcocystis*.

### 4. *Globidium* Infections

An organism originally identified as *Balbiania mucosa* was described in the submucosa of the large intestine of *Petrogale penicillatus*. Although it was later classified as *Sarcocystis mucosa* (Coutelen, 1933), Triffitt (1927) considered the parasite to be *Globidium mucosum*. Infection with *Globidium* sp. was also reported by Scott (1927) in two captive *M. rufogriseus frutica*.

## 5. *Coccidia* Infections

Several species of *Coccidia* have been reported from captive Macropodidae at the London, Khartov, and Philadelphia Zoological Gardens. They include *Eimeria macropodis* from *Macropus rufogriseus frutica* (Prasad, 1960; Triffitt, 1926; Wenyon and Scott, 1925), *Eimeria marsupialium* and *Eimeria fausti* from *M. giganteus* (Yakimoff and Matschoulsky, 1936), and *Eimeria rufusi* from *Macropus rufus* (Prasad, 1960). Unidentified species of *Coccidia* were also reported from *M. rufogriseus frutica* by Scott (1927) and from *Macropus giganteus* by Fox (1932).

Mykytowycz (1964a,b) surveyed wild populations of *Macropus rufus* and *Macropus giganteus* in New South Wales and southern Queensland, Australia. *Eimeria cunnamullensis* infection was most commonly observed among *M. giganteus*; however, *Eimeria gungahlinensis, Eimeria purchasei, Eimeria kongoni*, and *Eimeria hestermani* infections were also encountered. *Macropus rufus* was found to harbor *Eimeria wilcanniensis* and, less frequently, *Eimeria toganmainensis*. Infection with *Eimeria* sp. was reported from *Macropus parryi* by Johnston (1916). The incidence of infection with various types of *Coccidia* ranged from 6 to 14% among 523 *M. rufus* and 26 to 70% among 157 *M. giganteus* and appeared to occur principally during the hot seasons. Coccidial infections among livestock are said to be enhanced by concentrating susceptible animals on small areas of ground, particularly where sufficient moisture is present to provide good conditions for oocyst sporulation (Hagan and Bruner, 1957). In this regard, the sedentary habits and densely vegetated environment of *M. giganteus*, as contrasted with the nomadic behavior and plains habitat of *M. rufus*, may have been related to the higher incidence of coccidiosis among the former (Mykytowycz, 1964a). In addition, increased utilization of watering places may have been associated with the high infection rates observed during the hot seasons. The severity and frequency of coccidiosis, greatest in yearling kangaroos, appear to decline and disappear with increasing age. This finding is consonant with observations of others that age resistance or immunity to reinfection develops in most animals (Mykytowycz, 1964a; Hagan and Bruner, 1957).

Kangaroos comprising the wild populations studied by Mykytowycz (1964a) were considered healthy. None of the animals examined had pathologic lesions related to coccidial infections. However, five yearling kangaroos which were housed with adults died from occcidiosis (Mykytowycz, 1964a). This episode was preceded by extensive flooding.

Pathologic changes associated with coccidial infection were limited to the small intestine. Acute pseudomembranous inflammation, loss of epithelial cells, and thickening of submucosal tissue was observed (Triffitt, 1926).

### 6. *Hemogregarina* Infections

Welsh *et al.* (1910) described intra- and extraerythrocytic forms of a parasite, *Hemogregarina dasyuri*, obtained from one of 50 *Dasyurus viverrinus* in New South Wales. Welsh and Dalyell (1910) also identified *Hemogregarina peramelis* (extracorporeal forms) from *P. nasuta* in New South Wales. The same authors indicated that *Hemogregarina pertauri* infection had been described in *Petaurus* sp.

### 7. *Babesia* Infections

*Isoodon obesulus* was described by Mackerras (1959b) as a host for *Babesia thylacis* in Queensland. Infections were experimentally transmissible among bandicoots.

### 8. *Theileria* Infections

Mykytowycz (1964b) found one of 122 *Macropus rufus* from Queensland infected with *Theileria* sp. *Isoodon obesulus, Perameles nasuta*, and *Potorous tridactylus* were also reported as hosts for *Theileria peramelis* in south Queensland (Mackerras, 1959b). Experimental infections resulted in anemia characterized by anisocytosis, polychromasia, and the presence of nucleated erythrocytes.

### 9. *Hemobartonella* Infections

In Queensland, *Isoodon obesulus* was identified as a host for *Hemobartonella* sp. (Mackerras, 1959b).

### 10. *Hepatozoon* Infections

Several marsupial species are naturally susceptible to *Hepatozoon* infections. *Hepatozoon perameles* was identified in *Perameles nasuta* and *Isoodon obesulus* in New South Wales and Queensland (Mackerras, 1959b). Ectoparasites obtained from the bandicoots that were examined for the presence of *Hepatozoon* sp. were negative. In another survey, Mykytowycz (1964b) found that 24 of the 122 *M. rufus* examined in Queensland had positive peripheral blood smears for *Hepatozoon* sp. and *H. petauri,* and *H.*

*dasyuri* infections were described in *Petaurus norfolcensis* and *Dasyurus viverrinus* (Mackerras, 1959b). *Pseudocheirus peregrinus* and *Dasyuroides byrnei* in Queensland are also hosts for *Hemogregarina pseudocheirri* and *Hemogregarina dasyuroides* (Mackerras, 1959b).

### 11. *Trypanosoma* Infections

Twelve of 82 *Isoodon obesulus* examined in Queensland had *Trypanosoma thylacis* infections (Mackerras, 1959b). The same authors also described *T.* sp. in the lymph and subcutaneous tissues of *I. obesulus*.

### 12. *Amoeba* Infections

Hackel *et al.* (1953) observed organisms associated with gastric ulcers in a captive *M. rufus* with toxoplasmosis which they described as amebic trophozoites. The organism was not identified and may have been a saprophyte.

## E. Mycotic Diseases

### 1. Systemic Mycoses

Plimmer (1915) reported mycotic pneumonias which contributed to the deaths of captive kangaroos and wallabies. *Aspergillus fumigatus, Aspergillus glaucus*, and *Aspergillus niger* were isolated from several wallabies including *Macropus rufogriseus frutica* (Rewell, 1948a; Scott, 1927). The disease appeared to occur among groups of animals and was thought to be transmitted by contact or through inoculation of wounds by infected feces, feed, or soil. Sanitation of feed bins reduced its occurrence (Scott, 1927). Some of the literature does not clearly differentiate between diseases caused by *Nocardia* sp. and those associated with true fungus infections (see Section III,B,1).

## F. Nutritional Diseases

### Vitamin E Deficiency

Muscular dystrophy associated with vitamin E deficiency was described in captive *Setonix brachyurus* by Kakulas (1961) and *Dendrolagus* sp. by

Goss (1940, 1941). Kakulas (1963) demonstrated that clinical signs also could be induced in *S. brachyurus* by overcrowding and close confinement. In *Dendrolagus*, the disease was associated with the addition of fish oil to a vegetable and legume diet.

Clinical signs of muscular dystrophy begin insidiously with loss of weight and weakness of the hind limbs and progress rapidly to complete paralysis. Electromyograms of affected *S. brachyurus* are characterized by myotonia, and electrocardiographic (ECG) changes include flattening and inversions of T waves, depression of S–T segments, widening of QRS complexes, and ventricular extrasystoles (Durack *et al.*, 1969). ECG abnormalities are probably the result of focal myocardial lesions or serum potassium imbalances. The disease terminates in death 2 days to several weeks following the onset of clinical signs.

Necropsy findings are limited to the striated musculature. Atrophy, edema, opacity, and yellow-white discoloration of the femoral and pelvic muscles are observed grossly. The histologic appearance of individual skeletal muscle fibers is characterized by swelling, necrosis, fragmentation, and loss of fibrillar structure. Proliferation of sarcolemnal cells, infiltration by lymphocytes, polymorphonuclear leukocytes, and plasma cells, and focal myocardial necrosis are also observed. One *S. brachyurus* with myopathy which was examined had hyperkalemia (7.8 mEq/liter; normal = 4.3 mEq/liter).

Oral administration of 200 to 600 mg of vitamin E per day for several days is sufficient to reverse the clinical signs of muscular dystrophy in *S. brachyurus* within 2 to 3 weeks. Vitamin E therapy also prevents the occurrence of clinical signs in *S. brachyurus* which are housed in small enclosures. $\alpha$- Tocopherol (25 mg orally) has been successfully employed in the treatment of *Dendrolagus* sp. with muscular dystrophy.

### G. Neoplastic Diseases

Carcinoma of the stomach is common among captive *Macropodidae*. Malignant gastric papillomas have been observed in three *Macropus rufus*; metastases to the liver, spleen, and kidney were present in two (Fox, 1912, 1924; Boyd, 1929). Metastatic lesions in the same organs were associated with adenocarcinoma of the stomach in two other *M. rufus* (Ratcliffe, 1933). Plimmer (1912) reported gastric carcinoma in *M. robustus* and a kangaroo with metastases involving the lymph nodes, colon, and liver. Urbain *et al.* (1949) mentioned the stomach of *M. rufogriseus frutica* as the site of an unspecified tumor.

Carcinoma of the rectum and an unspecified tumor of the liver were reported in two *Dasyurus viverrinus* (Ratcliffe, 1933; Scott, 1927). Pale yellow, circumscribed nodular metastases in the liver, spleen, and lung were associated with adenocarcinoma of the intestine in *Dasyurus maculatus* (Fox, 1912). Two pulmonary neoplasms were reported: carcinoma with gastric metastases in *M. rufus* and adenocarcinoma in *Macrotis lagotis* (Fox, 1923s). Other reported neoplasms included carcinoma of the bladder in a Papuan flying phalanger (Fox, 1929) and squamous cell carcinoma of the perineum in *Sarcophilus harrisii* (Ratcliffe, 1933).

## H. Diseases of Specific Organ Systems

A variety of lesions involving specific organs or systems have been reported in Australian marsupials. The majority of the diseases occurred in zoological display specimens and were diagnosed at necropsy. Table VI lists the number of cases reported for each disease entity; data concerning total populations were not available.

### Arteriosclerosis and Atherosclerosis

The results of 365 necropsies reported by Fox (1933) indicated that 139 marsupials had arteriosclerosis. Affected families were Macropodidae, Peramelidae, and Dasyuridae. Statistically, these animals lived only two-thirds of their expected life spans in captivity. Lesions were generally found in the thoracic aorta and consisted of medial degeneration and calcium deposition. The intima was occasionally thickened. Coronary arteriosclerosis was observed in 16 to 27 Macropodidae examined by Ratcliffe (1965); aortic lesions, fibrosis, and infarction were also found. Individual cases of arteriosclerosis were reported in *Sarcophilus harrisii, Macropus* sp. (Fox, 1923q), and *Macropus rufogriseus frutica* (Heath and Whitehead, 1963).

Atherosclerosis, characterized by small aortic intimal plaques and nodules, was described in three wallabies (Rigg *et al*., 1960). Fat droplets were observed at the periphery of the plaques, and degenerative changes with calcium deposition occurred at the junction of the media and intima.

A high incidence of atherosclerosis was reported in a laboratory colony of rat kangaroos (*Potorous* sp. ?). Four of 10 animals died of the disease and each had elevated blood cholesterol levels (Anonymous, 1971).

## I. Ectoparasites

Table VII lists the arthropod parasites of Australasian marsupials and their geographic locations. Although infestations appear to be common, the

**TABLE VI**

Specific Anatomic Lesions Observed in Australasian Marsupials

| Lesion | Marsupial | Number of cases | References |
|---|---|---|---|
| Digestive System | | | |
| Dentigenous cyst | *Wallabia bicolor* | 1 | Hamerton (1939) |
| Molar malocclusion | Wombat | 1 | Beddard (1902) |
| Gastric ulcers | *Macropus rutogriseus frutica* | 1 | Hamerton (1939) |
| | Kangaroos | 8 | Fox (1923m) |
| Gastric impaction (hair ball) | *Dendrolagus matschiei* | 1 | Hamerton (1941) |
| Gastroenteritis | *Vombatus ursinus* | 1 | Goss (1941) |
| | *Macropus giganteus* | 1 | Goss (1941) |
| Enteritis | *Macropus giganteus* | 1 | Crisp (1860) |
| | *M. rufogriseus frutica* | 2 | Urbain *et al.* (1949) |
| | *M. rufogriseus frutica* | 1 | Hamerton (1935a) |
| | *M. rufus* | 1 | Hamerton (1933) |
| | *Trichosurus vulpecula* | 2 | Hamerton (1932, 1934a) |
| | *Pseudochireus peregrinus* | 1 | Hamerton (1941) |
| | Phalanger | 1 | Fox (1923n) |
| Duodenal ulcers | *Sarcophilus harrisii* | 2 | Hamerton (1935a, 1938) |
| | *Macropus eugenii* | 1 | Hamerton (1935a) |
| | *Wallabia bicolor* | 1 | Hamerton (1934a) |
| Impaction of the colon | *M. rufus* | 1 | Hamerton (1942) |
| | *Dendrolagus matschiei* | 1 | Hamerton (1941) |
| Colonic ulcers | *Trichosurus vulpecula* | 1 | Hamerton (1934a) |
| Intussusception | *Macropus giganteus* | 2 | Hamerton (1937, 1939) |
| | Rock wallaby (*Petrogale* sp.) | 1 | Scott (1927) |
| | *Sarcophilus harrisii* | 1 | Scott (1927) |
| Intestinal strangulation | *Macropus fuliginosus* | 1 | Garrod (1873) |
| Pancreatitis | Phalanger | 1 | Fox (1923n) |
| | *Sarcophilus harrisii* | 1 | Fox (1923n) |
| | *Pseudocheirus peregrinus* | 1 | Hamerton (1941) |

**TABLE VI (Cont.)**

| Lesion | Marsupial | Number of cases | References |
|---|---|---|---|
| Hepatitis | *Macropus robustus* | 1 | Hamerton (1942) |
| Cirrhosis | *M. rufus* | 1 | Hamerton (1935a) |
| | *Wallabia bicolor* | 1 | Fox (1914) |
| Fatty degeneration | *Petaurus breviceps* | 1 | Hamerton (1936) |
| Cholecystitis | *Sarcophilus harrisii* | 1 | Fox (1923o) |
| Choledochitis | *Wallabia bicolor* | 1 | Fox (1914) |
| Fat necrosis | Phalanger | 1 | Fox (1923n) |
| | *Sarcophilus harrisii* | 1 | Fox (1923n) |
| Peritonitis | Wombat | 1 | Beddard (1902) |
| Urogenital System | | | |
| Nephritis | *Wallabia bicolor* | 1 | Hamerton (1934a) |
| | *M. rufus* | 1 | Hamerton (1935a) |
| | *Petaurus breviceps* | 2 | Hamerton (1936, 1938) |
| | *Sarcophilus harrisii* | 1 | Fox (1923) |
| Renal calculi | Wallaby | 2 | Plimmer (1915, 1916) |
| | *Sarcophilus harrisii* | 1 | Fox (1923o) |
| Cystic calculi | *Dendrolagus inustus* | 1 | Hamerton (1933) |
| | *Sarcophilus harrisii* | 1 | Plimmer (1912) |
| Renal hypoplasia | *Wallabia bicolor* | 1 | Scott (1925) |
| Respiratory System | | | |
| Pneumonia (abscesses) | *Macropus robustus* | 1 | Fox (1923p) |
| | *Sarcophilus harrisii* | 1 | Fox (1923p) |
| Bronchopneumonia | *Trichosurus vulpecula* | 2 | Hamerton (1934a, 1936) |
| Acute lobar pneumonia | *Sarcophilus harrisii* | 1 | Hamerton (1936) |
| | *Macropus rufus* | 1 | Hamerton (1936) |
| | *M. robustus* | 1 | Hamerton (1936) |
| | *Thylogale thetis* | 1 | Hamerton (1936) |
| Pleural effusion and atelectasis | *Dendrolagus matschiei* | 1 | Hamerton (1936) |
| Purulent rhinitis | *Lasiorhinus latifrons* | 1 | H. Fox (1940) |

| | | | |
|---|---|---|---|
| Cardiovascular System | | | |
| Patent foramen ovale | Wallaby | 1 | Scott (1928) |
| Heart failure (right) | *Dendrolagus matschiei* | 1 | Hamerton (1936) |
| Atherosclerosis | Wallaby | 3 | Rigg *et al*. (1960) |
| | Rat kangaroo | 4(10)[a] | Anononmous (1971) |
| | *Dendrolagus inustus* | 1 | Hamerton (1936) |
| Arteriosclerosis | *Sarcophilus harrisii* | 1 | Fox (1923q) |
| | *Macropus* sp. | 2 | Fox (1923q) |
| | *M. rufogriseus frutica* | 1 | Heath and Whitehead (1963) |
| Musculoskeletal System | | | |
| Myositis (chronic) | *Wallabia bicolor* | 1 | Fox (1914) |
| Abscess of sphenomaxillary fossa | *Vombatus ursinus* | 1 | Hamerton (1945) |
| Tibiotarsal fractures | *Macropus rufogriseus frutica* | 1 | Brick (1970) |
| Fracture of leg | *Macropus giganteus* | 1 | Urbain *et al*. (1947) |
| Spondylosis | "Marsupials" | 16 | du Boulay *et al*. (1972) |
| Hematopoietic System | | | |
| Tonsilitis | *Macropus agilis* | 1 | Hamerton (1937) |
| Splenitis (chronic) | *W. bicolor* | 1 | Fox (1914) |
| Central Nervous System | | | |
| Posterior paralysis | *Macropus giganteus* | 1 | Rewell (1948c) |
| | *Sarcophilus harrisii* | 1 | Goss (1941) |
| Brain abscess | *Wallabia bicolor* | 1 | Hamerton (1942) |
| Organs of Special Sense | | | |
| Cataract, bilateral | *Sarcophilus harrisii* | 1 | Hamerton (1935a) |
| Middle ear infection | *Lasiorhinus latifrons* | 1 | H. Fox (1940) |
| | *Wallabia bicolor* | 1 | Hamerton (1942) |
| Endocrine System | | | |
| Thyroiditis (acute) | *Vombatus ursinus* | 1 | Hamerton (1935a) |
| Skin and Adnexae | | | |
| Albinism | *Echymipera kalubu* | 1 | Phillips and Wilson (1965) |
| | *Macropus rufogriseus frutica* | 1 | Hamerton (1935a) |
| Exfoliative dermatitis | *Wallabia bicolor* | 1 | Hamerton (1934a) |
| Inflammation of pouch | *Macropus rufus* × *M. fuliginosus* | 1 | Hamerton (1932) |
| Perineal abscess | *Sarcophilus harrisii* | 1 | Hamerton (1932) |

[a] The figure in parentheses indicates the total number of marsupials examined.

**TABLE VII**

Arthropod Parasites of Australasian Marsupials

| Parasite | Host | Location | References |
|---|---|---|---|
| Acarina | | | |
| Ticks | | | |
| *Argas gurneyi-warburtoni* | Kangaroo | *[a] | Seddon (1943) |
| *Boophilus australis* | Kangaroo | * | Seddon (1943) |
| *Haemaphysalis numerosa* | *Aepyprymnus rufescens* | N. and E. Queensland, Moreton Island | Derrick *et al*. (1939) |
| | *Isoodon obesulus* | N. and E. Queensland, Moreton Island | Heaslip (1941) |
| | *I. macrourus* | N. and E. Queensland, Moreton Island | Derrick (1944) |
| | *Perameles nasuta* | N. and E. Queensland, Moreton Island | Fenner (1946) |
| | *Thylogale stigmatica* | N. and E. Queensland, Moreton Island | Fenner (1946) |
| | *Trichosurus vulpecula* | N. and E. Queensland, Moreton Island | Fenner (1946) |
| | Bandicoots | N. and E. Queensland, Moreton Island | Fenner (1946) |
| *Haemaphysalis novaeguineae* | *Perameles* sp. | New Guinea | Hirst (1913) |
| | Kangaroo | New Guinea | Hirst (1913) |
| *Haemaphysalis spinigera novaeguineae* | *Isoodon macrourus* | N. Queensland | Derrick *et al*. (1939) |
| *Ixodes holocyclus* | *Aepyprymnus rufescens* | N. and E. Queensland, Moreton Island, New South Wales | Derrick *et al*. (1939) |
| | *Isoodon obesulus* | N. and E. Queensland, Moreton Island, New South Wales | Heaslip (1941) |
| | *I. macrourus* | N. and E. Queensland, Moreton Island, New South Wales | Smith (1942) |

| | | | |
|---|---|---|---|
| | *Perameles nasuta* | N. and E. Queensland, Moreton Island, New South Wales | Seddon (1943) |
| | *Isoodon obesulus* | N. and E. Queensland, Moreton Island, New South Wales | Fenner (1946) |
| | *Trichosurus vulpecula* | N. and E. Queensland, Moreton Island, New South Wales | Pope *et al.* (1957b) |
| | Bandicoots | N. and E. Queensland, Moreton Island, New South Wales | Mackerras (1959b) |
| *Ixodes tasmani* | *Aepyprymnus rufescens* | N. Queensland, New South Wales | Fenner (1946) |
| | *Isoodon obesulus* | N. Queensland, New South Wales | Pope *et al.* (1957) |
| | *Schoinobates volans* | N. Queensland, New South Wales | Moorehouse (1969) |
| | *Trichosurus vulpecula* | N. Queensland, New South Wales | Moorehouse (1969) |
| *Ornithodorus gurneyi* | Kangaroo | * | Hirst (1913) |
| Acarina | | | |
| Mites | | | |
| *Ascoschongastia accola* | *Phalanger vestitus* | New Guinea | Nadchatram (1970) |
| | *Phalanger* sp. | New Guinea | Nadchatram (1970) |
| | *Pseudochireus corrinae* | New Guinea | Nadchatram (1970) |
| | *Pseudochireus cupreus* | New Guinea | Nadchatram (1970) |
| | *Pseudochireus forbesi* | New Guinea | Nadchatram (1970) |
| | *Pseudochireus* sp. | New Guinea | Nadchatram (1970) |
| *Demodex* sp. | *Antechinus stuartii* | New South Wales | Nutting and Wooley (1965) |
| *Guntheria bipygalis* | *Echymipera kaluba* | Queensland, New Guinea | Gunther (1941a) |
| | *Isoodon macrourus* | Queensland, New Guinea | Heaslip (1941) |
| | *Peroryctes raffrayanus* | Queensland, New Guinea | Heaslip (1941) |
| *Heterolaelaps antipodidnum* | *Isoodon macrourus* | Queensland | Derrick *et al.* (1939) |
| *Hypoapsis perameles* | *Isoodon macrourus* | Queensland | Derrick *et al.* (1939) |
| *Laelaptidae* | *Echymipera rufescens* | Netherlands, New Guinea | Vogtman and Fitzwater (1951) |
| *Laelaps lagotisinus* | *Macrotis lagotis* | W. Australia | Hirst (1931) |

**TABLE VII (Cont.)**

| Parasite | Host | Location | References |
|---|---|---|---|
| *Leeuwenhoekia australiensis* | *Isoodon macrourus* | Queensland | Heaslip (1941) |
| *Listrophoroides expansus* | *Isoodon macrourus* | N. Queensland | Heaslip (1941) |
| *Mesolaelaps anomalus* | *Isoodon macrourus* | Queensland | Pope *et al.* (1957b) |
| *Myobia ensifera* | *Isoodon macrourus* | N. Queensland | Heaslip (1941) |
| *Neoschongastia cairnsensis* | *Isoodon macrourus* | N. Queensland | Fenner (1946) |
| | *Perameles nasuta* | N. Queensland | Fenner (1946) |
| *Neoschongastia coorongense* | *Isoodon macrourus* | N. Queensland | Heaslip (1941) |
| *Neoschongastia derricki* | *Isoodon macrourus* | N. Queensland | Heaslip (1941) |
| *Neoschongastia hirsti* | *Isoodon macrourus* | N. Queensland | Fenner (1946) |
| | *Trichosurus vulpecula* | N. Queensland | Fenner (1946) |
| *Neoschongastia impar* | *Echymipera rufescens* | Netherlands, New Guinea | Vogtman and Fitzwater (1951) |
| *Neoschongastia indica* | *Isoodon macrourus* | N. Queensland | Heaslip (1941) |
| *Neoschongastia isoodon* | *Isoodon macrourus* | Queensland | Derrick *et al.* (1939) |
| *Neoschongastia kallipypos* | *Isoodon macrourus* | Queensland | Derrick *et al.* (1939) |
| *Neoschongastia perameles* | *Echymipera rufescens* | N. Queensland, Netherlands, New Guinea | Fenner (1946) |
| | *Isoodon obesulus* | N. Queensland, Netherlands, New Guinea | Vogtman and Fitzwater (1951) |
| *Neoschongastia* sp. | *Perameles nasuta* | N. Queensland | Fenner (1946) |
| Parasid mites | *Aepyprymnus rufescens* | N. Queensland | Fenner (1946) |
| | *Isoodon obesulus* | N. Queensland | Fenner (1946) |
| | *Trichosurus vulpecula* | N. Queensland | Fenner (1946) |
| *Peramelaelaps bandicoota* | *Isoodon obesulus* | Queensland | Pope *et al.* (1957b) |
| *Schongastia taylori* | *Thylogale stigmatica* | New Guinea | Gunther (1941b) |
| *Trombicula deljensis* (= *akamushi*) | *Isoodon macrourus* | N. Queensland | Heaslip (1941) |

| | | | |
|---|---|---|---|
| *Trombicula minor* | *Isoodon obesulus* | N. Queensland | Heaslip (1941) |
| | *Isoodon macrourus* | N. Queensland | Fenner (1946) |
| | *Perameles nasuta* | N. Queensland | Fenner (1946) |
| | *Trichosurus vulpecula* | N. Queensland | Fenner (1946) |
| *Trombicula* sp. | *Isoodon obesulus* | N. Queensland | Heaslip (1941) |
| | *Isoodon macrourus* | N. Queensland | Fenner (1946) |
| | *Perameles nasuta* | N. Queensland | Fenner (1946) |
| | *Trichosurus vulpecula* | N. Queensland | Fenner (1946) |
| *Walachia* sp. | *Isoodon macrourus* | Queensland | Heaslip (1941) |
| Mallophaga | | | |
| Lice | | | |
| *Boopia bettongia* | *Aepyprymnus rufescens* | * | Hopkins (1949) |
| *Boopia dubia* | *Lasiorhinus latifrons* | * | Hopkins (1949) |
| *Boopia grandis* | *Macropus rufus* | * | Hopkins (1949) |
| *Boopia minuta* | *Macropus dorsalis* | * | Hopkins (1949) |
| *Boopia mjobergi* | *Macropus rufus* | * | Hopkins (1949) |
| *Boopia notafusca* | *Macropus giganteus* | Victoria | Harrison and Johnston (1916) |
| | *Macropus robustus* | Victoria | Hopkins (1949) |
| | *Wallabia bicolor* | Victoria | Hopkins (1949) |
| *Boopia phanocerata* | *Perameles nasuta* | New South Wales | Harrison and Johnston (1916) |
| | *Isoodon macrourus* | New South Wales | Hopkins (1949) |
| *Boopia spinosa* | *Wallabia bicolor* | Victoria | Harrison and Johnston (1916) |
| | *Phascogale tapoatafa* | Victoria | Hopkins (1949) |
| *Boopia tarsata* | *Vombatus ursinus* | London Zoo | Cummings (1916); Hopkins (1949) |
| | *Wallabia bicolor* | London Zoo | Hopkins (1949) |
| *Boopia uncinata* | *Dasyurus geoffroii* | New South Wales | Harrison and Johnston (1916) |
| | *Dasyurus hallucatus* | New South Wales | Hopkins (1949) |
| | *Dasyurus maculatus* | New South Wales | Hopkins (1949) |
| | *Dasyurus viverrinus* | New South Wales | Hopkins (1949) |
| | *Isoodon macrourus* | New South Wales | Hopkins (1949) |
| *Boopia* sp. | *Aepyprymnus rufescens* | N. Queensland | Fenner (1946) |
| | *Isoodon obesulus* | N. Queensland | Pope *et al.* (1957b) |
| | *Isoodon macrourus* | N. Queensland | Pope *et al.* (1957b) |
| | *Perameles nasuta* | N. Queensland | Pope *et al.* (1957b) |

**TABLE VII (Cont.)**

| Parasite | Host | Location | References |
|---|---|---|---|
| *Dendrolagia pygidialis* | *Dendrolagus lumholtzi* | * | Hopkins (1949) |
| *Heterodoxus forcipatus* | *Thylogale billardieri* | * | Hopkins (1949) |
| | *Thylogale stigmatica* | * | Hopkins (1949) |
| | *Thylogale thetis* | * | Hopkins (1949) |
| *Heterodoxus longitarsus* | *Macropus rufogriseus frutica* | New South Wales, Victoria, Queensland | Harrison and Johnston (1916) |
| | *Macropus dorsalis* | New South Wales, Victoria, Queensland | Hopkins (1949) |
| | *Macropus giganteus* | New South Wales, Victoria, Queensland | Hopkins (1949) |
| | *Macropus robustus* | New South Wales, Victoria, Queensland | Harrison and Johnston (1916) |
| | *Macropus rufogriseus* | New South Wales, Victoria, Queensland | Hopkins (1949) |
| | *Macropus rufus* | New South Wales, Victoria, Queensland | Hopkins (1949) |
| | *Thylogale stigmatica* | New South Wales, Victoria, Queensland | Hopkins (1949) |
| | *Wallabia bicolor* | New South Wales, Victoria, Queensland | Hopkins (1949) |
| | *Macropus agilis* | New South Wales, Victoria, Queensland | Hopkins (1949) |
| *Heterodoxus macropus* | *Wallabia bicolor* | Queensland, Victoria | Le Souef and Bullier (1902) |
| | Kangaroos | Queensland, Victoria | Hopkins (1949) |
| | Wallabies | Queensland, Victoria | Hopkins (1949) |
| *Heterodoxus spiniger* | *Wallabia bicolor* | * | Hopkins (1949) |
| *Latumcephalum lesouefi* | *Wallabia bicolor* | Victoria | Harrison and Johnston (1916) |

| | | | |
|---|---|---|---|
| *Latumcephalum macropus* | *Wallabia bicolor* | Queensland | Harrison and Johnston (1916) |
| | *Macropus dorsalis* | Queensland | Hopkins (1949) |
| *Paraboopia flava* | *Macropus robustus* | * | Hopkins (1949) |
| *Paraheterodoxus insignis* | *Aepyprymnus rufescens* | * | Harrison and Johnston (1916) |
| | *Wallabia bicolor* | * | Hopkins (1949) |
| *Phascogalia brevispinosa* | *Antechinus flavipes* | * | Hopkins (1949) |
| Siphonaptera | | | |
| Fleas | | | |
| *Acanthopsylla povida* | *Dendrolagus lumholtzi* | Queensland, New South Wales | Jordan and Rothschild (1922) |
| | *Thylogale thetis* | Queensland, New South Wales | Ferris (1924) |
| | *Petaurus breviceps* | Queensland, New South Wales | Ferris (1924) |
| | *Antechinus flavipes* | Queensland, New South Wales | Ferris (1924) |
| | *Pseudocheirus herbertensis* | Queensland, New South Wales | Ferris (1924) |
| | *Pseudocheirus peregrinus* | Queensland, New South Wales | Ferris (1924) |
| *Acanthopsylla rothschildi* | *Dasyurus maculatus* | Zoological Gardens, New South Wales | Jordan and Rothschild (1922) |
| *Acanthopsylla rothschildi nereis* | *Potorous tridactylus* | Clark Island, Bass Straits | Jordan and Rothschild (1922) |
| *Acanthopsylla rothschildi rainbowi* | *Dasyurus viverrinus* | Queensland, New South Wales | Jordan and Rothschild (1922) |
| *Acanthopsylla rothschildi victoriana* | *Antechinus swainsonii* | Victoria | Jordan and Rothschild (1922) |
| *Acanthopsylla saphes* | native "cat" (*Dasyurus* sp.) | Victoria | Jordan and Rothschild (1922) |
| *Choristopsylla ochi* | *Trichosurus vulpecula* | Tasmania | Jordan and Rothschild (1922) |
| *Choristopsylla thomasi* | *Acrobates pygmaeus* | * | Ferris (1924) |
| *Choristopsylla tristis* | *Acrobates pygmaeus* | Victoria | Jordan and Rothschild (1922) |
| | *Petaurus australis* | | Jordan and Rothschild (1922) |
| *Macroppsylla hercules* | *Antechinus flavipes* | New South Wales | Ferris (1924) |
| *Pygiopsylla congrua* | *Isoodon obesulus* | Tasmania | Jordan and Rothschild (1922) |
| *Pygiopsylla hilli* | *Bettongia penicillata* | W. Australia | Jordan and Rothschild (1922) |
| *Pygiopsylla hoplia* | *Dasyurus maculatus* | Queensland, New South Wales | Jordan and Rothschild (1922) |
| | *Perameles nasuta* | Queensland, New South Wales | Jordan and Rothschild (1922) |
| | *Isoodon obesulus* | Queensland, New South Wales | Jordan and Rothschild (1922) |
| | *Potorous tridactylus* | Queensland, New South Wales | Jordan and Rothschild (1922) |

**TABLE VII (Cont.)**

| Parasite | Host | Location | References |
| --- | --- | --- | --- |
| *Pygiopsylla rainbowi* | *Perameles nasuta* | Victoria, New South Wales | Jordan and Rothschild (1922) |
| | *Antechinus swainsonii* | Victoria, New South Wales | Jordan and Rothschild (1922) |
| *Pygiopsylla zethi* | *Bettongia gaimardi* | Victoria, New South Wales, Tasmania | Jordan and Rothschild (1922) |
| | *Dasyurus viverrinus* | Victoria, New South Wales, Tasmania | Jordan and Rothschild (1922) |
| | *Perameles nasuta* | Victoria, New South Wales, Tasmania | Jordan and Rothschild (1922) |
| | *Isoodon obesulus* | Victoria, New South Wales, Tasmania | Jordan and Rothschild (1922) |
| *Pygiopsylla* sp. | *Aepyprymnus rufescens* | N. Queensland | Derrick *et al.* (1939) |
| | *Isoodon obesulus* | N. Queensland | Fenner (1946) |
| | *Perameles nasuta* | N. Queensland | Pope *et al.* (1957b) |
| | *Trichosurus vulpecula* | N. Queensland | Pope *et al.* (1957b) |
| *Stephanocircus dasyuri* | *Dasyurus viverrinus* | Queensland, Victoria | Fleay (1935) |
| | *Isoodon macrourus* | Queensland, Victoria | Derrick *et al.* (1939) |
| *Stephanocircus* sp. | *Aepyprymnus rufescens* | N. Queensland | Fenner (1946) |
| | *Isoodon obesulus* | N. Queensland | Pope *et al.* (1957b) |
| | *Perameles nasuta* | N. Queensland | Pope *et al.* (1957b) |
| *Stivalins novaeguineae* | *Peroryctes raffrayanus* | New Guinea Admiralty Islands | Jordan and Rothschild (1922) |
| *Stivialius*-like sp. | *Aepyprymnus rufescens* | N. Queensland | Fenner (1946) |
| | *Perameles nasuta* | N. Queensland | Fenner (1946) |
| Diptera | | | |
| Flies | | | |
| *Hippobosca tasmanica* | *Macropus rufogriseus* | Tasmania | Wesche (1903) |
| *Oestrus macropi* (larvae) | *Macropus* sp. | New South Wales | Froggatt (1913) |
| *Ortholfersia raveni* | *Macropus* sp. | New South Wales | Ferris (1924) |
| *Ortholfersia tasmanica* | *Macropus rufogriseus* | N. Queensland, Tasmania | Ferris (1924) |
| | *Thylogale billardieri* | N. Queensland, Tasmania | Ferris (1924) |
| *Tachyomyia macropi* (larvae) | *Macropus rufus* | * | Mykytowycz (1964b) |

[a] Asterisk means location not given.

significance of the majority of the host–parasite relationships is not clear. The role of ticks as vectors of Q fever and north Queensland tick typhus was discussed previously (see Section III,C,1 and 2).

### 1. Demodicosis

Nutting and Wooley (1965) described demodicosis in a colony of *Antechinus stuartii* (marsupial mice) obtained from New South Wales. Cutaneous nodules (3–24 mm diameter) were observed in 13 of 114 animals housed under laboratory conditions 2 to 22 months after capture. The smooth surfaced, generally solid nodules were usually located anterior to the pouch, and on the outer aspect of the tibiatarsal joint or hind legs; less frequently the skin covering the angle of the jaw, base of the tail, top of the foot, or on the neck behind the ears was affected. Cut sections of the lesions were creamy white and appeared honeycombed; cavitations contained mites (*Demodex* sp.) in all stages of development. In other areas of skin, solitary mite invasions of single hair follicles were observed. Two *A. stuartii* obtained directly from the wild were examined: one had a low grade mite infestation and the other was negative. It was believed that dietary or environmental factors may have contributed to the development of grossly visible lesions in the captive animals. No marked clinical signs were observed in affected animals.

### 2. Tick Paralysis

A paralytic condition was induced in bandicoots (*Isoodon macrourus*) from tick-free regions by feeding larval, nymphal, and adult ticks (*Ixodes holocyclus*) on them (Smith, 1942). Clinical signs of tick paralysis occurred 7 to 14 days after exposure and included paralysis of the hind or all legs, respiratory paralysis, and recovery or death. Paralysis could be caused by as few as one tick.

## J. Helminth Parasites

Numerous species of helminths have been reported from a wide spectrum of wild and captive Australasian marsupials (Table VIII). Nematodes are most frequently encountered followed by the cestodes, trematodes, and acanthocephalids. Recently, comprehensive surveys of the parasites of Macropodidae appeared (Mackerras, 1959a; Mykytowycz, 1964b;

**TABLE VIII**

Helminth Parasites of Australasian Marsupials

| Name | Host | Habitat | Location | References |
|---|---|---|---|---|
| Nematodes | | | | |
| Oxyuroidea | | | | |
| *Austroxyuris finlaysoni* | *Schoinobates volans* | Intestine, cecum | Queensland, Australia | Johnston and Mawson (1938b) |
| *Passalurus parvus* | *Schoinobates volans* | Intestine, cecum | Queensland, Australia | Johnston and Mawson (1938b) |
| *Subulura* (2 species) | *Macrotis leucura, Perameles nasuta, P. bougainville* | Stomach, intestine | New South Wales, Central and South Australia | Johnston and Mawson (1939a,c, 1940a) |
| *Syphacia trichosuri* | *Trichosurus vulpecula* | Intestine | Queensland, Australia | Johnston and Mawson (1938b) |
| *Oxyuris* (3 species) | *Schoinobates volans, Potorous tridactylus* | Intestine, cecum | Queensland, Australia | Johnston and Mawson (1938b) |
| Ascaroidea | | | | |
| *Ascaris* sp. | *Perameles nasuta* | *[a] | New South Wales, Australia | Johnston (1909); Imperial Bureau (1933) |
| *Baylisascaris tasmaniensis* | *Sarcophilus harrisii, Dasyurus viverrinus, Dasyurus maculatus* | Stomach, duodenum | Tasmania | Sprent, (1970) |
| *Contracecum erraticum* | Wallaby | * | *[b] | Johnston and Mawson (1940b) |
| Rhabdidasoidea | | | | |
| *Strongyloides thylacis* | *Isoodon obesulus, Macropus rufus* | Stomach, small intestine | Queensland, Australia | Winter (1958); Mackerras (1959); Mykytowycz (1964b) |
| *Parastrongyloides* (2 species) | *Trichosurus vulpecula, Isoodon obesulus, Perameles nasuta* | Small intestine | Queensland, Australia | Mackerras (1959a) |
| Strongyloidea | | | | |
| *Strongylidae* | | | | |
| *Globocephaloides* (4 species) | *Macropus dorsalis, Thylogale thetis, Wallabia bicolor* | Duodenum, intestine | Queensland, Australia; London Zoological Gardens, England | Imperial Bureau (1933); Johnston and Mawson (1939a); Rewell (1948a) |

| | | | | |
|---|---|---|---|---|
| *Oesophagostomoides giltneri* | *Vombatus ursinus* | Large intestine | National Zoo, District of Columbia | Schwartz (1928) |
| *Zoniolaimus* (16 species) | *Wallabia bicolor, M. parryi, M. giganteus, M. dorsalis, M. fuliginosus, Thylogale billardieri, M. eugenii, M. robustus, Wallabia bicolor, Thylogale wilcoxi, M. rufus, M. irma, M. rufogriseus, Lagorchestes hirsutus, Onychogalea fraenata, Thylogale thetis, Potorous tridactylus* | Stomach | Queensland, New South Wales, Victoria, and West Australia | Johnston (1909); Imperial Bureau (1933); Johnston and Mawson (1939a,c, 1940b); Mykytowycz (1964b) |
| *Parazoniolaimus collaris* | *Macropus parma, Wallabia bicolor* | * | Victoria, Australia | Johnston and Mawson (1939c, 1940a) |
| *Buccostrongylus* (5 species) | *Macropus dorsalis, M. parryi, M. parma, M. rufogriseus, Thylogale thetis, T. stigmatica* | Stomach | Queensland and Victoria, Australia | Baylis (1934b); Johnston and Mawson (1939a,c) |
| *Cloacina* (32 species) | *Petrogale penicillata, Macropus rufus, M. robustus, M. agilis, M. giganteus, M. dorsalis, M. rufogriseus, Wallabia bicolor, Thylogale thetis, M. parma, M. parryi, Potorous tridactylus, Thylogale brunii, M. rufus, M. eugenii, Isoodon obesulus* | Stomach | Queensland, Victoria, West, Central, and South Australia; Dutch New Guinea; Bismarck Archipelago; and Philadelphia Zoological Gardens | Johnston (1909); Imperial Bureau (1933); Johnston and Mawson (1938a, 1939a,b,c, 1940a,b); Mykytowycz (1964b) |
| *Coronostrongylus coronatus* | *Macropus rufus, Macropus parma, M. rufogriseus, Thylogale thetis* | Stomach | Queensland and New South Wales, Australia | Johnston and Mawson (1939a, 1940a); Mykytowycz (1964b) |
| *Cyclostrongylus medioannulatus* (4 species) | *Thylogale thetis* | Stomach | Queensland, Australia | Johnston and Mawson (1940a) |

**TABLE VIII (Cont.)**

| Name | Host | Habitat | Location | References |
|---|---|---|---|---|
| *Labiostrongylus* (5 species) | *Macropus agilis, M. robustus, M. rufus, Wallabia bicolor, Petrogale penicillata* | Stomach | Queensland, New South Wales, West and South Australia; London Zoological Gardens, England | Baylis (1927); Imperial Bureau (1933); Johnston and Mawson (1938c); Rewell (1948c); Mykytowycz (1964b); Dudzinski and Mykytowycz (1965); Mykytowycz and Dudzinski (1965) |
| *Macropostrongylus* (9 species) | *Macropus agilis, M. robustus, M. dorsalis, Dorcopis veterum, M. rufus, M. irma* | Stomach, cecum | Queensland, New South Wales, West Australia; and Dutch New Guinea | Baylis (1927, 1940); Davey and Wood (1938); Johnston and Mawson (1940b); Mykytowycz (1964b) |
| *Pharyngostrongylus* (15 species) | *Macropus dorsalis, M. parma, M. rufogriseus, Thylogale billardieri, Wallabia bicolor, M. giganteus, M. parryi, M. robustus, M. rufus, Thylogale thetis, M. bernardi, Petrogale penicillata, M. eugenii, M. irma* | Esophagus, stomach | Queensland, Victoria, New South Wales, Central, West, and South Australia, London Zoological Gardens, England | Imperial Bureau (1933); Johnston and Mawson (1938b, 1939a,b,c, 1940a,b, 1942); Rewell (1948c) |
| *Rugopharynx* (=*Spirostrongylus*) (2 species) | *Wallabia bicolor* | Stomach | Queensland and New South Wales, Australia | Imperial Bureau (1933); Mykytowycz (1964b) |
| *Papillostrongylus labiatus* | *Macropus rufus, Macropus dorsalis* | Stomach | Queensland and New South Wales, Australia | Johnston and Mawson (1939a); Mykytowycz (1964b) |
| *Potorstrongylus finlaysoni* | *Potorous tridactylus* | Intestine | Victoria, Australia | Johnston and Mawson (1939c) |
| *Phascolostrongylus turleyi* | *Vombatus ursinus* | Colon | Philadelphia Zoological Gardens | Canavan (1931) |

| | | | | |
|---|---|---|---|---|
| Trichostrongylidae | | | | |
| *Austrostrongylus thylogale* | *Setonix brachyurus* | Intestine | Rottnest Island, Western Australia | Waring (1956) |
| *Austrostrongylus* (5 species) | *Wallabia bicolor, M. dorsalis, M. rufogriseus frutica* | Duodenum, upper small intestine, intestine | New South Wales and Queensland, Australia; Texas | Chandler (1924); Cameron (1926); Imperial Bureau (1933); Johnston and Mawson (1938b, 1940a) |
| *Filarinema* (3 species) | *Macropus rufus, Isoodon obesulus* | Stomach, small intestine, cecum, large intestine | Queensland and New South Wales, Australia | Johnston and Mawson (1938b); Mykytowycz (1964a) |
| *Nicollina* (3 species) | *Sarcophilus harrisii* | Intestine | Queensland, Australia; Scottish Zoological Park, Scotland | Cameron (1931); Imperial Bureau (1933) |
| *Trichostrongylus* (4 species) | *Isoodon obesulus, Dasyurus maculatus, Macropus robustus, M. rufogriseus frutica, Wallabia bicolor* | Stomach, duodenum, intestine | W. Australia; London Zoological Gardens, England | Imperial Bureau (1933); Johnston and Mawson (1939c); Pope *et al*. (1957b); Rewell (1948c) |
| Metastrongylidae | | | | |
| *Marsupostrongylus bronchialus* | *Isoodon obesulus* | Lungs | Queensland, Australia | Pope *et al*. (1957b) |
| Ancylostomatidae | | | | |
| *Hypodontus* (2 species) | *Macropus rufus, Thylogale thetis* | Small intestine | Queensland and New South Wales, Australia | Johnston and Mawson (1939a); Mykytowycz (1964b) |
| Spiruroidea | | | | |
| *Echimonema cincta* | *Dasyurus viverrinus, Isoodon obesulus, Perameles nasuta, Isoodon obesulus* | Intestine | Queensland and New South Wales, Australia | Imperial Bureau (1933); Johnston and Mawson (1939a, 1940a); Pope *et al*. (1957b) |
| *Cyliospirura* (=*Spirocerca*) | *Dasyurus viverrinus, D. maculatus* | Stomach; cysts in stomach and intestine | Queensland and New South Wales, Australia; Tasmania | Baylis (1927); Imperial Bureau (1933); Mawson (1968) |
| *Protospirura marsupialis* | *Trichosurus vulpecula* | * | Queensland and New South Wales, Australia | Baylis (1927); Imperial Bureau (1933); Johnston and Mawson (1938b, 1939b) |

TABLE VIII (Cont.)

| Name | Host | Habitat | Location | References |
|---|---|---|---|---|
| *Physaloptera* (6 species) | *Macrotis leucura, Sarcophilus harrisii, Perameles nasuta* | Stomach | New South Wales, Central Australia; Tasmania | Johnston and Mawson (1940a,b) |
| *Cyathospirura dasyuridis* | *Dasyurus viverrinus, D. maculatus* | * | New South Wales, Australia; Tasmania | Mawson (1968) |
| *Dirofilaria* (2 species) | *Macropus rufus* | Blood, knee joint | Queensland and New South Wales, Australia | Imperial Bureau (1933); Mykytowyzc (1964b) |
| *Filaria* (2 species) | *Macropus dorsalis, M. giganteus, M. parryi, M. antilopinus, M. rufogriseus frutica, Dendrolagus bennettianus, D. maculatus, Onychogalea fraenata, Potorous tridactylus, Petrogale penicillata, Trichosurus caninus, T. vulpecula* | Subcutaneous tissue, pericardium, peritoneal cavity, knee joint, liver, blood | Queensland, Victoria, and New South Wales, Australia; Tasmania; Heidelberg Zoo, Germany; and London Zoological Gardens, England | Crisp (1860); Johnston (1909, 1911, 1916); Leiper (1911); Imperial Bureau (1933); Johnston and Mawson (1938a) |
| *Setaria speleae* | *Macropus* sp. | * | Australia | Imperial Bureau (1933) |
| *Breinlia* (2 species) | *Dendrolagus inustus, Trichosurus vulpecula* | Peritoneal cavity | Queensland, Australia and London Zoological Gardens, England | Imperial Bureau (1933); Solomon (1933) |
| *Acanthocheilonema* (=*Dipetalonema*) (2 species) | *Macropus giganteus, M. antilopinus, Petrogale penicillata, Trichosurus vulpecula* | Peritoneal cavity, leg joints | Queensland, Australia | Imperial Bureau (1933) |
| *Dipetalonema* (10 species) | *Macropus dorsalis, M. robustus, M. parryi, M. fuliginosus, M. rufogriseus, M. giganteus,* | Subcutaneous tissue, lung, tail muscle, | Queensland, Victoria, New South Wales, and South Australia | Johnston and Mawson (1938a,b, 1939b,c, 1940a,b) |

| | | | | |
|---|---|---|---|---|
| | *Thylogale thetis, Wallabia bicolor, Onychogalea fraenata, Dendrolagus lumholtzi, D. benettianus, Perameles nasuta, Petrogale xanthopus, P. penicillata, Dasyurus viverrinus, Trichosurus vulpecula* | knee joint, aorta, and "body cavity" | | |
| Trichuroidea | | | | |
| *Trichuris peramelis* | *Perameles nasuta, Isoodon obesulus* | Intestine | North Queensland and New South Wales, Australia | Imperial Bureau (1933); Johnston and Mawson (1939b) |
| Trematodes | | | | |
| *Alaria* sp. | *Sarcophilus harrisii* | Intestine | Scottish Zoological Park, Scotland | Cameron (1931) |
| *Fasciola hepatica* | *Macropus giganteus, M. rufogriseus, Macropus eugenii, Vombatus ursinus, Trichosurus vulpecula,* kangaroo, wallaby | Bile ducts | Queensland and New South Wales, Australia | Johnston (1909, 1911); Boray (1969) |
| *Harmostomum dasyuri* | *Dasyurus viverrinus* | Intestine | New South Wales, Australia | Johnston (1913) |
| *Harmostomum simile* | *Isoodon obesulus* | Intestine | New South Wales, Australia | Johnston (1913) |
| *Mehlisia acuminata* | *Dasyurus viverrinus* | Intestine | New South Wales, Australia | Johnston (1913) |
| *Zonorchis* sp. | *Isoodon obesulus* | Pancreas | Queensland, Australia | Pope *et al*. (1957b) |
| "Trematodes" | *Echymipera kalubu, Dendrolagus inustus,* wallaby | Liver, bile ducts, peritoneum | Netherlands; New Guinea; London Zoological Gardens, England | Campbell (1891); Hamerton (1931); Vogtman and Fitzwater (1951) |
| Cestoda | | | | |
| Anoplocephalidae | | | | |
| *Anoplotaenia dasyuri* | *Sarcophilus harrisii* | * | London Zoological Gardens, England | Beddard (1911); Imperial Bureau (1933); Gregory and Munday (1975) |

**TABLE VIII (Cont.)**

| Name | Host | Habitat | Location | References |
|---|---|---|---|---|
| *Bancroftiella tenuis* | *Wallabia bicolor* | Intestine | Victoria, Australia | Imperial Bureau (1933) |
| *Bertiella aberrata* | *Pseudocheirus herbertensis* | * | * | Meggitt (1926) |
| *Bertiella edulis* | *Phalanger ursinus* | * | Celebes Islands, Australia | Imperial Bureau (1933) |
| *Bertiella kapul* | *Phalanger maculatus* | Intestine | Admiralty Islands | Baylis (1934a) |
| *Bertiella obesa* | *Phascolarctos* sp. | * | Queensland and New South Wales, Australia | Johnston (1916); Imperial Bureau (1933) |
| *Bertiella pellucida* | *Pseudocheirus lemuroides* | * | * | Meggitt (1926) |
| *Bertiella pseudochiri* | *Pseudocheirus herbertensis* | * | * | Meggitt (1926) |
| *Bertiella rigida* | *Phalanger* sp. | * | New Guinea | Imperial Bureau (1933) |
| *Bertiella sarasinorum* | *Phalanger ursinus* | * | * | Imperial Bureau (1933) |
| *Bertiella undulata* | *Pseudocheirus lemuroides* | * | * | Meggitt (1926) |
| *Cittotaenia bancrofti* | *Onychogalea fraenata* | * | Queensland, Australia | Lewis (1914); Johnston (1916) |
| *Cittotaenia villosa* | *Lagorchestes conspicillatus* | Stomach, intestine | Monte Bello Islands | Lewis (1914) |
| *Dasyurotaenia robustae* | *Sarcophilus harrisii* | Intestine | London Zoological Gardens, England | Beddard (1912); Johnston (1916) |
| *Hepatotaenia diaphana* | *Vombatus ursinus* | Liver | Tasmania; Bass Island Straits | Imperial Bureau (1933) |
| *Hepatotaenia fellicola* | *Macropus agilis* | Gall bladder | * | Imperial Bureau (1933) |
| *Hepatotaenia festiva* | *Macropus eugenii, M. giganteus, M. robustus, Onychogalea unguifera* | Gall bladder, bile ducts | Queensland and New South Wales, Australia | Imperial Bureau (1933); Mykytowycz (1964b) |
| *Moniezia bipapillosa* | *Vombatus ursinus* | * | New South Wales, Australia; and Philadelphia Zoological Gardens | Imperial Bureau (1933) |
| *Moniezia diaphana* | *Vombatus ursinus* | Liver | Tasmania | Johnston (1911) |

| | | | | |
|---|---|---|---|---|
| *Moniezia festiva* | *Macropus eugenii* | * | Queensland, Australia | Johnston (1916) |
| *Protogamotaenia bancrofti* | *Onychogalea fraenata, O. unguifera* | * | * | Imperial Bureau (1933) |
| *Protogamotaenia lagorchestis* | *Lagorchestes conspicillatus* | Stomach, intestine | Monte Bello Islands | Imperial Bureau (1933) |
| *Protogamotaenia zschokkei* | *Macropus* sp. | * | New Guinea | Imperial Bureau (1933) |
| *Prototaenia obesa* | *Phasolarctos cinereus* | Intestine | Queensland, Australia | Baylis (1934b) |
| *Triplotaenia mirabilis* | *Petrogale penicillata, Marcropus* sp. | Intestine | Australia | Imperial Bureau (1933) |
| Davaineida | | | | |
| *Raillietina macropa* | *Thylogale brunii* | Intestine | London Zoological Gardens, England | Imperial Bureau (1933) |
| Hymenolepididiae | | | | |
| *Hymenolepis perimelidarum* | *Isoodon macrourus* | * | * | Imperial Bureau (1933) |
| Linstowiidae | | | | |
| *Linstowia semoni* | *Perameles nasuta, Isoodon obesulus, Isoodon macrourus* | * | Queensland and New South Wales, Australia; Leningrad Zoological Gardens, USSR | Johnston (1911, 1916); Imperial Bureau (1933); Meggitt (1926) |
| *Linstowia echidnae* | *Isoodon obesulus* | * | * | Meggitt (1926); Imperial Bureau (1933) |
| Taeniidae | | | | |
| *Echinococcus granulosus* (larvae) | *Macropus dorsalis, M. eugenii, M. giganteus, M. robustus, Thylogale thetis, Wallabia bicolor* | Lungs, liver | Queensland and New South Wales, Australia | Johnston (1911, 1916); Meggitt (1926); Imperial Bureau (1933) |
| *Echinococcus polymorphus* (larvae) | *Macropus giganteus, M. dorsalis* | Lungs | Queensland, Australia; Cologne Zoo, France | Johnston (1909) |
| *Taenia bipapillosa* | *Vombatus ursinus* | * | Philadelphia Zoological Gardens | Johnston (1909) |

**TABLE VIII (Cont.)**

| Name | Host | Habitat | Location | References |
|---|---|---|---|---|
| *Taenia fimbriata* | *Macropus* sp. | * | Queensland, Australia | Johnston (1909) |
| *Taenis geophiloides* | *Phascolarctos cinereus* | * | Queensland, Australia | Imperial Bureau (1933) |
| *Taenia kreffti* | *Macropus* sp. | * | Queensland, Australia | Imperial Bureau (1933) |
| *Taenia mastersii* | *Macropus* sp., *M. eugenii* | * | Queensland, Australia | Johnston (1909, 1916); Imperial Bureau (1933) |
| *Taenia phalangistae* | *Trichosurus vulpecula* | Intestine | Queensland and New South Wales, Australia | Johnston (1909); Imperial Bureau (1933) |
| *Taenia semoni* | *Perameles nasuta* | * | St. Petersburg Zoological Gardens, USSR | Johnston (1911) |
| Diphyllobothriidae | | | | |
| *Bothriocephalus marginatus* | *Macropus* sp. | * | Queensland, Australia | Imperial Bureau (1933) |
| *Dithyridium cynocephali* (larvae) | *Thylacinus cynocephalus* | Skeletal and heart muscle | National Zoo, Washington, D.C. | Ransom (1907) |
| *Sparganum* sp. *(plerocercoids)* | *Dasyurus viverrinus, D. maculatus* | Body cavity, areolar tissue, axillae | New South Wales, Australia; London Zoological Gardens, England | Hamerton (1930); Imperial Bureau (1933) |
| "Cestodes" | *Echymipera kalubu* | * | Netherlands; New Guinea | Vogtman and Fitzwater (1951) |
| Acanthocephala | | | | |
| *Gigantorhynchus semoni* | *Perameles nasuta, Isoodon obesulus* | Intestine | Queensland and New South Wales, Australia | Johnston (1916); Imperial Bureau (1933) |
| *Gigantorhynchus* sp. | *Petrogale penicillata* | Intestine | Queensland and New South Wales, Australia | Johnston (1909); Imperial Bureau (1933) |
| *Moniliformis semoni* | *Isoodon obesulus* | * | Queensland, Australia | Pope *et al.* (1957b) |

[a] Asterisk means habitat not given.
[b] Asterisk means location not given.

Mykytowycz and Dudzinski, 1965). Most of the other literature cited contains descriptions of the parasites, but little information is available concerning their effects upon their hosts.

### 1. Nematodes

Mykytowycz (1964b) surveyed 454 wild *Macropus rufus* in three environmentally different areas of Australia. Nematode infections were commonly encountered, generally in young (1- to 2-year-old) animals. Small stomach worms (*Rugopharnynx, Zoniolaimus, Pharyngostrongylus, Coronostrongylus, Macropostrongylus,* and *Papillostrongylus* sp.), *Clocaina* sp., *Filarinema* sp., and *Labiostrongylus longispicularis* showed definite preferences for one or more of the six anatomical sections of the kangaroo stomach, apparently based on their different metabolic requirements. The kangaroo populations studied were not considered to be under any stress; they seemed to tolerate large parasite loads with no ill effects. However, under conditions of stress such as confinement, parasitism assumes a greater importance (Fiennes, 1968). Winter (1958) reported death resulting from *Strongyloides* infection in 15 of 20 pet kangaroos. Confinement of the hosts and environmental factors favoring the development of the parasite outside of the host were sufficient to induce overwhelming infections. Clinically, affected kangaroos became listless, ate poorly, and had intermittent diarrhea; death occurred in 1 to 3 days. Necropsy findings included edema of the lungs and dependent areas of the body, hydroperitoneum, hydropericardium, hydrothorax, and gastroenteritis. Adult *Strongyloides,* eggs, and larvae were found in the gastric mucosa. Strongyloidosis was also very common in wild bandicoots (*Isoodon obesulus*), though few worms were present (Mackerras, 1959a). A key to the classification of the nematode parasites of Australian marsupials has been published by Johnston and Mawson (1940c).

### 2. Trematodes

Few naturally occurring trematode infections were observed. They occurred chiefly among the Peramelidae and Dasyuridae.

Boray (1969) attempted to establish experimental *Fasciola hepatica* infections in several hosts. Wombats (*Vombatus ursinus*), wallabies (*Macropus eugenii*), and kangaroos (*Macropus giganteus*) appeared to possess some degree of natural resistance, although flukes did establish themselves in the bile ducts of the kangaroos. Administration of metacercariae to *Trichosurus vulpecula* produced chronic fascioliasis and severe anemia.

### 3. Cestodes

Tapeworm infections were most frequently reported among the Phalangeridae, but did not appear to be common. Attempts to establish *Taenia ovis* infections in *Macropus robustus* were unsuccessful (De Chaneet, 1971).

## REFERENCES

Adams, W. H., Emmons, W., and Brooks, J. E. (1970). *Am. J. Trop. Med. Hyg.* **19,** 311.
Alexander, A. D., Flyger, V., Herman, Y. F., and McConnell, S. J. (1972). *J. Wildl. Dis.* **8,** 119.
Andersen, F. L., and Reily, J. R. (1965). *J. Parasitol.* **51,** 913.
Anderson, C. R., and Roca-Garcia, M. (1947). *Am. J. Trop. Med.* **27,** 161.
Anonymous. (1971). *J. Am. Vet. Med. Assoc.* **158,** 1327.
Apgar, C. S. (1934). *Rep. Penrose Lab. Mus. Comp. Pathol.* p. 51.
Ash, L. R. (1962). *J. Parasitol.* **48,** 306.
Babero, B. B. (1957). *J. Parasitol.* **43,** 232.
Babero, B. B. (1960). *J. Parasitol.* **46,** 455.
Baer, J. G. (1927). *Abh. Senckenb. Naturforsch. Ges.* **40,** 375.
Baer, J. G. (1938). *Livro Jubilar Prof. Travassos* p. 75.
Barnes, R. D. (1968). *Lab. Anim. Care* **18,** 251.
Barr, T. R. B. (1961). *Diss. Abstr.* **21,** 3589.
Barr, T. R. B. (1963). *J. Wildl. Manage.* **27,** 53.
Barretto, M. P. (1940). *Arquiv. Zool.* **1,** 339.
Barretto, M. P. (1967). *Rev. Inst. Med. Trop. Sao Paulo* **9,** 313.
Barretto, M. P. (1968). *Rev. Bras. Biol.* **28,** 481.
Barretto, M. P., and Albuquerque, R. D. R. (1969). *Rev. Inst. Med. Trop. Sao Paulo.* **11,** 165.
Bates, M. (1944). *Am. J. Trop. Med.* **24,** 91.
Bates, M., and Roca-Garcia, M. (1946). *Am. J. Trop. Med.* **26,** 437.
Battey, Y. M., and Smith, D. J. W. (1964). *J. Hyg.* **62,** 485.
Baylis, H. A. (1927). *Ann. Mag. Nat. Hist.* **20,** 214.
Baylis, H. A. (1934a). *Ann. Mag. Nat. Hist. London* **14,** 412.
Baylis, H. A. (1934b). *Parasitology* **26,** 129.
Baylis, H. A. (1940). *Ann. Mag. Nat. Hist.* **6,** 313.
Beamer, P. D., Mohr, C. O., and Barr, T. R. B. (1960). *Am. J. Vet. Res.* **21,** 507.
Beddard, F. E. (1902). *Proc. Zool. Soc. London* **72,** 351.
Beddard, F. E. (1911). *Proc. Zool. Soc. London* **81,** 626.
Beddard, F. E. (1912). *Proc. Zool. Soc. London* **82,** 677.
Belem Virus Laboratory. (1967a). *U.S., Public Health Serv., Publ.* **1760,** 301–304.
Belem Virus Laboratory. (1967b). *U.S., Public Health Serv., Publ.* **1760,** 325–328.
Belem Virus Laboratory. (1967c). *U.S., Public Health Serv., Publ.* **1760,** 581–584.

Belem Virus Laboratory. (1967d). *U.S., Public Health Serv., Publ.* **1760,** 585–588.
Belem Virus Laboratory. (1967e). *U.S., Public Health Serv., Publ.* **1760,** 589–592.
Belem Virus Laboratory. (1967f). *U.S., Public Health Serv., Publ.* **1760,** 593–596.
Belem Virus Laboratory. (1967g). *U.S., Public Health Serv., Publ.* **1760,** 597–600.
Beveridge, W. I. B. (1934). *J. Pathol. Bacteriol.* **38,** 467.
Bigler, W. J. (1971). *J. Wildl. Dis.* **7,** 166.
Bigler, W. J., Lassing, E., Buff, E., Lewis, A. L., and Hoff, G. L. (1975). *J. Wildl. Dis.* **11,** 348.
Blair, W. R. (1916). *21st Ann. Rep. N.Y. Zool. Soc.* p. 77.
Blair, W. R. (1918). *23rd Ann. Rep. N.Y. Zool. Soc.* p. 65.
Blair, W. R. (1921). *26th Ann. Rep. N.Y. Zool. Soc.* p. 79.
Blair, W. R. (1924). *29th Ann. Rep. N.Y. Zool. Soc.* p. 50.
Bohls, S. W., and Schuhardt, V. T. (1933). *Tex. State J. Med.* **29,** 199.
Bollinger, A., and Bollinger, W. (1948). *Aust. J. Sci.* **10,** 182.
Boray, J. C. (1969). *Adv. Parasitol.* **7,** 95.
Bourne, G. (1934). *J. Proc. R. Soc. West. Aust.* **19,** 1.
Boyd, A. G. (1929). *Cornell Vet.* **19,** 33.
Bozeman, F. M., Shirai, A., Humphries, J. W., and Fuller, H. S. (1967). *Am. J. Trop. Med. Hyg.* **16,** 48.
Brick, J. O. (1970). *J. Am. Vet. Med. Assoc.* **157,** 650.
Brigham, G. D. (1936). *Publ. Health Rep.* **51,** 333.
Brooke, M. M., Norman, L., Allain, D., and Gorman, G. W. (1957). *Parasitology* **43,** 15.
Bugher, J. C., Boshell-Manrique, J., Roca-Garcia, M., and Gilmore, R. M. (1941). *Am. J. Trop. Med.* **21,** 309.
Byrd, E. E., and Ward, J. W. (1928). *J. Parasitol.* **28,** 28.
Byrd, E. E., and Ward, J. W. (1943). *J. Parasitol.* **29,** 217.
Bryd, E. E., Reiber, R. J., and Parker, M. V. (1942a). *J. Tenn. Acad. Sci.* **17,** 116.
Byrd, E. E., Reiber, R. J., and Parker, M. V. (1942b). *J. Tenn. Acad. Sci.* **17,** 130.
Caballero, E. (1951). *Ann. Inst. Biol. Univ. Mex.* **22,** 141.
Caballero, E., Bravo, C. M., and Cerecero, M. C. (1944). *Ann. Inst. Biol. Univ. Mex.* **15,** 59.
Calhoun, E. L., Mohr, C. O., and Alford, H. I., Jr. (1956). *Am. J. Hyg.* **63,** 127.
Cameron, T. W. M. (1926). *J. Helminthol.* **4,** 23.
Cameron, T. W. M. (1931). *J. Helminthol.* **9,** 153.
Campbell, H. J. (1891). *Guy's Hosp. Rep.* **48,** 19.
Canavan, W. P. N. (1929). *Parasitology* **21,** 63.
Canavan, W. P. N. (1931). *Parasitology* **23,** 196.
Chandler, A. C. (1924). *Parasitology* **16,** 160.
Chandler, A. C. (1932). *Proc. U.S. Natl. Mus.* **81,** 1.
Chandler, A. C., and Rausch, R. (1946). *Trans. Am. Microsc. Soc.* **65,** 328.
Chapman, J., and Chapman, A. A. (1935). *Tex. State J. Med.* **31,** 36.
Clark, F. H., and Shepard, C. C. (1963). *J. Bacteriol.* **86,** 1057.
Clark, H. C., and Dunn, L. H. (1931). *20th Annu. Rep. United Fruit Company, Med. Dep.* p. 131.
Clark, H. C., and Dunn, L. H. (1932). *Am. J. Trop. Med.* **12,** 49.
Clark, H. C., and Dunn, L. H. (1933). *Am. J. Trop. Med.* **13,** 273.
Clark, L. G., Kresse, J. I., Carbrey, E. A., Marshak, R. R., and Hollister, C. J. (1961). *J. Am. Vet. Med. Assoc.* **139,** 889.
Cobbold, S. T. (1874). *Proc. Zool. Soc. London,* **44,** 124.
Constantine, D. G. (1966). *Am. J. Vet. Res.* **27,** 16.
Constantine, D. G., and Woodall, D. F. (1966). *Am. J. Vet. Res.* **27,** 24.

Conti-Diaz, I. A., Turner, C., Tweedale, D., and Furcolow, M. L. (1970). *J. Parasitol.* **56,** 457.

Cook, I., and Pope, J. H. (1959). *Aust. J. Exp. Biol. Med. Sci.* **37,** 253.

Corkum, K. C. (1966). *J. Parasitol.* **52,** 444.

Correa, F. M. A., and Barretto, M. P. (1964). *Rev. Inst. Med. Trop. Sao Paulo* **6,** 157.

Correa, R. R. (1968). *Rev. Bras. Malariol. Doencas Trop.* **20,** 39.

Corson-White, E. P. (1922). *Arch. Intern. Med.* **30,** 620.

Coutelen, F. (1932). *C. R. Seances Soc. Biol. Ses Fil.* **110,** 1245.

Coutelen, F. (1933). *Ann. Parasitol. Hum. Comp.* **11,** 1.

Crisp, E. (1860). *Proc. Zool. Soc. London* **30,** 175.

Cummings, B. F. (1916). *Proc. Zool. Soc. London* **86,** 253.

Darling, S. T. (1910). *Bull. Soc. Pathol. Exot.* **3,** 513.

Davey, D. G., and Wood, W. A. (1938). *Parasitology* **30,** 258.

Deane, L. M. (1961). *Rev. Inst. Med. Trop. Sao Paulo* **3,** 15.

Deane, L. M. (1964). *Rev. Inst. Med. Trop. Sao Paulo* **6,** 225.

Deane, M. P., de Brito, T., and Deane, L. M. (1963). *Rev. Inst. Med. Trop. Sao Paulo* **5,** 225.

de Brito, T., and Deane, L. M. (1966). *Rev. Inst. Med. Trop. Sao Paulo* **8,** 79.

De Chaneet, G. (1971). *Aust. Vet. J.* **47,** 459.

Denton, J. F. (1944). *Proc. Helminthol. Soc. Wash.* **11,** 54.

De Roever-Bonnet, H. (1967). *Trop. Geogr. Med.* **19,** 221.

De Roever-Bonnet, H., Lelyveld, J., and Marinkelle, C. J. (1969). *Trop. Geogr. Med.* **21,** 451.

Derrick, E. H. (1944). *J. Hyg.* **43,** 357.

Derrick, E. H., Smith, D. J. W., Brown, H. E., and Freeman, M. (1939) *Med. J. Aust.* **1,** 150.

Derrick, E. H., Smith, D. J. W., and Brown, H. E. (1940). *Aust. J. Exp. Biol. Med. Sci.* **18,** 409.

Dias, E., and Martins, A. V. (1939). *Am. J. Trop. Med.* **19,** 103.

Dikerson, L. M. (1930). *Parasitology* **22,** 37.

Dikmans, G. (1931). *Proc. U.S. Natl. Mus.* **79,** 1.

Dikmans, G. (1943). *Proc. Helminthol. Soc. Wash.* **10,** 6.

Dobos-Kovacs, M., Meszaros, J., Pellerdy, L., and Balsai, A. (1975). *Acta Vet. Acad. Sci. Hung.* **24,** 293.

Doherty, R. L. (1967a). *Public Health Serv., U.S.,* **1760,** 345–348.

Doherty, R. L. (1967b). *U.S. Public Health Serv., Publ.* **1760,** 545–548.

Doherty, R. L., Carley, J. G., and Gorman, B. M. (1964). *Aust. J. Exp. Biol. Med. Sci.* **42,** 149.

Doherty, R. L., Whitehead, R. H., Wetters, E. J., Gorman, B. M., and Carley, J. G. (1970). *Trans. R. Soc. Trop. Med. Hyg.* **64,** 784.

de Boulay, G. H., Hime, J. M., and Verity, P. M. (1972). *Br. J. Radiol.* **45,** 841.

Dudzinski, M. L., and Mykytowycz, R. (1965). *Parasitology* **55,** 543.

d'Utra e Silva, O., and Arantes, J. B. (1917). *Mem. Inst. Oswaldo Cruz* **8,** 61.

Dunn, L. H., and Clark, H. C. (1933). *Am. J. Trop. Med.* **13,** 201.

Durack, D. T., Gubbay, S. S., and Kakulas, B. A. (1969). *Aust. J. Exp. Biol. Med. Sci.* **47,** 581.

Ekdahl, M. O. (1970). *N. Z. Vet. J.* **18,** 44.

Emanuel, M. L., and Mackerras, I. M. (1964). *J. Hyg.* **62,** 451.

Emmons, C. W., Rowley, D. A., Olson, B. J., Mattern, C. F. T., Bell, J. A., Powell, E., and Marcey, E. A. (1955). *Am. J. Hyg.* **61,** 40.

Emmons, R. W., and Lennette, E. H. (1969). *Science* **163,** 945.

Ernst, J. V., Cooper, C., Jr., and Chobotar, B. (1969). *Bull. Wildl. Dis. Assoc.* **5,** 406.

Evans, L. B., Wood, G. E., Flyger, V., Alexander, A. D., Yager, R. H., and Rubin, H. L. (1962). *Proc. Soc. Exp. Biol. Med.* **110,** 113.

Ewing, H. E., and Fox, I. (1943). *U.S., Dep. Agric. Misc. Publ.* **500**.
Farris, E. J. (1950). *In* "The Care and Breeding of Laboratory Animals" (E. J. Farris, ed.), pp. 256–267. Wiley, New York.
Feldman, D. B., Moore, J. A., Harris, M. W., and Self, J. L. (1972). *Lab. Anim. Sci.* **22,** 183.
Fenner, F. (1946). *Med. J. Aust.* **2,** 666.
Ferris, G. F. (1924). *Am. Mus. Novit.* **110,** 1.
Fiennes, R. N. T-W. (1968). *Proc. R. Soc. Med.* **61,** 161.
Flatt, R. E., Nelson, L. R., and Patton, N. M. (1971). *Lab. Anim. Sci.* **21,** 106.
Fleay, D. H. (1935). *J. Mammal.* **16,** 10.
Foster, A. O. (1939). *Trans. Am. Microsc. Soc.* **58,** 185.
Fox, H. (1912). *J. Pathol. Bacteriol.* **17,** 217.
Fox, H. (1914). *N. Y. Med. J.* **100,** 1209.
Fox, H. (1923a). *In* "Disease in Captive Wild Mammals and Birds," p. 520. Lippincott, Philadelphia, Pennsylvania.
Fox, H. (1923b). *In* "Disease in Captive Wild Mammals and Birds," p. 215. Lippincott, Philadelphia, Pennsylvania.
Fox, H. (1923c). *In* "Disease in Captive Wild Mammals and Birds," p. 253. Lippincott, Philadelphia, Pennsylvania.
Fox, H. (1923d). *In* "Disease in Captive Wild Mammals and Birds," p. 268. Lippincott, Philadelphia, Pennsylvania.
Fox, H. (1923e). *In* "Disease in Captive Wild Mammals and Birds," p. 278. Lippincott, Philadelphia, Pennsylvania.
Fox, H. (1923f). *In* "Disease in Captive Wild Mammals and Birds," p. 314. Lippincott, Philadelphia, Pennsylvania.
Fox, H. (1923g). *In* "Disease in Captive Wild Mammals and Birds," p. 296. Lippincott, Philadelphia, Pennsylvania.
Fox, H. (1923h). *In* "Disease in Captive Wild Mammals and Birds," pp. 285, 480. Lippincott, Philadelphia, Pennsylvania.
Fox, H. (1923i). *In* "Disease in Captive Wild Mammals and Birds," pp. 312, 480. Lippincott, Philadelphia, Pennsylvania.
Fox, H. (1923j). *In* "Disease in Captive Wild Mammals and Birds," p. 368. Lippincott, Philadelphia, Pennsylvania.
Fox, H. (1923k). *In* "Disease in Captive Wild Mammals and Birds," p. 623. Lippincott, Philadelphia, Pennsylvania.
Fox, H. (1923l). *In* "Disease in Captive Wild Mammals and Birds," pp. 149, 175, 632. Lippincott, Philadelphia, Pennsylvania.
Fox, H. (1923m). *In* "Disease in Captive Wild Mammals and Birds," p. 174. Lippincott, Philadelphia, Pennsylvania.
Fox, H. (1923n). *In* "Disease in Captive Wild Mammals and Birds," p. 252. Lippincott, Philadelphia, Pennsylvania.
Fox, H. (1923o). *In* "Disease in Captive Wild Mammals and Birds," pp. 278, 283. Lippincott, Philadelphia, Pennsylvania.
Fox, H. (1923p). *In* "Disease in Captive Wild Mammals and Birds," p. 157. Lippincott, Philadelphia, Pennsylvania.
Fox, H. (1923q). *In* "Disease in Captive Wild Mammals and Birds," p. 76. Lippincott, Philadelphia, Pennsylvania.
Fox, H. (1923r). *In* "Disease in Captive Wild Mammals and Birds," p. 567. Lippincott, Philadelphia, Pennsylvania.
Fox, H. (1923s). *In* "Disease in Captive Wild Mammals and Birds," p. 162. Lippincott, Philadelphia, Pennsylvania.

Fox, H. (1924). *J. Mammal.* **5,** 153.
Fox, H. (1926). *Rep. Penrose Lab. Mus. Comp. Pathol.,* p. 16.
Fox, H. (1927–1928). *J. Am. Vet. Med. Assoc.* **72,** 792.
Fox, H. (1929). *Rep. Penrose Lab. Mus. Comp. Pathol.*
Fox, H. (1932). *Rep. Penrose Lab. Mus. Comp. Pathol.,* p. 16.
Fox, H. (1933). *In* "Arteriosclerosis, a Survey of the Problem" (E. V. Cowdry, ed.), pp. 153–193. Macmillan, New York.
Fox, H. (1937). *Rep. Penrose Lab. Mus. Comp. Pathol.,* p. 26.
Fox, H. (1938). *Rep. Penrose Lab. Mus. Comp. Pathol.,* p. 28.
Fox, H. (1940). *Rep. Penrose Lab. Mus. Comp. Pathol.,* p. 10.
Fox, I. (1940a). "Fleas of the Eastern United States," p. 114. Iowa State Coll. Press, Ames, Iowa.
Fox, I. (1940b). "Fleas of the Eastern United States," p. 61. Iowa State Coll. Press, Ames, Iowa.
Francis, E. (1934). *Sci. Mon.* **38,** 476.
Francis, E. (1937). *Pub. Health Rep.* **52,** 103.
French, E. L. (1967). *Public Health Serv., Pub.* **1760,** 509–512.
Frerichs, W. M., and Holbrook, A. A. (1970). *J. Parasitol.* **56,** 130.
Froggatt, W. W. (1913). *Agric. Gaz. N. S. W.* **24,** 567.
Fuller, H. S. (1963). *Bull. Soc. Pathol. Exot.* **56,** 568.
Galton, M. M., Powers, D. K., McKeever, S., and Gorman, G. W. (1957). *Public Health Rep.* **72,** 431.
Galton, M. M., Menges, R. W., and Steele, J. H. (1958). *Ann. N.Y. Acad. Sci.* **70,** 427.
Galton, M. M., Gorman, G. W., and Shotts, E. B., Jr. (1960). *Public Health Rep.* **75,** 917.
Garrod, A. H. (1873). *Proc. Zool. Soc. London* **43,** 202.
Gilruth, J. A., and Bull, L. B. (1911). *Proc. R. Soc. Victoria* **24,** 432.
Giltner, L. T., and Shahan, M. S. (1936). *J. Am. Vet. Med. Assoc.* **88,** 363.
Gorman, G. W. McKeever, S., and Grimes, R. D. (1962). *Am. J. Trop. Med. Hyg.* **11,** 518.
Goss, L. J. (1940). *Zoologica (N.Y.)* **24,** 523.
Goss, L. J. (1941). *Zoologica (N.Y.)* **25,** 269.
Goss, L. J. (1942). *Cornell Vet.* **32,** 155.
Grayson, M. A., and Galindo, P. (1968). *Am. J. Epidemiol.* **88,** 80.
Grayson, M. A., and Galindo, P. (1969). *J. Am. Vet. Med. Assoc.* **155,** 2141.
Gregory, G. G., and Munday, B. L. (1975). *Int. J. Parasitol.* **5,** 187.
Griffith, S. A. (1928). *J. Hyg.* **28,** 198.
Grimes, J. E. (1967). *U.S., Public Health Serv., Publ.* **1760,** 813–816.
Gunther, C. E. M. (1941a). *Proc. Linn. Soc. N. S. W.* **66,** 155.
Gunther, C. E. M. (1941b). *Proc. Linn. Soc. N. S. W.* **66,** 391.
Gupta, B. N., and Feldman, D. B. (1975). *Lab. Anim. Sci.* **25,** 238.
Hackel, D. B., Kinney, T. D., and Wendt, W. (1953). *Lab. Invest.* **2,** 154.
Hagan, W. A., and Bruner, D. W. (1957). "The Infectious Diseases of Domestic Animals," 3rd ed. Cornell Univ. Press (Comstock), Ithaca, New York.
Haley, J. S. (1938). *Vet. Med.* **33,** 291.
Hamerton, A. E. (1929). *Proc. Zool. Soc. London* **99,** 49.
Hamerton, A. E. (1930). *Proc. Zool. Soc. London* **100,** 357.
Hamerton, A. E. (1931). *Proc. Zool. Soc. London* **101,** 527.
Hamerton, A. E. (1932). *Proc. Zool. Soc. London* **102,** 613.
Hamerton, A. E. (1933). *Proc. Zool. Soc. London* **103,** 451.
Hamerton, A. E. (1934a). *Proc. Zool. Soc. London* **104,** 389.
Hamerton, A. E. (1934b). *Trans. R. Soc. Trop. Med. Hyg.* **28,** 2.

Hamerton, A. E. (1935a). *Proc. Zool. Soc. London* **105,** 443.
Hamerton, A. E. (1935b). *Br. J. Tuberc.* **29,** 145.
Hamerton, A. E. (1936). *Proc. Zool. Soc. London* **106,** 659.
Hamerton, A. E. (1937). *Proc. Zool. Soc. London* **107,** 443.
Hamerton, A. E. (1938). *Proc. Zool. Soc. London* **108,** 489.
Hamerton, A. E. (1939). *Proc. Zool. Soc. London* **109,** 281.
Hamerton, A. E. (1941). *Proc. Zool. Soc. London* **111,** 151.
Hamerton, A. E. (1942). *Proc. Zool. Soc. London* **112,** 120.
Hamerton, A. E. (1944). *Proc. Zool. Soc. London* **114,** 307.
Hamerton, A. E. (1945). *Proc. Zool. Soc. London* **115,** 371.
Harrison, L., and Johnston, H. T. (1916). *Parasitology* **8,** 338.
Hayes, R. O., LaMotte, L. C., and Holden, P. (1967). *Am. J. Trop. Med.* **16,** 675.
Heaslip, W. G. (1941). *Med. J. Aust.* **1,** 380.
Heath, D., and Whitehead, T. P. (1963). *J. Comp. Pathol.* **73,** 439.
Henger, R., and Ratcliffe, H. (1927). *J. Parasitol.* **14,** 27.
Herrer, A., Thatcher, V. E., and Johnson, C. M. (1966). *J. Parasitol.* **52,** 954.
Herrer, A. Teleford, S. R., Jr., and Christensen, H. A. (1971). *Ann. Trop. Med. Parasitol.* **65,** 349.
Hill, O. W. C. (1951). *Proc. Zool. Soc. London* **121,** 641.
Hill, W. C. (1939). *Am. Midl. Nat.* **21,** 636.
Hime, J. M., and Jones, D. M. (1972). *J. Small Anim. Pract.* **13,** 91.
Hirst, S. (1931). *Proc. Zool. Soc. London* **20,** 325.
Hirst, S. A. (1931). *Proc. Zool. Soc. London* **101,** 561.
Holloway, H. L., Jr. (1966). *Bull. Wildl. Dis. Assoc.* **2,** 38.
Holloway, H. L., Jr., and Dowler, J. L. (1963). *Vet. J. Sci.* **14,** 203.
Hopkins, G. H. E. (1949). *Proc. Zool. Soc. London* **119,** 387.
Hubbert, W. T., and Rosen, M. N. (1970). *Am. J. Public Health* **60,** 1103.
Imperial Bureau of Agricultural Parasitology. (1933). *J. Helminthol.* **11,** 195.
Irons, J. V., Beck, O., and Murphy, J. N., Jr. (1946). *J. Bacteriol.* **51,** 609.
Irons, J. V., Eads, R. B., Johnson, C. W., Walker, O. L., and Norris, M. A. (1952). *J. Parasitol.* **38,** 1.
Iveson, J. B., and Bradshaw, S. D. (1973). *J. Hyg.* **71,** 423.
Jennings, W. L., Lewis, A. L., Sather, G. E., Hammond, W. M. D., and Bond, J. O. (1968). *Am. J. Trop. Med.* **17,** 781.
Johnston, S. J. (1913). *Proc. Linn. Soc. N. S. W.* **37,** 727.
Johnston, T. H. (1909). *Proc. Linn. Soc. N. S. W.* **24,** 514.
Johnston, T. H. (1911). *Proc. Linn. Soc. N. S. W.* **36,** 47.
Johnston, T. H. (1916). *Proc. R. Soc. Queensl.* **28,** 31.
Johnston, T. H. (1938a). *Trans. Proc. R. Soc. South Aust.* **62,** 263.
Johnston, T. H., and Mawson, P. M. (1938b). *Rec. South Aust. Mus.* **6,** 187.
Johnston, T. H., and Mawson, P. M. (1938c). *Trans. Proc. R. Soc. South. Aust.* **62,** 269.
Johnston, T. H., and Mawson, P. M. (1939a). *Trans. Proc. R. Soc. South Aust.* **63,** 121.
Johnston, T. H., and Mawson, P. M. (1939b). *Trans. Proc. R. Soc. South Aust.* **63,** 204.
Johnston, T. H., and Mawson, P. M. (1939c). *Trans. Proc. R. Soc. South Aust.* **63,** 307.
Johnston, T. H., and Mawson, P. M. (1940a). *Proc. Linn. Soc. N. S. W.* **65,** 468.
Johnston, T. H., and Mawson, P. M. (1940b). *Rec. South Aust. Mus.* **20,** 360.
Johnston, T. H., and Mawson, P. M. (1940c). *Trans. Proc. R. Soc. South Aust.* **64,** 363.
Johnston, T. H., and Mawson, P. M. (1942). *Rec. South Aust. Mus.* **21,** 110.
Jonkers, A. H., Spence, L., Downs, W. G., Aitken, T. H. G., and Tikasingh, E. S. (1968). *Am. J. Trop. Med. Hyg.* **17,** 276.

Jordan, K., and Rothschild, N. C. (1922). *Ectoparasites* **1,** 231.
Joyce, C. R., and Eddy, G. W. (1944). *Iowa State Coll. J. Sci.* **18,** 209.
Kakulas, B. A. (1961). *Nature (London)* **191,** 402.
Kakulas, B. A. (1963). *Nature (London)* **198,** 673.
Kaplan, E. H. (1964). *J. Parasitol.* **50,** 797.
Keaton, R., Nash, B. J., Murphy, N. J., Jr., and Irons, J. V. (1953). *Public Health Rep.* **68,** 28.
Keep, J. M. (1972). *Aust. Vet. J.* **48,** 119.
Kellogg, V. L. (1914). *Am. Nat.* **48,** 257.
Kofoid, C. A., and Donat, F. (1933). *Calif. West. Med.* **38,** 245.
Kokernot, R. H., Hayes, J., Will, R. L., Tempelis, C. H., Chan, D. H. M., and Radivojevic, B. (1969a). *Am. J. Trop. Med. Hyg.* **18,** 750.
Kokernot, R. H., Tempelis, C. H., Chan, D. H. M., Boyd, K. R., and Anderson, R. J. (1969b). *Am. J. Trop. Med. Hyg.* **18,** 768.
Kokernot, R. H., Calisher, C. H., Stannard, L. J., and Hayes, J. (1969c). *Am. J. Trop. Med. Hyg.* **18,** 789.
Kokernot, R. H., Radivojevic, B., and Anderson, R. J. (1969d). *Am. J. Vet. Res.* **30,** 2197.
Krull, W. H. (1933). *J. Parasitol.* **20,** 98.
Krupp, J. H. (1962). *J. Am. Vet. Med. Assoc.* **141,** 369.
Krupp, J. H., and Quillin, R. (1964). *Lab. Anim. Care* **14,** 189.
Laemmert, H. W., Jr. (1946). *Am. J. Trop. Med.* **26,** 33.
Laemmert, H. W., Jr. (1967). *U.S., Public Health Serv., Publ.* **1760,** 377–380.
Laemmert, H. W., Jr., Ferreira, L. de C., and Taylor, R. M. (1946). *Am. J. Trop. Med.* **26,** 23.
Lage, H. A. (1968). *Hospital (Rio de Janeiro)* **73,** 1955.
Lainson, R., and Shaw, J. J. (1969a). *Trans. R. Soc. Trop. Med. Hyg.* **63,** 408.
Lainson, R., and Shaw, J. J. (1969b). *Trans. R. Soc. Trop. Med. Hyg.* **63,** 738.
Lainson, R., and Shaw, J. J. (1970). *Trans. R. Soc. Trop. Med. Hyg.* **64,** 654.
La Plante, E. S., and Burrell, R. G. (1966). *Bull. Wildl. Dis. Assoc.* **2,** 10.
LaRue, G. R., and Bosma, N. J. (1958). *J. Parasitol.* **44,** 324.
Leigh, H. W. (1940). *Ill. Nat. Hist. Surv., Bull.* **21,** 185.
Leiper, R. T. (1911). *Proc. Zool. Soc. London* **81,** 620.
Lent, H., and Teixeira de Freitas, J. F. (1937). *Mem. Inst. Oswaldo Cruz* **32,** 221.
LeSouef, S. A., and Bullier, H. (1902). *Victorian Nat.* **18,** 159.
Lewis, R. C. (1914). *Proc. Zool. Soc. London* **84,** 419.
Little, M. D. (1960). *J. Parasitol.* **52,** 85.
Lombard, L. S., and Witte, E. J. (1959). *Cancer Res.* **19,** 127.
Lucas, N. (1925). *J. Pathol. Bacteriol.* **28,** 123.
McKeever, S. (1958). *J. Parasitol.* **44,** 324.
McKeever, S., Gorman, G. W., Chapman, J. F., Galton, M. M., and Powers, D. K. (1958a). *Am. J. Trop. Med. Hyg.* **7,** 646.
McKeever, S., Gorman, G. W., and Norman, L. (1958b). *J. Parasitol.* **44,** 583.
McKeever, S., Kaplan, W., and Ajello, L. (1958c). *Am. J. Vet. Res.* **19,** 973.
McKeever, S., Schubert, J. H., Moody, M. D., Gorman, G. W., and Chapman, J. F. (1958d). *J. Infect. Dis.* **103,** 120.
McKeever, S., Schubert, J. H., Gorman, G. W., and Grimes, R. D. (1959). *Am. J. Vet. Res.* **20,** 192.
Mackerras, M. J. (1959a). *Aust. J. Zool.* **7,** 87.
Mackerras, M. J. (1959b). *Aust. J. Zool.* **7,** 107.
Malek, E. A. (1970). *J. Parasitol.* **56,** 64.

Mandour, A. M. (1965). *J. Protozol.* **12,** 606.
Markham, N. P., and Markham, J. G. (1966). *J. Comp. Pathol.* **76,** 49.
Marx, M. B. (1969). *Am. J. Vet. Res.* **30,** 2003.
Mawson, P. M. (1968). *Parasitology* **58,** 75.
Mease, J. A., Jr. (1929). *J. Am. Vet. Med. Assoc.* **92,** 1042.
Meggitt, F. J. (1926). "The Cestodes of Mammals." Edward Goldston, London.
Menges, R. W., and Georg, L. K. (1957). *Public Health Rep.* **72,** 503.
Miles, J. A. R. (1967). *Public Health Serv., Publ.* **1760,** 749–752.
Miller, G. C., and Harkema, R. (1970). *Proc. Helminthol. Soc. Wash.* **37,** 36.
Miller, N. L., Frenkel, J. K., and Dubey, J. P. (1972). *J. Parasitol.* **58,** 928.
Mohr, C. O., and Morlan, H. B. (1959). *J. Parasitol.* **45,** 233.
Moore, M. J. (1904). *Vet. J.* **8,** 283.
Moore, T. D., Allen, A. M., Ganaway, J. R., and Sevy, C. E. (1971). *J. Infect. Dis.* **123,** 569.
Moorhouse, D. E. (1969). *Trans. R. Soc. Trop. Med. Hyg.* **63,** 22.
Moran, A. B. (1942). *Proc. U.S. Livestock Sanit. Assoc.* **46,** 441.
Morlan, H. B. (1949). *J. Parasitol.* **35,** 217.
Morlan, H. B., Hill, E. L., and Schubert, J. H. (1950). *Public Health Rep.* **65,** 57.
Munday, B. L. (1972). *J. Wildl. Dis.* **8,** 169.
Munday, B. L., and Corbould, A. (1973). *J. Wildl. Dis.* **9,** 72.
Mykytowycz, R. (1964a). *Parasitology* **54,** 105.
Mykytowycz, R. (1964b). *Parasitology* **54,** 677.
Mykytowycz, R., and Dudzinski, M. L. (1965). *Parasitology* **55,** 527.
Nadchatram, M. (1970). *J. Med. Entomol.* **7,** 1.
Nettles, V. F., Prestwood, A. K., and Davidson, W. R. (1975). *J. Wildl. Dis.* **11,** 419.
Noback, C. V. (1929). *34th Annu. Rep. Zool. Soc.* p. 47.
Noback, C. V. (1930). *35th Annu. Rep. Zool. Soc.* p. 50.
Norman, L., Brooke, M. M., Allain, D. S., and Gorman, G. W. (1959). *J. Parasitol.* **45,** 457.
Nutting, W., and Woolley, P. (1965). *Parasitology* **55,** 383.
Olsen, P. F., Shoemaker, J. P., and Turner, H. F. (1964). *J. Parasitol.* **50,** 599.
Packchanian, A. (1942). *Am. J. Trop. Med.* **22,** 623.
Paine, G. D. (1969). *J. Protozol.* **16,** 371.
Park, P. J. (1936). *Trans. Am. Microsc. Soc.* **55,** 49.
Petana, W. B. (1969). *Ann. Trop. Med. Parasitol.* **63,** 47.
Phillip, C. B. (1957). *In* "Bergey's Manual of Determinative Bacteriology" (R. S. Breed, E. G. D. Murray, and N. R. Smith, eds.), 7th ed., pp. 966–968. Williams & Wilkins, Baltimore, Maryland.
Phillips, C. J., and Wilson, N. (1965). *J. Mammal.* **46,** 698.
Pipkin, A. C., Sr. (1968). *J. Med. Entomol.* **5,** 107.
Plimmer, H. G. (1912). *Proc. Zool. Soc. London* **82,** 235.
Plimmer, H. G. (1915). *Proc. Zool. Soc. London* **85,** 123.
Plimmer, H. G. (1916). *Proc. Zool. Soc. London* **86,** 77.
Pope, J. H., Derrick, E. H., Cook, I. (1957a). *Aust. J. Exp. Biol. Med. Sci.* **35,** 467.
Pope, J. H., Bicks, V. A., and Cook, I. (1957b). *Aust. J. Exp. Biol. Med. Sci.* **35,** 481.
Potkay, S. (1970). *Lab. Anim. Care* **20,** 502.
Prasad, H. (1960). *Z. Parasitenkd.* **20,** 385.
Pray, L. L. (1921). *J. Mammal.* **2,** 109.
Pucak, G. J., Henderson, J. D., Jr., and Bullock, B. C. (1969). *J. Am. Vet. Med. Assoc.* **155,** 1228.
Ransom, B. H. (1907). *Trans. Am. Microsc. Soc.* **27,** 31.
Ratcliffe, H. L. (1933). *Am. J. Cancer Res.* **17,** 116.

Ratcliffe, H. L. (1965). *Ann. N.Y. Acad. Sci.* **127,** 715.
Ratcliffe, H. L., and Worth, C. B. (1951). *Am. J. Pathol.* **27,** 655.
Rausch, R., and Tiner, J. D. (1949). *Am. Midl. Nat.* **41,** 665.
Regendaz, R., and Kikuth, W. (1928). *C. R. Seances Soc. Biol. Ses Fil.* **98,** 1567.
Reilly, J. R., Hanson, L. E., and Ferris, D. H. (1970). *Am. J. Vet. Res.* **31,** 1443.
Rewell, R. E. (1948a). *Trans. R. Soc. Trop. Med. Hyg.* **42,** 17.
Rewell, R. E. (1948b). *J. Pathol. Bacteriol.* **60,** 155.
Rewell, R. E. (1948c). *Proc. Zool. Soc. London* **118,** 501.
Rieber, R. J., and Byrd, E. E. (1942). *J. Tenn. Acad. Sci.* **17,** 78.
Riemann, H. P., Behymer, D. E., Fowler, M. E., Schultz, T., Lock, A., Orthoefer, J. G., Silverman, S., and Franti, C. E. (1974). *J. Am. Vet. Med. Assoc.* **165,** 798.
Rigg, K. J., Finlayson, R., Symons, C., and Hill, K. R. (1960). *Proc. Zool. Soc. London* **130,** 157.
Robertson, A. (1930). *Trans. R. Soc. Trop. Med. Hyg.* **23,** 332.
Roca-Garcia, M. (1945). *Rev. Med. Vet., Bogata* **14,** 70.
Roca-Garcia, M. (1949). *J. Infect. Dis.* **85,** 275.
Roca-Garcia, M. (1967). *U.S., Public Health Serv. Publ.* **1760,** 405–408.
Roth, E. E., and Knieriem, B. B. (1958). *J. Am. Vet. Med. Assoc.* **132,** 97.
Roth, E. E., Linder, D., and Adams, W. V. (1961). *Am. J. Vet. Res.* **22,** 308.
Roth, E. E., Moore, M., Greer, B., Newman, K., Adams, W. V., and Sanford, G. E. (1963). *Zoonoses Res.* **2,** 91.
Rowlands, D. T., Jr., Vakilzadeh, J., Sherwood, B. F., and LeMay, J. C. (1970). *Am. J. Pathol.* **58,** 295.
Santa Rosa, C. A., Sulzer, C. R., Giorgi, W., da Silva, A. S., Yanaquita, R. M., and Lobao, A. O. (1975). *Am. J. Vet. Res.* **36,** 1365.
Sather, G. E., Lewis, A. L., Jennings, W., Bond, J. O., and Hammon, W. McD. (1970). *Am. J. Trop. Med. Hyg.* **19,** 319.
Schneider, C. R. (1967). *J. Parasitol.* **14,** 78.
Schnurrenberger, P. R., Held, L. J., Martin, R. J., Quist, K. O., and Galton, M. M. (1968). *J. Am. Vet. Med. Assoc.* **153,** 442.
Schwab, G. E., and Reeves, P. R. (1966). *J. Bacteriol.* **91,** 106.
Schwartz, B. (1928). *Proc. U.S. Natl. Mus.* **73,** 1.
Scorza, J. (1957). *Acta Biol. Venez.* **2,** 97.
Scott, H. H. (1925). *Proc. Zool. Soc. London* **95,** 1259.
Scott, H. H. (1926). *Proc. Zool. Soc. London* **96,** 231.
Scott, H. H. (1927). *Proc. Zool. Soc. London* **97,** 173.
Scott, H. H. (1928). *Proc. Zool. Soc. London* **98,** 81.
Scott, H. H. (1930). *Med. Res. Counc. (G.B.), Spec. Rep. Ser.* **149,** 1.
Seddon, H. R. (1943). *Proc. R. Soc. Queensl.* **54,** 1.
Seddon, H. R. (1952). *Aust. Dep. Health, Div. Vet. Hyg., Serv. Publ.* No. 8, pp. 92–93.
Seneviratna, P., Edward, A. G., and De Giusti, D. L. (1975). *Am. J. Vet. Res.* **36,** 337.
Shaw, J. J., and Lainson, R. (1969). *Parasitology* **59,** 223.
Sherwood, B. F., Rowlands, D. T., Jr., Hackel, D. B., and LeMay, J. C. (1968). *Am. J. Pathol.* **53,** 115.
Sherwood, B. F., Rowlands, D. T., Jr., and Hackel, D. B. (1969). *J. Am. Vet. Med. Assoc.* **155,** 1102.
Shope, R. E. (1967a). *U.S., Public Health Serv., Publ.* **1760,** 277–280.
Shope, R. E. (1967b). *U.S., Public Health Serv., Publ.* **1760,** 285–288.
Shope, R. E. (1967c). *U.S., Public Health Serv., Publ.* **1760,** 293–296.
Shope, R. E. (1967d). *U.S., Public Health Serv., Publ.* **1760,** 333–336.

Shope, R. E. (1967e). *U.S., Public Health Serv., Publ.* **1760,** 609–612.

Shope, R. E. (1967f). *U.S., Public Health Serv., Publ.* **1760,** 653–656.

Shope, R. E. (1967g). *U.S., Public Health Serv., Publ.* **1760,** 661–664.

Shope, R. E. (1967h). *U.S., Public Health Serv., Publ.* **1760,** 673–676.

Shope, R. E., and Whitman, L. (1967). *U.S., Public Health Serv., Publ.* **1760,** 677–680.

Shope, R. E., de Andrade, A. H. P., Bensabath, G., Causey, O. R., and Humphrey, P. S. (1966). *Am. J. Epidemiol.* **84,** 467.

Shotts, E. B., Jr., Andrews, C. L., and Harvey, T. W. (1975). *J. Am. Vet. Med. Assoc.* **167,** 587.

Smith, D. J. W. (1942). *Aust. J. Exp. Biol. Med. Sci.* **20,** 213.

Smith, D. J. W., and Derrick, E. H. (1940). *Aust. J. Exp. Biol. Med. Sci.* **18,** 1.

Snowdon, W. A. (1968). *Aust. J. Exp. Biol. Med. Sci.* **46,** 667.

Snyder, R. L., and Ratcliffe, H. L. (1966). *Cancer Res.* **26,** 514.

Solomon, S. G. (1933). *J. Helminthol.* **11,** 101.

Sprent, J. F. A. (1970). *Parasitology* **61,** 75.

Srihongse, S. (1969). *Am. J. Epidemiol.* **90,** 69.

Stabler, R. M., and Welch, K. (1961). *J. Parasitol.* **47,** 576.

Stanley, N. F., and Leak, P. J. (1963). *Am. J. Hyg.* **78,** 82.

Stanley, N. F., Leak, P. J., Grieve, G. M., and Perret, D. (1964). *Aust. J. Exp. Biol. Med. Sci.* **42,** 373.

Starr, L. E., Galton, M. M., Ammons, J., Le Maistre, C., and Menges, R. W. (1959). *Public Health Rep.* **74,** 271.

Stoner, D. (1945). *J. Mammal.* **26,** 192.

Stünzi, H. (1950). *Virchows Arch. Pathol. Anat. Physiol.* **318,** 97.

Sudia, W. D., Coleman, P. H., and Chamberlain, R. W. (1969). *Am. J. Trop. Med. Hyg.* **18,** 98.

Switzer, W. P., Mare, C. J., and Hubbard, E. D. (1966). *Am. J. Vet. Res.* **27,** 1134.

Syverton, J. T., and Berry, G. P. (1940). *Am. J. Hyg.* **32,** 19.

Szyfres, B., and Blood, B. D. (1964). *Trop. Geog. Med.* **16,** 263.

Taylor, D. L., and Burrell, R. (1968). *J. Immunol.* **101,** 1207.

Taylor, R. L., and Shacklette, M. H. (1962). *Am. J. Trop. Med. Hyg.* **11,** 796.

Taylor, R. M., ed. (1967). "Catalogue of Arthropod-borne Viruses of the World," Publ. No. 1760. U.S. Public Health Serv., Washington, D.C.

Teixeira de Freitas, J. F. (1937). *Mem. Inst. Oswaldo Cruz* **32,** 217.

Teixeira de Freitas, J. F., and Lent, H. (1936). *Mem. Inst. Oswaldo Cruz* **31,** 69.

Teleford, S. R., Herrer, A., and Christensen, H. A. (1972). *Ann. Trop. Med. Parasitol.* **66,** 173.

Tesh, R. B., Peralta, P. H., and Johnson, K. M. (1969). *Am. J. Epidemiol.* **90,** 255.

Tesh, R. B., Peralta, P. H., and Johnson, K. M. (1970). *Am. J. Epidemiol.* **91,** 216.

Teuscher, E. (1957). *Zentralbl. Veterinaermed.* **4,** 747.

Thatcher, V. E., Eisenmann, C., and Hertig, M. (1965). *J. Parasitol.* **51,** 842.

Thigpen, J. E., Moore, J. A., Gupta, B. N., and Feldman, D. B. (1975). *J. Am. Vet. Med. Assoc.* **167,** 590.

Thompson, G. P. (1939). *Ann. Mag. Nat. Hist.* **3,** 600.

Thompson, S. W., and Reed, T. H. (1957). *J. Am. Vet. Med. Assoc.* **131,** 545.

Thrasher, J. D. (1970). *Experientia* **26,** 871.

Thrasher, J. D., Barenfus, M., Rich, S. T., and Shope, D. V. (1971). *Lab. Anim. Sci.* **21,** 526.

Tikasingh, E. S. (1974). *Bull., Pan Am. Health Organ.* **8,** 232.

Tilden, E., and Williamson, W. (1957). *J. Am. Vet. Med. Assoc.* **131,** 526.

Toft, J. D., II, Pucak, G. J., and Bullock, B. C. (1973). *Lab. Anim. Sci.* **23,** 431.

Trainer, P. O., and Karstad, L. (1963). *Zoonoses Res.* **2,** 135.
Trapp, E., and Shope, R. E. (1967). *U.S., Public Health Serv., Publ.* **1760,** 249–252.
Travassos, L. (1942). *Rev. Bras. Biol.* **2,** 213.
Travassos, L. (1946). *Rev. Bras. Biol.* **6,** 499.
Travassos, L. (1948). *Proc. Int. Cong. Trop. Med. Malaria, 4th, 1947,* Vol. 1, p. 414.
Travis, B. V. (1941). *J. Parasitol.* **27,** 465.
Triffitt, M. J. (1926). *Protozoology* **2,** 31.
Triffitt, M. J. (1927). *Protozoology* **3,** 75.
Tucker, R., and Millar, R. (1953). *J. Comp. Pathol. Ther.* **63,** 143.
Urbain, A., Nouvel, J., and Bullier, P. (1947). *Bull. Mus. Nat. Hist. Nat., Paris* **19,** 137.
Urbain, A., Nouvel, J., Bullier, P., and Rinjard, J. (1949). *Bull. Mus. Nat. Hist. Nat., Paris* **21,** 178.
Vakilzadeh, J., Rowlands, D. T., Jr., Sherwood, B. F., and LeMay, J. C. (1970). *J. Infect. Dis.* **122,** 89.
Vakilzadeh, J., Sherwood, B. F., Hackel, D. B., and LeMay, J. C. (1971). *Lab. Anim. Sci.* **21,** 224.
Van Cleave, H. J. (1953). *Ill. Biol. Monogr.* **23,** 1.
Vaz, Z., and Perier, C. (1934). *J. Wash. Acad. Sci.* **24,** 54.
Viral and Rickettsial Disease Laboratory. (1967). *U.S., Public Health Serv., Publ.* **1760,** 237–240.
Vogtman, D. B., and Fitzwater, W. D., Jr. (1951). *J. Mammal.* **32,** 211.
Volk, J. J. (1938). *J. Parasitol.* **24,** 547.
Volkmar, F. (1932). *Am. Fur Breeder* **5,** 11.
von Haffner, K. (1913). *Abh. Senckenb. Naturforsch. Ges.* **40,** 361.
von Janicki, C. (1904). *J. R. Microsc. Soc.* [2] **5,** 533.
von Rieth, H. (1966). *Mykosen* **8–9,** 37.
Waddell, M. B., and Taylor, R. M. (1948). *Am. J. Trop. Med.* **28,** 87.
Walton, B. C., and Walls, K. W. (1964). *Am. J. Trop. Med. Hyg.* **13,** 53.
Waring, H. (1956). *Aust. J. Sci.* **18,** 66.
Watts, P. S., and McLean, S. J. (1956). *J. Comp. Pathol.* **66,** 159.
Wellde, B. T., Briggs, N. T., and Sadun, E. H. (1966). *Mil. Med.* **131,** 859.
Welsh, D. A., and Dalyell, E. J. (1910). *J. Pathol. Bacteriol.* **14,** 547.
Welsh, D. A., Dalyell, E. J., and Burfitt, M. B. (1910). *J. Pathol. Bacteriol.* **14,** 542.
Wenyon, C. M., and Scott, H. H. (1925). *Trans. R. Soc. Trop. Med. Hyg.* **19,** 7.
Wesche, W. (1903). *Ann. Mag. Nat. Hist.* **11,** 384.
Wilner, B. I. (1969). "A Classification of the Major Groups of Human and Other Animal Viruses," 4th ed. Burgess, Minneapolis, Minnesota.
Winter, H. (1958). *Aust. Vet. J.* **34,** 118.
Wong, Y. W., Dorsey, C. D., Hunphreys, M. J., and Hansler, W. J. (1970). *Health Lab. Sci.* **7,** 117.
Wood, F. D., and Wood, S. F. (1941). *Am. J. Trop. Med.* **21,** 335.
Yager, R. H., Gochenour, W. S., Jr., Alexander, A. D., and Wetmore, P. W. (1953). *Proc. Soc. Exp. Biol. Med.* **84,** 589.
Yakimoff, W. L., and Matschoulsky, S. N. (1936). *J. Parasitol.* **22,** 514.
Zeledon, R., Solano, G., Saenez, S. G., and Swartzwelder, J. C. (1970). *J. Parasitol.* **56,** 38.
Zimmerman, W. J., and Hubbard, E. D. (1969). *Am. J. Epidemiol.* **90,** 84.

# 9

# A Selected Bibliography of New World Marsupials

DON HUNSAKER II

Allen, J. A. (1900). Note on the generic names *Didelphis* and *Philander*. *Bull. Am. Mus. Nat. Hist.* **13,** 185–190.

Allen, J. A. (1901). A preliminary study of the North American opossums of the genus *Didelphis*. *Bull. Am. Mus. Nat. Hist.* **14,** 149–188.

Altman, D. (1968). Bemerkungen uber ein Wollopossum, *Caluromys laniger* (Desmarest). *Zool. Garten* (*Leipzig*) **35,** 22–29.

Anderson, S., and Jones, J. K., Jr., eds. (1967). "Recent Mammals of the World—Synopsis of Families." Ronald Press, New York.

Asdell, S. A. (1964). "Patterns of Mammalian Reproduction," 2nd ed. Cornell Univ. Press, Ithaca, New York.

Audubon, J. J., and Bachman, J. (1851). "Quadrupeds of North America," Vol. 2. New York.

Azzali, G., and Dio, L. D. (1965). The lymphatic system of *Didelphis azarae* and *Didelphys marsupialis*. *Am. J. Anat.* **116,** 449–470.

Babero, B. B. (1960). Further studies on helminths of the opossum (*Didelphis virginiana*) with a description of new species from this host. *J. Parasitol.* **46,** 455–463.

Baer, J. G. (1927). Die Cestoden der Saugetiere Brasiliens. *Abh. Senckenb. Naturforsch. Ges.* **40,** 375.

Barker, P. R., Bergman, H. C., Hunter, W. H., Lawrence, W. H., Steers, C. W., Wright, L. N., Willis, R. A., and Barrow, D. M. (1967). "Marsupial Biomodule Evaluation Study," Report no. SID66-1647. U.S. Defense Documentation Center, Cameron Station, Alexandria, Virginia.

Barnes, R. D. (1968a). Small marsupials as experimental animals. *Lab Anim. Care* **18,** 251–257.

Barnes, R. D. (1968b). *Marmosa mitis,* a small marsupial for experimental biology. *In* "Animal Models for Biomedical Research," Publ. No. 1594, pp. 88–97. Nat. Acad. Sci., Washington, D.C.

Barnes, R. D. (1977). Special anatomy of *Marmosa robinsoni. In* "The Biology of Marsupials" (D. Hunsaker, ed.), Chapter 7, this volume. Academic Press, New York.

Barnes, R. D., and Barthold, S. W. (1969). Reproduction and breeding behavior in an experimental colony of *Marmosa mitis* Bangs (Didelphidae). *J. Reprod. Fertil.* **6,** 477–482.

Barnes, R. D., and Wolf, H. G. (1971). The husbandry of *Marmosa mitis* as a laboratory animal. *Int. Zoo Yearb.* **11,** 50–54.
Barr, T. R. B. (1961). Experimental rabies in the opossum (*D. marsupialis virginiana* Kerr). *Diss. Abstr.* **21,** 3589.
Barr, T. R. B. (1963). Infectious diseases in the opossum: A review. *J. Wildl. Manage.* **27,** 53–71.
Barton, B. S. (1823). Facts, observations, and conjectures relative to the generation of the opossum of North America. In a letter from Prof. Barton to Mons. Roume of Paris. *Ann. Philos.* [N. S.] **6,** 349–354.
Bates, M. (1944). Experiments with the virus of yellow fever in marsupials with special reference to brown and grey masked opossums. *Am. J. Trop. Med.* **24,** 91.
Bates, M., and Roca-Garcia, M. (1946). Experiments with various Colombian marsupials and primates in laboratory cycles of yellow fever. *Am. J. Trop. Med.* **26,** 437.
Bauman, T. R., and Turner, C. W. (1966). L-thyroxine secretion rates and litrilodothyronine equivalents in the opossum, *Didelphis virginiana. Gen. Comp. Endocrinol.* **6,** 109–113.
Bautista, N.S., and Matzke, H. A. (1965). A degeneration study of the course and extent of the pyramidal tract of the opossum. *J. Comp. Neurol.* **124,** 367–376.
Beach, F. A. (1939). Maternal behavior of the pouchless marsupial *Marmosa cinerea. J. Mammal.* **20,** 315–322.
Beamer, P. D., Mohr, C. O., and Barr, T. R. B. (1960). Resistance of the opossum to rabies virus. *Am. J. Vet. Res.* **21,** 507–510.
Beebe, W. (1923). Jacking for yapocks. *Zool. Soc. N.Y. Bull.* **26,** 150–156.
Bensley, R. R. (1914). The thyroid gland of the opossum. *Anat. Rec.* **8,** 431–440.
Bergman, H. C., and Housley, C. (1968). Chemical analysis of American opossum (*Didelphis virginiana*) milk. *Comp. Biochem. Physiol.* **25,** 213–218.
Bergquist, E. H. (1970). Output pathways of hypothalamic mechanisms for sexual aggressive and other motivated behavior in opossums. *J. Comp. Physiol. Psychol.* **70,** 389–398.
Bick, Y. A. E. (1970). Comparison of the effects of LSD, heliotrine, and X-irradiation on chromosome breakage and the effects of LSD on the rate of cell division. *Nature* (*London*) **226,** 1165–1167.
Bick, Y. A. E., and Brown, J. K. (1969). Diploidy and contact inhibition in long-established marsupial cell line (HPKI). *Cytobios* **2,** 123.
Biggers, J. D. (1966). Reproduction in male marsupials. *Symp. Zool. Soc. London* **15,** 251–280.
Biggers, J. D. (1967). Notes on reproduction of the woolly opossum (*Caluromys derbianus*) in Nicaragua. *J. Mammal.* **48,** 678–680.
Biggers, J. D., and DeLamater, E. D. (1965). Marsupial spermatozoa; Pairing in the epididymis of the American forms. *Nature* (*London*) **208,** 402–404.
Biggers, J. D., Fritz, H. I., Hare, W. C. D., and McFeely, R. A. (1965). Chromosomes of American marsupials. *Science* **148,** 1602–1603.
Blair, W. F. (1936). An opossum dies of cold and hunger. *J. Mammal.* **17,** 410.
Block, M. (1960). Wound healing in the newborn opossum (*Didelphis virginiana*). *Nature* (*London*) **187,** 340.
Block, M. (1964). The blood forming tissues and blood of the newborn opossum (*Didelphis virginiana*) L. Normal development through about the one hundredth day of life. *Ergeb. Anat. Entwicklungsgesch.* **37,** 237–366.
Bodian, D. (1939). Studies on the diencephalon of the Virginia opossum. *J. Comp. Neurol.* **71,** 259–323.
Bodian, D. (1963). Cytological aspects of neurosecretion in opossum neurohypophysis. *Bull. Johns Hopkins Hosp.* **131,** 57–93.

Boggs, J. R. (1969). The general and agonistic behavior of the mouse opossum, *Marmosa robinsoni.* Master's Thesis, San Diego State University, San Diego, California.

Bolliger, A., and Gross, R. (1960). Nutrition of the marsupial suckling. *Aust. Sci.* **22,** 292–294.

Bremer, J. L. (1904). On the lung of the opossum. *Am. J. Anat.* **3,** 67–73.

Bringham, G. D. (1936). Susceptibility of the opossum (*D. virginiana*) to the virus of endemic typhus fever. *Public Health Rep.* **51,** 333–337.

Brinkley, B. R., and Humphrey, R. M. (1969). Evidence for subchromatial organization in marsupial chromosomes. *J. Cell Biol.* **42,** 827–831.

Brocke, R. H. (1970). Ecological inferences from oxygen consumption data on the opossum. *Bull. Ecol. Soc. Am.* **51,** 29.

Bryant, B. J. (1977). The development of the lymphatic and immunohematopoietic systems. *In* "The Biology of Marsupials" (D. Hunsaker, ed.), Chapter 6, this volume. Academic Press, New York.

Bugher, J. C., Boshell-Manrique, J., Roca-Garcia, M., and Gilmore, R. M. (1941). The susceptibility to yellow fever of the vertebrates of Eastern Colombia. *Am. J. Trop. Med.* **21,** 309–333.

Burns, R. K., and Burns, L. M. (1957). Observations on the breeding of the American opossum in Florida. *Rev. Suisse Zool.* **64,** 595–605.

Butler, P. M., and Zielan-Jaworowska, Z. (1973). Is *Deltatheridium* a marsupial? *Nature (London)* **245,** 105–106.

Byrd, E. C., Reiber, R. J., and Parker, M. V. (1942). Mammalian trematodes. I. Trematodes from the opossum, *D. virginiana* Kerr. *J. Tenn. Acad. Sci.* **17,** 130–142.

Cabrera, A. (1958). Catalogo de los mamiferos de America del Sur. 1. *Rev. Mus. Argent. Cienc. Nat.* "*Bernardino Rivadavia*" *Inst. Nac. Invest. Ciene, Nat., Cienc. Geol.* **4,** 1–307.

Cabrera, A., and Yepes, J. (1940). "Mamiferos Sud-Americanos." Compania Argentina de Editores, Buenos Aires.

Cadzow, G. (1974). Observations of pouch environment and behavior of pouch young of the Virginia opossum. Master's Thesis, California State University, Long Beach.

Chandler, A. C. (1932). Notes on the helminth parasites of the opossum in southeast Texas with descriptions of four new species. *Proc. U.S. Natl. Mus.* **81,** 1–15.

Chase, E. B. (1939). The reproductive system of the male opossum. *Didelphis virginiana,* and its experimental modifications. *J. Morphol.* **65,** 215–239.

Christensen, A. K., and Fawcett, D. W. (1961). The normal free structure of opossum testicular interstitial cells. *J. Biophys. Biochem. Cytol.* **9,** 653–670.

Clemens, W. A. (1968). Origin and early evolution of marsupials. *Evolution* **22,** 1–18.

Coghill, G. E. (1939). Studies on rearing the opossum (*Didelphis virginiana*). *Ohio J. Sci.* **39,** 239–249.

Collins, L. R. (1973). "Monotremes and Marsupials, A Reference for Zoological Institutions," Smithson. Inst. Publ. No. 4888. Random House (Smithsonian Inst. Press), New York.

Cook, B., and Nalbandov, A. V. (1968). The effect of some pituitary hormones on progesterone synthesis *in vitro* by the luteinized ovary of the common opossum. *J. Reprod. Fertil.* **15,** 267–275.

Coues, E. (1972). On the osteology and myology of *D. virginiana. Mem. Boston Soc. Nat. Hist.* **2,** 41–149.

Crandall, L. S. (1964). "The Management of Wild Animals in Captivity." Univ. of Chicago Press, Chicago, Illinois.

Davis, D. E. (1945). The annual cycle of plants, mosquitoes, birds, and mammals in two Brazilian forests. *Ecol. Monogr.* **15,** 243–295.

Davis, J. A., Jr. (1966). Maverick opossums. *Anim. Kingdom* **69,** 112–117.

Davis, P. J., and Jurgelski, W. (1973). Thyroid hormone binding in opossum serum: Evidence for polymorphism and relationship to haptoglobin polymorphism. *Endocrinol.* **92,** 822–832.

de Almeida, M. O., and Rocha, A. (1932). Sur la composition de l'airde la poche mammaire du gamba (*Didelphis*). *C. R. Seances Soc. Biol. Ses Fil.* **109,** 131.

de Buffon, G. L. L. (1763). Le sarigue ou l'opossum. *In* "Historie Naturelle Générale et Particulière, *avec la description* du cabinet du roi" (G. L. L. de Buffon and L. J. M. Daubenton, eds.), Vol. 10, pp. 279–310. L'Imprinabrie Royal, Paris.

DeLamater, E. D., and Biggers, J. D. (1964). Spermatogenesis and spermateliosis in the opossum (*Didelphis virginiana*) by histochemical and electron microscopical methods. *J. Histochem. Cytochem.* **12,** 35–36.

de Miranda-Ribeiro, A. (1936). Didelphia au mammalia—ovovipera. Marsupials, didelphos, pedimanos ou metatherios. *Rev. Mus. Paulista* **20,** 245–427.

Dickerson, L. M. (1930). A new variety of *Harmostomum opisthotrias* from the North American opossum, *Didelphis virginiana,* with a discussion of its possible bearing on the origin of its host. *Parasitology* **22,** 37–46.

Doutt, K. (1954). The swimming of the opossum, *Didelphis marsupialis virginiana. J. Mammal.* **35,** 581–583.

Downing, S. C. (1948). A provisional check-list of the mammals of Ontario. *R. Ont. Mus. Zool., Misc. Publ.* **2,** 11.

Ebner, F. F. (1967). Afferent connections to neocortex in the opossum (*Didelphis virginiana*). *J. Comp. Neurol.* **129,** 241–268.

Eisenberg, J., and Maliniak, E. (1967). Breeding the murine opossum *Marmosa* sp. in captivity. *Int. Zoo Yearb.* **7,** 78–79.

Eisenberg, J., and Thorington, R. W. (1973). A preliminary analysis of a neotropical fauna. *Biotropica* **9,** 150–161.

Eisentraut, M. (1970). Beitrag zur Fortpfzanzungsbiologie der Zwerbeutelratte *Marmosa murina* (Didelphidae, Marsupialia). *Z. Saeugetierkd.* **35,** 159–173.

Ellsworth, A. F. (1976). "North American Opossum [An Anatomical Atlas]." Robert E. Krieger Pub. Inc., Huntington, New York.

Enders, A. C., and Enders, R. K. (1969). The placenta of the four-eyed opossum (*Philander opossum*). *Anat. Rec.* **165,** 431–450.

Enders, R. K. (1935). Mammalian life histories from Barro Colorado Island, Panama. *Bull. Mus. Comp. Zool.* **78,** 383–502.

Enders, R. K. (1937). Panniculus carnosus and formation of the pouch in *Didelphis. J. Morphol.* **61,** 1–26.

Enders, R. K. (1966). Attachment, nursing, and survival of young in some didelphids. *Symp. Zool. Soc. London* **15,** 195–203.

Farber, J. P., and Tenney, S. M. (1970), Development and ventilatory control mechanisms in the Virginia opossum. *Physiologist* **13,** 194.

Farber, J. P., and Tenney, S. M. (1971). The pouch gas of the Virginia opossum (*Didelphis virginiana*). *Resp. Physiol.* **11,** 335–345.

Farris, E. J. (1950). The opossum. *In* "Care and Breeding of Laboratory Animals" (E. J. Farris, ed.), pp. 256–267. Wiley, New York.

Feldman, D. B., and Self, J. L. (1973). Establishment of a helminth free opossum colony. *Lab. Anim. Sci.* **23,** 855–857.

Feldman, D. B., Moore, J. A., Harris, M. W., and Self, J. L. (1972). Characteristics of common helminths of the Virginia opossum (*Didelphis virginiana*) from North Carolina. *Lab. Anim. Sci.* **22,** 183.

Fitch, H. S., and Sandidge, L. L. (1953). Ecology of the opossum on a natural area in northeastern Kansas. *Publ. Mus. Nat. Hist. Univ. Kans.* **7**, 305–338.

Fitch, H. S., and Shirer, H. W. (1970). A radiotelemetric study of spatial relationships in the opossum. *Am. Midl. Nat.* **84**, 170–186.

Fleming, T. H. (1972). Aspects of the population dynamics of three species of opossums in the Panama Canal Zone. *J. Mammal.* **53**, 619–623.

Fleming, T. H. (1973). The reproductive cycles of three species of opossums and other mammals in the Panama Canal Zone. *J. Mammal.* **54**, 439–455.

Foster, A. O. (1939). Some helminths of the woolly opossum in Panama. *Trans. Am. Microsc. Soc.* **58**, 185.

Francq, E. N. (1969). Behavioral aspects of feigned death in the opossum *Didelphis marsupialis. Am. Mild. Nat.* **81**, 556–568.

Francq, E. N. (1970). Electrocardiograms of the opossum, *Didelphis marsupialis*, during feigned death. *J. Mammal.* **51**, 395.

Friedman, H. (1964). Taming of the Virginia opossum. *Nature (London)* **201**, 323–324.

Friedman, H. (1967). Colour vision in the Virginia opossum. *Nature (London)* **213**, 835–836.

Friedman, H., and Marshall, D. (1965). Position reversal training in the Virginia opossum: Evidence for the acquisition of a learning set. *Q. J. Exp. Psychol.* **17**, 250–254.

Fritz, H. I. (1971). Maintenance of the common opossum in captivity. *Int. Zoo Yearb.* **11**, 46–49.

Gardner, A. L. (1973). "Systematics of the genus *Didelphis*," Spec. Publ. Mus. Texas Tech. Univ., Lubbock.

Gilmore, R. M. (1943). Mammalogy in an epidemiology study of jungle yellow fever in Brazil. *J. Mammal.* **24**, 144–162.

Goldman, E. A. (1920). Mammals of Panama. *Smithson. Misc. Collect.* **69**, No. 5.

Goldman, E. A. (1951). Biological investigations in Mexico. *Smithson. Misc. Collect.* **115**.

Gregory, W. K. (1922). On the "habitus" and "heritage" of *Caenolestes. J. Mammal.* **3**, 106–114.

Grinnell, J. (1915). The Tennessee opossum has arrived in California. *Calif. Fish Game* **1**, 114–116.

Guilday, J. E. (1958). The prehistoric distribution of the opossum. *J. Mammal.* **39**, 39–43.

Hall, E. R., and Kelson, K. R. (1952). Comments on the taxonomy and geographic distribution of some North American marsupials, insectivores, and carnivores. *Publ. Mus. Nat. Hist. Univ. Kan.* **9**, 319–342.

Hall, E. R., and Kelson, K. R. (1959). "The mammals of North America," Vol. 1. Ronald Press, New York.

Hamilton, W. J., Jr. (1951). The food of the opossum in New York State. *J. Wildl. Manage.* **15**, 258–264.

Hamilton, W. J., Jr. (1958). Life history and economic relation of the opossum (*Didelphis marsupialis virginiana*) in New York State. *N.Y., Agric. Exp. Stn., Ithaca, Mem.* **354**, 1–48.

Hamilton, W. J., Jr. (1963). Success story of the opossum. *Nat. Hist. (N.Y.)* **72**, 16–25.

Harrison, L., and Johnston, H. T. (1916). Mallophaga from marsupials. *Parasitology* **8**, 338.

Hartman, C. G. (1920). Studies in the development of the opossum *Didelphis virginiana* L. (The phenomena of parturition). *Anat. Rec.* **19**, 251–261.

Hartman, C. G. (1922). Breeding habits, development, and birth of the opossum. *Smithson. Inst. Annu. Rep.* (1921), pp. 347–364.

Hartman, C. G. (1924). The oestrous cycle in the opossum. *Am. J. Anat.* **32**, 353–421.

Hartman, C. G. (1928). The breeding season of the opossum (*D. virginiana*) and the rate of intra-uterine and post-natal development. *J. Morphol. Physiol.* **46,** 143–215.

Hartman, C. G. (1952). "Possums." Univ. of Texas Press, Austin.

Hayes, T. G. (1968). Studies of a primitive mammalian spleen, the opossum (*Didelphis virginiana*). *J. Morphol.* **124,** 445–450.

Hayman, D. L., Kirsch, J. A. W., Martin, P. G., and Waller, P. F. (1971). Chromosomal and serological studies of the Caenolestidae and their implications for marsupial evolution. *Nature* (*London*) **321,** 194–195.

Herrick, C. J. (1921). The brain of *Caenolestes obscurus. Field Mus. Nat. Hist. Publ., Zool. Ser.* **14,** 157–162.

Hershkovitz, P. (1969). The evolution of mammals on southern continents. *Q. Rev. Biol.* **44,** 1–70.

Hershkovitz, P. (1949). Generic names of the four-eyed pouch opossum and the woolly opossum (Didelphidae). *Proc. Biol. Soc. Wash.* **62,** 11–12.

Hibbard, C. W., Ray, D. E., Savage, D. E., Taylor, D. W., and Guilday, J. E. (1965). The Quarternary of the United States. *In* "Quaternary Mammals of North America" (H. E. Wright and D. G. Frey, eds.), pp. 509–525. Princeton Univ. Press, Princeton, New Jersey.

Higginbotham, A. C., and Koon, W. E. (1955). Temperature regulation in the Virginia opossum. *Am. J. Physiol.* **181,** 69–71.

Hildebrand, M. (1961). Body proportions of didelphid (and some other) marsupials, with emphasis on variability. *Am. J. Anat.* **109,** 239–249.

Hill, J. P. (1918). Some observations on the early development of *Didelphys aurita. Q. J. Microsc. Sci.* **63,** 91–139.

Holloway, H. L., Jr., and Dowler, J. L. (1963). The helminths of opossums in Western Virginia. *Va. J. Sci.* **14,** 203.

Hoversland, A. S., Dhindsa, D. S., Murphy, W. S., Parer, J. T., and Metcalfe, J. (1973). Respiratory properties and 2,3-diphosphoglycerate (DPG) concentration in blood of the adult opossum. *Physiologist* **16,** 348.

Hoversland, A. S., Dhindsa, D. S., Murphy, W. S., Parer, J. T., and Metcalfe, J. (1975). Oxygen transport and hemodynamics in unanesthetized American opossum (*Didelphis virginiana*). *Comp. Biochem. Physiol.* **50A,** 519–525.

Hudson, G. E. (1932). On the food habits of *Marmosa. J. Mammal.* **13,** 159.

Hunsaker, D. (1970) "Ecological Notes on the Mouse Opossum, *Marmosa mitis,*" Marsupial Newsletter, II. University of California, Davis.

Hunsaker, D. (1977). Ecology of the New World Marsupials. *In* "The Biology of Marsupials" (D. Hunsaker, ed.), Chapter 3, this volume. Academic Press, New York.

Hunsaker, D., and Shupe, D. V. (1977). Behavior of New World Marsupials. *In* "The Biology of Marsupials" (D. Hunsaker, ed.), Chapter 5, this volume. Academic Press, New York.

James, W. T. (1937). An experimental study of the defense mechanism in the opossum with emphasis on natural behavior and its relation to mode of life. *J. Genet. Psychol.* **51,** 95–100.

James, W. T. (1955). The behavior of the opossum in the Guthrie–Horton puzzle box. *J. Genet. Psychol.* **87,** 203–206.

James, W. T. (1958). Conditioned responses in the opossum. *J. Genet. Psychol.* **93,** 179–183.

James, W. T. (1959). Behavior of the opossum in the Fink–Arrow maze. *J. Genet. Psychol.* **94,** 199–203.

James, W. T. (1960). A study of visual discrimination in the opossum. *J. Genet. Psychol.* **97,** 127–130.

James, W. T., and Turner, W. W. (1963). Experimental study of maze learning in young opossums. *Psychol. Rep.* **13,** 921–922.

Jenkins, F. A., Jr. (1971). Limb posture and locomotion in the Virginia opossum (*Didelphis marsupialis*) and in other non-cursorial mammals. *J. Zool.* **165,** 303–315.

Jewett, S. G., and Dobyns, H. W. (1929). The Virginia opossum in Oregon. *J. Mammal.* **10,** 351.

Johnson, J. (1977). Central nervous systems in marsupials. *In* "The Biology of Marsupials" (D. Hunsaker, ed.), Chapter 4, this volume. Academic Press, New York.

Jones, M. L. (1963). Mammals in captivity. Unpublished ms. (rev. ed.).

Jurgelski, W., Jr., and Porter, M. (1974). III. Breeding the opossum in captivity: Methods. *Lab. Anim. Sci.* **24,** 412–425.

Jurgelski, W., Jr. (1971). A method of administering test materials to the newborn opossum (*Didelphis marsupialis virginiana* Kerr). *Lab. Anim. Sci.* **21,** 748–751.

Jurgelski, W., Jr. (1974). The opossum (*Didelphis virginiana* Kerr) as a biomedical model. I. Research perspective, husbandry, and laboratory techniques for the opossum. *Lab. Anim. Sci.* **24,** 376–403.

Jurgelski, W., Jr., Forsythe, W., Dahl, D., Thomas, L. D., Moore, J. A., Kotin, P., Falk, H. L., and Vogel, F. S. (1974). The opossum (*Didelphis virginiana* Kerr) as a biomedical model. II. Breeding the opossum in captivity: Facility design. *Lab Anim. Sci.* **24,** 404–411.

Kennard, F. H. (1925). The Virginia opossum in Massachusetts and New Hampshire. *J. Mammal.* **6,** 196.

Kerr, R. (1792). "The Animal Kingdom or Zoological system of the Celebrated Sir Charles Linnaeus. Class I Mammalia: Containing a complete systematic description, arrangement, and nomenclature, of all the known species and varieties of the mammalia, or animals which give suck to their young; being a translation of that part of the Systema Natura, as lately published, with great improvements, by Professor Gmelin of Goettingen. Together with numerous additions from more recent zoological writers, and illustrated with copperplates." A. Strahan and T. Cadell, London, and W. Creech, Edinburgh.

Kingsbury B. F. (1940). The development of the pharyngeal derivatives of the opossum (*Didelphis virginiana*) with special reference to the thymus. *Am. J. Anat.* **67,** 393–435.

Kirsch, J. A. W. (1968). Prodromus of the comparative serology of marsupials. *Nature (London)* **217,** 418–420.

Kirsch, J. A. W. (1977). The classification of marsupials. *In* "The Biology of Marsupials" Hunsaker, ed.), Chapter 1, this volume. Academic Press, New York.

Krause, W. J., and Leeson, C. R. (1973). The postnatal development of the respiratory system of the opossum. I. Light and scanning electron microscopy. *Am. J. Anat.* **137,** 337–356.

Krupp, J. H. (1962). Treatment of opossums with *Physaloptera* infections. *J. Am. Vet. Med. Assoc.* **141,** 369–370.

Krupp, J. H. (1966). Parasitic diseases of the opossum. *Lab. Anim. Dig.* **4,** 12–13.

Krupp, J. H., and Quillin, R. A. (1964). Review of the use of the opossum for research-husbandry, experimental techniques and routine health measures. *Lab. Anim. Care* **14,** 189–194.

Ladman, A. J. (1967). The fine structure of the ductuli efferentes of the opossum. *Anat. Rec.* **157,** 559–576.

Laemmert, H. W., Jr. (1946). Studies on susceptibility of marsupials to different strains of yellow fever virus. *Am. J. Trop. Med.* **26,** 33.

Langworthy, O. R. (1925). The development of progression and posture in young opossums. *Am. J. Physiol.* **74,** 1–13.

Langworthy, O. R. (1927). Correlated physiological and morphological studies of the development of electrically responsive areas in the cerebral cortex of the opossum. *Contrib. Embryol. Carnegie Inst.* **19,** 149–175.

Langworthy, O. R. (1928). The behavior of pouch young opossums correlated with the myelination of tracts in the nervous system. *J. Comp. Neurol.* **46,** 201–247.

La Plante, G. S., and Burrell, R. (1966). Bacterial endocarditis in opossums. *Bull. Wildl. Dis. Assoc.* **2,** 10–12.

La Plante, G. S., Burrell, R., Watne, A. L., Taylor, D. L., and Zimmerman, B. (1969). Skin allograft studies in the pouch young of the opossum. *Transplantation* **7,** 67–72.

La Via, M., Rowlands, D., and Block, M. (1963). Antibody formation in embryos. *Science* **140,** 1219–1220.

Lay, D. W. (1942). Ecology of the opossum in eastern Texas. *J. Mammal.* **23,** 147–159.

Layne, J. M. (1951). The use of the tail by an opossum. *J. Mammal.* **32,** 464–465.

Layne, J. M., and Benton, A. H. (1954). Some speeds of small animals. *J. Mammal.* **35,** 103–104.

Lende, R. A. (1963a). Sensory representation in the cerebral cortex of the opossum (*Didelphis virginiana*). *J. Comp. Neurol.* **121,** 395–403.

Lende, R. A. (1963b). Motor representation in the cerebral cortex of the opossum (*Didelphis virginiana*). *J. Comp. Neurol.* **121,** 405–415.

Lillegraven, J. A. (1969). Latest cretaceous mammals of upper part of Edmonton formation of Alberta, Canada, and review of marsupial–placental dichotomy in mammalian evolution. *Univ. Kans. Paleontol. Contrib.* **50,** 1–122.

Lillegraven, J. A. (1974). Biogeographical considerations of the marsupial–placental dichotomy. *Annu. Rev. Ecol. Syst.* **5,** 263–283.

Lillegraven, J. A. (1975). Biological considerations of the marsupial–placental dichotomy. *Evolution* **29,** 707–722.

Llewellyn, L. M., and Dale, E. H. (1964). Notes on the ecology of the opossums in Maryland. *J. Mammal.* **45,** 113–122.

Loo, Y. T. W. (1930). The forebrain of the opossum *Didelphis virginiana. J. Comp. Neurol.* **51,** 13–64.

Lowrance, E. W. (1949). Variability and growth of the opossum skeleton. *J. Morphol.* **85,** 569–593.

Lustic, S., and Lustic, D. D. (1972). Energetics in the opossum, *Didelphis marsupialis virginiana. Comp. Biochem. Physiol. A* **43,** 643–647.

McCrady, E., Jr. (1938). The embryology of the opossum. *Am. Anat. Mem.* **16,** 1–233.

McKeever, S., Schubert, J. H., Moody, M. D., Gorman, G. W., and Chapman, J. F. (1958). Natural occurrence of tularemia in marsupials, carnivores, lagomorphs, and large rodents in southwestern Georgia and northwestern Florida. *J. Infect. Dis.* **103,** 120–126.

McManus, J. J. (1967). Observations on sexual behavior in the opossum, *Didelphis marsupialis. J. Mammal.* **48,** 486–487.

McManus, J. J. (1969). Temperature regulation in the opossum, *Didelphis marsupialis virginiana. J. Mammal.* **50,** 550–558.

McManus, J. J. (1970). Behavior of captive opossums, *Didelphis marsupialis virginiana. Am. Midl. Nat.* **84,** 144–169.

McManus, J. J. (1971). Activity of captive *Didelphis marsupialis. J. Mammal.* **52,** 846–848.

McManus, J. J. (1974). *Didelphis virginiana. Mamm. Species* **40,** 1–6.

Magne de la Croix, P. (1936). The evolution of locomotion in mammals. *J. Mammal.* **17,** 51–54.

Major, P. C., and Burrell, R. (1971). Induction of acquired tolerence in neonatal opossums, *J. Immunol.* **106,** 1621–1690.

Maller, O., Clark, J. M., and Kare, M. R. (1965). Short term caloric regulation in the adult opossum (*Didelphis virginiana*). *Proc. Soc. Exp. Biol. Med.* **118,** 275–277.

Mann, G. F. (1955). Monito del Monte. *Dromiciops australis* Philippe. *Invest. Zool. Chil.* **2,** 159–166.

Mann, G. F. (1958). Reproduction de *Dromiciops australis* (Marsupialia: Didelphydae). *Invest. Zool. Chil.* **4,** 209–213.

Marshall Graves, J. A. (1967). DNA synthesis in chromosomes of cultural leucocytes from two marsupial species. *Exp. Cell Res.* **46,** 37–57.

Martan, J., and Allen, J. M. (1962). The cytology and chemical organization of the prostrate epithelium of *Didelphis virginiana* Kerr. *J. Exp. Zool.* **159,** 209–230.

Martin, C. J. (1903). Thermal adjustment and respiratory exchange in monotremes and marsupials—a study in the development of homoeothermism. *Philos. Trans. R. Soc. London, Ser. B* 1–37.

Martin, G. F., and King, J. S. (1968). The basilar pontine gray of the opossum (*Didelphis virginiana*). *J. Comp. Neurol.* **133,** 447–462.

Martin, G. F., Dom, R., King, J. S., RoBards, M., and Watson, C. R. R. (1975). The inferior olivary nucleus of the opossum (*Didelphis marsupialis virginiana*). Its organization and connections. *J. Comp. Neurol.* **160,** 507–534.

Martinez-Esteve, P. (1937). Le cycle sexuel vaginal chez le marsupial *Didelphis azarae. C. R. Soc. Biol. Paris* **124,** 502–504.

Martinez-Esteve, P. (1942). Observations on the history of the opossum ovary. *Contrib. Embryol. Carnegie Inst.* **189,** 19–26.

Mays, A., and Loew, F. M. (1968). Hemograms of laboratory confined opossums (*Didelphis virginiana*). *J. Am. Vet. Med. Assoc.* **153,** 800–802.

Mease, J. A., Jr. (1929). Tularemia from opossums. *J. Am. Vet. Med. Assoc.* **92,** 1042.

Mihailoff, G. A., and King, J. S. (1974). The basilar pontine gray of the opossum: A correlated light and electron microscopic analysis. *J. Comp. Neurol.* **159,** 521–552.

Mizell, M., and Isaaca, J. J. (1970). Induced regeneration of hind limbs in the newborn opossum. *Am. Zool.* **10,** 141–155.

Monodolfi, E., and Podilla, G. M. (1957). Contribucion al conocimiento del "Perrito de Agua" (*Chironectes minimes* Zimmerman). *Mem. Soc. Cienc. Nat. La Salle* **17,** 141–149.

Moore, C. R. (1955). Opossum taking refuge under water. *J. Mammal.* **36,** 559–561.

Moore, J. C. (1939). Modification of sexual development in the opossum by sex hormones (steroids). *Proc. Soc. Exp. Biol. Med.* **40,** 544–546.

Morrison, P. R. (1946). Temperature regulation in three Central American mammals. *J. Cell. Comp. Physiol.* **27,** 125–137.

Morrison, P. R., and McNab, B. K. (1962). Daily torpor in a Brazilian murine opossum (*Marmosa*). *Comp. Biochem. Physiol.* **6,** 57–68.

Morrison, P. R., and Petajan, J. H. (1962). The development of temperature regulation in the opossum, *Didelphis marsupialis virginiana. J. Physiol. Zool.* **35,** 52–65.

Nardone, R. M., Wilber, C. G., and Musacchia, X. J. (1955). Electrocardiogram of opossum during exposure to cold. *Am. J. Physiol.* **181,** 352–356.

Nelson, L. R., and Lende, R. A. (1965). Interhemispheric responses in the opossum. *J. Neurophysiol.* **28,** 189–199.

Nelson, O., and White, E. L. (1941). A method for inducing ovulation in the anestrous opossum (*Didelphis virginiana*). *Anat. Rec.* **81,** 529.

New, D. A. T., and Mizell, M. (1972). Opossum fetuses grown in culture. *Science* **175,** 523–536.

Norton, A. C., Beran, A. V., and Misrahy, G. A. (1964). Electroencephalograph during feigned sleep in the opossum. *Nature* (*London*) **204,** 162–163.

Obenchain, J. B. (1925). The brains of the South American marsupials *Caenolestes* and *Orolestes. Field Mus. Nat. His. Publ. Zool. Ser.* **14,** 173–232.

Osgood, W. H. (1921). A monographic study of the American marsupial *Caenolestes. Field Mus. Nat. Hist. Publ., Zool. Ser.* **14,** 1–156.

Osgood, W. H. (1924). Review of living caenolestids with description of a new genus from Chile. *Field Mus. Nat. Hist. Publ. Zool. Ser.* **14,** 163–172.

Osgood, W. H. (1943). Mammals of Chile. *Field Mus. Nat. Hist. Publ., Zool. Ser.* **20,** 1–268.

Oswaldo-Cruz, E. and Rocha-Miranda, C. E. (1968). The brain of the opossum (*Didelphis marsupialis*): A cytoarchitectonic atlas in stereotaxic coordinates. Inst. de Biofisica, Univ. Fed. do Rio de Janeiro. 1–99.

Padykula, H. A. (1976). Cellular mechanisms involved in cyclic stromal renewal of the uterus. III. Cells of the immune response. *Anat. Rec.* **184,** 49–72.

Padykula, H. A., and Taylor, J. M. (1976). Cellular mechanisms involved in cyclic stromal renewal of the uterus. I. The opossum, *Didelphis virginiana. Anat. Rec.* **184,** 5–26.

Parmelee, A. H., Jr., Brooks, M. B., and Stone, R. S. (1960). Effect of high oxygen concentration on opossum pouch young (*Didelphis virginiana*). *Anat. Rec.* **136,** 73–78.

Petajan, J. H., and Morrison, P. R. (1962). Physical and physiological factors modifying the development of temperature regulation in the opossum. *J. Exp. Zool.* **149,** 45–57.

Petajan, J. H., Morrison, P. R., and Akert, K. (1962). Localization of central nervous control of temperature regulation in the opossum. *J. Exp. Zool.* **150,** 225–231.

Peterson, R., and Downing, S. C. (1956). Distributional records of the opossum in Ontario. *J. Mammal.* **37,** 431–434.

Petrides, G. A. (1949). Sex and age determination in the opossum. *J. Mammal.* **30,** 364–378.

Petty, C., Bethlenfalvay, N. C., and Bageant, T. (1975). Spectrophotometric measurement of hemoglobin oxygen saturation in the opossum *Didelphis virginiana, Comp. Biochem. Physiol.* **50A,** 273–276.

Phillips, C. J., and Jones, J. K. (1968). Additional comments on reproduction in the woolly opossum (*Caluromys derbianus*) in Nicaragua. *J. Mammal.* **49,** 320–321.

Phillips, C. J., and Jones, J. K. (1969). *Notes on the reproduction and development* in *The four-eyed opossum, Philander opossum,* in Nicaragua. *J. Mammal.* **50,** 345–348.

Pine, R. H. (1973). Anatomical and nomenclature notes on opossums. *Proc. Biol. Soc. Wash.* **86,** 391–402.

Plakke, R. K. (1970). Urea, electrolyte and total solution excretion following water deprivation in the opossum (*Didelphis marsupialis virginiana*). *Comp. Biochem. Physiol.* **34,** 325–332.

Plakke, R. K., and Pfeiffer, E. W. (1965). Influence of plasma urea on urine concentration in the opossum (*Didelphis marsupialis virginiana*). *Nature* (*London*) **207,** 866–867.

Potkay, S. (1970). Diseases of the opossum (*Didelphys marsupialis*): A review. *Lab. Anim. Care* **20,** 502–511.

Potkay, S. (1977). Diseases of marsupials. *In* "The Biology of Marsupials" (D. Hunsaker, II, ed.), Chapter 8, this volume. Academic Press, New York.

Pray, L. L. (1921). Opossum carries leaves with its tail. *J. Mammal.* **2,** 109–110.

Pressman, T. G., and Doolittle, J. H. (1966). Taste preference in the Virginia opossum. *Psychol. Rep.* **18,** 875–878.

Pubols, B. H., Jr., Pubols, L. M., DiPette, D. J., and Sheely, J. C. (1976). Opossum somatic sensory cortex: A microelectrode mapping study. *J. Comp. Neurol.* **165,** 229–246.

Pucak, G. J., Henderson, J. D., Jr., and Bullock, B. C. (1969). *Pasteurella* multocide septicemia in a colony of *Marmosa mitis. J. Am. Vet. Med. Assoc.* **155,** 1228.

Reig, O. A., and Bianchi, N. O. (1969). The occurrence of an intermediate didelphid karyotype in the short-tailed opossum (Genus *Monodelphis*). *Experientia* **25,** 1210–1211.

Reig, O. A., and Sonnenscheim, C. (1970). The chromosomes of *Marmosa fuscata* Thomas, from northern Venezuela (Marsupialia, Didelphidae). *Experientia* **26,** 199–201.

Reiss, J. A., Neu, R. L., Kajii, T., and Gardner, L. I. (1970). Late DNA replication in the sex chromosomes of *Didelphis virginiana. Experientia* **26,** 917–918.

Renfree, M. B. (1974). Ovariectomy during gestation in the American opossum, *Didelphis marsupialis virginiana*, *J. Reprod. Fert.* **39,** 127–130.

Renfree, M. B. (1975). Uterine proteins in the marsupial, *Didelphis marsupialis virginiana*, during gestation, *J. Reprod. Fert.* **42,** 163–166.

Renfree, M. B., and Fox, D. J. (1975). Pre- and postnatal development of lactate and malate dehydrogenases in the marsupial *Didelphis marsupialis virginiana*, *Comp. Biochem. Physiol.* **52B,** 347–350.

Reynolds, H. C. (1945). Some aspects of the life history and ecology of the opossum in central Missouri. *J. Mammal.* **26,** 361–379.

Reynolds, H. C. (1952). Studies on reproduction in the opossum (*Didelphis virginiana*). *Univ. Calif., Berkeley, Publ. Zool.* **52,** 223–284.

Reynolds, H. C. (1953). The opossum. *Sci. Am.* **188,** 88–94.

Riese, W. (1945). Structure and function of the brain of the opossum (*Didelphis virginiana*) at the time of birth. *J. Mammal.* **26,** 148–153.

Rink, R., and Miller, J. A., Jr. (1967). Temperature, weight (=age), and resistance to asphyxia in pouch-young opossums. *Cryobiology* **4,** 24–29.

Roberts, W. W., Steinberg, M. L., and Means, L. W. (1967). Hypothalamic mechanisms for sexual aggressive and other motivated behaviors in the opossum, *Didelphis virginiana*. *J. Comp. Physiol. Psychol.* **64,** 1–15.

Roberts, W. W., Bergquist, E. H., and Robinson, T. C. L. (1969). Thermoregulatory grooming and sleep-like relaxation induced by local warming of preoptic area and anterior hypothalamus in opossum. *J. Comp. Physiol. Psychol.* **67,** 182–188.

Roca-Garcia, M. (1949). Viruses of the lymphogranuloma—psittacosis group isolated from opossums in Colombia. *J. Infect. Dis.* **85,** 275.

Rothstein, R., and Hunsaker, D., II. (1972). Baseline hematology and blood chemistry of the South American woolly opossum, *Caluromys derbianus*. *Lab. Anim. Sci.* **22,** 227–232.

Rowlands, D. T., Jr., and Dudley, M. A. (1968). The isolation of immunoglobulins of the adult opossum (*Didelphis virginiana*). *J. Immunol.* **100,** 736–743.

Rowlands, D. T., Jr., LaVia, M. F., and Block, M. H. (1964). The blood forming tissues and blood of the newborn opossum (*Didelphys virginiana*). II. Onteogenesis of antibody formation to flagella of *Salmonella typhi*. *J. Immunol.* **93,** 157–164.

Sanderson, G. C. (1961). Estimating opossums by marking young. *J. Wildl. Manage.* **25,** 20–27.

Sands, J. L. (1960). The opossum in New Mexico. *J. Mammal.* **41,** 393.

Scheffer, V. B. (1943). The opossum settles in Washington state, *Murrelet* **24,** 27–28.

Schneider, L. K. (1971). Evidence for genetic activity in the sex chromosomes of the male opossum (*Didelphis virginiana*). *Anat. Rec.* **169,** 420.

Schneider, L. K. (1973). Cell cycle determination of phytohemagglutinin—simulated lymphocytes from the opossum, *Didelphis virginiana*. *Experientia* **29,** 468–469.

Schneider, L. K. (1977). Marsupial chromosomes, cell cycles, and cytogenetics. *In* "The Biology of Marsupials" (D. Hunsaker, ed.), Chapter 2, this volume. Academic Press, New York.

Scott, W. J. (1938). Gas transport by the blood of the opossum *Didelphis virginiana*. *J. Cell. Comp. Physiol.* **12,** 391–401.

Seiler, B. W. (1828). Einge bemerktungen uber die erste geburt des kangaruembryo und seine ernahrung in dem beutel. *Isis Oken* **12,** 475–477.

Shaver, E. L. (1972). The chromosomes of the opossum, *Didelphis virginiana*. *Can. J. Genet. Cytol.* **4,** 62–68.

Sherwood, B. F., Rowlands, D. T., Jr., and Hackel, D. B. (1969a). Pulmonary ademomatosis in opossums (*Didelphis virginiana*). *J. Am. Vet. Med. Assoc.* **155,** 1102.

Sherwood, B. F., Rowlands, D. T., Jr., and Hackel, D. B. (1969b). The opossum, *Didelphis virginiana,* as a laboratory animal. *Lab. Anim. Care* **19,** 494–499.

Shirer, H. W., and Fitch, H. S. (1970). Comparison from radio movements and denning habits of raccoon, striped skunk, and opossum in northeastern Kansas. *J. Mammal.* **51,** 491–503.

Silvette, H., and Britton, S. W. (1938). Renal function in the opossum and the mechanism of cortico-adrenal and post pituitary action. *Am. J. Physiol.* **123,** 630–639.

Simpson, G. G. (1970). The Argyrolagidae, extinct South American marsupials. *Bull Mus. Comp. Zool.* **139,** 1–86.

Sinha, A. K., Kakati, S., and Pathak, S. (1972). Exclusive localization of C-bands within opossum sex chromosomes. *Exp. Cell Res.* **75,** 265–268.

Slaughter, B. H. (1968). Earliest known marsupials. *Science* **162,** 254–255.

Smith, L. (1941). Observation on the nest building behavior of the opossum. *J. Mammal.* **22,** 201–202.

Sorokin, S. P. (1965). On the cytology and cytochemistry of the opossum's bronchial glands. *Am. J. Anat.* **117,** 311–338.

Stieglitz, W. D., and Klimstra, W. D. (1962). Dietary pattern of the Virginia opossum, *Didelphis marsupialis virginiana* Kerr, late summer-winter, southern Illinois. *Trans. Ill. Acad. Sci.* **55,** 198–208.

Stoner, D. (1939). Remarks on abundance and range of the opossum. *J. Mammal.* **20,** 205–251.

Szyfres, B., and Blood, B. D. (1964). *Leptospira paidjan* isolated from opossums in Argentina. *Trop. Geog. Med.* **16,** 263.

Talice, R. V., and Lagomarsino, J. C. (1959). Comportamiento Sexual y nacimientos en cautividad de la "comadreja overa": *Didelphis azarae. Congr. Sudamer. Zool.* **5,** 81–96.

Tamar, H. (1961). Taste reception in the opossum and the bat. *Physiol. Zool.* **34,** 86–91.

Tate, G. H. H. (1933). A systematic revision of the marsupial genus *Marmosa. Bull. Am. Mus. Nat. Hist.* **66,** 6–236.

Taylor, D. L., and Burrell, R. (1968). The immunologic responses of the North American opossum (*Didelphis virginiana*). *J. Immunol.* **101,** 1207–1216.

Thomas, O. (1888). "Catalogue of the Marsupialia and Monotremata in the Collection of the British Museum." London.

Thrasher, J. D. (1969a). Preliminary estimates of the mean duration of the S-phase in the opossum. *Didelphis virginiana. Exp. Cell Res.* **57,** 441–442.

Thrasher, J. D. (1969b). Preliminary observations on mate compatibility in *Marmosa mitis. Lab. Anim. Care* **19,** 67–70.

Thrasher, J. D., Barenfus, M., Rich, S. T., and Shupe, D. V. (1971). The colony management of *Marmosa mitis,* the pouchless opossum. *Lab. Anim. Sci.* **21,** 526–536.

Tilley, M. W., Doolittle, J. H., and Mason, D. J. (1966). Spontaneous alternation in the Virginia opossum. *Psychol. Rep.* **19,** 593–594.

Timmons, E. H., and Marques, P. A. (1969). Blood chemical and hematological studies in the laboratory-confined, unanesthetized opossum, *Didelphis virginiana. Lab. Anim. Care* **19,** 342–344.

Toft, J. D., II, Pucak, G. J., and Bullock, B. C. (1973). Basal-cell tumor in a Virginia opossum (*Didelphis marsupialis*). *Lab. Anim. Sci.* **23,** 431.

Travassos, L. (1946). Sobre un "metastrongylidae" das cavidades nasais de *Didelphis marsupialis aurita* Wied. *Rev. Bras. Biol.* **6,** 499.

Tyndale-Biscoe, H. (1973). "Life of Marsupials." Am. Elsevier, New York.

Tyndale-Biscoe, H., and Mackenzie, R. B. (1976). Reproduction in *Didelphis marsupialis* and *D. albiventris* in Colombia. *J. Mammal.* **57,** 249–265.

Vakilzadek, J., Sherwood, B. F., Hackel, D. B., and LeMay, J. C. (1971). Experimental study of pulmonary adenomas in the opossum (*Didelphis virginiana*). *Lab. Anim. Sci.* **21**, 224–228.

Van Twyver, H., and Allison, T. (1970). Sleep in the opossum *Didelphis marsupialis*. *Electroencephalogr. Clin. Neurophysiol.* **29**, 181–189.

Verts, B. J. (1963). Movements and populations of opossums in a cultivated area. *J. Wildl. Manage.* **27**, 127–129.

von Haffner, K. (1913). Linguatuliden als parasiten von *Marmosa*. *Abh. Senckenb. Naturforsch. Ges.* **40**, 361.

Waddell, M. B., and Taylor, R. M. (1948). Studies on cyclic passage of yellow fever virus in South American mammals and mosquitoes. IV. Marsupials (*Metachirus nudicaudatus* and *Marmosa*) in combination with *Aedes aegypti* as vector. *Am. J. Trop. Med.* **28**, 87.

Wade, D., and Neeley, P. (1949). The heart and attached vessels of the opossum, a marsupial. *J. Mammal.* **30**, 111–116.

Walen, K. H. (1965). Spatial relationships in the replication of chromosomal DNA. *Genetics* **51**, 915–929.

Walker, E. P. (1975). "Mammals of the World," 3rd ed. Johns Hopkins Press, Baltimore, Maryland.

Washburn, S. L. (1946). The sequence of epiphyseal union in the opossum. *Anat. Rec.* **95**, 353–363.

Weber, C. G. (1955). Electrocardiographic studies on the opossum. *J. Mammal.* **36**, 284–286.

Wemyss, C. T., Jr. (1953). A preliminary study of marsupial relationships as indicated by the precipitin test. *Zoologica* (*N.Y.*) **38**, 173–181.

Wheeler, R. S. (1943). Normal development of the pituitary in the opossum and its response to hormonal treatments. *J. Morphol.* **73**, 43–87.

Wiedorn, W. S. (1954). A new experimental animal for psychiatric research. The opossum (*Didelphis virginiana*). *Science* **119**, 360–361.

Wiseman, G. L., and Hendrickson, G. O. (1950). Notes on the life history and ecology of the opossum in southeast Iowa. *J. Mammal.* **31**, 331–337.

Yeager, L. E. (1936). Winter daytime dens of opossums. *J. Mammal.* **17**, 410.

Zetek, J. (1930). The water opossum, *Chironectes panamensis* Goldman. *J. Mammal.* **11**, 420–421.

# Appendix: Notes on Nomenclature

JOHN A. W. KIRSCH

Through the courtesy of the editors, I have been able to read the complete manuscript and bring the scientific nomenclature into conformity with a recent complete list of living marsupial species (Kirsch and Calaby, 1976). Such lists are not, of course, ever final, but there are some advantages in achieving a degree of consistency. Kirsch and Calaby do not give synonyms, nor do they employ subspecies, and disadvantages of using their nomenclature may include possible confusion when readers wish to look into original sources, where they may find a different, and confusing, name employed. To overcome this problem, I append a list which gives the synonyms recorded while correcting the manuscript; I emphasize that this listing is not a formal synonomy: in some cases the original nomenclature may not even have been legal, and for the most part, trivial or obvious spelling variants are not listed.

A perusal of the list will reveal that there have been notable vicissitudes in generic nomenclature, particularly in Didelphidae—discussed in Chapters 1 and 3—and Macropodidae. The history of generic nomenclature in Macropodinae is taken up in detail in Ride (1962). Most of the wallabies and kangaroos were placed in *Macropus* by Thomas (1888); present distinctions are based mainly on karyotypic differences, as recommended by Calaby (1966): with a few exceptions typical *Macropus* have $2n = 16$, while *Thylogale* have $2n = 22$, and the (now) monotypic *Wallabia* has $2n = 10$ in the female and $2n = 11$ in the male.

Fashions also change in specific nomenclature, but increasingly in marsupial systematics they change for good biological reasons. Thus, Gardner has shown (1973) that *Didelphis virginiana* is unquestionably a distinct species, in spite of many years of synonomy with *D. marsupialis*. In most older references it is possible to discern which species is meant by the locality from which the specimens came, but since *D. virginiana* and *D. marsupialis* are sympatric from Costa Rica to Mexico, this will frequently be impossible without voucher specimens.

That a specific distinction of this sort may be biologically meaningful—and therefore worth recording—is illustrated by an exactly parallel case among Australian marsupials: *Macropus giganteus* (gray kangaroo) was formerly thought to be a single wide-ranging species, including the dark phase *M. g. melanops*, until it was shown that it in fact comprised two species with considerable geographic overlap (Kirsch and Poole, 1967; Kirsch and Poole, 1972). At least one reason Mykytowycz (1964) found peculiar distributions of *Eimeria* in populations of *M. giganteus* (see Chapter 8) seems to be that he was sampling in an area within the zone of sympatry.

Subspecies are not normally used in Australia, and I have reduced these to the appropriate species when originally given as subspecies or full species in the text, except in the case of the Bennett's Wallaby, which is a strongly marked form of *Macropus rufogriseus* found in Tasmania and properly referred to as *M. rufogriseus fruticus*.

In the case of American marsupials, on the other hand, I have frequently allowed subspecies to stand when (a) subspecies are commonly used, or (b) the subspecies designation was given as a full species in the original reference cited. However, I have adopted what I believe to be current subspecies usage.

Finally, I should remark that corrections have been made on the copyedited manuscripts; some discrepancies may have crept in with the authors' final revisions.

| Current name | Synonyms and/or included subspecies |
|---|---|
| Family Didelphidae | |
| *Caluromys derbianus* | Includes *Caluromys* (or *Philander*) *laniger pallidus* |
| *Caluromys lanatus* | *Caluromys laniger, Philander laniger* |
| *Caluromys philander* | *Philander philander, Didelphis philander* |
| *Didelphis albiventris* | *Didelphis azarae, D. marsupialis azarae;* includes *D. albiventris paraguayensis* |
| *Didelphis marsupialis* | Sometimes may mean *Didelphis virginiana;* includes *D. marsupialis aurita, D.m. caucae* (in turn including *D. m. etensis, D. m. karkinophaga, D. m. yucatanensis*) |
| *Lutreolina crassicaudata* | *Metachirus crassicaudatus* |
| *Marmosa cinerea* | Includes *Marmosa demararae* |

| Current name | Synonyms and/or included subspecies |
|---|---|
| *Marmosa elegans* | Includes *Marmosa venusta* |
| *Marmosa grisea* | *Marmosa marmota* |
| *Marmosa impavida* | Includes *Marmosa caucae* |
| *Marmosa pusilla* | Includes *Marmosa pallidor* |
| *Marmosa robinsoni* | *Marmosa mitis;* includes *M. simonsi* |
| *Metachirus nudicaudatus* | *Philander nudicaudatus* |
| *Monodelphis brevicaudata* | *Didelphis brevicaudatus* |
| *Monodelphis domestica* | *Didelphis domesticus* |
| *Philander opossum* | *Metachirops opossum, Metachirops quica* |
| Family Caenolestidae | |
| *Lestoros inca* | *Orolestes incae* |
| Family Dasyuridae | |
| *Antechinus flavipes* | *Phascogale flavipes, P. flaviceps* |
| *Antechinus swainsonii* | *Phascogale swainsoni* |
| *Dasyurus maculatus* | *Dasyurops maculatus* |
| *Dasyurus viverrinus* | *Dasyurus quoll, Dasyurinus viverrinus* |
| *Sarcophilus harrisii* | *Sarcophilus ursinus, Dasyurus ursinus* |
| Family Peramelidae | |
| *Echymipera kalubu* | *Echymipera doreyana, E. cockerelli* |
| *Isoodon macrourus* | *Thylacis macrourus, T. macroura;* includes *Isoodon torosus, Thylacis torosus, Perameles torosus* |
| *Isoodon obesulus* | *Thylacis obesulus, T. obesula, Perameles obesula* |
| *Perameles bougainville* | *Perameles myosura* |
| Family Thylacomyidae | |
| *Macrotis lagotis* | *Thylacomys lagotis* |
| *Macrotis leucura* | *Macrotis minor, Thylacomys minor, Peragale minor, P. leucura* |
| Family Petauridae | |
| *Pseudocheirus peregrinus* | Includes *Pseudocheirus laniginosus* |
| *Schoinobates volans* | *Petauroides volans* |
| Family Macropodidae | |
| *Bettongia gaimardi* | *Bettongia cuniculus* |
| *Macropus* | *Halmaturus, Protemnodon, Thylogale, Wallabia* |
| *Macropus dorsalis* | *Protemnodon dorsalis, Wallabia dorsalis* |
| *Macropus eugenii* | *Protemnodon eugenii, Thylogale eugenii;* includes *T. derbyanus, T. flindersi* |
| *Macropus fuliginosus* | *Macropus giganteus melanops, M. melanops;* includes *M. ocydromus;* at times *M. giganteus* may be *M. fuliginosus* |
| *Macropus giganteus* | *Macropus canguru;* includes *M. major, M. tasmaniensis;* may sometimes mean *M. fuliginosus* |
| *Macropus irma* | *Protemnodon irma, Wallabia irma* |
| *Macropus parryi* | *Macropus elegans, Wallabia elegans* |
| *Macropus rufogriseus* | *Macropus ruficollis, Wallabia ruficollis, Wallabia rufogrisea* |
| *Macropus r. fruticus* | *Macropus bennetti, Wallabia bennetti* |
| *Macropus rufus* | *Megaleia rufa* |

| Current name | Synonyms and/or included subspecies |
|---|---|
| *Petrogale penicillata* | *Macropus penicillata;* includes *Petrogale inornata, P. lateralis, P. pearsoni, P. herberti* |
| *Potorous tridactylus* | *Hypsyprymnus tridactylus;* includes *Hypsyprymnus apicalis, Macropus apicalis, Potorous apicalis* |
| *Thylogale brunii* | *Macropus brunii* |
| *Thylogale stigmatica* | *Macropus stigmaticus;* includes *Macropus coxeni, Thylogale coxeni* |
| *Thylogale thetis* | *Macropus thetidis* |
| *Wallabia bicolor* | *Macropus bicolor, M. ualabatus, Protemnodon bicolor* |
| Family Vombatidae | |
| *Lasiorhinus latifrons* | *Phascolomys* (or *Phascolomis) latifrons* |
| *Vombatus ursinus* | *Phascolomys* (or *Phascolomis) ursinus, P. tasmaniensis;* includes *P. mitchelli, P. hirsutus, V. hirsutus* |

## REFERENCES


Calaby, J. H. (1966). *CSIRO Div. Wildlife Res. Techn. Pap.* **No. 10,** 1.

Gardner, A. L. (1973). *Spec. Pub. Mus. Texas Tech. Univ.,* **No. 4,** 1.

Kirsch, J. A. W., and Calaby, J. H. (1976). *In* "The Biology of Marsupials" (B. Stonehouse and D. Gilmore, eds.), pp. 9–26. Macmillan, London.

Kirsch, J. A. W., and Poole, W. E. (1967). *Nature* (*London*) **215,** 1097.

Kirsch, J. A. W., and Poole, W. E. (1972). *Aust. J. Zool.* **20,** 315.

Mykytowycz, R. (1964). *Parasitology* **54,** 105.

Ride, W. D. L. (1962). *Aust. J. Sci.* **24,** 367.

Thomas, O. (1888). "Catalogue of the Marsupialia and Monotremata in the Collection of the British Museum (Natural History)." British Museum, London.

# Index

**W**



**X**



**Y**



A 7
B 8
C 9
D 0
E 1
F 2
G 3